Herbert Amann
Joachim Escher

Analysis III

Translated from the German
by Silvio Levy and Matthew Cargo

Birkhäuser
Basel · Boston · Berlin

Authors:

Herbert Amann
Institut für Mathematik
Universität Zürich
Winterthurerstr. 190
8057 Zürich
Switzerland
e-mail: herbert.amann@math.uzh.ch

Joachim Escher
Institut für Angewandte Mathematik
Universität Hannover
Welfengarten 1
30167 Hannover
Germany
e-mail: escher@ifam.uni-hannover.de

Originally published in German under the same title by Birkhäuser Verlag, Switzerland
© 2001 by Birkhäuser Verlag

2000 Mathematics Subject Classification: 28-01, 28A05, 28A20, 28A25, 28B05, 58-01, 58A05, 28A10, 58C35

Library of Congress Control Number: 2008939525

Bibliographic information published by Die Deutsche Bibliothek
Die Deutsche Bibliothek lists this publication in the Deutsche Nationalbibliografie;
detailed bibliographic data is available in the Internet at <http://dnb.ddb.de>.

ISBN 3-7643-7479-2 Birkhäuser Verlag, Basel – Boston – Berlin

© 2009 Birkhäuser Verlag AG
Basel · Boston · Berlin
P.O. Box 133, CH-4010 Basel, Switzerland
Part of Springer Science+Business Media
Layout and LATEX: Gisela Amann, Zürich
Printed on acid-free paper produced of chlorine-free pulp. TCF ∞
Printed in Germany

ISBN 978-3-7643-7479-2
9 8 7 6 5 4 3 2 1

e-ISBN 978-3-7643-7480-8
www.birkhauser.ch

Foreword

This third volume concludes our introduction to analysis, wherein we finish laying the groundwork needed for further study of the subject.

As with the first two, this volume contains more material than can treated in a single course. It is therefore important in preparing lectures to choose a suitable subset of its content; the remainder can be treated in seminars or left to independent study. For a quick overview of this content, consult the table of contents and the chapter introductions.

This book is also suitable as background for other courses or for self study. We hope that its numerous glimpses into more advanced analysis will arouse curiosity and so invite students to further explore the beauty and scope of this branch of mathematics.

In writing this volume, we counted on the invaluable help of friends, colleagues, staff, and students. Special thanks go to Georg Prokert, Pavol Quittner, Olivier Steiger, and Christoph Walker, who worked through the entire text critically and so helped us remove errors and make substantial improvements. Our thanks also goes out to Carlheinz Kneisel and Bea Wollenmann, who likewise read the majority of the manuscript and pointed out various inconsistencies.

Without the inestimable effort of our "typesetting perfectionist", this volume could not have reached its present form: her tirelessness and patience with TEX and other software brought not only the end product, but also numerous previous versions, to a high degree of perfection. For this contribution, she has our greatest thanks.

Finally, it is our pleasure to thank Thomas Hintermann and Birkhäuser for their usual flexibility and friendly cooperation.

Zürich and Hannover, July 2001 H. Amann and J. Escher

Foreword to the English translation

We are again much obliged to Silvio Levy and Matt Cargo for their careful and accurate translation of this last part of the original German treatise. Special thanks go to Thomas Hempfling from Birkhäuser Verlag for rendering possible this translation so that our analysis course is now available to a larger audience.

Zürich and Hannover, January 2009   H. Amann and J. Escher

Contents

Chapter X   Integration theory

Chapter IX

Elements of measure theory

In this chapter, we treat the general theory of the measure of lines, areas, volumes, and sets in even higher dimensional spaces. The theory is guided by elementary geometrical facts. In particular, we will assign the length measure to intervals, the area measure (length times width) to rectangles, and the volume measure (length times width times height) to rectangular boxes.

Naturally, we do not only want to assign measure these elementary domains, that is Cartesian products of intervals, but we also want to measure much more general sets. To do this, it is natural to break up a given set into a disjoint collection of such elementary domains; then the sum of their measures determines the measure of the set. Here it will be of fundamental significance that this process admits not only finite but also countably infinite divisions of the original set. In this way, we will see that we can assign a measure to every open subset of $\mathbb{R}^n$, and this measure has the natural properties expected of it, for example, independence on the location of the set in its space. In addition, will be able to measure not only open sets, but also, for example, closed sets or any set that can be suitably approximated by open sets. It will turn out, however, that it is not possible "to measure" every subset of $\mathbb{R}^n$ in this way.

To introduce measures in practice, however, we will follow another path, which is substantially more general and technically simpler. Only at the end of it will we then encounter a characterization of measurable sets in $\mathbb{R}^n$. Our general approach takes us into the realm of abstract measure theory, and besides its relative simplicity, it has the advantage of providing other measures that have nothing to do with the original geometric ideas. This more general theory will be needed in the last chapter of this volume. It is also important in probability theory, physics, and many mathematical applications.

Section 1 is devoted to σ-algebras. A σ-algebra is a collection of sets that constitutes the domain of definition of a measure. If the underlying set has a topology, then the Borel σ-algebra, which is determined by open sets, has a prominent significance. We show, among other facts, that the Borel σ-algebra of a topological product is determined in all cases of practical relevance by the product of open sets.

Section 2 focuses on the fundamental properties of general measures. Also, we prove that every measure space has a completion, that is, a certain natural minimal extension.

In Sections 3 and 4, we construct the most important measures for applications, namely, those developed by Lebesgue, Stieltjes, and Hausdorff. Here we apply the approach suggested by Carathéodory, which uses as its superstructure the idea of outer measure.

The last section of the chapter is devoted to a detailed study of the Lebesgue measure. First, we characterize the σ-algebra of Lebesgue measurable sets as the completion of the Borel σ-algebra. After that, we study the behavior of the Lebesgue measure under maps, which leads to its invariance under rigid motions and, in particular, translations. Finally, we see how the Lebesgue measure stands out among all locally finite Borel measures and is of fundamental significance in the construction of non-Lebesgue measurable sets.

1 Measurable spaces

In Chapter VI we used the Cauchy–Riemann notion of integral to assign an area to the region between the graph of a sufficiently regular function and its abscissa. Our goal now is to specify a largest possible class of domains in $\mathbb{R}^n$ that can meaningfully be assigned a "generalized area", or *content*. That is, we seek a subset $\mathcal{A}$ of $\mathfrak{P}(\mathbb{R}^n)$ and a map $\mu \colon \mathcal{A} \to [0, \infty)$ such that for $A \in \mathcal{A}$, the number $\mu(A)$ can be interpreted as the content of A. This function μ must of course satisfy certain rules — those that might be expected based on the case of areas of plane domains. For example, the content of the union of two disjoint domains should equal the sum of their individual contents. Also, the content of a domain should be independent of its overall location in space. The concept of content will be gradually clarified, and we will prove in Section 5 that it is not possible to assign (nontrivially) a content, or "measure", to all subsets of $\mathbb{R}^n$: it cannot be the case that $\mathcal{A} = \mathfrak{P}(\mathbb{R}^n)$.

In this section

- X, X_1 and X_2 are nonempty sets.

σ-algebras

We start with the axiomatic introduction of those collections of sets on which "measures" will be defined later: A subset $\mathcal{A}$ of $\mathfrak{P}(X)$ is called a **σ-algebra over X** if it satisfies the properties

(i) $X \in \mathcal{A}$;

(ii) $A \in \mathcal{A} \Rightarrow A^c \in \mathcal{A}$;

(iii) $(A_j) \in \mathcal{A}^{\mathbb{N}} \Rightarrow \bigcup_{j \in \mathbb{N}} A_j \in \mathcal{A}$.

If $\mathcal{A}$ is a σ-algebra over X, one calls $(X, \mathcal{A})$ a **measure space**, and every $A \in \mathcal{A}$ is said to be **$\mathcal{A}$-measurable**.

1.1 Remark Suppose $\mathcal{A}$ is a σ-algebra, $(A_j) \in \mathcal{A}^{\mathbb{N}}$, and $m \in \mathbb{N}$. Then each of the sets

$$\emptyset , \quad A_0 \setminus A_1 , \quad \bigcup_{j=0}^{m} A_j , \quad \bigcap_{j=0}^{m} A_j , \quad \bigcap_{j \in \mathbb{N}} A_j$$

also belongs to $\mathcal{A}$.

Proof Set

$$B_k := \begin{cases} A_k , & k \le m , \\ A_m , & k > m , \end{cases}$$

so that $(B_k) \in \mathcal{A}^{\mathbb{N}}$ and therefore

$$\bigcup_{k \in \mathbb{N}} B_k = \bigcup_{j=0}^{m} A_j \in \mathcal{A} .$$

The remaining statements follow from de Morgan's laws — see Proposition I.2.7(iii). ∎

We say a collection of sets $\mathcal{S} \subset \mathfrak{P}(X)$ is **closed under finite set operations** if

$$A \in \mathcal{S} \Rightarrow A^c \in \mathcal{S} \tag{1.1}$$

and if, for every finite family $A_0, \ldots, A_m$, the union $\bigcup_{j=0}^{m} A_j$ also belongs to $\mathcal{S}$. If $\mathcal{S}$ satisfies (1.1) and $\bigcup_{j=0}^{\infty} A_j$ belongs to $\mathcal{S}$ for every sequence (A_j) in $\mathcal{S}$, we say $\mathcal{S}$ is **closed under countable set operations**. These definitions are justified because, by de Morgan's laws, $\mathcal{S}$ also contains $\bigcap_{j=0}^{m} A_j$ (finite case) and $\bigcap_{j=0}^{\infty} A_j$ (countable case).

We say $\mathcal{S}$ is an **algebra over X** if these properties are satisfied:

(i) $X \in \mathcal{S}$;

(ii) $A \in \mathcal{S} \Rightarrow A^c \in \mathcal{S}$;

(iii) $A, B \in \mathcal{S} \Rightarrow A \cup B \in \mathcal{S}$.

1.2 Remarks   Suppose $\mathcal{S} \subset \mathfrak{P}(X)$ contains X.

(a) $\mathcal{S}$ is an algebra if and only if $\mathcal{S}$ is closed under finite set operations.

(b) $\mathcal{S}$ is a σ-algebra if and only if $\mathcal{S}$ is closed under countable set operations. In this case, $\mathcal{S}$ is also an algebra.

(c) Suppose $\mathcal{S}$ is an algebra and for every disjoint sequence[1] $(B_j) \in \mathcal{S}^{\mathbb{N}}$ we have $\bigcup_{j \in \mathbb{N}} B_j \in \mathcal{S}$. Then $\mathcal{S}$ is a σ-algebra.

Proof   Suppose $(A_k) \in \mathcal{S}^{\mathbb{N}}$. We recursively set

$$B_0 := A_0 \quad \text{and} \quad B_{j+1} := A_{j+1} \setminus \bigcup_{k=0}^{j} A_k \text{ for } j \in \mathbb{N} .$$

Then (B_j) is a disjoint sequence with $\bigcup_k A_k = \bigcup_j B_j$. By assumption, the union $\bigcup_j B_j$ lies in $\mathcal{S}$, and the claim follows. ∎

1.3 Examples   **(a)** $\{\emptyset, X\}$ and $\mathfrak{P}(X)$ are σ-algebras.

(b) $\{A \subset X \; ; \; A \text{ or } A^c \text{ is countable}\}$ is a σ-algebra.

(c) $\{A \subset X \; ; \; A \text{ or } A^c \text{ is finite}\}$ is an algebra, and it is a σ-algebra if and only if X is finite.

(d) Suppose A is a nonempty set, and suppose $\mathcal{A}_\alpha$ is a σ-algebra over X for every $\alpha \in \mathsf{A}$. Then $\bigcap_{\alpha \in \mathsf{A}} \mathcal{A}_\alpha$ is a σ-algebra over X.

(e) Suppose Y is a nonempty set and let $f \in Y^X$. Further let $\mathcal{A}$ and $\mathcal{B}$ be σ-algebras over X and Y, respectively. Then

$$f^{-1}(\mathcal{B}) := \{ f^{-1}(B) \; ; \; B \in \mathcal{B} \} \quad \text{and} \quad f_*(\mathcal{A}) := \{ B \subset Y \; ; \; f^{-1}(B) \in \mathcal{A} \}$$

are respectively σ-algebras over X and Y. One says $f^{-1}(\mathcal{B})$ is the **inverse image of $\mathcal{B}$ under f** and $f_*(\mathcal{A})$ is the **image** (or **push-forward**) of $\mathcal{A}$ under f.

---

[1]We agree upon simplified language as follows. A sequence $(A_j) \in \mathcal{S}^{\mathbb{N}}$ is **disjoint** if $A_j \cap A_k = \emptyset$ for all $j, k \in \mathbb{N}$ with $j \neq k$.

Proof We verify only part (e) and leave the rest to the reader.

Obviously Y belongs to $f_*(\mathcal{A})$. For $B \in f_*(\mathcal{A})$, the set $f^{-1}(B)$ belongs to $\mathcal{A}$. Due to Proposition I.3.8 (ii′) and (iv′), we have

$$f^{-1}(B^c) = [f^{-1}(B)]^c \quad \text{and} \quad f^{-1}\left(\bigcup_j B_j\right) = \bigcup_j f^{-1}(B_j).$$

Therefore, if B lies in $f_*(\mathcal{A})$, so does B^c, and similarly if $B_j \in f_*(\mathcal{A})$ for $j \in \mathbb{N}$ then $\bigcup_{j \in \mathbb{N}} B_j \in f_*(\mathcal{A})$. ∎

The Borel σ-algebra

Let $\mathcal{S}$ be a nonempty subset of $\mathfrak{P}(X)$. Then

$$\mathcal{A}_\sigma(\mathcal{S}) := \bigcap \{ \mathcal{A} \subset \mathfrak{P}(X) \, ; \, \mathcal{A} \supset \mathcal{S}, \, \mathcal{A} \text{ is a } \sigma\text{-algebra over } X \}$$

is the **σ-algebra generated by $\mathcal{S}$**, and $\mathcal{S}$ is a **generating set** for $\mathcal{A}_\sigma(\mathcal{S})$.

1.4 Remarks (a) $\mathcal{A}_\sigma(\mathcal{S})$ is well defined and is the smallest σ-algebra containing $\mathcal{S}$.

Proof This follows from Examples 1.3(a) and (d). ∎

(b) If $\mathcal{S}$ is a σ-algebra, then $\mathcal{A}_\sigma(\mathcal{S}) = \mathcal{S}$.

(c) From $\mathcal{S} \subset \mathcal{T}$ follows $\mathcal{A}(\mathcal{S}) \subset \mathcal{A}(\mathcal{T})$.

(d) For $\mathcal{S} = \{A\}$, we have $\mathcal{A}_\sigma(\mathcal{S}) = \{\emptyset, A, A^c, X\}$. ∎

Let $X := (X, \mathcal{T})$ be a topological space. Since $\mathcal{T}$ is nonempty, it generates a well defined σ-algebra, called the **Borel σ-algebra** of X and denoted by $\mathcal{B}(X)$. The elements of $\mathcal{B}(X)$ are the **Borel subsets** of X. As a shorthand, we write $\mathcal{B}^n := \mathcal{B}(\mathbb{R}^n)$.

A subset A of X is called a **G_δ** (or **G_δ-set**) if there exist open sets O_j with $A = \bigcap_{j \in \mathbb{N}} O_j$, that is, if A is an intersection of countably many open sets in X. The set A is called an **F_σ** (or **F_σ-set**) if is a countable union of closed subsets of X. Therefore A is a F_σ-set if and only if A^c is a G_δ-set.[2]

1.5 Examples (a) For $\mathcal{F} := \{ A \subset X \, ; \, A \text{ is closed} \}$, we have $\mathcal{B}(X) = \mathcal{A}_\sigma(\mathcal{F})$.

(b) Every G_δ-set and every F_σ-set is a Borel set.

(c) Every closed interval I is an F_σ and a G_δ.

Proof Suppose $I = [a, b]$ with $-\infty < a \leq b < \infty$. It is clear that I is a F_σ-set. Because $[a, b] = \bigcap_{k \in \mathbb{N}^\times} (a - 1/k, b + 1/k)$, we see I is also a G_δ-set. The cases $I = [a, \infty)$ and $I = (-\infty, a]$ with $a \in \mathbb{R}$ are treated analogously. The case $I = \mathbb{R}$ is clear. ∎

---

[2]The symbols F_σ and G_δ are explained as follows: F stands for fermé (French for closed) and σ for somme (a sum of sets being another name for their union); G stands for Gebiet (German for domain, an old-fashioned term for an open set) and δ for Durchschnitt (intersection).

(d) Suppose $Y \subset X$ with $Y \neq \emptyset$ and $Y \neq X$. Further suppose $\mathcal{T} := \{\emptyset, X\}$ is the **trivial topology** on X. Then Y is neither an F_σ-set nor a G_δ-set in $(X, \mathcal{T})$.

(e) $\mathbb{Q}$ is an F_σ but not a G_δ in $\mathbb{R}$.

Proof Since $\mathbb{Q}$ is countable, it is clearly an F_σ (see Corollary III.2.18). For the second statement, assume that $\mathbb{Q}$ is a G_δ and choose open sets Q_j, for $j \in \mathbb{N}$, such that $\mathbb{Q} = \bigcap_j Q_j$. Since $\mathbb{Q} \subset Q_j$ for $j \in \mathbb{N}$ and in view of Proposition I.10.8, every Q_j is open and dense in $\mathbb{R}$. Now it follows from Exercise V.4.4 that $\mathbb{Q}$ is uncountable, which is wrong. ∎

The second countability axiom

Let $(X, \mathcal{T})$ be a topological space. We call $\mathcal{M} \subset \mathcal{T}$ a **basis** of $\mathcal{T}$ if for every $O \in \mathcal{T}$ there is an $\mathcal{M}' \subset \mathcal{M}$ with $O = \bigcup\{M \subset X \; ; \; M \in \mathcal{M}'\}$, that is, if every open set can be expressed as a union of sets from $\mathcal{M}$. A topological space $(X, \mathcal{T})$ is **second countable** (or satisfies the **second countability axiom**) if $\mathcal{T}$ has a countable basis. Finally $(X, \mathcal{T})$ is called a **Lindelöf space** if every open cover of X has a countable subcover. Obviously every compact space is Lindelöf.

1.6 Remarks **(a)** $\mathcal{M} \subset \mathcal{T}$ is a basis of $\mathcal{T}$ if and only if for every point $x \in X$ and every neighborhood U of x there exists $M \in \mathcal{M}$ such that $x \in M \subset U$.

Proof (i) "$\Rightarrow$" Suppose $\mathcal{M}$ is a basis of $\mathcal{T}$, and take $x \in X$ and $U \in \mathcal{U}(x)$. By assumption, there exists $O \in \mathcal{T}$ such that $x \in O \subset U$. Further, there is an $\mathcal{M}' \subset \mathcal{M}$ such that $O = \bigcup\{M \subset X \; ; \; M \in \mathcal{M}'\}$. Thus we have found an $M \in \mathcal{M}' \subset \mathcal{M}$ such that $x \in M \subset O \subset U$.

(ii) "$\Leftarrow$" Suppose $O \in \mathcal{T}$. For every $x \in O$, O is a neighborhood of x. Therefore by assumption, there is an $M_x \in \mathcal{M}$ such that $x \in M_x \subset O$, and we find that

$$O = \bigcup_{x \in O} \{x\} \subset \bigcup_{x \in O} M_x \subset O \, ,$$

that is, $O = \bigcup_{x \in O} M_x$. ∎

(b) Any topological space that satisfies the second countability axiom also satisfies the first (see Remark III.2.29(c)).

Proof This follows immediately from (a). ∎

(c) The converse of (b) is false.

Proof Let X be uncountable. Then $(X, \mathfrak{P}(X))$ satisfies the first countability axiom, because any $x \in X$ admits $\{\{x\}\}$ as a neighborhood basis. But $(X, \mathfrak{P}(X))$ cannot be second countable, because any basis of $\mathfrak{P}(X)$ must contain the set $\{\{x\} \; ; \; x \in X\}$ and is therefore uncountable. ∎

1.7 Lemma *Suppose X is a metric space and $A \subset X$ is dense in X. Further let $\mathcal{M} := \{\mathbb{B}(a, r) \; ; \; a \in A, \; r \in \mathbb{Q}^+\}$. Then every open set in X can be written as a union of sets from $\mathcal{M}$.*

Proof Suppose O is open in X and $x \in O$. Then there is an $\varepsilon > 0$ with $\mathbb{B}(x, \varepsilon_x) \subset O$. Because A is dense in X and $\mathbb{Q}$ is dense in $\mathbb{R}$, there is an $a_x \in A$ such that $d(x, a_x) < \varepsilon_x/4$ and an $r_x \in \mathbb{Q}^+$ such that $r_x \in (\varepsilon_x/4, \varepsilon_x/2)$. The triangle inequality then yields

$$x \in \mathbb{B}(a_x, r_x) \subset \mathbb{B}(x, \varepsilon_x) \subset O ,$$

and it follows that $O = \bigcup_{x \in O} \mathbb{B}(a_x, r_x)$. ∎

1.8 Proposition *Let X be a metric space. The following statements are equivalent:*

(i) *X satisfies the second countability axiom.*

(ii) *X is a Lindelöf space.*

(iii) *X is separable.*

Proof "(i)⇒(ii)" Suppose $\mathcal{M}$ is a countable basis and $\{ O_\alpha \; ; \; \alpha \in A \}$ is an open cover of X. By assumption, for every $\alpha \in A$ there is a sequence $(U_{\alpha,j})_{j \in \mathbb{N}}$ in $\mathcal{M}$ such that $O_\alpha = \bigcup_{j \in \mathbb{N}} U_{\alpha,j}$. Now, the collection $\mathcal{M}' := \{ U_{\alpha,j} \; ; \; \alpha \in A, \; j \in \mathbb{N} \}$ covers X and is countable (since $\mathcal{M}' \subset \mathcal{M}$), so we can arrange it in the form $\mathcal{M}' = \{ M_j \; ; \; j \in \mathbb{N} \}$. By construction, there exists for every $j \in \mathbb{N}$ an $\alpha_j \in A$ such that $M_j \subset O_{\alpha_j}$. Therefore $\{ O_{\alpha_j} \; ; \; j \in \mathbb{N} \}$ is a countable subcover of $\{ O_\alpha \; ; \; \alpha \in A \}$.

"(ii)⇒(iii)" For every $n \in \mathbb{N}^\times$ we know $\mathcal{U}_n := \{ \mathbb{B}(x, 1/n) \; ; \; x \in X \}$ is an open cover of X. By assumption, for every $n \in \mathbb{N}^\times$, there are points $x_{n,k} \in X$, for $k \in \mathbb{N}$, such that $\mathcal{V}_n := \{ \mathbb{B}(x_{n,k}, 1/n) \; ; \; k \in \mathbb{N} \}$ is a subcover of $\mathcal{U}_n$. According to Proposition I.6.8, $D := \{ x_{n,k} \; ; \; n \in \mathbb{N}^\times, \; k \in \mathbb{N} \}$ is countable. Now take $x \in X$, $\varepsilon > 0$, $n > 1/\varepsilon$. Since $\mathcal{V}_n$ covers X, there is an $x_{n,k} \in D$ such that $x \in \mathbb{B}(x_{n,k}, 1/n)$. Therefore D is dense in X.

"(iii)⇒(i)" A separable set is second countable by Lemma 1.7. ∎

1.9 Corollary

(i) *Suppose X is a separable metric space and A is countable and dense in X. Then*

$$\mathcal{B}(X) = \mathcal{A}_\sigma\bigl(\{ \mathbb{B}(a, r) \; ; \; a \in A, \; r \in \mathbb{Q}^+ \}\bigr) .$$

(ii) *Suppose $X \subset \mathbb{R}^n$ is not empty. Then the metric space X has a countable basis.*

Proof (i) Define $\mathcal{S} := \{ \mathbb{B}(a, r) \; ; \; a \in A, \; r \in \mathbb{Q}^+ \}$, and let $\mathcal{T}$ denote the topology of X. Lemma 1.7 implies $\mathcal{T} \subset \mathcal{A}_\sigma(\mathcal{S})$, and we find with Remarks 1.4(b) and (c) that

$$\mathcal{B}(X) = \mathcal{A}_\sigma(\mathcal{T}) \subset \mathcal{A}_\sigma\bigl(\mathcal{A}_\sigma(\mathcal{S})\bigr) = \mathcal{A}_\sigma(\mathcal{S}) .$$

The inclusion $\mathcal{A}_\sigma(\mathcal{S}) \subset \mathcal{B}(X)$ follows from $\mathcal{S} \subset \mathcal{T}$ and Remark 1.4(a).

(ii) This follows from Exercise V.4.13 and Proposition 1.8. ∎

For general topological spaces we have the following result.

1.10 Corollary   *Suppose X is a topological space with a countable basis. Then X is separable and Lindelöf.*

Proof   (i) Suppose $\{ B_j \; ; \; j \in \mathbb{N} \}$ is a basis for X. For every $j \in \mathbb{N}$, we select $b_j \in B_j$ and set $D := \{ b_j \; ; \; j \in \mathbb{N} \}$. Obviously D is countable. Now suppose $x \in X$ and U is an open neighborhood of x. Then there is an $I \subset \mathbb{N}$ such that $U = \bigcup_{i \in I} B_i$. Therefore $U \cap D \neq \emptyset$, that is, D is dense in X.

(ii) The argument showing "(i)$\Rightarrow$(ii)" in Proposition 1.8 implies X is Lindelöf.   ∎

Generating the Borel σ-algebra with intervals

We give $\mathbb{R}^n$ the **natural** (product) **ordering**, that is, for $a, b \in \mathbb{R}^n$, the relation $a \leq b$ holds if and only if $a_k \leq b_k$ for $1 \leq k \leq n$.

A subset J of $\mathbb{R}^n$ is called an **interval** in $\mathbb{R}^n$ if there are (ordinary) intervals $J_k \subset \mathbb{R}$ with $1 \leq k \leq n$ such that $J = \prod_{k=1}^n J_k$. For $a, b \in \mathbb{R}^n$ with $a \leq b$ we use the notation

$$(a, b) := \prod_{k=1}^n (a_k, b_k) \;, \qquad [a, b] := \prod_{k=1}^n [a_k, b_k] \;,$$

$$(a, b] := \prod_{k=1}^n (a_k, b_k] \;, \qquad [a, b) := \prod_{k=1}^n [a_k, b_k) \;.$$

If $a \leq b$ is not satisfied, we set

$$(a, b) := [a, b] := (a, b] := [a, b) := \emptyset \;.$$

In analogy with the one dimensional case, we call (a, b) an **open interval** and $[a, b]$ a **closed interval** in $\mathbb{R}^n$. Obviously open and closed intervals in $\mathbb{R}^n$ are respectively open and closed subsets of $\mathbb{R}^n$. We denote the set of open intervals in $\mathbb{R}^n$ by $\mathbb{J}(n)$.

Suppose Y is a set and E is a property which is either true or false for $y \in Y$. When the identity of Y is clear from the context, we use the notation

$$[E] := \big[E(y) \big] := \big\{ y \in Y \; ; \; E(y) \text{ is true} \big\} \;.$$

For example, the set $[x_k \geq \alpha]$, where $k \in \{1, \dots, n\}$ and $\alpha \in \mathbb{R}$, is the closed half-space

$$H_k(\alpha) := \{ x \in \mathbb{R}^n \; ; \; x_k \geq \alpha \}$$

in $\mathbb{R}^n$. If $f \in Y^X$, we set

$$\big[E(f) \big] := \big\{ x \in X \; ; \; E\big(f(x) \big) \text{ is true} \big\} \;.$$

Thus, for $f \in \mathbb{R}^X$, we have for instance $[f > 0] = \big\{ x \in X \; ; \; f(x) > 0 \big\}$.

The following theorem shows that the Borel σ-algebra over $\mathbb{R}^n$ is generated merely by the set of half-spaces with rational coordinates.

1.11 Theorem *Define*

$$
\begin{aligned}
\mathcal{A}_{\mathbb{Q}} &:= \mathcal{A}_{\sigma}\big(\{\,(a,b)\ ;\ a,b \in \mathbb{Q}^n\,\}\big), \\
\mathcal{A}_0 &:= \mathcal{A}_{\sigma}\big(\{\,H_k(\alpha)\ ;\ 1 \le k \le n,\ \alpha \in \mathbb{Q}\,\}\big), \\
\mathcal{A}_1 &:= \mathcal{A}_{\sigma}\big(\{\,H_k(\alpha)\ ;\ 1 \le k \le n,\ \alpha \in \mathbb{R}\,\}\big).
\end{aligned}
$$

Then

$$
\mathcal{B}^n = \mathcal{A}_{\sigma}\big(\mathbb{J}(n)\big) = \mathcal{A}_{\mathbb{Q}} = \mathcal{A}_0 = \mathcal{A}_1 .
$$

Proof Because every closed half-space belongs to $\mathcal{B}^n$, it follows that

$$
\mathcal{A}_0 \subset \mathcal{A}_1 \subset \mathcal{B}^n . \tag{1.2}
$$

Now take $a, b \in \mathbb{R}^n$ with $a \le b$. For $k \in \{1, \dots, n\}$, we have

$$
[x_k < b_k] = [x_k \ge b_k]^c = H_k(b_k)^c \in \mathcal{A}_1
$$

and

$$
[x_k > a_k] = \bigcup_{j=1}^{\infty} [x_k \ge a_k + 1/j] \in \mathcal{A}_1 ,
$$

because $\mathcal{A}_1$ is a σ-algebra. This implies

$$
(a,b) = \prod_{k=1}^{n} (a_k, b_k) = \bigcap_{k=1}^{n} \big([x_k < b_k] \cap [x_k > a_k]\big) \in \mathcal{A}_1 .
$$

For $a, b \in \mathbb{Q}^n$, this shows that (a,b) belongs to $\mathcal{A}_0$. In view of (1.2) and the inclusion $\{\,(a,b)\ ;\ a,b \in \mathbb{Q}^n\,\} \subset \mathbb{J}(n)$, it follows that

$$
\mathcal{A}_{\mathbb{Q}} \subset \mathcal{A}_{\sigma}\big(\mathbb{J}(n)\big) \subset \mathcal{A}_1 \subset \mathcal{B}^n \quad \text{and} \quad \mathcal{A}_{\mathbb{Q}} \subset \mathcal{A}_0 \subset \mathcal{B}^n . \tag{1.3}
$$

Finally $\mathbb{B}_{\infty}^n(c,r) = \prod_{k=1}^{n}(c_k - r, c_k + r)$ belongs to $\mathcal{A}_{\mathbb{Q}}$ for every $c \in \mathbb{Q}^n$ and $r \in \mathbb{Q}^+$. Thus Corollary 1.9(i) yields

$$
\mathcal{B}^n = \mathcal{A}_{\sigma}\big(\{\,\mathbb{B}_{\infty}^n(c,r)\ ;\ c \in \mathbb{Q}^n,\ r \in \mathbb{Q}^+\,\}\big) \subset \mathcal{A}_{\mathbb{Q}} ,
$$

which, together with (1.3), implies the claim. ∎

Bases of topological spaces

The topology of a set X is uniquely determined by specifying a basis. It is easy to see that not every nontrivial collection of sets $\mathcal{M} \subset \mathfrak{P}(X)$ can be the basis of a topology. The next theorem characterizes which ones can.

1.12 Theorem *A collection of sets $\mathcal{M} = \{ M_\alpha \subset X \; ; \; \alpha \in \mathsf{A} \}$ with $\bigcup_{\alpha \in \mathsf{A}} M_\alpha = X$ is a basis for a topology on X, called the* **topology generated by $\mathcal{M}$**, *if and only if for every $(\alpha, \beta) \in \mathsf{A} \times \mathsf{A}$ and every $x \in M_\alpha \cap M_\beta$, there exists $\gamma \in \mathsf{A}$ with $x \in M_\gamma \subset M_\alpha \cap M_\beta$.*

Proof "$\Rightarrow$" Suppose $\mathcal{T}$ is a topology on X and $\mathcal{M} = \{ M_\alpha \subset X \; ; \; \alpha \in \mathsf{A} \}$ is a basis of $\mathcal{T}$. Take $\alpha, \beta \in \mathsf{A}$ and $x \in M_\alpha \cap M_\beta$. Then $M_\alpha \cap M_\beta$ is an open neighborhood of x. Because $\mathcal{M}$ is a basis of $\mathcal{T}$, we can express $M_\alpha \cap M_\beta$ as a union of sets from $\mathcal{M}$; in particular, there is $\gamma \in \mathsf{A}$ such that $x \in M_\gamma \subset M_\alpha \cap M_\beta$.

"$\Leftarrow$" Suppose $\mathcal{M}$ is a collection of sets with the given properties, and set $\mathcal{T}(\mathcal{M}) := \{ \bigcup_{\alpha \in \mathsf{A}'} M_\alpha \; ; \; \mathsf{A}' \subset \mathsf{A} \}$. Obviously $\emptyset$, X, and any union of sets from $\mathcal{T}(\mathcal{M})$ belong to $\mathcal{T}(\mathcal{M})$.

Suppose $O_1, O_2 \in \mathcal{T}(\mathcal{M})$ and define $O := O_1 \cap O_2$. We check that O belongs to $\mathcal{T}(\mathcal{M})$, and we may as well assume that O is nonempty. From the definition of $\mathcal{T}(\mathcal{M})$, there are $\mathsf{A}_j \subset \mathsf{A}$ such that $O_j = \bigcup_{\alpha \in \mathsf{A}_j} M_\alpha$ for $j = 1, 2$. To every $x \in O$ we therefore find $\alpha_j(x) \in \mathsf{A}$ for $j = 1, 2$ with $x \in M_{\alpha_1(x)} \cap M_{\alpha_2(x)}$. Further, there is by assumption an $\alpha(x) \in \mathsf{A}$ such that

$$x \in M_{\alpha(x)} \subset M_{\alpha_1(x)} \cap M_{\alpha_2(x)} \subset O \; .$$

Therefore $O = \bigcup_{x \in O} M_{\alpha(x)}$, that is, O belongs to $\mathcal{T}(\mathcal{M})$.

A simple induction argument now shows that the intersection of any finite number of sets in $\mathcal{T}(\mathcal{M})$ lies in $\mathcal{T}(\mathcal{M})$. ∎

The product topology

Suppose $\mathcal{T}_1$ and $\mathcal{T}_2$ are topologies of X. If $\mathcal{T}_1 \subset \mathcal{T}_2$, we say that $\mathcal{T}_1$ is **coarser** than $\mathcal{T}_2$ (and $\mathcal{T}_2$ is **finer** then $\mathcal{T}_1$).

1.13 Remarks **(a)** $\{\emptyset, X\}$ is the coarsest and $\mathfrak{P}(X)$ is the finest topology of X, that is, $\{\emptyset, X\} \subset \mathcal{T} \subset \mathfrak{P}(X)$ for every topology of X.

(b) Suppose $\mathcal{M} \subset \mathfrak{P}(X)$ is a basis for a topology $\mathcal{T}(\mathcal{M})$. Then $\mathcal{T}(\mathcal{M})$ is the coarsest topology on X that contains $\mathcal{M}$. In other words, if $\mathcal{T}$ a topology of X with $\mathcal{M} \subset \mathcal{T}$, then $\mathcal{T} \supset \mathcal{T}(\mathcal{M})$.

(c) If $\mathcal{T}_0$ is a topology on X, then $\mathcal{T}_0$ is a basis for itself: $\mathcal{T}(\mathcal{T}_0) = \mathcal{T}_0$.

(d) Suppose $\mathcal{M}_j \subset \mathfrak{P}(X)$ is a basis of $\mathcal{T}_j$ for $j = 1, 2$ with $\mathcal{M}_1 \subset \mathcal{M}_2$. Then $\mathcal{T}_1 \subset \mathcal{T}_2$. ∎

Suppose $(X_1, \mathcal{T}_1)$ and $(X_2, \mathcal{T}_2)$ are topological spaces and $(O_j, U_j) \in \mathcal{T}_1 \times \mathcal{T}_2$ for $j = 1, 2$. Obviously,

$$(O_1 \times U_1) \cap (O_2 \times U_2) = (O_1 \cap O_2) \times (U_1 \cap U_2) \; .$$

Hence Theorem 1.12 shows that

$$T_1 \boxtimes T_2 := \left\{ O_1 \times O_2 \subset X_1 \times X_2 \; ; \; (O_1, O_2) \in T_1 \times T_2 \right\}$$

is a basis for a topology $T := T(T_1 \boxtimes T_2)$ on $X_1 \times X_2$, which we call the **product topology of T_1 and T_2** (on $X_1 \times X_2$). The topological space $(X_1 \times X_2, T)$ is the **topological product** of (X_1, T_1) and (X_2, T_2). Unless explicitly stated otherwise, we always provide $X_1 \times X_2$ with the product topology.

1.14 Remarks (a) The product topology is the coarsest topology on $X_1 \times X_2$ that contains $T_1 \boxtimes T_2$.

(b) The product topology is the coarsest topology on $X_1 \times X_2$ for which the projections $\mathrm{pr}_j : X_1 \times X_2 \to X_j$ for $j = 1, 2$ are continuous.

Proof (i) For $O_1 \in T_1$ and $O_2 \in T_2$, we have

$$\mathrm{pr}_1^{-1}(O_1) = O_1 \times X_2 \quad \text{and} \quad \mathrm{pr}_2^{-1}(O_2) = X_1 \times O_2 \ .$$

Therefore the projections pr_1 and pr_2 are continuous with respect to $T := T(T_1 \boxtimes T_2)$.

(ii) Denote by $\widetilde{T}$ a topology of $X_1 \times X_2$ for which pr_1 and pr_2 are continuous. For every $V \in T$, there is an index set A and pairs $(O_\alpha, U_\alpha) \in T_1 \times T_2$, for $\alpha \in \mathsf{A}$, such that $V = \bigcup_{\alpha \in \mathsf{A}} O_\alpha \times U_\alpha$. By Theorem III.2.20 and Remark III.2.29(e), the sets $\mathrm{pr}_1^{-1}(O_\alpha)$ and $\mathrm{pr}_2^{-1}(U_\alpha)$ belongs to $\widetilde{T}$. Since $O_\alpha \times U_\alpha = \mathrm{pr}_1^{-1}(O_\alpha) \cap \mathrm{pr}_2^{-1}(U_\alpha)$, this shows that $V \in \widetilde{T}$, that is, $T \subset \widetilde{T}$. ∎

(c) Suppose $\mathcal{M}_j \subset \mathfrak{P}(X_j)$ is a basis of T_j for $j = 1, 2$. Then $\mathcal{M}_1 \boxtimes \mathcal{M}_2$ is a basis of the product topology of $X_1 \times X_2$.

(d) Suppose (X_j, d_j) are metric spaces for $j = 1, 2$, and let T_j denote the topology induced on X_j by d_j. Further let $T(d_1 \vee d_2)$ be the topology induced on $X_1 \times X_2$ by the product metric $d_1 \vee d_2$; see Example II.1.2(e). Then

$$T(T_1 \boxtimes T_2) = T(d_1 \vee d_2) \ ,$$

that is, the product topology induced by d_1 and d_2 coincides with the topology induced by the product metric $d_1 \vee d_2$.

Proof $T(d_1 \vee d_2)$ is a topology of $X_1 \times X_2$ satisfying

$$T_1 \boxtimes T_2 \subset T(d_1 \vee d_2) \subset T(T_1 \boxtimes T_2) \ ,$$

by Exercise III.2.6 and Theorem 1.12. The claim then follows from (a). ∎

(e) The definitions and results above have obvious generalizations to products of finitely many topological spaces. We leave the formulations and proofs to the reader. ∎

Product Borel σ-algebras

Suppose $(X_1, \mathcal{A}_1)$ and $(X_2, \mathcal{A}_2)$ are measurable spaces. $\mathcal{A}_1 \boxtimes \mathcal{A}_2$ need not be a σ-algebra over $X_1 \times X_2$, as can be seen already in simple cases (see Exercise 15). We therefore define the **product σ-algebra** $\mathcal{A}_1 \otimes \mathcal{A}_2$ of $\mathcal{A}_1$ and $\mathcal{A}_2$ as the smallest σ-algebra over $X_1 \times X_2$ that contains $\mathcal{A}_1 \boxtimes \mathcal{A}_2$:

$$\mathcal{A}_1 \otimes \mathcal{A}_2 := \mathcal{A}_\sigma(\mathcal{A}_1 \boxtimes \mathcal{A}_2) .$$

The next proposition shows how to obtain a generating set for $\mathcal{A}_1 \otimes \mathcal{A}_2$ from generating sets for $\mathcal{A}_1$ and $\mathcal{A}_2$.

1.15 Proposition   For $\mathcal{S}_j \subset \mathfrak{P}(X_j)$ with $X_j \in \mathcal{S}_j$ for $j = 1, 2$ we have

$$\mathcal{A}_\sigma(\mathcal{S}_1) \otimes \mathcal{A}_\sigma(\mathcal{S}_2) = \mathcal{A}_\sigma(\mathcal{S}_1 \boxtimes \mathcal{S}_2) .$$

Proof   Putting $\mathcal{A}_j := \mathcal{A}_\sigma(\mathcal{S}_j)$, we clearly have $\mathcal{A}_\sigma(\mathcal{S}_1 \boxtimes \mathcal{S}_2) \subset \mathcal{A}_1 \otimes \mathcal{A}_2$. To show the converse inclusion, define

$$\widetilde{\mathcal{A}}_j := (\mathrm{pr}_j)_* \big(\mathcal{A}_\sigma(\mathcal{S}_1 \boxtimes \mathcal{S}_2) \big) \quad \text{for } j = 1, 2 .$$

Because $X_2 \in \mathcal{S}_2$, we have $\mathcal{S}_1 \subset \widetilde{\mathcal{A}}_1$; likewise $\mathcal{S}_2$ is a subset of $\widetilde{\mathcal{A}}_2$. Example 1.3(e) now shows that $\widetilde{\mathcal{A}}_j \supset \mathcal{A}_j$ for $j = 1, 2$. Given $A_1 \times A_2 \in \mathcal{A}_1 \boxtimes \mathcal{A}_2$ we then conclude that

$$A_1 \times X_2 = (\mathrm{pr}_1)^{-1}(A_1) \in \mathcal{A}_\sigma(\mathcal{S}_1 \boxtimes \mathcal{S}_2) ,$$
$$X_1 \times A_2 = (\mathrm{pr}_2)^{-1}(A_2) \in \mathcal{A}_\sigma(\mathcal{S}_1 \boxtimes \mathcal{S}_2) ,$$

so $A_1 \times A_2 = (A_1 \times X_2) \cap (X_1 \times A_2) \in \mathcal{A}_\sigma(\mathcal{S}_1 \boxtimes \mathcal{S}_2)$. This implies that $\mathcal{A}_1 \otimes \mathcal{A}_2 = \mathcal{A}_\sigma(\mathcal{A}_1 \boxtimes \mathcal{A}_2)$ is contained in $\mathcal{A}_\sigma(\mathcal{S}_1 \boxtimes \mathcal{S}_2)$. $\blacksquare$

This theorem, too, generalizes easily to the case of finitely many measurable spaces, the product σ-algebra being defined in the obvious way.

Let $(X_1, \mathcal{T}_1)$ and $(X_2, \mathcal{T}_2)$ be topological spaces. Two σ-algebras arise naturally on $X_1 \times X_2$: the product σ-algebra $\mathcal{B}(X_1) \otimes \mathcal{B}(X_2)$ of the Borel σ-algebras $\mathcal{B}(X_1)$ and $\mathcal{B}(X_2)$, and the Borel σ-algebra $\mathcal{B}(X_1 \times X_2)$ of the topological product $X_1 \times X_2$. We now study the question of how the two are related.

1.16 Proposition   Suppose X_1 and X_2 are topological spaces. Then

$$\mathcal{B}(X_1) \otimes \mathcal{B}(X_2) \subset \mathcal{B}(X_1 \times X_2) .$$

Proof   Let $\mathcal{T}_j$ be the topology of X_j. The product topology $\mathcal{T}$ on $X_1 \times X_2$ contains $\mathcal{T}_1 \boxtimes \mathcal{T}_2$. Thus

$$\mathcal{A}_\sigma(\mathcal{T}_1 \boxtimes \mathcal{T}_2) \subset \mathcal{A}_\sigma(\mathcal{T}) = \mathcal{B}(X_1 \times X_2) .$$

Moreover, Proposition 1.15 shows that $\mathcal{B}(X_1) \otimes \mathcal{B}(X_2) = \mathcal{A}_\sigma(\mathcal{T}_1 \boxtimes \mathcal{T}_2)$, from which the claim follows. $\blacksquare$

Exercise 19 shows by example that the inclusion opposite that of Proposition 1.16 need not hold: in general, $\mathcal{B}(X_1 \times X_2) \neq \mathcal{B}(X_1) \otimes \mathcal{B}(X_2)$. The next result is thus of particular importance.

1.17 Theorem *For topological spaces X_1 and X_2 satisfying the second countability axiom,*

$$\mathcal{B}(X_1 \times X_2) = \mathcal{B}(X_1) \otimes \mathcal{B}(X_2) \ .$$

Proof Let $\mathcal{M}_j$ be a countable basis for the topology $\mathcal{T}_j$ of X_j. By Remark 1.14(c) and Proposition I.6.9, $\mathcal{M}_1 \boxtimes \mathcal{M}_2$ is a countable basis of the product topology $\mathcal{T} := \mathcal{T}(\mathcal{T}_1 \boxtimes \mathcal{T}_2)$. Thus every $O \in \mathcal{T}$ can be represented as a countable union of sets from $\mathcal{M}_1 \boxtimes \mathcal{M}_2$. Therefore $\mathcal{T} \subset \mathcal{B}(X_1) \otimes \mathcal{B}(X_2)$, hence $\mathcal{B}(X_1 \times X_2) \subset \mathcal{B}(X_1) \otimes \mathcal{B}(X_2)$. The claim follows from Proposition 1.16. ∎

1.18 Corollary $\mathcal{B}^m \otimes \mathcal{B}^n = \mathcal{B}^{m+n}$ and $\mathcal{B}^m = \underbrace{\mathcal{B}^1 \otimes \cdots \otimes \mathcal{B}^1}_{m}$ for $m, n \in \mathbb{N}^\times$.

Proof This follows from Remark 1.14(e), Corollary 1.9(ii), Theorem 1.17, and the appropriate generalizations to the case of m factors. ∎

Measurability of sections

For $C \subset X \times Y$ and $(a, b) \in X \times Y$, the sets

$$C_{[a]} := \{ y \in Y \ ; \ (a, y) \in C \} \ ,$$
$$C^{[b]} := \{ x \in X \ ; \ (x, b) \in C \}$$

are called **sections** of C (at $a \in X$ and $b \in Y$, respectively).

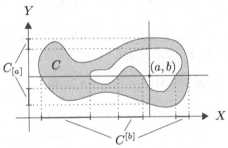

Cross sections of measurable sets are measurable:

1.19 Proposition *Let $(X, \mathcal{A})$ and $(Y, \mathcal{B})$ be measurable spaces and suppose $C \in \mathcal{A} \otimes \mathcal{B}$. Then $C_{[x]} \in \mathcal{B}$ and $C^{[y]} \in \mathcal{A}$, for any $x \in X$ and $y \in Y$.*

Proof (i) We define

$$\mathcal{C} := \{ C \in \mathcal{A} \otimes \mathcal{B} \ ; \ C_{[x]} \in \mathcal{B}, \ C^{[y]} \in \mathcal{A}, \ (x, y) \in X \times Y \}$$

and show that $\mathcal{C}$ is a σ-algebra over $X \times Y$.

Obviously $X \times Y$ belongs to $\mathcal{C}$. For $C \in \mathcal{C}$ and $(x, y) \in X \times Y$, we have

$$(C^c)_{[x]} = (C_{[x]})^c \in \mathcal{B} \quad \text{and} \quad (C^c)^{[y]} = (C^{[y]})^c \in \mathcal{A} \ ,$$

so C^c belongs to $\mathcal{C}$. Finally every sequence (C_j) in $\mathcal{C}$ satisfies

$$\left(\bigcup_j C_j\right)_{[x]} = \bigcup_j (C_j)_{[x]} \quad \text{and} \quad \left(\bigcup_j C_j\right)^{[y]} = \bigcup_j (C_j)^{[y]} \,,$$

so $\bigcup_j C_j$ also belongs to $\mathcal{C}$.

(ii) For $A \times B \in \mathcal{A} \boxtimes \mathcal{B}$ and $(x, y) \in X \times Y$, we have

$$(A \times B)_{[x]} = \begin{cases} B, & x \in A, \\ \emptyset, & x \in A^c, \end{cases} \qquad (A \times B)^{[y]} = \begin{cases} A, & y \in B, \\ \emptyset, & y \in B^c. \end{cases}$$

Thus $\mathcal{A} \boxtimes \mathcal{B}$ is contained in $\mathcal{C}$, and therefore so is $\mathcal{A} \otimes \mathcal{B}$. But $\mathcal{C} \subset \mathcal{A} \otimes \mathcal{B}$ by construction, so everything is proved. ∎

Exercises

1 Prove the statements in Examples 1.3(a)–(d).

2 Let $\mathcal{S}, \mathcal{S}' \subset \mathfrak{P}(X)$ be nonempty. Prove or disprove: $\mathcal{A}_\sigma(\mathcal{S}) = \mathcal{A}_\sigma(\mathcal{S}')$ implies $\mathcal{S} = \mathcal{S}'$.

3 Suppose $(X_1, \mathcal{A}_1)$ and $(X_2, \mathcal{A}_2)$ are measurable spaces. A subset $F \subset X_1 \times X_2$ is called a **mosaic** in $X_1 \times X_2$ if there is an $m \in \mathbb{N}$ and $R_0, \ldots, R_m \in \mathcal{A}_1 \boxtimes \mathcal{A}_2$ such that $R_j \cap R_k = \emptyset$ for $j \neq k$ and $F = \bigcup_{j=0}^m R_j$.
Prove:

(a) $\mathcal{F} := \{ F \subset X_1 \times X_2 \,;\, F \text{ is a mosaic in } X_1 \times X_2 \}$ is an algebra on $X_1 \times X_2$.

(b) $\mathcal{A}_\sigma(\mathcal{F}) = \mathcal{A}_1 \otimes \mathcal{A}_2$.

4 Suppose (A_j) is a sequence in $\mathfrak{P}(X)$ and define

$$\varlimsup_j A_j := \bigcap_{j=0}^\infty \bigcup_{k=j}^\infty A_k \,, \qquad \varliminf_j A_j := \bigcup_{j=0}^\infty \bigcap_{k=j}^\infty A_k \,.$$

We call $\varlimsup_j A_j$ the **limit superior** (or limsup) and $\varliminf_j A_j$ the **limit inferior** (liminf) of (A_j).

(a) Describe the sets $\varlimsup_j A_j$ and $\varliminf_j A_j$.

(b) Prove or disprove: $\varliminf_j A_j \subset \varlimsup_j A_j$, $\varlimsup_j A_j \subset \varliminf_j A_j$.

5 A sequence $(A_j) \in \mathfrak{P}(X)^\mathbb{N}$ is said to be **convergent** if $\varliminf_j A_j = \varlimsup_j A_j$. This common value is called the **limit** of (A_j) and is written $\lim_j A_j$.
Verify:

(a) Any increasing sequence (A_j) converges, with limit $\bigcup_j A_j$. Any decreasing sequence (A_j) converges, with limit $\bigcap_j A_j$.

(b) (A_j) converges to A if and only if (χ_{A_j}) converges pointwise to χ_A.

6 Suppose $(X, \mathcal{A})$ and $(Y, \mathcal{B})$ are measurable spaces. A map $f \colon X \to Y$ is said to be **$\mathcal{A}$-$\mathcal{B}$-measurable**[3] if $f^{-1}(\mathcal{B}) \subset \mathcal{A}$. If X and Y are topological spaces, a $\mathcal{B}(X)$-$\mathcal{B}(Y)$-measurable function is also called **Borel measurable**.

(a) Suppose $\mathcal{S} \subset \mathfrak{P}(Y)$ satisfies $\mathcal{A}_\sigma(\mathcal{S}) = \mathcal{B}$. Prove that $f \in Y^X$ is $\mathcal{A}$-$\mathcal{B}$-measurable if and only if $f^{-1}(\mathcal{S}) \subset \mathcal{A}$.

(b) Suppose X and Y are topological spaces. Prove that every continuous map from X to Y is Borel measurable.

7 Let $(X, \mathcal{A})$ be a measurable space and let $Y \subset X$. Show that $\mathcal{A} \,|\, Y := \{\, A \cap Y \,;\, A \in \mathcal{A} \,\}$ is a σ-algebra over Y. We call it **the σ-algebra induced on Y by $\mathcal{A}$**.

8 Suppose $(X, \mathcal{A})$, $(Y, \mathcal{B})$ and $(Z, \mathcal{C})$ are measurable spaces. Check:

(a) If $f \in Y^X$ and $g \in Z^Y$ are measurable, then $g \circ f \in Z^X$ is also measurable.

(b) If $f \in Y^X$ is measurable and $A \subset X$, then $f \,|\, A \in Y^A$ is $(\mathcal{A} \,|\, A)$-$\mathcal{B}$-measurable.

9 Suppose Y is a topological space and $X \subset Y$. Show that $\mathcal{B}(X) = \mathcal{B}(Y) \,|\, X$.

10 Suppose d_1 and d_2 are equivalent metrics on X and let $X_j := (X, d_j)$ for $j = 1, 2$. Show that $\mathcal{B}(X_1) = \mathcal{B}(X_2)$.

11 Show that if a topological space X has a countable basis $\mathcal{M}$, then $\mathcal{B}(X) = \mathcal{A}_\sigma(\mathcal{M})$.

12 Suppose $(X, \mathcal{T})$ is a Hausdorff space and there is a sequence $(K_j) \in X^{\mathbb{N}}$ of compact sets such that $X = \bigcup_j K_j$. Verify that $\mathcal{B}(X) = \mathcal{A}_\sigma(\mathcal{K})$, where $\mathcal{K}$ is the set of all compact subsets of X.

13 Suppose the topological spaces X and Y satisfy $\mathcal{B}(X \times Y) = \mathcal{B}(X) \otimes \mathcal{B}(Y)$. Show that for every nonempty $Z \subset Y$, we have $\mathcal{B}(X \times Z) = \mathcal{B}(X) \otimes \mathcal{B}(Z)$. (Hint: Check that $\mathcal{A} := \{\, M \subset X \times Y \,;\, (X \times Z) \cap M \in \mathcal{B}(X) \otimes \mathcal{B}(Z) \,\}$ is a σ-algebra over $X \times Y$ containing $\mathcal{B}(X) \otimes \mathcal{B}(Y)$. Further note Remark III.2.29(h).)

14 Suppose X_j and Y_j are topological spaces and $f_j \colon X_j \to Y_j$ is Borel measurable for $j = 1, 2$. Define

$$f_1 \times f_2 \colon X_1 \times X_2 \to Y_1 \times Y_2 \quad \text{by} \quad (x_1, x_2) \mapsto \big(f_1(x_1), f_2(x_2)\big) \,.$$

Check that $(f_1 \times f_2)^{-1}\big(\mathcal{B}(Y_1) \otimes \mathcal{B}(Y_2)\big) \subset \mathcal{B}(X_1) \otimes \mathcal{B}(X_2)$.

15 Take $A \subset X$ and define $\mathcal{A} := \{\emptyset, A, A^c, X\}$. Under what conditions on A is $\mathcal{A} \boxtimes \mathcal{A}$ a σ-algebra over $X \times X$?

16 Suppose $\mathcal{S} \subset \mathfrak{P}(X)$. Show that

$$\mathcal{A}_\sigma(\mathcal{S}) = \bigcup \{\, \mathcal{A}_\sigma(\mathcal{C}) \,;\, \mathcal{C} \subset \mathcal{S} \text{ is countable} \,\} \,.$$

(Hint: The collection of sets on the right is a σ-algebra over X and contains $\mathcal{S}$.)

17 For $A \subset X$, let $\mathcal{A}_A := \{\, B \subset X \,;\, A \subset B \text{ or } A \subset B^c \,\}$. Prove:

(i) $\mathcal{A}_A$ is a σ-algebra over X.

(ii) If $\mathcal{S} \subset \mathcal{A}_A$ then $\mathcal{A}_\sigma(\mathcal{S}) \subset \mathcal{A}_A$.

---

[3]If $\mathcal{A}$ and $\mathcal{B}$ are implicit from the context, we say for short that f is **measurable**.

18   Let $X = (X, \mathcal{T})$ be a topological space. Show that if the diagonal

$$\Delta_X = \big\{ (x, y) \in X \times X \; ; \; x = y \big\}$$

belongs to $\mathcal{B}(X) \otimes \mathcal{B}(X)$, there is an injection of X in $\mathbb{R}$.

(Hint: (i) By Exercise 16, there exist $\mathcal{S}_j = \{ A_{k,j} \; ; \; k \in \mathbb{N} \} \subset \mathcal{T}$ for $j = 1, 2$ with $\Delta_X \in \mathcal{A}_\sigma(\mathcal{S}_1 \boxtimes \mathcal{S}_2)$. Set $\mathcal{A}_j := \mathcal{A}_\sigma(\mathcal{S}_j)$ for $j = 1, 2$. Then $\mathcal{A}_1 \otimes \mathcal{A}_2 \supset \mathcal{A}_\sigma(\mathcal{S}_1 \boxtimes \mathcal{S}_2)$, which, with $\Delta_X \in \mathcal{A}_\sigma(\mathcal{S}_1 \boxtimes \mathcal{S}_2)$ and Proposition 1.19, implies that $\{x\} \in \mathcal{A}_1$ for every $x \in X$.
(ii) Define $\varphi := \sum_{k=0}^{\infty} 3^{-k} \chi_{A_{k,1}} \in B(X, \mathbb{R})$ and take $x, y \in X$ such that $\varphi(x) = \varphi(y)$. By Theorem II.7.11, either $\{x, y\} \subset A_{k,1}$ or $\{x, y\} \subset A_{k,1}^c$ for every $k \in \mathbb{N}$. Then Exercise 17 shows that either $\{x, y\} \subset A$ or $\{x, y\} \subset A^c$ for every $A \in \mathcal{A}_1$. By (i), this is only possible when $x = y$.)

19   Suppose $X := \mathfrak{P}(\mathbb{R})$ and let $\mathcal{T}$ be a Hausdorff topology on X. Then $\mathcal{B}(X) \otimes \mathcal{B}(X) \neq \mathcal{B}(X \times X)$.

(Hint: Because $\mathcal{T}$ satisfies the Hausdorff condition, the diagonal Δ_X is closed in $X \times X$. If $\mathcal{B}(X) \otimes \mathcal{B}(X) = \mathcal{B}(X \times X)$, Exercise 18 says there is an injection of $\mathfrak{P}(\mathbb{R})$ in $\mathbb{R}$. This contradicts Theorem I.6.5.)

2 Measures

We will now introduce the concept of a measure and study its general properties, those that follow more or less immediately from the definition. The resulting rules form the foundation for our deeper exploration of measure and integration theory.

In the following,
- X is a nonempty set and $[0, \infty] := \mathbb{R}^+ \cup \{\infty\}$.

Recall also the facts set forth in Sections I.10 and II.5 about the arithmetic and topology of $\overline{\mathbb{R}}$.

Set functions

Suppose $\mathcal{C}$ is a collection of subsets of X with $\emptyset \in \mathcal{C}$. Let φ be a map (or **set function**) from $\mathcal{C}$ into $[0, \infty]$ with $\varphi(\emptyset) = 0$. We say that φ is $\boldsymbol{\sigma}$**-subadditive** if for every sequence (A_j) in $\mathcal{C}$ such that $\bigcup_j A_j \in \mathcal{C}$,[1]

$$\varphi\left(\bigcup_j A_j\right) \leq \sum_j \varphi(A_j) . \tag{2.1}$$

We say a map φ of $\mathcal{C}$ in $[0, \infty]$ or $\mathbb{K}$ such that $\varphi(\emptyset) = 0$ is $\boldsymbol{\sigma}$**-additive** if

$$\varphi\left(\bigcup_j A_j\right) = \sum_j \varphi(A_j) \tag{2.2}$$

for every *disjoint* sequence (A_j) in $\mathcal{C}$ such that $\bigcup_j A_j \in \mathcal{C}$. If (2.1) holds for every finite collection $A_0, \ldots, A_m$ of subsets of $\mathcal{C}$ such that $\bigcup_j A_j \in \mathcal{C}$, we say that φ is **subadditive**. Likewise, if (2.2) holds for every finite collection $A_0, \ldots, A_m$ of pairwise disjoint subsets of $\mathcal{C}$ such that $\bigcup_j A_j \in \mathcal{C}$, we say φ is **additive**. Finally we say $\varphi \colon \mathcal{C} \to [0, \infty]$ is $\boldsymbol{\sigma}$**-finite** if X belongs to $\mathcal{C}$ and there is a sequence (A_j) in $\mathcal{C}$ such that $\bigcup_j A_j = X$ and $\varphi(A_j) < \infty$ for $j \in \mathbb{N}$. If $\varphi(X) < \infty$, we say φ is **finite**.

2.1 Remarks (a) Every σ-additive set function is additive; every σ-subadditive set function is subadditive.

(b) Suppose φ is a σ-additive map from $\mathcal{C}$ into $[0, \infty]$ [or into $\mathbb{K}$]. If $(A_j) \in \mathcal{C}^{\mathbb{N}}$ are disjoint and $\bigcup_j A_j \in \mathcal{C}$, the series $\sum_j \varphi(A_j)$ converges **absolutely** in $[0, \infty]$ [or in $\mathbb{K}$], that is, it can be reordered arbitrarily with no effect in its sum.

Proof This follows from (2.2), since $\varphi(\bigcup_j A_j)$ does not depend on the ordering on A_j. ∎

(c) The map

$$\mathfrak{P}(X) \to [0, \infty] , \quad A \mapsto \begin{cases} 1 , & A \neq \emptyset , \\ 0 , & A = \emptyset , \end{cases}$$

is σ-subadditive; it is σ-additive if and only if X has a single element. ∎

---

[1] Here and below, $\bigcup_j A_j$ is understood to mean $\bigcup_{j=0}^{\infty} A_j$, etc.

Measure spaces

Suppose $\mathcal{A}$ is a σ-algebra over X and $\mu: \mathcal{A} \to [0, \infty]$ is σ-additive. We say that μ is a (positive) **measure** on X (or on $\mathcal{A}$)[2], and we call $(X, \mathcal{A}, \mu)$ a **measure space**. If $\mu(X) = 1$, we also call μ a **probability measure** and $(X, \mathcal{A}, \mu)$ a **probability space**.

2.2 Examples (a) For fixed $a \in X$, define

$$\delta_a(A) := \left\{ \begin{array}{ll} 1 , & a \in A , \\ 0 , & a \notin A , \end{array} \right.$$

for $A \subset X$. Then $\delta_a : \mathfrak{P}(X) \to [0, \infty]$ is a probability measure, the **Dirac measure on X at a** (or with support a).

(b) For $A \subset X$ define $\mathcal{H}^0(A) := \mathrm{Num}(A)$. Then $\mathcal{H}^0 : \mathfrak{P}(X) \to [0, \infty]$ is a measure, the **counting measure** on X. It is finite [or σ-finite] if and only if X is finite [or countable].

(c) For $A \subset X$, let $\mu(A) := 0$ for $A = \emptyset$ and let $\mu(A) := \infty$ otherwise. Then $(X, \mathfrak{P}(X), \mu)$ is a measure space.

(d) Let $(X, \mathcal{A}, \mu)$ be a measure space and take $A \in \mathcal{A}$. Then $(A, \mathcal{A} \,|\, A, \mu \,|\, A)$ is also a measure space. ∎

Properties of measures

We gather here the most important rules for working with measures.

2.3 Proposition Let $(X, \mathcal{A}, \mu)$ be a measure space. For $A, B \in \mathcal{A}$ and $(A_j) \in \mathcal{A}^{\mathbb{N}}$, we have:

  (i) $\mu(A \cup B) + \mu(A \cap B) = \mu(A) + \mu(B)$.
 (ii) $\mu(B \backslash A) = \mu(B) - \mu(A)$ if $A \subset B$ and $\mu(A) < \infty$.
(iii) $A \subset B \Rightarrow \mu(A) \leq \mu(B)$, that is, μ is increasing.[3]
 (iv) $\mu(A_k) \uparrow \mu(\bigcup_j A_j)$ if $A_0 \subset A_1 \subset A_2 \subset \cdots$.
  (v) $\mu(A_k) \downarrow \mu(\bigcap_j A_j)$ if $A_0 \supset A_1 \supset A_2 \supset \cdots$ with $\mu(A_0) < \infty$.
 (vi) $\mu(\bigcup_j A_j) \leq \sum_j \mu(A_j)$, that is, μ is σ-subadditive.

Proof (i) From $A \cup B = A \cup (B \backslash A)$ and $A \cap (B \backslash A) = \emptyset$ it follows by Remark 2.1(a) that

$$\mu(A \cup B) = \mu(A) + \mu(B \backslash A) . \tag{2.3}$$

---

[2] The specification of $\mathcal{A}$ is actually superfluous, as $\mathcal{A}$ is the domain of definition of μ.
[3] This refers to the natural ordering of $\mathcal{A}$ induced by $(\mathfrak{P}(X), \subset)$; see Examples I.44(a) and (b). Instead of increasing, we may also say **monotone**.

Analogously, we get from $B = (A \cap B) \cup (B \backslash A)$ and $(A \cap B) \cap (B \backslash A) = \emptyset$ that

$$\mu(A \cap B) + \mu(B \backslash A) = \mu(B) . \tag{2.4}$$

By adding (2.3) and (2.4) we find

$$\mu(A \cup B) + \mu(A \cap B) + \mu(B \backslash A) = \mu(A) + \mu(B) + \mu(B \backslash A) .$$

If $\mu(B \backslash A)$ is finite, the claim follows. If $\mu(B \backslash A) = \infty$, we get $\mu(A \cup B) = \mu(B) = \infty$ from (2.3) and (2.4), and the claim is again verified.

(ii) Since $A \subset B$ we have $B = A \cup (B \backslash A)$; but A and $B \backslash A$ are disjoint, so $\mu(B) = \mu(A) + \mu(B \backslash A)$. By assumption, $\mu(A) < \infty$, and we find $\mu(B) - \mu(A) = \mu(B \backslash A)$.

(iii) As in (ii) we have $\mu(B) = \mu(A) + \mu(B \backslash A)$ and thus $\mu(B) \geq \mu(A)$.

(iv) We set $A_{-1} := \emptyset$ and $B_k := A_k \backslash A_{k-1}$ for $k \in \mathbb{N}$. By assumption, (B_k) is a disjoint sequence in $\mathcal{A}$ with $\bigcup_{k=0}^{\infty} B_k = \bigcup_{j=0}^{\infty} A_j$ and $\bigcup_{k=0}^{m} B_k = A_m$. The σ-additivity of μ therefore implies

$$\mu\Big(\bigcup_j A_j\Big) = \mu\Big(\bigcup_k B_k\Big) = \lim_{m \to \infty} \sum_{k=0}^{m} \mu(B_k) = \lim_{m \to \infty} \mu\Big(\bigcup_{k=0}^{m} B_k\Big) = \lim_{m \to \infty} \mu(A_m) .$$

(v) If (A_k) is a decreasing sequence in $\mathcal{A}$, then $(A_0 \backslash A_k)$ is increasing. Further,

$$A_0 \backslash \Big(\bigcap_k A_k\Big) = A_0 \cap \Big(\bigcap_k A_k\Big)^c = \bigcup_k \Big(A_0 \cap A_k^c\Big) = \bigcup_k (A_0 \backslash A_k) .$$

Using (ii) and (iv), we get

$$\mu(A_0) - \mu\Big(\bigcap_k A_k\Big) = \mu\Big(A_0 \backslash \Big(\bigcap_k A_k\Big)\Big) = \mu\Big(\bigcup_k (A_0 \backslash A_k)\Big)$$
$$= \lim_{m \to \infty} \mu(A_0 \backslash A_m) = \mu(A_0) - \lim_{m \to \infty} \mu(A_m) ,$$

from which the claim follows.

(vi) Set $B_0 := A_0$ and $B_k := A_k \backslash \big(\bigcup_{j=0}^{k-1} A_j\big)$ for $k \in \mathbb{N}^\times$. The sequence (B_k) in $\mathcal{A}$ satisfies $\bigcup_k B_k = \bigcup_k A_k$ and $B_k \subset A_k$ for $k \in \mathbb{N}$. From (iii) and the σ-additivity of μ, we have

$$\mu\Big(\bigcup_k A_k\Big) = \mu\Big(\bigcup_k B_k\Big) = \sum_k \mu(B_k) \leq \sum_k \mu(A_k) . \quad \blacksquare$$

2.4 Remarks   (a) Parts (iv) and (v) of Proposition 2.3 express the **continuity of measures from below** and **from above**, respectively.

(b) Parts (i)–(iii) clearly remain true when $\mathcal{A}$ is an algebra and $\mu : \mathcal{A} \to [0, \infty]$ is additive.

(c) If $\mathcal{S}$ is an algebra over X and $\mu : \mathcal{S} \to [0, \infty]$ is additive, monotone, and σ-finite, there is a disjoint sequence (B_k) in $\mathcal{S}$ such that $\bigcup_k B_k = X$ and $\mu(B_k) < \infty$ for $k \in \mathbb{N}$.

Proof   Because of the σ-finiteness of μ, there is a sequence (A_j) in $\mathcal{S}$ with $\bigcup_j A_j = X$ and $\mu(A_j) < \infty$. Setting $B_0 := A_0$ and $B_k := A_k \setminus \bigcup_{j=0}^{k-1} A_j$ for $k \in \mathbb{N}^\times$, we find easily that (B_k) has the stated properties. ∎

Null sets

Suppose $(X, \mathcal{A}, \mu)$ is a measure space. A set $N \in \mathcal{A}$ with $\mu(N) = 0$ is said to be **μ-null**. We denote the set of all μ-null sets by $\mathcal{N}_\mu$. A measure μ or measure space $(X, \mathcal{A}, \mu)$ is called **complete** if any subset of a μ-null set lies in $\mathcal{A}$.

2.5 Remarks   (a) For $M \in \mathcal{A}$ and $N \in \mathcal{N}_\mu$ such that $M \subset N$, we have $M \in \mathcal{N}_\mu$.

Proof   This follows from the monotony of μ. ∎

(b) Countable unions of μ-null sets are μ-null.

Proof   This follows from the σ-subadditivity of μ. ∎

(c) A measure μ is complete if and only if *every* subset of a μ-null set is μ-null.

Proof   This is a consequence of (a). ∎

(d) If $\mathcal{A} = \mathfrak{P}(X)$, then μ is complete. For example, the Dirac measure and the counting measure are complete. ∎

We denote by

$$\mathcal{M}_\mu := \{\, M \subset X \; ; \; \exists\, N \in \mathcal{N}_\mu \text{ such that } M \subset N \,\}$$

the set of all subsets of μ-null sets. Clearly μ is complete if and only if $\mathcal{M}_\mu$ is contained in $\mathcal{A}$. Thus, for an incomplete measure space,[4]

$$\overline{\mathcal{A}}_\mu := \{\, A \cup M \; ; \; A \in \mathcal{A},\ M \in \mathcal{M}_\mu \,\}$$

does augment $\mathcal{A}$. The next proposition shows that $\overline{\mathcal{A}}_\mu$ is a σ-algebra admitting a complete measure that agrees with μ on $\mathcal{A}$.

---

[4]Corollary 5.29 will show that there are incomplete measure spaces.

2.6 Proposition *Suppose $(X, \mathcal{A}, \mu)$ is a measure space.*

(a) *For $A \in \mathcal{A}$ and $M \in \mathcal{M}_\mu$, define $\bar{\mu}(A \cup M) := \mu(A)$. Then $\bar{\mu}$ is a well defined set function on $\bar{\mathcal{A}}_\mu$ and $(X, \bar{\mathcal{A}}_\mu, \bar{\mu})$ is a complete measure space with $\bar{\mu} \supset \mu$ (that is, $\bar{\mu}$ extends μ).*

(b) *If $(X, \mathcal{B}, \nu)$ is a complete measure space with $\nu \supset \mu$, then $\nu \supset \bar{\mu}$.*

Proof (i) We first show that $\bar{\mathcal{A}}_\mu$ is a σ-algebra. Manifestly X belongs to $\bar{\mathcal{A}}_\mu$. Suppose $A_0 \in \bar{\mathcal{A}}_\mu$. Then there are $A \in \mathcal{A}$, $N \in \mathcal{N}_\mu$ and $M \subset N$ such that $A_0 = A \cup M$. From $M \subset N$, it follows that $M^c = N^c \cup (N \cap M^c)$, and we find

$$A_0^c = A^c \cap M^c = (A^c \cap N^c) \cup (A^c \cap N \cap M^c) \,.$$

Because A and N belong to $\mathcal{A}$, so does $A^c \cap N^c$. Moreover $A^c \cap N \cap M^c$ lies in $\mathcal{M}_\mu$, because it is contained in the μ-null set N. Therefore $A_0^c \in \bar{\mathcal{A}}_\mu$. Finally, let (B_j) be a sequence in $\bar{\mathcal{A}}_\mu$. There are sequences (A_j) in $\mathcal{A}$, (N_j) in $\mathcal{N}_\mu$, and (M_j) in $\mathfrak{P}(X)$ such that $M_j \subset N_j$ and $B_j = A_j \cup M_j$ for $j \in \mathbb{N}$. Because $\bigcup N_j$ is a μ-null set that contains $\bigcup M_j$ and because $\mathcal{A}$ is a σ-algebra, we have

$$\bigcup B_j = \left(\bigcup A_j \right) \cup \left(\bigcup M_j \right) \in \bar{\mathcal{A}}_\mu \,.$$

(ii) We show that the set function $\bar{\mu}: \bar{\mathcal{A}}_\mu \to [0, \infty]$ is well defined. Take $A_1, A_2 \in \mathcal{A}$ and $M_1, M_2 \in \mathcal{M}_\mu$ with $A_1 \cup M_1 = A_2 \cup M_2$, and suppose N is a μ-null set with $M_2 \subset N$. Then $A_1 \subset A_1 \cup M_1 \subset A_2 \cup N$, and Proposition 2.3 yields

$$\mu(A_1) \leq \mu(A_2 \cup N) = \mu(A_2) + \mu(N) - \mu(A_2 \cap N) = \mu(A_2) \,.$$

Analogously, we show $\mu(A_2) \leq \mu(A_1)$. Therefore $\bar{\mu}$ is well defined.

(iii) Suppose A_0 is a $\bar{\mu}$-null set and take $B \subset A_0$. There exist $A, N \in \mathcal{N}_\mu$ and $M \subset N$ such that $A_0 = A \cup M$. Therefore $B \subset A_0 \subset A \cup N$, and thus B belongs to $\mathcal{M}_\mu \subset \bar{\mathcal{A}}_\mu$, that is, $\bar{\mu}$ is complete.

(iv) By construction $\bar{\mu}$ is an extension of μ, and it is easy to see that $\bar{\mu}$ is also σ-additive. This proves (a).

(v) Suppose $(X, \mathcal{B}, \nu)$ is a measure space with $\mathcal{B} \supset \mathcal{A}$ and $\nu \,|\, \mathcal{A} = \mu$. Then $\mathcal{N}_\mu \subset \mathcal{N}_\nu$, hence also $\mathcal{M}_\mu \subset \mathcal{M}_\nu$. If ν is complete, so that $\mathcal{M}_\nu \subset \mathcal{B}$, we obtain $\mathcal{M}_\mu \subset \mathcal{B}$, and therefore $\bar{\mathcal{A}}_\mu \subset \mathcal{B}$. This proves (b). ∎

Part (b) of Proposition 2.6 says that $(X, \bar{\mathcal{A}}_\mu, \bar{\mu})$ is the minimal complete extension of $(X, \mathcal{A}, \mu)$. We call $(X, \bar{\mathcal{A}}_\mu, \bar{\mu})$ and $\bar{\mu}$ the **completion** of $(X, \mathcal{A}, \mu)$; we also say $\bar{\mu}$ is the completion of μ. An important example of this construction will surface in Theorem 5.8.

Exercises

1 For $A \subset X$, let $\mathcal{A} := \mathcal{A}_\sigma(\{A\})$. Put $\mu(\emptyset) := 0$ and $\mu(B) := \infty$ otherwise. Show that $(X, \mathcal{A}, \mu)$ is a complete measure space.

2 Suppose $(X, \mathcal{A}, \mu)$ is a measure space and take $A_1, \ldots, A_n \in \mathcal{A}$ for $n \in \mathbb{N}^\times$. Show that

$$\mu\left(\bigcup_{j=1}^n A_j\right) = \sum_{k=1}^n (-1)^{k+1} \sum_{1 \le j_1 < \cdots < j_k \le n} \mu\left(\bigcap_{\ell=1}^k A_{j_\ell}\right) .$$

3 Find in the measure space $(\mathbb{N}, \mathfrak{P}(\mathbb{N}), \mathcal{H}^0)$ a decreasing sequence $(A_j) \in \mathfrak{P}(\mathbb{N})^\mathbb{N}$ for which $\lim_j \mathcal{H}^0(A_j)$ exists but $\mathcal{H}^0(\bigcap_j A_j) \ne \lim_j \mathcal{H}^0(A_j)$.

4 Suppose $(X, \mathcal{A})$ is a measurable space and $\mu : \mathcal{A} \to [0, \infty]$ is additive and continuous from below. Prove that $(X, \mathcal{A}, \mu)$ is a measure space.

5 Let $(X, \mathcal{A}, \mu)$ be a measure space and take $B \in \mathcal{A}$. For $A \in \mathcal{A}$, set $\mu_B(A) := \mu(A \cap B)$. Show that $(X, \mathcal{A}, \mu_B)$ is a measure space.

6 Let $(X, \mathcal{A}, \mu)$ be a measure space and consider a sequence $(A_j) \in \mathcal{A}^\mathbb{N}$. Prove the following statements:

(a) $\mu\left(\varliminf_j A_j\right) \le \varliminf_j \mu(A_j)$.

(b) $\mu\left(\varlimsup_j A_j\right) \ge \varlimsup_j \mu(A_j)$ if there exists $k \in \mathbb{N}$ such that $\mu\left(\bigcup_{j=k}^\infty A_j\right) < \infty$.

(c) If there is a $k \in \mathbb{N}$ such that $\mu\left(\bigcup_{j=k}^\infty A_j\right) < \infty$ and the sequence (A_j) converges, then $\mu(\lim_j A_j) = \lim_j \mu(A_j)$.

7 Show that for every measure space $(X, \mathcal{A}, \mu)$ we have $(X, \overline{\mathcal{A}}_\mu, \bar{\mu}) = (X, \overline{[\overline{\mathcal{A}_\mu}]_{\bar{\mu}}}, \overline{[\bar{\mu}]})$.

8 Suppose $(X, \mathcal{A}, \mu)$ and $(X, \mathcal{A}, \nu)$ are finite measure spaces. Prove or disprove that

$$(X, \overline{\mathcal{A}}, \bar{\mu}) = (X, \overline{\mathcal{A}}, \bar{\nu}) \iff \mathcal{N}_\mu = \mathcal{N}_\nu .$$

9 Suppose $(X, \mathcal{A})$ is a measure space and $\mathcal{N} \subset \mathcal{A}$ satisfies

  (i) $\emptyset \in \mathcal{N}$;

  (ii) $(A_j) \in \mathcal{N}^\mathbb{N} \Rightarrow \bigcup_j A_j \in \mathcal{N}$;

  (iii) $(A \in \mathcal{A},\ B \in \mathcal{N},\ A \subset B) \Rightarrow A \in \mathcal{N}$.

Construct a measure μ on $(X, \mathcal{A})$ such that $\mathcal{N}_\mu = \mathcal{N}$.

10 Suppose X is uncountable and $\mathcal{A} := \{ A \subset X \ ;\ A \text{ or } A^c \text{ is countable} \}$. For $A \in \mathcal{A}$, set $\mu(A) := 0$ if A is countable and $\mu(A) := \infty$ otherwise. Show that $(X, \mathcal{A}, \mu)$ is a complete measure space.

11 Suppose $(X, \mathcal{A}, \mu)$ is a measure space. We call $A \in \mathcal{A}$ a **μ-atom** if $\mu(A) > 0$ and, for every $B \in \mathcal{A}$ such that $B \subset A$, either $\mu(B) = 0$ or $\mu(A \backslash B) = 0$.

(a) Prove:

  (i) Let A be a μ-atom and take $B \in \mathcal{A}$ with $B \subset A$. Then either $\mu(B) = \mu(A)$ or $\mu(B) = 0$.

  (ii) Suppose that $A \in \mathcal{A}$ satisfies $0 < \mu(A) < \infty$, and that for every $B \in \mathcal{A}$ with $B \subset A$, either $\mu(B) = \mu(A)$ or $\mu(B) = 0$. Then A is a μ-atom.

  (iii) Suppose μ is σ-finite and $A \in \mathcal{A}$ is a μ-atom. Then $\mu(A) < \infty$.

(b) Determine all atoms of the counting measure $\mathcal{H}^0$. Repeat for the measures of Exercises 1 and 10.

12 Suppose $(X, \mathcal{A}, \mu)$ is a complete measure space. Let $A \in \mathcal{A}$ be a μ-atom and suppose $B \subset A$. Is B measurable? Justify your answer.

13 Suppose $(X, \mathcal{A}, \mu)$ is a complete measure space, and take $A, N \in \mathcal{A}$ with $\mu(A) > 0$ and $\mu(N) = 0$. Show that $\mu(A \cap N^c) > 0$.

3   Outer measures

Until now, all measures we've encountered have been of the banal variety. None of them would do for measuring, say, surface areas, if we want the result to agree with the familiar geometric one in the simplest case of a rectangle!

In this section, we lay the foundation for the introduction of more interesting classes of measures. We first construct set functions, called "outer measures", that are defined on *all* subsets of a given set and have some, although not all, the properties of measures. We will see important examples thereof. In later sections we then obtain many actual measures as appropriately chosen restrictions of outer measures.

As always, we suppose

- X is a nonempty set.

The construction of outer measures

A map $\mu^* : \mathfrak{P}(X) \to [0, \infty]$ such that $\mu^*(\emptyset) = 0$ is called an **outer measure** on X if it is increasing and σ-subadditive. A subset $\mathcal{K}$ of $\mathfrak{P}(X)$ is said to be a **conforming cover** for X if it contains the empty set as well as elements K_j, for $j \in \mathbb{N}$, such that $X = \bigcup_j K_j$.

3.1 Remarks   (a) Any outer measure on X is already defined on all of $\mathfrak{P}(X)$.

(b) Every measure defined on $\mathfrak{P}(X)$ is an outer measure on X.

Proof   This follows from Proposition 2.3(vi). ■

(c) For $A \subset X$ set

$$\mu^*(A) := \begin{cases} 0, & A = \emptyset, \\ 1, & A \neq \emptyset. \end{cases}$$

Then μ^* is an outer measure on X, and it is a measure if and only if X has a single point.

(d) $\{\emptyset, X\}$ is a conforming cover for X.

(e) For each $a, b \in \mathbb{R}^n$ let $A(a, b)$ be some subset of $\mathbb{R}^n$ with $(a, b) \subset A(a, b) \subset [a, b]$. Then $\{ A(a, b) \; ; \; a, b \in \mathbb{R}^n \}$ is a conforming cover for $\mathbb{R}^n$; in particular, so is $\mathbb{J}(n)$.

(f) If $(X, \mathcal{T})$ is a topological space, $\mathcal{T}$ is a conforming cover for X.

(g) Suppose X is a separable metric space and $\mathcal{T}$ is the corresponding topology. For any $\varepsilon > 0$, the set $\{ O \in \mathcal{T} \; ; \; \operatorname{diam}(O) < \varepsilon \}$ is a conforming cover for X.

Proof   By Proposition 1.8, X is Lindelöf, which implies the claim. ■

The next theorem allows the systematic construction of outer measures.

3.2 Theorem *Suppose $\mathcal{K}$ is a conforming cover for X and $\nu : \mathcal{K} \to [0, \infty]$ satisfies $\nu(\emptyset) = 0$. For $A \subset X$, set*

$$\mu^*(A) := \inf\left\{ \sum_{j=0}^{\infty} \nu(K_j) \; ; \; (K_j) \in \mathcal{K}^{\mathbb{N}}, \; A \subset \bigcup_j K_j \right\}.$$

*Then μ^* is an outer measure on X, said to be **induced** by $(\mathcal{K}, \nu)$.*

Proof It is clear that $\mu^* : \mathfrak{P}(X) \to [0, \infty]$ is increasing and that $\mu^*(\emptyset) = 0$. To check σ-subadditivity, suppose (A_j) is a sequence in $\mathfrak{P}(X)$. For every $\varepsilon > 0$ and every $j \in \mathbb{N}$, there is a sequence $(K_{j,k})_{k \in \mathbb{N}}$ in $\mathcal{K}$ with

$$A_j \subset \bigcup_k K_{j,k} \quad \text{and} \quad \sum_k \nu(K_{j,k}) \le \mu^*(A_j) + \varepsilon / 2^{j+1}.$$

Then $\bigcup_j A_j \subset \bigcup_j \bigcup_k K_{j,k}$ and we get

$$\mu^*\left(\bigcup_j A_j \right) \le \sum_j \sum_k \nu(K_{j,k})$$
$$\le \sum_j \left(\mu^*(A_j) + \varepsilon / 2^{j+1} \right) = \left(\sum_j \mu^*(A_j) \right) + \varepsilon.$$

Because $\varepsilon > 0$ is arbitrary, σ-subadditivity follows. ∎

The Lebesgue outer measure

For $a, b \in \mathbb{R}^n$, the **n-dimensional volume** of the interval (a, b) in $\mathbb{R}^n$ is defined as

$$\text{vol}_n(a, b) := \left\{ \begin{array}{ll} \prod_{j=1}^{n} (b_j - a_j), & a \le b, \\ 0 & \text{otherwise}. \end{array} \right.$$

If (a, b) is nonempty, this coincides with the product of the n edge lengths of (a, b) — more specifically, with the everyday notion of length of an interval, area of a rectangle, and volume of a parallelepiped, for $n = 1, 2$, and 3 respectively.

3.3 Proposition *For $A \subset \mathbb{R}^n$, let*

$$\lambda_n^*(A) := \inf\left\{ \sum_{j=0}^{\infty} \text{vol}_n(I_j) \; ; \; I_j \in \mathbb{J}(n), \; j \in \mathbb{N}, \; \bigcup_{j=0}^{\infty} I_j \supset A \right\}.$$

*Then λ_n^* is an outer measure on $\mathbb{R}^n$, called **n-dimensional Lebesgue outer measure**. For $a, b \in \mathbb{R}^n$ and $(a, b) \subset A \subset [a, b]$, we have $\lambda^*(A) = \text{vol}_n(a, b)$.*

Proof (i) The first claim follows from Remark 3.1(e) and Theorem 3.2.

(ii) Suppose $a, b \in \mathbb{R}^n$, and set $I_0 := (a, b)$ and $I_j := \emptyset$ for $j \in \mathbb{N}^\times$. Obviously this defines a sequence of intervals in $\mathbb{R}^n$ such that $(a, b) \subset \bigcup_j I_j$. Therefore

$$\lambda_n^*((a, b)) \leq \sum_j \mathrm{vol}_n(I_j) = \mathrm{vol}_n(a, b) . \tag{3.1}$$

(iii) Let $\mathcal{A}_0$ be the set of all subsets[1] J of $[a, b]$ such that $a, b \in \mathbb{R}^n$ with $a_k = b_k$ for some $k \in \{1, \ldots, n\}$. Clearly, for every $J \in \mathcal{A}_0$ and $\varepsilon > 0$, there exists $I_\varepsilon \in \mathbb{J}(n)$ such that $J \subset I_\varepsilon$ and $\mathrm{vol}_n(I_\varepsilon) < \varepsilon$. Thus $\lambda_n^*(J) = 0$ for $J \in \mathcal{A}_0$. Now, given $a, b \in \mathbb{R}^n$, there are $2n$ "faces" $J_j \in \mathcal{A}_0$ such that

$$[a, b] = (a, b) \cup \bigcup_{j=1}^{2n} J_j .$$

From this and Remark 2.1(a), it follows that

$$\lambda_n^*([a, b]) \leq \lambda_n^*((a, b)) + \sum_{j=1}^{2n} \lambda_n^*(J_j) = \lambda_n^*((a, b)) .$$

For $(a, b) \subset A \subset [a, b]$ we conclude from the monotony of λ_n^* that

$$\lambda_n^*((a, b)) = \lambda_n^*(A) = \lambda_n^*([a, b]) . \tag{3.2}$$

(iv) Suppose (I_j) is a sequence in $\mathbb{J}(n)$ such that $[a, b] \subset \bigcup_j I_j$. Since $[a, b]$ is compact, there exists $N \in \mathbb{N}$ such that $[a, b] \subset \bigcup_{j=0}^N I_j$. Thus, by Exercise 1 below,

$$\mathrm{vol}_n(a, b) \leq \sum_{j=0}^N \mathrm{vol}_n(I_j) \leq \sum_{j=0}^\infty \mathrm{vol}_n(I_j) ,$$

and we find by taking the infimum that $\mathrm{vol}_n(a, b) \leq \lambda_n^*([a, b])$. Together with (3.1) and (3.2), this yields the claim. ■

For $a, b \in \mathbb{R}^n$, suppose $J(a, b)$ is an interval in $\mathbb{R}^n$ such that $(a, b) \subset J(a, b) \subset [a, b]$. Then Proposition 3.3 shows that

$$\lambda_n^*(J(a, b)) = \mathrm{vol}_n(a, b) . \tag{3.3}$$

For this reason, we set

$$\mathrm{vol}_n J(a, b) := \lambda_n^*(J(a, b))$$

and again we call $\mathrm{vol}_n J(a, b)$ the **n-dimensional volume** of the interval $J(a, b)$. Formula (3.3) says, informally, that the boundary faces do not contribute to the volume of an n-dimensional box.

---

[1] J itself need not be an interval in $\mathbb{R}^n$.

Any interval of the form $J = [a, b)$, for $a, b \in \mathbb{R}^n$, is said to be **left closed**, and one of the form $J = (a, b]$ is called **right closed**. We denote the set of all left closed intervals in $\mathbb{R}^n$ by $\mathbb{J}_\ell(n)$, and that of right closed intervals by $\mathbb{J}_r(n)$. By $\bar{\mathbb{J}}(n)$ we denote the set of intervals in $\mathbb{R}^n$ that are bounded and closed (on both sides).

The next result shows that, in the definition of the Lebesgue outer measure, we can use left, right, or both-sided closed intervals instead of open intervals.

3.4 Proposition *Suppose $A \subset \mathbb{R}^n$ and $\mathbb{J} \in \{\mathbb{J}_\ell(n), \mathbb{J}_r(n), \bar{\mathbb{J}}(n)\}$. Then*

$$\lambda_n^*(A) = \inf\left\{ \sum_{j=0}^{\infty} \mathrm{vol}_n(J_j) \; ; \; J_j \in \mathbb{J}, \; j \in \mathbb{N}, \; \bigcup_{j=0}^{\infty} J_j \supset A \right\} .$$

Proof We consider the case $\mathbb{J} = \mathbb{J}_\ell(n)$. For $J = (a, b) \in \mathbb{J}(n)$, let $\ell J := [a, b)$.

If a sequence $(J_j) \in \mathbb{J}(n)^{\mathbb{N}}$ covers A, so does the sequence $(\ell J_j) \in \mathbb{J}^{\mathbb{N}}$. Hence there are no fewer sequences in $\mathbb{J}$ covering A than there are in $\mathbb{J}(n)$. So (3.3) and the definition of $\lambda_n^*(A)$ imply

$$\inf\left\{ \sum_{j=0}^{\infty} \mathrm{vol}_n(J_j) \; ; \; J_j \in \mathbb{J}, \; j \in \mathbb{N}, \; \bigcup_j J_j \supset A \right\}$$

$$\leq \inf\left\{ \sum_{j=0}^{\infty} \mathrm{vol}_n(\ell J_j) \; ; \; J_j \in \mathbb{J}(n), \; j \in \mathbb{N}, \; \bigcup_j J_j \supset A \right\} = \lambda_n^*(A) . \tag{3.4}$$

Suppose (J_j) is a sequence in $\mathbb{J}$ that covers A, and take $\varepsilon > 0$. For $a_j, b_j \in \mathbb{R}^n$ and $J_j = [a_j, b_j)$, set

$$J_j^\varepsilon := \left(a_j - \varepsilon(b_j - a_j), b_j\right) \quad \text{for } j \in \mathbb{N} .$$

Then (J_j^ε) is a sequence in $\mathbb{J}(n)$ that covers A, and

$$\sum_{j=0}^{\infty} \mathrm{vol}_n(J_j^\varepsilon) = \sum_{j=0}^{\infty} (1 + \varepsilon)^n \, \mathrm{vol}_n(J_j) = \left(\sum_{j=0}^{\infty} \mathrm{vol}_n(J_j)\right)(1 + \varepsilon)^n .$$

From this, it follows that

$$\lambda_n^*(A) = \inf\left\{ \sum_{j=0}^{\infty} \mathrm{vol}_n(I_j) \; ; \; I_j \in \mathbb{J}(n), \; j \in \mathbb{N}, \; \bigcup_j I_j \supset A \right\}$$

$$\leq \inf\left\{ \sum_{j=0}^{\infty} \mathrm{vol}_n(J_j^\varepsilon) \; ; \; J_j \in \mathbb{J}, \; j \in \mathbb{N}, \; \bigcup_j J_j \supset A \right\}$$

$$= \inf\left\{ \sum_{j=0}^{\infty} \mathrm{vol}_n(J_j) \; ; \; J_j \in \mathbb{J}, \; j \in \mathbb{N}, \; \bigcup_j J_j \supset A \right\}(1 + \varepsilon)^n .$$

Since $\varepsilon > 0$ is arbitrary, we see that

$$\lambda_n^*(A) \leq \inf\left\{ \sum_{j=0}^{\infty} \mathrm{vol}_n(J_j) \; ; \; J_j \in \mathbb{J}, \; j \in \mathbb{N}, \; \bigcup_j J_j \supset A \right\} .$$

Now the claim follows from (3.4). Obvious modifications achieve the proof for the cases $\mathbb{J} = \mathbb{J}_r(n)$ and $\mathbb{J} = \bar{\mathbb{J}}(n)$. ∎

The Lebesgue–Stieltjes outer measure

Let $F\colon \mathbb{R} \to \mathbb{R}$ be increasing and continuous from the left. We say that F is a **measure-generating function**. If $\lim_{x\to-\infty} F(x) = 0$ and $\lim_{x\to\infty} F(x) = 1$, we also say F is a **(probability) distribution function**. If F is a measure-generating function, we set

$$\nu_F\big([a,b)\big) := \begin{cases} F(b) - F(a)\,, & a < b\,, \\ 0\,, & a \geq b\,, \end{cases}$$

for $a, b \in \mathbb{R}$. Because F is increasing, ν_F is an increasing map into $\mathbb{R}$ from the set of intervals of the form $[a,b)$ with $a, b \in \mathbb{R}$.

3.5 Proposition *Suppose F is a measure-generating function, and for $A \subset \mathbb{R}$ let*

$$\mu_F^*(A) := \inf\Big\{ \sum_{j=0}^{\infty} \nu_F(I_j)\,;\ I_j = [a_j, b_j),\ a_j, b_j \in \mathbb{R} \text{ with } A \subset \bigcup_{j=0}^{\infty} I_j \Big\}\,.$$

*Then μ_F^* is an outer measure on $\mathbb{R}$, the **Lebesgue–Stieltjes outer measure arising from F**. For $-\infty < a < b < \infty$, we have $\mu_F^*\big([a,b)\big) = F(b) - F(a)$.*

Proof (i) That μ_F^* is an outer measure follows from Remark 3.1(e) and Theorem 3.2.

(ii) Suppose $a, b \in \mathbb{R}$ with $a < b$. We set $I_0 := [a, b)$ and $I_j := \emptyset$ for $j \in \mathbb{N}^\times$. Then $[a, b) \subset \bigcup_j I_j$ and

$$\mu_F^*\big([a,b)\big) \leq \sum_{j=0}^{\infty} \nu_F(I_j) = \nu_F(I_0) = F(b) - F(a)\,. \tag{3.5}$$

(iii) Now let $I_j := [a_j, b_j)$ for $j \in \mathbb{N}$ be such that $[a, b) \subset \bigcup_j I_j$, and take $\varepsilon > 0$. Because F is continuous from the left, there are positive numbers c and c_j such that

$$F(b) - F(b - c) < \varepsilon/2\,, \quad F(a_j) - F(a_j - c_j) < \varepsilon 2^{-(j+2)} \quad \text{for } j \in \mathbb{N}\,, \tag{3.6}$$

and $[a, b{-}c] \subset \bigcup_j (a_j{-}c_j, b_j)$. Because $[a, b{-}c]$ is compact, there is an N such that $[a, b{-}c] \subset \bigcup_{j=0}^{N} (a_j{-}c_j, b_j)$. Now the monotony of F implies that

$$F(b - c) - F(a) \leq \sum_{j=0}^{N}\big(F(b_j) - F(a_j - c_j)\big) \leq \sum_{j=0}^{\infty}\big(F(b_j) - F(a_j - c_j)\big)\,.$$

Together with (3.6), this yields

$$F(b) - F(a) = F(b) - F(b - c) + F(b - c) - F(a)$$
$$\leq \sum_{j=0}^{\infty}\big[F(b_j) - F(a_j) + \varepsilon 2^{-(j+2)}\big] + \varepsilon/2 \leq \sum_{j=0}^{\infty}\big[F(b_j) - F(a_j)\big] + \varepsilon\,.$$

This is true for every $\varepsilon > 0$, so $F(b) - F(a) \leq \sum_{j=0}^{\infty} \nu_F(I_j)$, which gives the desired equality in view of (3.5). ∎

3.6 Remarks (a) In the case $F(x) := x$ for $x \in \mathbb{R}$, we have $\mu_F^* = \lambda_1^*$.

Proof This follows from Proposition 3.4. ∎

(b) If one replaces "continuous from the left" in the definition of a measure-generating function by "continuous from the right", Proposition 3.5 remains true if one replaces all left closed intervals by right closed ones. ∎

Hausdorff outer measures

Suppose X is a separable metric space, and let $\mathcal{T}$ be the topology induced by the metric. For $s > 0$, $\varepsilon > 0$, and $A \subset X$, we set

$$\mathcal{H}_\varepsilon^s(A) := \inf\left\{ \sum_{j=0}^\infty [\operatorname{diam} O_j]^s \; ; \; O_j \in \mathcal{T}, \; \operatorname{diam}(O_j) < \varepsilon, \; A \subset \bigcup_{j=0}^\infty O_j \right\}.$$

According to Remark 3.1(g) and Theorem 3.2, $\mathcal{H}_\varepsilon^s$ is an outer measure on X. Also $\mathcal{H}_{\varepsilon_1}^s \leq \mathcal{H}_{\varepsilon_2}^s$ for $\varepsilon_1 > \varepsilon_2$, because for ε_1 there are more sets available for covering than for ε_2. Therefore (see Proposition II.5.3)

$$\mathcal{H}_*^s(A) := \lim_{\varepsilon \to 0+} \mathcal{H}_\varepsilon^s(A) = \sup_{\varepsilon > 0} \mathcal{H}_\varepsilon^s(A)$$

exists for all $s > 0$ and $A \subset X$. We call $\mathcal{H}_*^s$ the **s-dimensional Hausdorff outer measure** on X. For completeness, we define the 0-dimensional Hausdorff (outer) measure by $\mathcal{H}_*^0 := \mathcal{H}^0$, where $\mathcal{H}^0$ is the counting measure on X.

3.7 Proposition *For every $s \geq 0$, $\mathcal{H}_*^s$ is an outer measure on X.*

Proof The case $s = 0$ is covered by Remark 3.1(b) since $\mathcal{H}^0$ is a measure — see Example 2.2(b). So suppose $s > 0$. Obviously $\mathcal{H}_*^s$ is an increasing map from $\mathfrak{P}(X)$ into $[0, \infty]$ with $\mathcal{H}_*^s(\emptyset) = 0$. To show σ-subadditivity, let (A_j) be a sequence in $\mathfrak{P}(X)$. Because $\mathcal{H}_\varepsilon^s(A)$ is an outer measure on X for every $\varepsilon > 0$, we have

$$\mathcal{H}_\varepsilon^s\left(\bigcup_j A_j\right) \leq \sum_j \mathcal{H}_\varepsilon^s(A_j) \leq \sum_j \mathcal{H}_*^s(A_j).$$

Taking the limit $\varepsilon \to 0$ we obtain the claim. ∎

Exercises

1 Prove:

(a) $I, J \in \mathbb{J}(n) \Rightarrow I \cap J \in \mathbb{J}(n)$.

(b) Suppose $I_0, \ldots, I_k \in \mathbb{J}(n)$ and I is an interval such that $I \subset \bigcup_{j=0}^n I_j$. Then $\operatorname{vol}_n(I) \leq \sum_{j=0}^k \operatorname{vol}_n(I_j)$. (Prove this without using Proposition 3.3.)

2 (a) Let μ be a measure on the Borel σ-algebra $\mathcal{B}^1$ and suppose $\mu((-\infty, x))$ is finite for $x \in \mathbb{R}$. Further let

$$F_\mu(x) := \mu((-\infty, x)) \quad \text{for } x \in \mathbb{R}.$$

Show that F_μ is a measure-generating function with $\lim_{x \to -\infty} F_\mu(x) = 0$.

(b) Determine F_{δ_0}, where δ_0 denotes the Dirac measure on $(\mathbb{R}, \mathcal{B}^1)$ with support at 0.

3 Suppose $f : \mathbb{R} \to [0, \infty)$ in improperly integrable and

$$F_f(x) := \int_{-\infty}^{x} f(\xi)\, d\xi \quad \text{for } x \in \mathbb{R} .$$

Verify that F_f is a measure-generating function for which $\mu_{F_f}^*\big([a,b)\big) = \int_a^b f(\xi)\, d\xi$ when $-\infty < a < b < \infty$.

4 Suppose $A \subset \mathbb{R}^n$. Prove:

(a) $\mathcal{H}_*^s(A) = \lim_{\varepsilon \to 0+} \inf\big\{ \sum_{k=0}^{\infty} [\operatorname{diam}(A_k)]^s \ ; \ A_k \subset \mathbb{R}^n, \ \operatorname{diam}(A_k) \leq \varepsilon, \ k \in \mathbb{N}, \ A \subset \bigcup_k A_k \big\}$.

(b) If $f : A \to \mathbb{R}^m$ is Lipschitz continuous with Lipschitz constant λ, then

$$\mathcal{H}_*^s\big(f(A)\big) \leq \lambda^s \mathcal{H}_*^s(A) .$$

(c) For every isometry $\varphi : \mathbb{R}^n \to \mathbb{R}^n$, we have $\mathcal{H}_*^s\big(\varphi(A)\big) = \mathcal{H}_*^s(A)$. Thus the Hausdorff outer measure on $\mathbb{R}^n$ is invariant under isometries, that is, an invariant of motion.[2]

(d) Suppose $\bar{n} > n$ and $\overline{\mathcal{H}}_*^s$ is the Hausdorff outer measure on $\mathbb{R}^{\bar{n}}$. Then $\overline{\mathcal{H}}_*^s(A) = \mathcal{H}_*^s(A)$. That is, the Hausdorff outer measure is independent of the dimension of the ambient $\mathbb{R}^n$.

5 Suppose $A \subset \mathbb{R}^n$ and $0 \leq s < t < \infty$. Show these facts:

(a) $\mathcal{H}_*^s(A) < \infty \Rightarrow \mathcal{H}_*^t(A) = 0$.

(b) $\mathcal{H}_*^t(A) > 0 \ \Rightarrow \mathcal{H}_*^s(A) = \infty$.

(c) $\inf\big\{ s > 0 \ ; \ \mathcal{H}_*^s(A) = 0 \big\} = \sup\big\{ s \geq 0 \ ; \ \mathcal{H}_*^s(A) = \infty \big\}$. The unique number

$$\dim_H(A) := \inf\big\{ s > 0 \ ; \ \mathcal{H}_*^s(A) = 0 \big\}$$

is called the **Hausdorff dimension** of A.

6 Let A, B, and A_j, for $j \in \mathbb{N}$, be subsets of $\mathbb{R}^n$. Prove:

(a) $0 \leq \dim_H(A) \leq n$.

(b) If A is open and not empty, then $\dim_H(A) = n$.

(c) $A \subset B \Rightarrow \dim_H(A) \leq \dim_H(B)$.

(d) $\dim_H\big(\bigcup_j A_j\big) = \sup_j\big\{\dim_H(A_j)\big\}$.

(e) If A is countable, it has Hausdorff dimension 0.

(f) $\dim_H\big(f(A)\big) \leq \dim_H(A)$ for any Lipschitz continuous function $f : A \to \mathbb{R}^n$.

(g) The Hausdorff dimension of A is independent of that of the ambient $\mathbb{R}^n$.

7 Suppose $A \subset \mathbb{R}^n$ and $B \subset \mathbb{R}^m$. Show then that $\dim_H(A \times B) = \dim_H(A) + \dim_H(B)$.

8 Suppose $I \subset \mathbb{R}$ is a perfect compact interval and $\gamma \in C(I, \mathbb{R}^n)$ is an injective rectifiable path with image Γ. Then $\dim_H(\Gamma) = 1$.

9 Verify that setting $\mu^*(A) := \lambda_1^*\big(\operatorname{pr}_1(A)\big)$ for $A \subset \mathbb{R}^2$ defines an outer measure on $\mathbb{R}^2$.

---

[2]By Exercises VII.9.1 and VII.9.2, every isometry φ of $\mathbb{R}^n$ is an **affine map** — that is, of the form $\varphi(x) = Tx + a$ with $T \in O(n)$ and $a \in \mathbb{R}^n$ — and can be interpreted as a rigid motion.

10 Suppose $\{\mu_j^* \; ; \; j \in \mathbb{N}\}$ is a family of outer measures on X. Then

$$\mu^* : \mathfrak{P}(X) \to [0, \infty] \; , \quad A \mapsto \sum_{j=0}^{\infty} \mu_j^*(A)$$

is an outer measure on X.

11 Show that for every $A \subset \mathbb{R}^n$ there is a G_δ-set G such that $A \subset G$ and $\lambda_n^*(A) = \lambda_n^*(G)$.

4   Measurable sets

In this section we finish the process of constructing measures on a set. We start with an outer measure and restrict it to an appropriate collection of subsets. By choosing this subset skillfully, we end up with a complete measure space. This technique, which goes back to Carathéodory, is then applied to the examples of the last sections to obtain the most important measures for applications — in particular, the Lebesgue measure.

Motivation

The key point in Carathéodory's construction is the definition of measurable sets. It's a convenient definition for the proof of the main theorems, but not one that is immediately grasped by intuition. Therefore we first give a heuristic motivation.

Suppose A is a bounded subset of $\mathbb{R}^n$. If (I_j) is a sequence of open intervals such that $\bigcup I_j \supset A$, then $\sum_{j=0}^{\infty} \mathrm{vol}_n(I_j)$ represents an approximate value for $\lambda_n^*(A)$, which becomes closer to $\lambda_n^*(A)$ as $\bigcup_j I_j$ better approximates the set A. By Proposition 3.4, we can replace open intervals by left-open ones; in particular, we can choose finitely many pairwise disjoint intervals whose union contains A. The set A is approximated "from the outside" by a *mosaic*, a shape whose boundary is piecewise parallel to the coordinate hyperplanes. In this sense, we regard

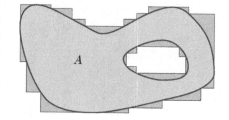

$$\lambda_n^*(A) := \inf\left\{ \sum_{j=0}^{\infty} \mathrm{vol}_n(I_j) \; ; \; I_j \in \mathbb{J}(n), \; j \in \mathbb{N}, \; A \subset \bigcup_{j=0}^{\infty} I_j \right\}$$

as an "approximation from the outside" to the volume of A.

Now instead of A consider the set $D \backslash A$, where D is a bounded superset of A in $\mathbb{R}^n$. Approximating $D \backslash A$ from the outside by a mosaic, as above, we get an approximation of A "from the inside". It is therefore natural to define the *inner* measure of A (relative to D) by

$$\lambda_{n,*}^D(A) := \lambda_n^*(D) - \lambda_n^*(D \backslash A) .$$

Now it is reasonable to expect a special role for those subsets A of $\mathbb{R}^n$ whose outer measures agree with their inner measures relative to every bounded superset D, that is, those satisfying

$$\lambda_n^*(A) = \lambda_{n,*}^D(A) \quad \text{for } D \subset \mathbb{R}^n \text{ and } D \supset A ,$$

where again D is bounded. This corresponds to the equality

$$\lambda_n^*(D) = \lambda_n^*(A) + \lambda_n^*(D\backslash A) \quad \text{for } D \subset \mathbb{R}^n \text{ and } D \supset A , \tag{4.1}$$

from which we can now drop the requirement that A and D be bounded. Thus (4.1) singles out precisely those sets A for which the Lebesgue outer measure behaves additively with respect to the disjoint decomposition $A \cup (D\backslash A)$ of D, for every $D \subset \mathbb{R}^n$ with $D \supset A$.

The σ-algebra of μ^*-measurable sets

Suppose μ^* is an outer measure on X. If we replace $\mathbb{R}^n$ by X and λ_n^* by μ^*, equation (4.1) is meaningful for every $A \subset X$. Because outer measures are subadditive, we can also replace the equality sign in (4.1) by $\geq$. We then reach the key definition: A subset A of X is μ^*-**measurable** if, for every $D \subset X$,

$$\mu^*(D) \geq \mu^*(A \cap D) + \mu^*(A^c \cap D) .$$

We denote the set of all μ^*-measurable subsets of X by $\mathcal{A}(\mu^*)$. If $N \subset X$ has $\mu^*(N) = 0$, we say the set N is μ^*-**null** (or of μ^* measure zero).

4.1 Remarks (a) Every μ^*-null set is μ^*-measurable.

Proof Take Suppose $D \subset X$ and $N \subset X$ with $\mu^*(N) = 0$. The monotony of μ^* gives $0 \leq \mu^*(N \cap D) \leq \mu^*(N) = 0$. Thus $N \cap D$ is μ^*-null, and it follows that

$$\mu^*(N \cap D) + \mu^*(N^c \cap D) = \mu^*(N^c \cap D) \leq \mu^*(D) .$$

Therefore N is μ^*-measurable. ∎

(b) For $A \subset X$, these statements are equivalent:

(i) $A \in \mathcal{A}(\mu^*)$.

(ii) $\mu^*(D) \geq \mu^*(A \cap D) + \mu^*(A^c \cap D)$ for all $D \subset X$ such that $\mu^*(D) < \infty$.

(iii) $\mu^*(D) = \mu^*(A \cap D) + \mu^*(A^c \cap D)$ for all $D \subset X$.

Proof The implications "(i)$\Rightarrow$(ii)" and "(iii)$\Rightarrow$(i)" are obvious.

"(ii)$\Rightarrow$(iii)" Suppose $D \subset X$. The subadditivity of μ^* gives

$$\mu^*(D) = \mu^*((A \cap D) \cup (A^c \cap D)) \leq \mu^*(A \cap D) + \mu^*(A^c \cap D) . \tag{4.2}$$

If $\mu^*(D) < \infty$ then (iii) follows from (4.2) and (ii). In the case $\mu^*(D) = \infty$, the statement is likewise correct due to (4.2). ∎

The next theorem shows that the set of all μ^*-measurable sets forms a σ-algebra and the restriction of the outer measures μ^* to this σ-algebra is a complete measure. This is the important **Carathéodory extension theorem**, which allows the construction of nontrivial measures.

4.2 Theorem *Suppose μ^* is an outer measure on X. Then $\mathcal{A}(\mu^*)$ is a σ-algebra on X, and $\mu := \mu^* | \mathcal{A}(\mu^*)$ is a complete measure on $\mathcal{A}(\mu^*)$, the measure on X induced by μ^*.*

Proof   (i) Obviously $\emptyset$ belongs to $\mathcal{A}(\mu^*)$. Also, A^c lies in $\mathcal{A}(\mu^*)$ if A does, because the notion of μ^*-measurability is symmetric in A and A^c.

   (ii) Take $A, B \in \mathcal{A}(\mu^*)$ and $D \subset X$. Then

$$\mu^*(D) \geq \mu^*(A \cap D) + \mu^*(A^c \cap D) . \tag{4.3}$$

Because B is μ^*-measurable, we have

$$\mu^*(A^c \cap D) \geq \mu^*(B \cap A^c \cap D) + \mu^*(B^c \cap A^c \cap D) .$$

Thus (4.3) and the subadditivity of μ^* give

$$\mu^*(D) \geq \mu^*\big((A \cap D) \cup (B \cap A^c \cap D)\big) + \mu^*(B^c \cap A^c \cap D) .$$

Noting that

$$(A \cap D) \cup (B \cap A^c \cap D) = \big[A \cup (B \cap A^c)\big] \cap D = (A \cup B) \cap D$$

and $(A \cup B)^c = A^c \cap B^c$, we see that

$$\mu^*(D) \geq \mu^*\big((A \cup B) \cap D\big) + \mu^*\big((A \cup B)^c \cap D\big) .$$

Therefore $A \cup B$ is μ^*-measurable, and $\mathcal{A}(\mu^*)$ is an algebra over X.

   (iii) Let (A_j) be a disjoint sequence in $\mathcal{A}(\mu^*)$. Because A_0 is μ^*-measurable, Remark 4.1(b) results in

$$\mu^*\big((A_0 \cup A_1) \cap D\big) = \mu^*\big(((A_0 \cup A_1) \cap D) \cap A_0\big) + \mu^*\big(((A_0 \cup A_1) \cap D) \cap A_0^c\big) ,$$

and from $A_0 \cap A_1 = \emptyset$, it follows that

$$\mu^*\big((A_0 \cup A_1) \cap D\big) = \mu^*(A_0 \cap D) + \mu^*(A_1 \cap D) .$$

By complete induction, we get

$$\mu^*\left(\left(\bigcup_{j=0}^{m} A_j\right) \cap D\right) = \sum_{j=0}^{m} \mu^*(A_j \cap D) \quad \text{for } m \in \mathbb{N} . \tag{4.4}$$

After setting for short $A := \bigcup_j A_j$, the monotony of μ^* shows that

$$\mu^*(A \cap D) \geq \sum_{j=0}^{m} \mu^*(A_j \cap D) \quad \text{for } m \in \mathbb{N} .$$

For $m \to \infty$, we get the inequality $\mu^*(A \cap D) \geq \sum_{j=0}^{\infty} \mu^*(A_j \cap D)$. Together with the σ-subadditivity of μ^*, this implies

$$\mu^*(A \cap D) = \sum_{j=0}^{\infty} \mu^*(A_j \cap D) . \tag{4.5}$$

Because, by (ii), $\mathcal{A}(\mu^*)$ is an algebra over X, we have for every $m \in \mathbb{N}$ that

$$\mu^*(D) = \mu^*\left(\left(\bigcup_{j=0}^{m} A_j\right)^c \cap D\right) + \mu^*\left(\left(\bigcup_{j=0}^{m} A_j\right) \cap D\right) .$$

The monotony of μ^* and (4.4) then give

$$\mu^*(D) \geq \mu^*(A^c \cap D) + \sum_{j=0}^{m} \mu^*(A_j \cap D) ,$$

so we find using (4.5) that

$$\mu^*(D) \geq \mu^*(A^c \cap D) + \sum_{j=0}^{\infty} \mu^*(A_j \cap D) = \mu^*(A^c \cap D) + \mu^*(A \cap D)$$

as $m \to \infty$. Therefore A is μ^*-measurable, and Remark 1.2(c) implies that $\mathcal{A}(\mu^*)$ is a σ-algebra.

(iv) To see that $\mu^* \,|\, \mathcal{A}(\mu^*)$ is a measure on $\mathcal{A}(\mu^*)$, it suffices to set $D = X$ in (4.5). Finally, the monotony of μ^* and Remark 4.1(a) shows that this measure is complete. ∎

If μ is the measure on X induced by μ^*, sets in $\mathcal{A}(\mu^*)$ are called **μ-measurable**, naturally enough, while μ^*-null sets and **μ-null** sets coincide.

Lebesgue measure and Hausdorff measure

We now apply Theorem 4.2 to the outer measures discussed in Propositions 3.3, 3.5, and 3.7.

- The measure on $\mathbb{R}^n$ induced by λ_n^* is called **n-dimensional Lebesgue measure** on $\mathbb{R}^n$ and is denoted by λ_n. We call λ_n-measurable sets **Lebesgue measurable**.

- If $F : \mathbb{R} \to \mathbb{R}$ is a measure-generating function, we call the measure on $\mathbb{R}$ generated by μ_F^* is the **Lebesgue–Stieltjes measure** on $\mathbb{R}$ induced by F. We denote it μ_F.

- Suppose X is a separable metric space and $s > 0$. The measure on X induced by $\mathcal{H}_*^s$ is the **s-dimensional Hausdorff measure** on X; it is denoted by $\mathcal{H}^s$.

Metric measures

Granted that Theorem 4.2 guarantees that the restriction of μ^* to $\mathcal{A}(\mu^*)$ is a measure, it says nothing about how many or how few sets belong to $\mathcal{A}(\mu^*)$. For metric spaces, we will now specify a sufficient condition such that at least all Borel sets are μ-measurable.

Suppose $X = (X, d)$ is a metric space and μ^* is an outer measure on X. If

$$\mu^*(A \cup B) = \mu^*(A) + \mu^*(B)$$

for all $A, B \subset X$ separated by a positive distance, that is, for which[1] $d(A, B) > 0$, we say μ^* and the measure on $\mathcal{A}(\mu^*)$ induced by μ^* are **metric**.

The next theorem shows that the σ-algebra induced by a metric outer measure contains the Borel σ-algebra. Conversely, one can show that an outer measure μ^* whose σ-algebra of μ^*-measurable sets contains the Borel σ-algebra is a metric measure; see Exercise 1.

4.3 Theorem   *Suppose μ^* is a metric outer outer measure on X. Then $\mathcal{A}(\mu^*) \supset \mathcal{B}(X)$.*

Proof (i) Because $\mathcal{A}(\mu^*)$ is a σ-algebra and because the Borel σ-algebra is generated by open sets, if suffices to verify that every open set is μ^*-measurable.

(ii) Take O open in X and $D \subset X$ such that $\mu^*(D) < \infty$. We will show that

$$\mu^*(D) \geq \mu^*(O \cap D) + \mu^*(O^c \cap D) .$$

From Remark 4.1(b), it follows that $O \in \mathcal{A}(\mu^*)$.

We set

$$O_n := \left\{ x \in X ; \ d(x, O^c) > 1/n \right\}$$

and

$$A_n := \left\{ x \in X ; \ 1/(n+1) < d(x, O^c) \leq 1/n \right\}$$

for $n \in \mathbb{N}^\times$. Clearly $d(O_n, O^c) \geq 1/n > 0$. For $x \in A_k$, we have

$$1/(k+1) < d(x, O^c) \leq d(x, z) \leq d(x, y) + d(y, z) \quad \text{for } z \in O^c \text{ and } y \in X .$$

Because this holds for every $z \in O^c$, we have

$$1/(k+1) \leq d(x, y) + d(y, O^c) \leq d(x, y) + 1/(k+2) \quad \text{for } y \in A_{k+2} ,$$

---

[1] See Example III.3.9(c).

and therefore

$$d(A_k, A_{k+2}) \geq \frac{1}{k+1} - \frac{1}{k+2} > 0 \quad \text{for } k \in \mathbb{N}^\times . \tag{4.6}$$

(iii) Because μ^* is a metric outer measure, it follows from (4.6) by complete induction that

$$\sum_{j=1}^n \mu^*(A_{2j-i} \cap D) = \mu^* \left(\left(\bigcup_{j=1}^n A_{2j-i} \right) \cap D \right) \leq \mu^*(D) \quad \text{for } n \in \mathbb{N}^\times \text{ and } i = 0,1 .$$

From this we get

$$\sum_{k=1}^\infty \mu^*(A_k \cap D) \leq 2\mu^*(D) < \infty .$$

In particular, we find that the series remainders $r_n := \sum_{k=n}^\infty \mu^*(A_k \cap D)$ form a null sequence. It is also clear that $O \backslash O_n = \bigcup_{j=n}^\infty A_j$. The σ-subadditivity of μ^* hence gives

$$0 \leq \mu^* \big((O \backslash O_n) \cap D \big) \leq \sum_{j=n}^\infty \mu^*(A_j \cap D) = r_n .$$

Therefore $\big(\mu^* ((O \backslash O_n) \cap D) \big)_{n \in \mathbb{N}^\times}$ is also a null sequence.

(iv) Clearly

$$\mu^*(O \cap D) \leq \mu^*(O_n \cap D) + \mu^* \big((O \backslash O_n) \cap D \big) . \tag{4.7}$$

Because $d(O_n \cap D, O^c \cap D) \geq d(O_n, O^c) \geq 1/n$ and since μ^* is an outer measure, we have

$$\mu^*(O_n \cap D) + \mu^*(O^c \cap D) = \mu^* \big((O_n \cup O^c) \cap D \big) \leq \mu^*(D) .$$

From this and (4.7), we conclude that

$$\mu^*(O \cap D) + \mu^*(O^c \cap D) \leq \mu^*(D) + \mu^* \big((O \backslash O_n) \cap D \big) \quad \text{for } n \in \mathbb{N}^\times .$$

Taking $n \to \infty$, we find the desired inequality. ∎

4.4 Examples (a) λ_n^* is a metric outer measure on $\mathbb{R}^n$. Therefore every Borel set is Lebesgue measurable.

Proof Suppose $A, B \subset \mathbb{R}^n$ with $d(A,B) > 0$, and let $\delta := d(A,B)/2$. According to Proposition 3.4, given $\varepsilon > 0$ there is a sequence (I_j) in $\mathbb{J}_\ell(n)$ such that $\bigcup_j I_j \supset A \cup B$ and $\sum_j \mathrm{vol}_n(I_j) \leq \lambda_n^*(A \cup B) + \varepsilon$. By cutting along coordinate hyperplanes, we can write each I_j as a disjoint finite union of left closed intervals all having diameter less than δ. Thus we lose no generality in assuming that $\mathrm{diam}(I_j) < \delta$ for every $j \in \mathbb{N}$. Because $d(A,B) = 2\delta$, this means that for each $j \in \mathbb{N}$ we have either $I_j \cap A = \emptyset$ or

$I_j \cap B = \emptyset$; in other words, there exist subsequences (I_k') and (I_ℓ'') of (I_j), covering A and B respectively, and such that $I_k' \cap I_\ell'' = \emptyset$ for $k, \ell \in \mathbb{N}$. Thus

$$\lambda_n^*(A) \le \sum_k \mathrm{vol}_n(I_k') \quad \text{and} \quad \lambda_n^*(B) \le \sum_\ell \mathrm{vol}_n(I_\ell'') ,$$

and we get

$$\lambda_n^*(A) + \lambda_n^*(B) \le \sum_k \mathrm{vol}_n(I_k') + \sum_\ell \mathrm{vol}_n(I_\ell'') \le \sum_j \mathrm{vol}_n(I_j)$$

$$\le \lambda_n^*(A \cup B) + \varepsilon .$$

Since $\varepsilon > 0$ was arbitrary, the claim follows using the subadditivity of λ_n^*. ∎

(b) For a measure-generating function $F : \mathbb{R} \to \mathbb{R}$, the corresponding Lebesgue-Stieltjes outer measure μ_F^* on $\mathbb{R}$ is metric.

Proof This follows by a simple modification of the proof of (a). ∎

(c) The Hausdorff outer measure $\mathcal{H}_*^s$ on $\mathbb{R}^n$ is metric for every $s > 0$. Every $A \in \mathcal{B}^n$ is $\mathcal{H}^n$-measurable.

Proof This also follows in analogy to the proof of (a). ∎

Exercises

1 Suppose X is a metric space and μ^* is an outer measure on X. Prove that if $\mathcal{A}(\mu^*)$ contains all Borel sets, μ^* is metric.

2 Let $(X, \mathcal{A}, \nu)$ be a measure space. Denote by μ^* the outer measure on X induced by $(\mathcal{A}, \nu)$ and by μ the measure on X induced by μ^*. Show that μ is an extension of ν. Are they equal?

3 Prove the statements in Examples 4.4(b) and (c).

4 Let μ^* be an outer measure on X, and define $\mu_* : \mathfrak{P}(X) \to [0, \infty]$, the **inner measure on X induced by μ^***, by

$$\mu_*(A) := \sup\{\mu^*(D) - \mu^*(D \setminus A) \; ; \; D \subset X, \; \mu^*(D \setminus A) < \infty\} \quad \text{for } A \subset X .$$

Show that $\mu_*(A) = \mu^*(A)$ for $A \in \mathcal{A}(\mu^*)$.

5 Suppose $I \subset \mathbb{R}$ is a perfect, compact interval, and $\gamma \in C(I, \mathbb{R}^n)$ is an injective rectifiable path in $\mathbb{R}^n$ with image Γ. Show that $\mathcal{H}^1(\Gamma) = L(\gamma)$.

6 Set $A_0 := [0, 1]^2 \subset \mathbb{R}^2$. Partition A_0 into a 4×4 array of identical squares and remove twelve of these squares according to the sketch below, so that exactly one closed square remains in every row and every column. This forms the set A_1. Repeat this procedure for every remaining square to get A_2, which consists of sixteen squares. Generally, obtain A_{k+1} from A_k by applying this subdivision and then removing subsquares from A_k. The intersection of all the A_k, that is, $A := \bigcap_{k=0}^\infty A_k$, is called **Cantor dust**.

Show that $1 \leq \mathcal{H}^1(A) \leq \sqrt{2}$ and $\dim_H(A) = 1$.

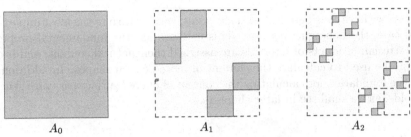

$$A_0 \qquad\qquad\qquad A_1 \qquad\qquad\qquad A_2$$

(Hint: To bound $\mathcal{H}^1(A)$ from above, use the covers suggested by the construction of A. For the lower bound, consider $\mathrm{pr}_1 : A \to \mathbb{R}$ and apply Exercises 5 and 3.6(f).)

7 Show that the Cantor set[2] C from Exercise III.3.8 satisfies

(i) $\dim_H(C) = \log 2 / \log 3 =: s$ and $1/2 \leq \mathcal{H}^s(C) \leq 1$;

(ii) $\lambda_1(C) = 0$.

(Hints for (i): The upper bound for $\mathcal{H}^s(C)$ is obtained much like the one for $\mathcal{H}^1(A)$ in Exercise 6. For the lower bound, a compactness argument shows that one need only consider coverings by finitely many open intervals. If $\{ I_i ; 0 \leq i \leq N \}$ is such a cover, choose for every i the integer k such that $3^{-(k+1)} \leq \mathrm{diam}(I_i) < 3^{-k}$. Then I_i can intersect at most one interval from C_k (Exercise III.3.8). For $j \geq k$, the cover I_i intersects at most $2^{j-k} = 2^j 3^{-sk} \leq 2^j (3\,\mathrm{diam}(I_i))^s$ intervals from C_j. Now choose j large enough that $3^{-(j+1)} \leq \mathrm{diam}(I_i)$ for all i. Then count intervals.)

8 Suppose $F : \mathbb{R} \to \mathbb{R}$ is a measure-generating function and let μ_F be the Lebesgue–Stieltjes measure induced by F. For $a \in \mathbb{R}$, calculate $\mu_F(\{a\})$.

9 Suppose $(\mathbb{R}, \mathcal{B}^1, \mu)$ is a locally finite[3] measure space. Prove:

(i) There is a measure-generating function F such that $\mu = \mu_F \,|\, \mathcal{B}^1$, that is, μ is the **Borel–Stieltjes measure** induced by F. This F is unique up to an additive constant.

(ii) Define

$$\mathcal{F}_0 := \big\{ F : \mathbb{R} \to \mathbb{R} \; ; \; F \text{ is measure-generating with } F(0) = 0 \big\} .$$

Then $F \mapsto \mu_F \,|\, \mathcal{B}^1$ is a bijection from $\mathcal{F}_0$ to the set of locally finite measures on $\mathcal{B}^1$.

(Hint for (i): Consider $F(t) := \mu([0, t))$ for $t \geq 0$ and $F(t) := -\mu([t, 0))$ for $t < 0$.)

10 Suppose $F : \mathbb{R} \to \mathbb{R}$ is a measure-generating function with the following properties: F is constant on each interval (a_k, a_{k+1}), where the numbers a_k, for $k \in \mathbb{Z}$, satisfy $\lim_{k \to \pm\infty} a_k = \pm\infty$; moreover F has at each a_k a jump discontinuity of height $p_k \geq 0$. Show that $\mathcal{A}(\mu_F) = \mathfrak{P}(\mathbb{R})$ and calculate $\mu_F(A)$ for $A \subset \mathbb{R}$.

---

[2]The Cantor set and Cantor dust are examples of **fractals**. (see for example [Fal90]).

[3]If X is a topological space and $\mu : \mathcal{A} \to [0, \infty]$ is a measure with $\mathcal{A} \supset \mathcal{B}(X)$, we say μ is **locally finite** if every $x \in X$ has an open neighborhood $U \ni x$ such that $\mu(U) < \infty$.

5 The Lebesgue measure

Until now we have considered general measures; we now turn to the most important special case, the Lebesgue measure. This measure has the fundamental property that Cartesian products of intervals are assigned their natural content, and it can therefore be used to calculate the content of more general shapes. In addition, it forms the foundation for calculating the content of curved surfaces or more general manifolds, as we shall see in later chapters.

The Lebesgue measure space

The σ-algebra $\mathcal{A}(\lambda_n^*)$ generated by the n-dimensional Lebesgue outer measure is called the **σ-algebra of Lebesgue measurable subsets** of $\mathbb{R}^n$ and will be denoted by $\mathcal{L}(n)$. Accordingly, Lebesgue null subsets of $\mathbb{R}^n$ (that is, λ_n^*-null or equivalently λ_n-null sets) are said to have **Lebesgue measure zero** (the use of this expression implies membership in $\mathcal{L}(n)$). If necessary, we speak also of **Lebesgue n-measure**.

In the next theorem, we list some first properties of the **Lebesgue measure space** $(\mathbb{R}^n, \mathcal{L}(n), \lambda_n)$.

5.1 Theorem

(i) $(\mathbb{R}^n, \mathcal{L}(n), \lambda_n)$ is a complete, σ-finite measure space.

(ii) $\mathcal{B}^n \subset \mathcal{L}(n)$, that is, every Borel subset of $\mathbb{R}^n$ is Lebesgue measurable.

(iii) For $a, b \in \mathbb{R}^n$, any set A satisfying $(a, b) \subset A \subset [a, b]$ belongs to $\mathcal{L}(n)$, and

$$\lambda_n(A) = \mathrm{vol}_n(a, b) = \prod_{j=1}^{n} (b_j - a_j) .$$

(iv) Every compact subset of $\mathbb{R}^n$ is Lebesgue measurable and has finite measure.

(v) A set $N \subset \mathbb{R}^n$ has Lebesgue measure zero if and only if for every $\varepsilon > 0$ there is a sequence (I_j) in $\mathbb{J}(n)$ such that $\bigcup_j I_j \supset N$ and $\sum_j \lambda_n(I_j) < \varepsilon$.

(vi) Every countable subset of $\mathbb{R}^n$ has Lebesgue measure zero.

Proof (i) Theorem 4.2 and Proposition 3.3 show that $(\mathbb{R}^n, \mathcal{L}(n), \lambda_n)$ is a complete measure space. Because $\mathbb{R}^n = \bigcup_{j=1}^{\infty} (j\mathbb{B}_\infty)$ and $\lambda_n(j\mathbb{B}_\infty) = (2j)^n$, it is σ-finite.

(ii) This follows from Theorem 4.3 and Example 4.4(a).

(iii) For $M := A \backslash (a, b)$, we have $M \subset N := [a, b] \backslash (a, b) \in \mathcal{B}^n$. Therefore part (ii) and Proposition 2.3 imply that N has Lebesgue measure zero, since $\lambda_n(N) = \lambda_n([a, b]) - \lambda_n((a, b)) = 0$. Now (i) says λ_n is complete, so M also has Lebesgue measure zero. Therefore $A = (a, b) \cup M$ belongs to $\mathcal{L}(n)$, and since (a, b) is disjoint from M, we get $\lambda_n(A) = \lambda_n((a, b)) = \mathrm{vol}_n(a, b)$.

(iv) follows from (ii) and (iii). Statement (v) is an immediate consequence of the definition of the Lebesgue outer measure. To see (vi), use the obvious fact that any one-point has Lebesgue measure zero. ∎

5.2 Example Any subset of $\mathbb{R}^n$ confined to a single coordinate hyperplane has Lebesgue measure zero.

Proof Since λ_n is complete, it suffices to verify that every coordinate hyperplane is λ_n-null. We consider the case $H := \mathbb{R}^{n-1} \times \{0\}$ (an obvious variant of the argument works for any other coordinate hyperplane).

Take $\varepsilon > 0$, and for $k \in \mathbb{N}^\times$ define $\varepsilon_k := \varepsilon(2k)^{-n+1} 2^{-(k+2)}$ and

$$J_k(\varepsilon) := (-k, k)^{n-1} \times (-\varepsilon_k, \varepsilon_k) \in \mathbb{J} .$$

Then $\mathrm{vol}_n(J_k(\varepsilon)) = \varepsilon 2^{-(k+1)}$, and thus $\sum_{k=1}^\infty \mathrm{vol}_n(J_k(\varepsilon)) = \varepsilon/2 < \varepsilon$. Because $(J_k(\varepsilon))$ covers H, the equality $\lambda_n(H) = 0$ follows from Theorem 5.1(v). $\blacksquare$

Corollary 5.23 below will show that every subset of $\mathbb{R}^n$ contained in a proper affine subspace has Lebesgue measure zero.

The Lebesgue measure is regular

We now prove several basic approximation results, but first we collect some terminology about measures on topological spaces.

Let X be a topological space and $(X, \mathcal{A}, \mu)$ a measure space with $\mathcal{B}(X) \subset \mathcal{A}$. We say $(X, \mathcal{A}, \mu)$ and μ are **regular** if, for every $A \in \mathcal{A}$,

$$\mu(A) = \inf\{ \mu(O) ; \ O \subset X \text{ is open with } O \supset A \}$$
$$= \sup\{ \mu(K) ; \ K \subset X \text{ is compact with } K \subset A \} .$$

If every $x \in X$ has an open neighborhood U such that $\mu(U) < \infty$, we say $(X, \mathcal{A}, \mu)$ and μ are **locally finite**. Finally, if $\mathcal{B}(X) = \mathcal{A}$, we call μ the **Borel measure** on X. In particular $\beta_n := \lambda_n \,|\, \mathcal{B}^n$ is called the **Borel–Lebesgue measure** on $\mathbb{R}^n$.

5.3 Remarks (a) If μ is locally finite, then every compact set $K \subset X$ has an open neighborhood U such that $\mu(U) < \infty$.

Proof Because μ is locally finite, every $x \in K$ has an open neighborhood U_x such that $\mu(U_x) < \infty$. Since K is compact, there are $x_0, \ldots, x_m \in K$ with $K \subset U := \bigcup_{j=0}^m U_{x_j}$, and we get $\mu(U) \le \sum_{j=0}^m \mu(U_{x_j}) < \infty$. $\blacksquare$

(b) Suppose X is locally compact.[1] Then μ is locally finite if and only if every compact set $K \subset X$ satisfies $\mu(K) < \infty$.

Proof This follows immediately from (a). $\blacksquare$

5.4 Theorem *The Lebesgue measure is regular.*

---

[1] A topological space is said to be **locally compact** if it is Hausdorff and every point has a compact neighborhood.

Proof Let $A \in \mathcal{L}(n)$.

(i) For every $\varepsilon > 0$, there is a sequence (I_j) in $\mathbb{J}(n)$ such that

$$A \subset \bigcup_j I_j \quad \text{and} \quad \sum_j \mathrm{vol}_n(I_j) < \lambda_n(A) + \varepsilon .$$

The open set $O := \bigcup_j I_j$ therefore satisfies

$$\lambda_n(A) \leq \lambda_n(O) \leq \sum_j \lambda_n(I_j) = \sum_j \mathrm{vol}_n(I_j) < \lambda_n(A) + \varepsilon . \tag{5.1}$$

Because this is true for every $\varepsilon > 0$,

$$\lambda_n(A) = \inf\{\lambda_n(O) \; ; \; O \subset \mathbb{R}^n \text{ is open with } O \supset A \} .$$

(ii) To verify that

$$\lambda_n(A) = \sup\{\lambda_n(K) \; ; \; K \subset \mathbb{R}^n \text{ is compact with } K \subset A \}$$

we consider first the case of a Lebesgue measurable set B that is bounded. Then there is a compact set $C \subset \mathbb{R}^n$ such that $B \subset C$. Using (i) we find for every $\varepsilon > 0$ an open set $O \subset \mathbb{R}^n$ containing $C \setminus B$ and for which $\lambda_n(O) < \lambda_n(C \setminus B) + \varepsilon$. Because $\lambda_n(B) < \infty$ if follows from Proposition 2.3(ii) that

$$\lambda_n(O) < \lambda_n(C) - \lambda_n(B) + \varepsilon . \tag{5.2}$$

The compact set $K := C \setminus O$ satisfies $K \subset B$ and $C \subset K \cup O$. Thus (5.2) shows

$$\lambda_n(C) \leq \lambda_n(K \cup O) \leq \lambda_n(K) + \lambda_n(O) < \lambda_n(K) + \lambda_n(C) - \lambda_n(B) + \varepsilon ,$$

which implies the inequality $\lambda_n(B) - \varepsilon < \lambda_n(K)$. Therefore

$$\lambda_n(B) = \sup\{\lambda_n(K) \; ; \; K \subset \mathbb{R}^n \text{ is compact with } K \subset B \}$$

for every bounded Lebesgue set B.

(iii) We can assume that $\lambda_n(A)$ is positive. There exists $\alpha > 0$ such that $\alpha < \lambda_n(A)$. With $B_j := A \cap \mathbb{B}^n(0, j)$, the continuity of the Lebesgue measure from below shows that $\lambda_n(A) = \lim_j \lambda_n(B_j)$. Thus there is $k \in \mathbb{N}$ such that $\lambda_n(B_k) > \alpha$. Because B_k is bounded, we find due to (ii) a compact set K such that $K \subset B_k \subset A$ and $\lambda_n(K) > \alpha$. Therefore

$$\sup\{\lambda_n(K) \; ; \; K \subset \mathbb{R}^n \text{ is compact with } K \subset A \} > \alpha .$$

The claim now follows because $\alpha < \lambda_n(A)$ is arbitrary. ∎

5.5 Corollary *Suppose $A \in \mathcal{L}(n)$. Then there is an F_σ-set F and a G_δ-set G such that $F \subset A \subset G$ and $\lambda_n(F) = \lambda_n(A) = \lambda_n(G)$. If A is bounded, G can be chosen to be bounded.*

Proof (i) We prove only the first statement, the second being clear. We start with the case $\lambda_n(A) < \infty$. By Theorem 5.4 there is for every $k \in \mathbb{N}^\times$ a compact set K_k and an open set O_k such that $K_k \subset A \subset O_k$ and

$$\lambda_n(A) - 1/k \le \lambda_n(K_k) \le \lambda_n(A) \le \lambda_n(O_k) \le \lambda_n(A) + 1/k . \tag{5.3}$$

Setting $F := \bigcup_k K_k$ and $G := \bigcap_k O_k$, we have the inclusions $F \subset A \subset G$, and Proposition 2.3(ii) applied to (5.3) gives for every $k \in \mathbb{N}^\times$

$$\lambda_n(A \backslash F) \le \lambda_n(A \backslash K_k) \le 1/k , \quad \lambda_n(G \backslash A) \le \lambda_n(O_k \backslash A) \le 1/k .$$

Hence $\lambda_n(A \backslash F) = \lambda_n(G \backslash A) = 0$, and the claim follows from Proposition 2.3(ii).

(ii) If instead $\lambda_n(A) = \infty$, Theorem 5.3 provides for every $k \in \mathbb{N}$ a compact set K_k such that $K_k \subset A$ and $k \le \lambda_n(K_k)$. The F_σ-set $F := \bigcup_k K_k$ and the G_δ-set $G := \mathbb{R}^n$ satisfy the desired equations. $\blacksquare$

Theorem 5.4 implies that we can approximate the measure of a Lebesgue measurable subset of $\mathbb{R}^n$ to arbitrary precision by the measure of a suitably chosen open superset. By the next proposition, the Lebesgue measure of an open set is itself the limit of the values obtained by adding up the volumes of finitely many disjoint intervals $[a, b)$, chosen so their union approximates the open set. This is the method for calculating content described in the introduction to this chapter.

5.6 Proposition *Every open subset O in $\mathbb{R}^n$ can be represented as the union of a disjoint sequence (I_j) of intervals of the form $[a, b)$ with $a, b \in \mathbb{Q}^n$. Then*

$$\lambda_n(O) = \sum_{j=0}^{\infty} \mathrm{vol}_n(I_j) .$$

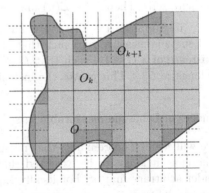

Proof For $k \in \mathbb{N}$, define

$$\mathcal{W}_k := \left\{ a + [0, 2^{-k}\mathbf{1}_n) \; ; \; a \in 2^{-k}\mathbb{Z}^n \right\}$$

with $\mathbf{1}_n := (1, \ldots, 1) \in \mathbb{R}^n$. In other words, every $W \in \mathcal{W}_k$ is a cube (aligned with the coordinate hyperplanes) whose sides have length 2^{-k} and whose "lower left corners" lie on points of the grid $2^{-k}\mathbb{Z}^n$. Obviously $\mathcal{W}_k$ is a countable disjoint cover of $\mathbb{R}^n$. If O_k is the union of those cubes in $\mathcal{W}_k$ that lie entirely in O, an application of Proposition I.6.8 concludes the proof, because

$$O = O_0 \cup (O_1 \backslash O_0) \cup (O_2 \backslash (O_0 \cup O_1)) \cup \cdots \blacksquare$$

A characterization of Lebesgue measurability

Let X be a topological space. A subset M of X is **σ-compact** if there is a sequence (K_j) of compact subsets such that $M = \bigcup_j K_j$.

5.7 Theorem   *For $A \subset \mathbb{R}^n$, these statements are equivalent:*

(i) *A is Lebesgue measurable.*

(ii) *There is a σ-compact subset S of $\mathbb{R}^n$ and a set N of Lebesgue measure zero such that $A = S \cup N$.*

Proof   "(i)$\Rightarrow$(ii)" Since the measure space $(\mathbb{R}^n, \mathcal{L}(n), \lambda_n)$ is σ-finite, there is a sequence (A_j) in $\mathcal{L}(n)$ such that $A = \bigcup_j A_j$ and $\lambda_n(A_j) < \infty$ for $j \in \mathbb{N}$. The proof of Corollary 5.5 shows that for every $j \in \mathbb{N}$ there is a σ-compact subset S_j of $\mathbb{R}^n$ such that $S_j \subset A_j$ and $\lambda_n(S_j) = \lambda_n(A_j)$. Therefore $N_j := A_j \setminus S_j$ is a set of Lebesgue measure zero such that $A_j = S_j \cup N_j$. Thus $S := \bigcup_j S_j$ is σ-compact, $N := \bigcup_j N_j$ has Lebesgue measure zero, and $A = S \cup N$.

"(ii)$\Rightarrow$(i)" Every σ-compact subset of $\mathbb{R}^n$ is Lebesgue measurable, because $\mathcal{B}^n \subset \mathcal{L}(n)$. Sets of Lebesgue measure zero are likewise Lebesgue measurable. ∎

In Corollary 5.29 we will show that the Borel–Lebesgue measure is not complete. With the help of Theorem 5.7 we can right away determine its completion.

5.8 Theorem   *The Lebesgue measure λ_n is the completion of the Borel–Lebesgue measure $(\mathbb{R}^n, \mathcal{B}^n, \beta_n)$.*

Proof   (i) Given $A \in \overline{\mathcal{B}^n_{\beta_n}}$, take $B, N \in \mathcal{B}^n$ and $M \subset \mathbb{R}^n$ such that $A = B \cup M$, $M \subset N$ and $\lambda_n(N) = 0$. The completeness of λ_n shows that M has Lebesgue measure zero. Then $\mathcal{B}^n \subset \mathcal{L}(n)$ implies $A = B \cup M \in \mathcal{L}(n)$, that is, $\overline{\mathcal{B}^n_{\beta_n}} \subset \mathcal{L}(n)$.

(ii) Suppose $A \in \mathcal{L}(n)$. By Theorem 5.7, there exist $B \in \mathcal{B}^n$ and a set M of Lebesgue measure zero such that $A = B \cup M$. Also, Corollary 5.5 secures the existence of a $G \in \mathcal{B}^n$ such that $M \subset G$ and $\lambda_n(G) = \lambda_n(M) = 0$. Therefore A belongs to $\overline{\mathcal{B}^n_{\beta_n}}$. This proves the inclusion $\mathcal{L}(n) \subset \overline{\mathcal{B}^n_{\beta_n}}$. ∎

Images of Lebesgue measurable sets

We will see in Theorem 5.28 that not every subset of $\mathbb{R}^n$ is Lebesgue measurable. Therefore it is not to be expected that measurable sets have measurable images under arbitrary maps.[2] For locally Lipschitz continuous functions, however, the measurability of these images can be guaranteed outright. To show this, we first consider null sets.

---

[2] In the following, we will deal almost exclusively with the Lebesgue measure and will omit the qualifier "Lebesgue" if no confusion is to be feared.

5.9 Theorem *Suppose $N \subset \mathbb{R}^n$ has Lebesgue n-measure zero and f is an element of $C^{1-}(N, \mathbb{R}^m)$, where $m \geq n$. (That is, f is locally Lipschitz continuous.) Then $f(N)$ has Lebesgue m-measure zero.*

Proof (i) We first assume that $f : N \to \mathbb{R}^m$ is (globally) Lipschitz continuous. Then there is an $L > 0$ such that

$$|f(x) - f(y)|_\infty \leq L\,|x - y|_\infty \quad \text{for } x, y \in N \ . \tag{5.4}$$

Suppose $0 < \varepsilon < L^m$. Because N has measure zero, we can find by Proposition 3.4 a sequence (I_k) in $\mathbb{J}_\ell(n)$ that covers N and satisfies $\sum_{k=0}^\infty \lambda_n(I_k) < \varepsilon / L^m$. We can take the edge lengths to be e rational. By subdivision, we can also assume without losing generality that every I_k is a cube, of side length a_k, say. By (5.4), then, $f(N \cap I_k)$ is contained in a cube $J_k \subset \bar{\mathbb{J}}(m)$ of side length $L a_k$. The n-volume of these cubes is

$$\lambda_m(J_k) = (La_k)^m = L^m \lambda_n(I_k)^{m/n} \quad \text{for } k \in \mathbb{N} \ .$$

Thus we find

$$f(N) = \bigcup_k f(N \cap I_k) \subset \bigcup_k J_k \tag{5.5}$$

and

$$\sum_{k=0}^\infty \lambda_m(J_k) = L^m \sum_{k=0}^\infty \lambda_n(I_k)^{m/n} \leq L^m \sum_{k=0}^\infty \lambda_n(I_k) < \varepsilon \ , \tag{5.6}$$

where the $\leq$ estimate relies on the assumption $m \geq n$: we have

$$\lambda_n(I_k) \leq \sum_{j=0}^\infty \lambda_n(I_j) < \varepsilon / L^m < 1 \ ,$$

hence $\lambda_n(I_k)^{m/n} \leq \lambda_n(I_k)$. Since all this holds for every $\varepsilon \in (0, L^m)$, we see from (5.5) and (5.6) that $f(N)$ has m-measure zero.

(ii) Now suppose f is only locally Lipschitz continuous. Every $x \in N$ has an open neighborhood U_x such that $f\,|\,(N \cap U_x)$ is Lipschitz continuous. From Corollary 1.9(ii) and Proposition 1.8, it follows that N is a Lindelöf space. Thus there is a countable subcover $\{V_j \ ; \ j \in \mathbb{N}\}$ of the open cover $\{U_x \cap N \ ; \ x \in N\}$ of N. Because Lebesgue measure is complete, every V_j has n-measure zero. Therefore part (i) implies that $f(V_j)$ has m-measure zero, and the claim then follows from the equality $f(N) = \bigcup_j f(V_j)$ via Remark 2.5(b). ∎

5.10 Corollary *Suppose U is open in $\mathbb{R}^n$ and $f \in C^1(U, \mathbb{R}^m)$ with $m \geq n$. If $N \subset U$ has Lebesgue n-measure zero, then $f(N)$ has Lebesgue m-measure zero.*

Proof This follows from Theorem 5.9 and Remark VII.8.12(b). ∎

5.11 Remarks (a) In the situation of Theorem 5.9, if we just ask that $f : N \to \mathbb{R}^m$ be continuous, the conclusion need not be true. Control over expansion is essential.

Proof For $N := [0,1] \times \{0\} \subset \mathbb{R}^2$, we have $\lambda_2(N) = 0$. Denoting by γ the parametrization of the Peano curve of Exercise VIII.1.8, we have $\gamma \in C(N, \mathbb{R}^2)$ and $\gamma(N) = \overline{\mathbb{B}}^2$ but $\lambda_2(\gamma(N)) > 2$, because $\overline{\mathbb{B}}^2$ contains a square of edge length $\sqrt{2}$ aligned with the axes. ∎

(b) Again in the situation of Theorem 5.9, the conclusion fails if instead of $m \geq n$ we have $m < n$.

Proof For $N := (0,1) \times \{0\} \subset \mathbb{R}^2$ and $f := \mathrm{pr}_1 \in C^\infty(N, \mathbb{R})$, we have $\lambda_2(N) = 0$ and $\lambda_1(f(N)) = \lambda_1((0,1)) = 1$. ∎

Since the continuous image of a σ-compact set is σ-compact, Theorem 5.9 and the characterization of Lebesgue sets in Theorem 5.7 easily lead to an invariance result for Lebesgue measurability:

5.12 Theorem *Suppose $A \in \mathcal{L}(n)$ and let $f \in C^{1-}(A, \mathbb{R}^m)$ be a locally Lipschitz map from A into $\mathbb{R}^m$, where $m \geq n$. Then $f(A)$ belongs to $\mathcal{L}(m)$.*

Proof By Theorem 5.7, there is a σ-compact subset S of $\mathbb{R}^n$ and a set N of Lebesgue n-measure zero such that $A = S \cup N$. Then $f(S)$ is a σ-compact subset of $\mathbb{R}^m$. According to Theorem 5.9, $f(N)$ has m-measure zero. By Theorem 5.7, $f(A) = f(S) \cup f(N)$ thus belongs to $\mathcal{L}(m)$. ∎

5.13 Corollary *Let U be open in $\mathbb{R}^n$. Suppose that $f \in C^1(U, \mathbb{R}^m)$ with $m \geq n$ and $A \in \mathcal{L}(n)$ with $A \subset U$. Then $f(A)$ belongs to $\mathcal{L}(m)$.*

Proof This follows from Theorem 5.12 and Remark VII.8.12(b). ∎

5.14 Remarks (a) In the situation of Theorem 5.12, if we just ask that $f : N \to \mathbb{R}^m$ be continuous, the conclusion need not be true.

Proof Let C be the Cantor set of Exercise III.3.8. In Exercise 17 you will show the existence of a homeomorphism $g : [0,1] \to [0,2]$ mapping C to a set of measure 1. Any set of positive measure has a nonmeasurable subset, by Theorem 5.28; hence we can fix a nonmeasurable $B \subset C$. But the inverse image $A := g^{-1}(B)$ is measurable, because λ_1 is complete, $A \subset C$, and C has measure zero (Exercise 4.7). Now take $f := g \,|\, A$. ∎

(b) Again in the situation of Theorem 5.12, the conclusion fails if instead of $m \geq n$ we have $m < n$.

Proof For $V \in \mathbb{R} \backslash \mathcal{L}(1)$ let $A := V \times \{0\}$. Then Example 5.2 and the completeness of the Lebesgue measure imply that A belongs to $\mathcal{L}(2)$. Also $f := \mathrm{pr}_1 \,|\, A$ is Lipschitz continuous, but $f(A) = V$ is not λ_1-measurable. ∎

(c) A subset A of $\mathbb{R}^n$ is Lebesgue measurable if and only if every $x \in A$ has an open neighborhood U_x in $\mathbb{R}^n$ such that $A \cap U_x$ is Lebesgue measurable. That is, measurability is a local property.

Proof The implication "$\Rightarrow$" is clear. For the converse, choose for each $x \in A$ an open neighborhood $U_x \ni x$ such that $A \cap U_x \in \mathcal{L}(n)$. Then $A \subset \bigcup_{x \in A} U_x$, and since A is Lindelöf (Corollary 1.9(ii) and Proposition 1.8), there is a countable set $\{ x_j \in A \,;\, j \in \mathbb{N} \}$ such that the U_{x_j} still cover A. Therefore $A = \bigcup_j (A \cap U_{x_j})$ belongs to $\mathcal{L}(n)$. ∎

The Lebesgue measure is translation invariant

We now turn to the task of checking that the Lebesgue measure of a set is independent of its position in space. As a first step, we show that it is invariant under translations. Given a vector $a \in \mathbb{R}^n$, the **translation** by a is the map

$$\tau_a : \mathbb{R}^n \to \mathbb{R}^n \,, \quad x \mapsto x + a \,. \tag{5.7}$$

5.15 Remark By taking map composition as multiplication, $\mathfrak{T} := \{ \tau_a \,;\, a \in \mathbb{R}^n \}$ becomes a commutative group, the **translation group** of $\mathbb{R}^n$. The map $a \mapsto \tau_a$ is an isomorphism from the additive group $(\mathbb{R}^n, +)$ to the translation group $\mathfrak{T}$. ∎

5.16 Lemma *The Borel and Lebesgue σ-algebras over $\mathbb{R}^n$ are translation invariant, that is, $\tau_a(\mathcal{B}^n) = \mathcal{B}^n$ and $\tau_a(\mathcal{L}(n)) = \mathcal{L}(n)$ for $a \in \mathbb{R}^n$.*

Proof (i) For $a \in \mathbb{R}^n$, the map τ_{-a} is a continuous map of $\mathbb{R}^n$ to itself. Therefore τ_{-a} is Borel measurable according to Exercise 1.6. Thus

$$\tau_a(\mathcal{B}) = (\tau_{-a})^{-1}(\mathcal{B}) \subset \mathcal{B} \,. \tag{5.8}$$

Replacing a by $-a$, we also get $\tau_{-a}(\mathcal{B}) \subset \mathcal{B}$. Using (5.8) we conclude that

$$\mathcal{B} = \tau_a \circ \tau_{-a}(\mathcal{B}) = \tau_a(\tau_{-a}(\mathcal{B})) \subset \tau_a(\mathcal{B}) \subset \mathcal{B} \,, \tag{5.9}$$

which proves that $\tau_a(\mathcal{B}) = \mathcal{B}$.

(ii) Because τ_a is a smooth map of $\mathbb{R}^n$ onto itself, $\tau_a(\mathcal{L}(n)) = \mathcal{L}(n)$ for $a \in \mathbb{R}^n$ by Theorem 5.12 and the group property. ∎

5.17 Theorem *The Lebesgue and Borel–Lebesgue measures are translation invariant: For $a \in \mathbb{R}^n$, we have $\lambda_n = \lambda_n \circ \tau_a$ and $\beta_n = \beta_n \circ \tau_a$.*

Proof Obviously $\mathbb{J}(n)$ and $\mathrm{vol}_n : \mathbb{J}(n) \to \mathbb{R}$ are translation invariant. Therefore the Lebesgue outer measure is also translation invariant, and the claim follows from Lemma 5.16 and the definitions of λ_n and β_n. ∎

Suppose O is open in $\mathbb{R}^n$ and nonempty. One easily checks that $O - O$ is a neighborhood of 0. The next theorem shows that this in fact holds for every Lebesgue measurable set with positive measure. Intuitively this means that such sets are never "too thin". (Compare also Exercise 12.)

5.18 Theorem (Steinhaus)   *For every $A \in \mathcal{L}(n)$ such that $\lambda_n(A) > 0$, the set $A - A$ is a neighborhood of 0.*

Proof Suppose $A \in \mathcal{L}(n)$ with $\lambda_n(A) > 0$. By replacing A with $A \cap k\mathbb{B}^n$ for a suitable $k \in \mathbb{N}^\times$, we can assume that $\lambda_n(A) < \infty$.

The regularity of λ_n ensures the existence of a compact set K and an open set O such that $K \subset A \subset O$ and

$$0 < \lambda_n(K) < \lambda_n(O) < 2\lambda_n(K) . \tag{5.10}$$

Because $K \subset O$ is compact, we have $\delta := d(K, O^c) > 0$; see Example III.3.9(c).

We claim that any $x \in \delta\mathbb{B}^n$ lies in $K - K$ (hence also in $A - A$, which is what we need). For suppose to the contrary that K and $x + K$ are disjoint, with $x \in \delta\mathbb{B}^n$. Because λ_n is additive and translation invariant, this gives

$$\lambda_n\big(K \cup (x + K)\big) = \lambda_n(K) + \lambda_n(x + K) = 2\lambda_n(K) . \tag{5.11}$$

At the same time, $x + K \subset O$, by the definition of δ, and hence $K \cup (x + K) \subset O$. Thus (5.11) implies $\lambda_n(O) \geq 2\lambda_n(K)$, in conflict with (5.10). ∎

A characterization of Lebesgue measure

The next theorem shows in particular that Lebesgue measure is determined up to normalization by translation invariance.

5.19 Theorem   *Let μ be a translation invariant locally finite measure on $\mathcal{B}^n$ or $\mathcal{L}(n)$. Then $\mu = \alpha_n \beta_n$ or $\mu = \alpha_n \lambda_n$, respectively, where $\alpha_n := \mu\big([0,1)^n\big)$.*

Proof (i) As a first step we will show

$$\mu\big([a, b)\big) = \alpha_n \operatorname{vol}_n\big([a, b)\big) \quad \text{for } a, b \in \mathbb{R}^n .$$

First assume $n = 1$ and set $g(s) := \mu\big([0, s)\big)$ for $s > 0$. Then $g \colon (0, \infty) \to (0, \infty)$ is increasing, and the translation invariance of μ implies, for $s, t \in (0, \infty)$,

$$\begin{aligned}
g(s + t) = \mu\big([0, s + t)\big) &= \mu\big([0, s) \cup [s, s + t)\big) \\
&= \mu\big([0, s)\big) + \mu\big([s, s + t)\big) = \mu\big([0, s)\big) + \mu\big([0, t)\big) \\
&= g(s) + g(t) .
\end{aligned}$$

Exercise 5 then shows that $g(s) = sg(1)$ for $s > 0$. Since $s = \operatorname{vol}_1\big([0, s)\big)$ and $\alpha_1 = \mu\big([0, 1)\big)$, this implies

$$\mu\big([0, s)\big) = g(s) = sg(1) = \operatorname{vol}_1\big([0, s)\big)\alpha_1 \quad \text{for } s > 0 ,$$

and we find

$$\mu\big([\alpha, \beta)\big) = \mu\big([0, \beta - \alpha)\big) = \alpha_1 \operatorname{vol}_1\big([0, \beta - \alpha)\big) = \alpha_1 \operatorname{vol}_1\big([\alpha, \beta)\big) \tag{5.12}$$

for $\alpha, \beta \in \mathbb{R}$.

To treat the case $n \geq 2$, we fix $a', b' \in \mathbb{R}^{n-1}$ and set

$$\mu_1([\alpha, \beta)) := \mu([\alpha, \beta) \times [a', b')) \quad \text{for } \alpha, \beta \in \mathbb{R} .$$

Exercise 7 and (5.12) imply

$$\mu_1([\alpha, \beta)) = \mu_1([0, 1)) \operatorname{vol}_1([\alpha, \beta)) \quad \text{for } \alpha, \beta \in \mathbb{R} .$$

Take $a = (a_1, \ldots, a_n) \in \mathbb{R}^n$, $b = (b_1, \ldots, b_n) \in \mathbb{R}^n$, and set $a' = (a_2, \ldots, a_n)$, $b' = (b_2, \ldots, b_n)$. Then

$$\begin{aligned}
\mu([a, b)) &= \mu([a_1, b_1) \times [a', b')) = \mu_1([a_1, b_1)) \\
&= \operatorname{vol}_1([a_1, b_1)) \mu_1([0, 1)) \\
&= \operatorname{vol}_1([a_1, b_1)) \mu([0, 1) \times [a', b')) .
\end{aligned}$$

A simple induction argument now gives

$$\mu([a, b)) = \mu([0, 1)^n) \prod_{j=1}^n \operatorname{vol}_1([a_j, b_j)) = \alpha_n \operatorname{vol}_n([a, b)) .$$

(ii) Suppose $A \in \mathcal{B}^n$ [or $A \in \mathcal{L}(n)$] and let (I_k) be a sequence in $\mathbb{J}_\ell(n)$ that covers A. It follows from (i) that

$$\mu(A) \leq \sum_k \mu(I_k) = \alpha_n \sum_k \lambda_n(I_k) .$$

Therefore we get from Proposition 3.4 that

$$\mu(A) \leq \alpha_n \lambda_n^*(A) = \alpha_n \lambda_n(A) .$$

(iii) Now suppose $B \in \mathcal{B}^n$ [or $B \in \mathcal{L}(n)$] is bounded. There exists $I \in \mathbb{J}_\ell(n)$ such that $B \subset I \subset \bar{I}$. Because $\bar{I}$ is compact and μ is locally finite, Remark 5.3(a) says that $\mu(B) < \infty$. Moreover $\lambda_n(B) < \infty$ by Theorem 5.1(iv), so Proposition 2.3(ii) yields

$$\mu(I \setminus B) = \mu(I) - \mu(B) \quad \text{and} \quad \lambda_n(I \setminus B) = \lambda_n(I) - \lambda_n(B) ,$$

and we find with (ii) that

$$\mu(I) - \mu(B) = \mu(I \setminus B) \leq \alpha_n \lambda_n(I \setminus B) = \alpha_n \lambda_n(I) - \alpha_n \lambda_n(B) .$$

From (i) we have $\mu(I) = \alpha_n \lambda_n(I)$, and the inequality $\mu(B) \geq \alpha_n \lambda_n(B)$ follows. Together with (ii), we therefore get $\mu(B) = \alpha_n \lambda_n(B)$ for every bounded $B \in \mathcal{B}^n$ [or $B \in \mathcal{L}(n)$].

(iv) Finally take an arbitrary $A \in \mathcal{B}^n$ [or $A \in \mathcal{L}(n)$] and set $B_j := A \cap \mathbb{B}^n(0, j)$ for $j \in \mathbb{N}$. The sequence (B_j) is increasing and covers A; moreover each B_j is a bounded Borel [or Lebesgue] set in $\mathbb{R}^n$. Applying (iii) and Proposition 2.3(iv), we see that

$$\mu(A) = \lim_j \mu(B_j) = \alpha_n \lim_j \lambda_n(B_j) = \alpha_n \lambda_n(A) ,$$

which finishes the proof. ∎

5.20 Remark   In the theorem just proved we cannot relax the assumption that μ is locally finite.

Proof   Clearly the counting measure $\mathcal{H}^0$ on Borel or Lebesgue-measurable sets is translation invariant. However, it is not a multiple of either measure. ∎

The Lebesgue measure is invariant under rigid motions

Theorem 5.19 allows a comparison of n-dimensional Lebesgue and Hausdorff measures. For this we need a lemma:

5.21 Lemma   The n-dimensional Hausdorff measure $\mathcal{H}^n$ on $\mathbb{R}^n$ is locally finite, and satisfies $\mathcal{H}^n\big([0,1)^n\big) > 0$.

Proof   (i) From Theorem 4.3 and Example 4.4(c), we know that every Borel set is $\mathcal{H}^n$-measurable. Suppose $K \subset \mathbb{R}^n$ is compact and take $\varepsilon > 0$. Choose $a > 0$ such that $K \subset [-a,a]^n$ and $m \in \mathbb{N}$ such that $m \geq 2a\sqrt{n}/\varepsilon$. Subdivide $[-a,a]^n$ into m^n subcubes W_j of length $2a/m$. Then $\mathrm{diam}(W_j) = 2a\sqrt{n}/m \leq \varepsilon$, and therefore

$$\sum_{j=1}^{m^n} \big[\mathrm{diam}(W_j)\big]^n = (2a)^n n^{n/2} \ .$$

Exercise 3.4 shows that $\mathcal{H}_*^n(K) \leq \mathcal{H}_*^n\big([-a,a]^n\big) \leq (2a)^n n^{n/2}$. Now we obtain from Remark 5.3(b) that $\mathcal{H}^n$ is locally finite.

(ii) It remains to verify that $\mathcal{H}^n\big([0,1)^n\big) > 0$. Take $\varepsilon > 0$ and let (U_j) be a sequence of open sets in $\mathbb{R}^n$ covering $[0,1)^n$ and such that $\mathrm{diam}(U_j) < \varepsilon$. For each $j \in \mathbb{N}$, there is $I_j \in \mathbb{J}(n)$ such that every edge length of I_j is bounded by $2\,\mathrm{diam}(U_j)$ and such that $U_j \subset I_j$. It follows that

$$1 = \lambda_n\big([0,1)^n\big) \leq \sum_j \mathrm{vol}_n(I_j) \leq 2^n \sum_j \big[\mathrm{diam}(U_j)\big]^n \ ,$$

and hence $2^{-n} \leq \mathcal{H}_\varepsilon^n\big([0,1)^n\big)$. This implies $\mathcal{H}^n\big([0,1)^n\big) \geq 2^{-n} > 0$. ∎

5.22 Corollary   The n-dimensional Hausdorff measure $\mathcal{H}^n$ on $\mathbb{R}^n$ is an extension of $\alpha_n \lambda_n$ with $\alpha_n := \mathcal{H}^n\big([0,1)^n\big)$; that is, every $A \in \mathcal{L}(n)$ is $\mathcal{H}^n$-measurable, and $\mathcal{H}^n(A) = \alpha_n \lambda_n(A)$.

Proof   (i) Lemma 5.21, Exercise 3.4, and the $\mathcal{H}^n$-measurability of Borel sets show that $\mathcal{H}^n$ is a locally finite translation invariant measure on $\mathcal{B}^n$. By Theorem 5.19, then, $\mathcal{H}^n \,|\, \mathcal{B}^n = \alpha_n \beta_n$.

(ii) Suppose N is a set of Lebesgue measure zero and $\varepsilon > 0$. Then there is a sequence (I_j) in $\mathbb{J}(n)$ such that $\sum_j \mathrm{vol}_n(I_j) < \varepsilon$ and $N \subset \bigcup_j I_j$. From (i), it follows that

$$\mathcal{H}_*^n(I_j) = \mathcal{H}^n(I_j) = \alpha_n \lambda_n(I_j) \ ,$$

and we find

$$\mathcal{H}_*^n(N) \leq \mathcal{H}_*^n\left(\bigcup_j I_j\right) \leq \sum_j \mathcal{H}_*^n(I_j) = \alpha_n \sum_j \mathrm{vol}_n(I_j) < \alpha_n \varepsilon \ .$$

Therefore N is $\mathcal{H}^n$-null.

(iii) Suppose $A \in \mathcal{L}(n)$. According to Theorem 5.7, we can write $A = S \cup N$, where $S \in \mathcal{B}^n$ and N has Lebesgue measure zero. Therefore A is $\mathcal{H}^n$-measurable. And it follows from (i) and (ii) that

$$\mathcal{H}^n(A) \leq \mathcal{H}^n(S) + \mathcal{H}^n(N) = \mathcal{H}^n(S) = \alpha_n \lambda_n(S) \leq \alpha_n \lambda_n(A) \ ,$$
$$\alpha_n \lambda_n(A) = \alpha_n \lambda_n(S) = \mathcal{H}^n(S) \leq \mathcal{H}^n(A) \ ,$$

which together show that $\mathcal{H}^n(A) = \alpha_n \lambda_n(A)$. ∎

5.23 Corollary *The Lebesgue and Borel–Lebesgue measures are invariants of motion, that is, they are preserved under isometries. In symbols, any isometry φ of $\mathbb{R}^n$ satisfies $\lambda_n = \lambda_n \circ \varphi$ and $\beta_n = \beta_n \circ \varphi$.*

Proof Let φ be an isometry of $\mathbb{R}^n$ and take $A \in \mathcal{L}(n)$ [or $A \in \mathcal{B}^n$]. Since φ and φ^{-1} are Lipschitz continuous by Conclusion VI.2.4(b), we obtain from Theorem 5.12 [or Exercise 1.6(b)] that $\varphi(A) \in \mathcal{L}(n)$ [or $\varphi(A) \in \mathcal{B}^n$]. Next, $\mathcal{H}_*^n$ is invariant under isometries, by Exercise 3.4(c); hence Lemma 5.21 and Corollary 5.22 show that

$$\alpha_n \lambda_n(\varphi(A)) = \mathcal{H}^n(\varphi(A)) = \mathcal{H}^n(A) = \alpha_n \lambda_n(A) \ . \quad ∎$$

5.24 Remarks **(a)** Though Corollary 5.22 talks of an extension, in fact the domains of $\mathcal{H}^n$ and λ_n coincide. Moreover the proportionality constant $\alpha_n = \mathcal{H}^n([0,1)^n)$ equals $2^n/\omega_n$, where $\omega_n = \pi^{n/2}/\Gamma((n/2)+1)$. For proofs of these statements, see [Rog70, Theorem 30 and subsequent remark].

(b) There are true extensions of the Lebesgue measure on $\mathbb{R}^n$ that are invariant under isometries; see [Els99]. ∎

The substitution rule for linear maps

Let φ be an isometry of $\mathbb{R}^n$ with $\varphi(0) = 0$. Exercises VII.9.1 and VII.9.2 show that φ is an automorphism and $|\det \varphi| = 1$. Therefore it follows from Corollary 5.23 that

$$\lambda_n(\varphi(A)) = |\det \varphi| \lambda_n(A) \quad \text{for } A \in \mathcal{L}(n) \ .$$

Our goal now is to extend this formula from isometries to arbitrary linear maps $T \in \mathcal{L}(\mathbb{R}^n)$. In the next chapter, we will obtain an even more far-reaching generalization, in which φ is replaced by a C^1 diffeomorphism and λ_n is replaced by the Lebesgue integral.

5.25 Theorem *Suppose $T \in \mathcal{L}(\mathbb{R}^n)$ is a linear map. Then*

$$\lambda_n\big(T(A)\big) = |\det T|\,\lambda_n(A) \quad \text{for } A \in \mathcal{L}(n) .\tag{5.13}$$

Proof We know that T is Lipschitz continuous; see Conclusion VI.2.4(b). Hence, by Theorem 5.12, $T(A)$ is Lebesgue measurable for every $A \in \mathcal{L}(n)$.

(i) If T is not an automorphism of $\mathbb{R}^n$, then $\det T = 0$ and $T(A)$ lies in an $(n-1)$-dimensional hyperplane of $\mathbb{R}^n$. Since λ_n is an invariant of motion, we can assume that $T(A)$ lies in a coordinate hyperplane. Then Example 5.2 shows that $T(A)$ has Lebesgue n-measure zero, proving (5.13) in this case.

(ii) Suppose instead that $T \in \mathcal{L}\mathrm{aut}(\mathbb{R}^n)$, and define $\mu(A) := \lambda_n\big(T(A)\big)$ for $A \in \mathcal{L}(n)$. It is not hard to verify that μ is a locally finite translation invariant measure on $\mathcal{L}(n)$. Theorem 5.19 then says that $\mu = \mu\big([0,1)^n\big)\lambda_n$; this will imply (5.13) if we show that

$$\lambda_n\big(T([0,1)^n)\big) = |\det T| .\tag{5.14}$$

(iii) Let the ordered n-tuple $[Te_1,\ldots,Te_n]$ be a permutation of the standard basis $[e_1,\ldots,e_n]$ of $\mathbb{R}^n$. Then $T\big([0,1)^n\big) = [0,1)^n$ and $|\det T| = 1$. Therefore (5.14) holds and therefore so does (5.13).

(iv) Let $\alpha \in \mathbb{R}^\times$ and define T by

$$Te_j = \begin{cases} \alpha e_1 , & j = 1 , \\ e_j , & j \in \{2,\ldots,n\} . \end{cases}$$

Then $|\det T| = |\alpha|$ and

$$T\big([0,1)^n\big) = \begin{cases} [0,\alpha) \times [0,1)^{n-1} , & \alpha > 0 , \\ (\alpha,0] \times [0,1)^{n-1} , & \alpha < 0 . \end{cases}$$

Again (5.14), and consequently (5.13), are satisfied.

(v) Finally suppose $n \geq 2$ and set

$$Te_j = \begin{cases} e_1 + e_2 , & j = 1 , \\ e_j , & j \in \{2,\ldots,n\} . \end{cases}$$

Then $\det T = 1$, and

$$T\big([0,1)^n\big) = \big\{ (y_1,\ldots,y_n) \in \mathbb{R}^n \,;\, 0 \leq y_1 \leq y_2 < y_1 + 1,\ y_j \in [0,1) \text{ for } j \neq 2 \big\} .$$

Setting $B_1 := \big\{ y \in T\big([0,1)^n\big) \,;\, y_2 < 1 \big\}$ and $B_2 := T\big([0,1)^n\big) \setminus B_1$, we see that $B_1 \cup (B_2 - e_2) = [0,1)^n$ and $B_1 \cap (B_2 - e_2) = \emptyset$.

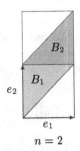

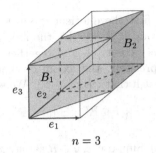

$$n = 2 \qquad n = 3$$

The translation invariance of λ_n then yields

$$\lambda_n\big(T([0,1)^n)\big) = \lambda_n(B_1 \cup B_2) = \lambda_n(B_1) + \lambda_n(B_2)$$
$$= \lambda_n(B_1) + \lambda_n(B_2 - e_2) = \lambda_n\big(B_1 \cup (B_2 - e_2)\big) = \lambda_n\big([0,1)^n\big) .$$

Once more (5.14) and hence (5.13) hold.

(vi) Now consider an arbitrary automorphism T of $\mathbb{R}^n$. The normal form theorem of linear algebra (see § 2.6 in [Koe83]) says that T can be written as a composition $T = T_1 \circ \cdots \circ T_k$ of maps $T_1, \ldots, T_k \in \mathcal{L}aut(\mathbb{R}^n)$, each having the form of one of the linear maps treated in (iii)–(v). Therefore

$$\lambda_n\big(T(A)\big) = \lambda_n\big((T_1 \circ T_2 \circ \cdots \circ T_k)(A)\big)$$
$$= |\det T_1| \, \lambda_n\big((T_2 \circ \cdots \circ T_k)(A)\big) = \cdots$$
$$= |\det T_1| \cdot \cdots \cdot |\det T_k| \, \lambda_n(A) = |\det T| \, \lambda_n(A)$$

for $A \in \mathcal{L}(n)$. ∎

5.26 Remarks (a) Suppose $[t_1, \ldots, t_n] \in \mathbb{R}^{n \times n}$ are the columns of the matrix $[T]$ representation $T \in \mathcal{L}(\mathbb{R}^n)$ with respect to the canonical basis. Then

$$T([0,1)^n) = \{\, x_1 t_1 + \cdots + x_n t_n \ ; \ 0 \le x_j < 1, \ 1 \le j \le n \,\} = P(t_1, \ldots, t_n)$$

is the parallelepiped formed by the vectors $t_1, \ldots, t_n$. Theorem 5.25 says that $|\det T|$ is the volume, or Lebesgue n-measure, of the parallelepiped $P(t_1, \ldots, t_n)$.

(b) For $r \ge 0$, we have
$$\lambda_n(r\mathbb{B}^n) = r^n \lambda_n(\mathbb{B}^n) .$$

Proof Letting $T := r1_n$, we have $\det T = r^n$ and $T(\mathbb{B}^n) = r\mathbb{B}^n$. The conclusion follows from Theorem 5.25. ∎

Sets without Lebesgue measure

We now turn to the question of whether the σ-algebra $\mathcal{L}(n)$ coincides with the power set of $\mathbb{R}^n$ or whether there are sets that are not Lebesgue measurable. To answer this, we must resort to the axiom of choice (see Remarks I.6.10 and 5.31(d)). In this connection, Steinhaus's theorem (Theorem 5.18) will prove very useful.

We consider first the quotient group $\mathbb{R}^n/\mathbb{Q}^n$ of the Abelian group $(\mathbb{R}^n, +)$ modulo the subgroup $(\mathbb{Q}^n, +)$. By the axiom of choice, we can choose a representative from every coset $[x]$, and together these representatives form a subset R of $\mathbb{R}^n$. More precisely, the axiom of choice ensures the existence of a map $\varphi \colon \mathbb{R}^n/\mathbb{Q}^n \to \mathbb{R}^n$ such that $\varphi([x]) \in [x]$. We set

$$R := \mathrm{im}(\varphi) = \left\{ x \in \mathbb{R}^n ; \ \exists\, [u] \in \mathbb{R}^n/\mathbb{Q}^n \text{ with } x = \varphi([u]) \right\} . \tag{5.15}$$

5.27 Remarks (a) Suppose $x, y \in R$ satisfy $x - y \in \mathbb{Q}^n$. Then $x = y$.

Proof For $x \in R$ and $[u]$ as in (5.15) we have $x = \varphi([u]) \in [u]$, so $[u] = [x]$, so $x = \varphi([x])$. Similarly, $y = \varphi([y])$. But $x - y \in \mathbb{Q}^n$ implies $[x] = [y]$, hence $x = \varphi([x]) = \varphi([y]) = y$. ∎

(b) For every $B \subset R$, we have $(B - B) \cap \mathbb{Q}^n = \{0\}$.

Proof This follows from (a). ∎

5.28 Theorem For every $A \in \mathcal{L}(n)$ with $\lambda_n(A) > 0$, there is a $B \subset A$ such that $B \notin \mathcal{L}(n)$.

Proof Fix $A \in \mathcal{L}(n)$ with $\lambda_n(A) > 0$, and define $B := \left\{ b \in R ; \ [b] \cap A \neq \emptyset \right\}$. By demanding in the construction of R that $b \in R$ lie in A whenever $[b] \cap A \neq \emptyset$, we ensure that B is a subset of A. Suppose B is Lebesgue-measurable; then it has measure zero, for otherwise Theorem 5.18 would imply that $B - B$ is a neighborhood of 0 in $\mathbb{R}^n$, contradicting Remark 5.27(b). Because λ_n is translation invariant, every set $B + r$ with $r \in \mathbb{Q}^n$ also has measure zero.

Now consider $a \in A$ and take $b := \varphi([a]) \in R$. Then $b \in [a]$, hence $a \in [b]$. Therefore $a \in [b] \cap A$, that is, $b \in B$ and

$$A \subset \bigcup_{b \in B} [b] = \bigcup_{r \in \mathbb{Q}^n} (B + r) .$$

The completeness of λ_n implies that A also has measure zero, in contradiction to our assumption. ∎

5.29 Corollary The Borel–Lebesgue measure space $(\mathbb{R}^n, \mathcal{B}^n, \beta_n)$ is not complete.

Proof (i) We consider first the case $n = 1$. Let C be Cantor set and $f \colon [0,1] \to [0,1]$ the Cantor function of Exercise III.3.8. Being compact, C is a Borel set, and Exercise 4.7 tells us that it has measure zero. Exercise 17 below shows that the map $g \colon [0,1] \to [0,2]$ given by $g(x) = x + f(x)$ is a homeomorphism and that $g(C)$ has measure 1. By Theorem 5.28, $g(C)$ contains a set B that is not even Lebesgue measurable. We claim that $N_1 := g^{-1}(B) \subset C$ is not Borel measurable. Indeed, since g is a homeomorphism, g^{-1} is a Borel measurable map, and we have

$$B = g(N_1) = (g^{-1})^{-1}(N_1) .$$

If N_1 were Borel measurable, B would be as well, contrary to assumption.

(ii) In the case $n \geq 2$, let $A := C \times \mathbb{R}^{n-1}$ and $N_n := N_1 \times \mathbb{R}^{n-1}$. Then Corollary 1.18 and Exercise 1 show that A has Borel n-measure zero. If we were to assume that $N_n \in \mathcal{B}^n$, it would follow, by Corollary 1.18 and Proposition 1.19, that N_1 belongs to $\mathcal{B}^1$, in conflict with (i). This finishes the proof. ∎

5.30 Corollary   *The Borel σ-algebra is a proper σ-subalgebra of the Lebesgue σ-algebra.*

Proof   Let A and N_n be as in the proof of Corollary 5.29. Then $\lambda_n(A) = 0$ and the completeness of λ_n implies N_n belongs to $\mathcal{L}(n)$, that is, $N_n \in \mathcal{L}(n) \backslash \mathcal{B}^n$. ∎

5.31 Remarks   **(a)** Suppose $(X, \leq)$ is an ordered set. A nonempty subset Y of X is said to be **totally ordered** if any two elements from Y are comparable to one another, that is, if $(x, y) \in Y \times Y$ always implies $(x \leq y) \vee (y \leq x)$. An element m of X is **maximal** if $x \geq m$ implies $x \leq m$, that is, if X has no element strictly larger than[3] m. **Zorn's lemma** reads: *If X is an ordered set and every totally ordered subset of X has an upper bound, then X has a maximal element.* One can show (see Theorem II.2.1 in [Dug66]) that Zorn's lemma is equivalent to the axiom of choice.

(b) Suppose $\mathcal{V}$ is a nontrivial vector space over a field. Then $\mathcal{V}$ has a basis.

Proof   For a proof (using Zorn's lemma), we refer to Proposition (1.10) in the appendix of [Art93]. ∎

(c) Suppose $B \subset \mathbb{R}$ is a basis of the vector space $\mathbb{R}$ over $\mathbb{Q}$. Take $b_0 \in B$, and let $M := \operatorname{span}(B \backslash \{b_0\})$. Then M is not Lebesgue measurable.

Proof   Assume that M belongs to $\mathcal{L}(1)$. Then $\lambda_1(M) > 0$, because otherwise it would follow from the translation invariance of λ_1 that $M + rb_0$ is a λ_1-null set for every $r \in \mathbb{Q}$. However, because

$$\bigcup_{r \in \mathbb{Q}} (M + rb_0) = \operatorname{span}(B) = \mathbb{R} ,$$

this is not possible. Thus Theorem 5.18 shows that $M - M$ is a neighborhood of 0 in $\mathbb{R}$; in particular, there exists $r_0 \in \mathbb{Q}$ such that $r_0 \neq 0$ and $r_0 b_0 \in M - M$. Because $M = M - M$, there are $k \in \mathbb{N}^\times$, $r_j \in \mathbb{Q}$, and $b_j \in B$ for $j = 1, \ldots, k$ such that $r_0 b_0 = \sum_{j=1}^{k} r_j b_j$, which contradicts the linear independence of B over $\mathbb{Q}$. ∎

(d) In the proof of Theorem 5.28, we have explicitly used the axiom of choice. The proof of (c), too, rests on the axiom of choice; see (a) and (b). In fact, one can show (see [BS79], [Sol70]) that it is not possible in principle to specify a set that is not Lebesgue measurable if one works in an axiomatic set theory not containing the axiom of choice. ∎

---

[3] Note that a set can generally have multiple maximal elements.

Exercises

1 Show that

$$\mathcal{L}(m) \boxtimes \mathcal{L}(n) \subset \mathcal{L}(m+n) \quad \text{and} \quad \lambda_m(A)\lambda_n(B) = \lambda_{m+n}(A \times B)$$

for $A \times B \in \mathcal{L}(m) \boxtimes \mathcal{L}(n)$.

(Hint: Consider first the case of open sets A in $\mathbb{R}^m$ and B in $\mathbb{R}^n$ and use Proposition 5.6 and Theorem II.8.10. For $A \times B \in \mathcal{L}(m) \boxtimes \mathcal{L}(n)$, note Corollary 5.5.)

2 Show that $\mathcal{B}^m \otimes \mathcal{B}^n \subset \mathcal{L}(m) \otimes \mathcal{L}(n) \subset \mathcal{L}(m+n)$ and that these inclusions are proper.

3 Suppose M is an m-dimensional C^1 submanifold of $\mathbb{R}^n$. Prove that M has Lebesgue n-measure zero if $m < n$.

4 Verify that for $A \in \mathcal{L}(n)$, we have

$$\lambda_n(A) = \sup\{ \lambda_n(B) \; ; \; B \subset \mathbb{R}^n \text{ is closed and } B \subset A \} \, .$$

5 Suppose $g : (0,\infty) \to \mathbb{R}$ satisfies $g(s+t) = g(s) + g(t)$ for $s, t \in (0,\infty)$. Prove that if g is increasing or bounded on bounded sets, then $g(s) = sg(1)$ for $s > 0$.

6 Let

$$S := \{ g \in \mathbb{R}^{\mathbb{R}} \; ; \; g(s+t) = g(s) + g(t), \; s, t \in \mathbb{R}, \; \exists \, s_0 \in \mathbb{R} : g(s_0) \neq s_0 g(1) \} \, .$$

Prove:

(a) For every $g \in S$, the graph of g is dense in $\mathbb{R}^2$;

(b) $S \neq \emptyset$.

(Hint for (b): Define g using a basis of the $\mathbb{Q}$-vector space $\mathbb{R}$.)

7 For $n \geq 2$, let μ be a translation invariant locally finite measure on $\mathcal{B}^n$ [or $\mathcal{L}(n)$]. If $A \in \mathcal{B}^1$ [or $A \in \mathcal{L}(1)$] and $a', b' \in \mathbb{R}^{n-1}$, let

$$\mu_1(A) := \mu(A \times [a', b')) \, .$$

Show that μ_1 is a translation invariant locally finite measure on $\mathcal{B}^1$ [or $\mathcal{L}(1)$].

8 Let B be a basis of the $\mathbb{Q}$-vector space $\mathbb{R}$. Prove or disprove that B is finite.

9 Define $M := \{ \log p \; ; \; p \in \mathbb{N} \text{ is prime} \}$. Prove:

(a) M is linearly independent over $\mathbb{Q}$.

(b) M is not a basis of $\mathbb{R}$.

10 If B is a Lebesgue measurable basis of $\mathbb{R}$ over $\mathbb{Q}$, then B has Lebesgue measure zero.

11 Verify that the Cantor set C satisfies $C + C = [0, 2]$.

12 Show there is a set A of Lebesgue measure zero such that $A - A$ is a neighborhood of 0.[4]

---

[4]Compare Theorem 5.18.

13 Show that there is a Lebesgue measurable basis of the $\mathbb{Q}$-vector space $\mathbb{R}$.
(Hint: Let C be the Cantor set, and define

$$\mathcal{A} := \{\, M \subset C \ ; \ M \text{ is linearly independent over } \mathbb{Q} \,\} \,.$$

Then $\mathcal{A}$ has a maximal element B, which spans $\mathbb{R}$ in view of Exercise 11.)

14 Suppose R is as in (5.15). Verify that R does not belong to $\mathcal{L}(n)$.

15 Set $G := \mathbb{Q} + \sqrt{2}\,\mathbb{Z}$, $G_1 := \mathbb{Q} + 2\sqrt{2}\,\mathbb{Z}$ and $G_2 := G\backslash G_1$. Prove:

(a) G and G_1 are subgroups of the Abelian group $(\mathbb{R}, +)$.

(b) Take $\varphi \colon \mathbb{R}/G \to \mathbb{R}$ such that $\varphi([x]) \in [x]$, and let $R := \mathrm{im}(\varphi)$. Set $A := R + G_1$. Then $(A - A) \cap G_2 = \emptyset$.

16 Prove that there is a subset A of $\mathbb{R}$ such that every Lebesgue measurable set contained in either A or A^c has measure zero.
(Hint: Show with the help of Theorem 5.18 that the set A of Exercise 15 has the desired properties.)

17 Let $f \colon [0,1] \to [0,1]$ be the Cantor function of Exercise III.3.8. Prove:

(a) There is a set $N \subset [0,1]$ of measure zero such that f is differentiable at every point of $[0,1]\backslash N$, with derivative zero.

(b) The map $g \colon [0,1] \to [0,2]$, $x \mapsto x + f(x)$ is a homeomorphism.

(c) $\lambda_1\big(g(C)\big) = 1$.

18 Verify:

(a) Every finite-dimensional normed vector space is locally compact.

(b) Any open subset and any closed subset of a locally compact space is locally compact.

(c) A locally compact space is σ-compact if and only if it is separable.

(d) Any open subset and any closed subset of a σ-compact locally compact metric space is is σ-compact.

19 Suppose $F \colon \mathbb{R} \to \mathbb{R}$ is measure-generating. Then the Lebesgue–Stieltjes measure on $\mathbb{R}$ induced by F is regular.

20 Suppose X is a metric space. Check that

$$\mathcal{B}(X) = \mathcal{A}_\sigma\big(\{\, f^{-1}(0) \ ; \ f \in C(X, \mathbb{R}) \,\}\big) \,.$$

(There exist nonmetrizable topological spaces for which $\mathcal{B}(X)$ is strictly bigger than $\mathcal{A}_\sigma\big(\{\, f^{-1}(0) \ ; \ f \in C(X, \mathbb{R}) \,\}\big)$; see [Flo81, 11.1.2].)

21 Suppose X is a topological space, and denote by $(X, \mathcal{A}, \mu)$ a regular measure space with $\mathcal{A} \supset \mathcal{B}(X)$. Further let $A \in \mathcal{A}$, $\mathcal{C} := \mathcal{A}\,|\,A$, and $\nu := \mu\,|\,\mathcal{C}$. Verify that $(A, \mathcal{C}, \nu)$ is regular.

Chapter X

Integration theory

Having made acquaintance in the last chapter with the fundamentals of measure theory, we will now turn to the theory of integration. In the first part of the chapter we study integrals over general measure spaces, while in the second half we take advantage of the special properties of the Lebesgue measure.

Integration with respect to arbitrary measures is not only important in many applications, but it will also be essential in the last chapter, when the underlying set is not "flat" but rather a manifold. This is why even an introductory text such as ours must deal with the subject.

In Section 1, we introduce μ-measurable functions and investigate their basic properties. A position of keen interest in analysis is held by natural measures with respect to which every continuous function is measurable. An example is the class of Radon measures, which we introduce in this section and which we will encounter again in Chapter XII.

In analysis, and not only there, it will be increasingly important to be able to deal with vector-valued functions, that is, maps with values in a Banach space. We have already worked along these lines in the first two volumes, and you will have noticed that the resulting exposition gains not only in elegance but, in many cases, also in simplicity. The same situation obtains regarding integration theory. Hence we have resolved from the outset to develop the theory in terms of vector-valued functions, and we therefore treat the Bochner–Lebesgue integral. This is possible with no significant extra effort. One of the few exceptions is the proof that a vector-valued function is μ-measurable if and only if it is measurable in the usual sense and is μ-almost separable valued. Of course, you could ignore this result and consider only scalar-valued functions. But this is not recommended, as it would cause you to miss out on an important and efficient addition to your toolkit.

Besides vector-valued maps, we will investigate in some detail functions with values in the extended number line $[0, \infty]$. This is primarily for technical reasons; in later sections it will save us from having to always single out special cases.

In Section 2, we introduce the general Bochner–Lebesgue integral, and do so via the $\mathcal{L}_1$-completion of the space of simple functions. This approach not only extends essentially unchanged to vector-valued functions, but also lays the foundation for the proof of Lebesgue's convergence theorem. We treat the latter, as well as other important convergence theorems, in Section 3.

Section 4 is devoted to the elementary theory of Lebesgue spaces. We prove their completeness and show that they become Banach spaces if we identify functions that agree almost everywhere. Because this identification is in our experience a source of difficulties for beginners, we make a meticulous distinction throughout the chapter between equivalence classes of functions and their respective representatives.

Although up to this point, we have considered integrals with respect to an arbitrary measure, we treat in Section 5 the special case of Lebesgue measure in $\mathbb{R}^n$. We show that the one-dimensional Lebesgue integral is an extension of the Cauchy–Riemann integral for absolutely integrable functions. This puts us in the position to bring what we learned about integrals in Volume II into the framework of the general theory. This is of particular significance in the context of Fubini's theorem, which gives a reduction procedure for evaluating higher-dimensional integrals.

Section 6 treats Fubini's theorem. We have decided not to prove it for arbitrary product measure spaces, but rather only for the Lebesgue measure space. This simplifies the presentation considerably and is in practice sufficient for all the needs of analysis — once strengthened by a extension to product manifolds, to be treated in Chapter XII.

The proof of Fubini's theorem in the vector-valued case requires delicate measurability arguments. For this reason, we study first the scalar case. We prove the vector-valued version at the end of Section 6 and exhibit some important applications. On first reading, this part may be skipped, because its results are not used in any essential way afterward, and also because the reader will probably become acquainted at some later point with the Hahn–Banach theorem of functional analysis: with its help Fubini's theorem for vector-valued functions is easily deduced from the scalar version.

Section 7 studies the convolution. This operation allows us to prove with extraordinary efficiency some fundamental approximation theorems, such as the theorem on smooth partitions of unity, which plays in important role in the final chapter. In the second half of the section we address the significance of the convolution and the approximation theorems in analysis and mathematical physics, offering a first glimpse of the very important generalization of the classical differential calculus known as the theory of distributions.

Besides the convergence theorems of Lebesgue and Fubini, the transformation theorem forms the third pillar of the entire integral calculus. It will be proved in Section 8, where we also discuss its more basic applications.

In the last section, we illustrate the power of the theory just developed by proving several basic facts about the Fourier transform. Like the second half of Section 7, this part affords a look at a related area of analysis which you may later encounter in more advanced studies.

1 Measurable functions

Suppose $(X, \mathcal{A}, \mu)$ is a measure space and $A \in \mathcal{A}$. An analogy with elementary geometrical constructions leads one to define the integral over X of the characteristic function χ_A with respect to the measure μ as $\int_X \chi_A \, d\mu := \mu(A)$. Obviously this only makes sense if A belongs to $\mathcal{A}$. The function $f = \chi_A$ must therefore be "compatible" in this sense with the underlying measure space $(\mathcal{A}, \mu)$. For more complicated functions, a suitable approximation argument makes it possible to generalize this notion of compatibility between functions and measures, leading to the concept of the measurability of functions.

In this section denote by

- $(X, \mathcal{A}, \mu)$ a complete, σ-finite measure space;
  $E = (E, |\cdot|)$ a Banach space.

Simple functions and measurable functions

Suppose E is a property that is either true or false of each point in X. We say that E holds **μ-almost everywhere**, or **for μ-almost every** $x \in X$, if there exists a μ-null set N such that $E(x)$ is true for every $x \in N^c$. "Almost every" and "almost everywhere" are both abbreviated "a.e."

1.1 Examples (a) For $f, g \in \mathbb{R}^X$, we write "$f \geq g$ μ-a.e." if there is a μ-null set N such that $f(x) \geq g(x)$ for every $x \in N^c$.

(b) Suppose $f_j, f \in E^X$ for $j \in \mathbb{N}$. Then (f_j) converges to f μ-a.e. if and only if there is a μ-null set N such that $f_j(x) \to f(x)$ for $x \in N^c$.

(c) A function $f \in E^X$ is bounded μ-a.e. if and only if there is a μ-null set N and an $M \geq 0$ such that $|f(x)| \leq M$ for every $x \in N^c$.

(d) If E holds μ-a.e., the set $\{ x \in X \; ; \; E(x) \text{ is not true} \}$ is μ-null.

Proof This follows from the completeness of $(X, \mathcal{A}, \mu)$. ∎

(e) Suppose $(X, \mathcal{B}, \nu)$ is an incomplete measure space. Then there is a property E of X that holds ν-almost everywhere for which $\{ x \in X \; ; \; E(x) \text{ is not true} \}$ is however not a ν-null set.

Proof Because $(X, \mathcal{B}, \nu)$ is not complete, there is a ν-null set N and an $M \subset N$ such that $M \notin \mathcal{B}$. If $f := \chi_M$, then $f = 0$ ν-almost everywhere, but $\{ x \in X \; ; \; f(x) \neq 0 \} = M$ is not a ν-null set. ∎

We say $f \in E^X$ is **μ-simple**[1] if $f(X)$ is finite, $f^{-1}(e) \in \mathcal{A}$ for every $e \in E$, and $\mu\big(f^{-1}(E \setminus \{0\})\big) < \infty$. We denote by $\mathcal{S}(X, \mu, E)$ the set of all μ-simple functions

---

[1] If the identity of the measure space is clear, we call functions **simple** instead of μ-simple; similarly in the case of μ-measurable functions, about to be introduced.

from X to E.[2]

A function $f \in E^X$ is said to be **μ-measurable** if there is a sequence (f_j) in $\mathcal{S}(X, \mu, E)$ such that $f_j \to f$ μ-almost everywhere as $j \to \infty$. We set[3]

$$\mathcal{L}_0(X, \mu, E) := \{ f \in E^X ; f \text{ is } \mu\text{-measurable} \} .$$

1.2 Remarks (a) We have the inclusions of vector subspaces

$$\mathcal{S}(X, \mu, E) \subset \mathcal{L}_0(X, \mu, E) \subset E^X .$$

(b) For $j = 0, \ldots, m$, where $m \in \mathbb{N}$, consider $e_j \in E$ and $A_j \in \mathcal{A}$ such that $\mu(A_j) < \infty$. Then $f := \sum_{j=0}^{m} e_j \chi_{A_j}$ belongs to $\mathcal{S}(X, \mu, E)$. We call this the **normal form** of f if

$$\begin{aligned}
e_j &\neq 0 \quad \text{for } j = 0, \ldots, m , \\
e_j &\neq e_k \quad \text{for } j \neq k , \\
A_j \cap A_k &= \emptyset \quad \text{for } j \neq k .
\end{aligned}$$

(c) Every simple function has a unique normal form, and[4]

$$\mathcal{S}(X, \mu, E) = \Big\{ \sum_{j=1}^{m} e_j \chi_{A_j} ; m \in \mathbb{N}, e_j \in E \backslash \{0\}, A_j \in \mathcal{A},$$
$$\mu(A_j) < \infty, A_j \cap A_k = \emptyset \text{ for } j \neq k \Big\} .$$

Proof Suppose $f \in \mathcal{S}(X, \mu, E)$. Then there is an $m \in \mathbb{N}$ and pairwise distinct elements $e_0, \ldots, e_m$ in E such that $f(X) \backslash \{0\} = \{e_0, \ldots, e_m\}$. Setting $A_j := f^{-1}(e_j)$, we have $A_j \in \mathcal{A}$ such that $\mu(A_j) < \infty$ and $A_j \cap A_k = \emptyset$ for $j \neq k$. One checks easily that

$$\sum_{j=0}^{m} e_j \chi_{A_j}$$

is the unique normal form of f. The second part now follows from (b). $\blacksquare$

(d) Suppose $f \in E^X$ and $g \in \mathbb{K}^X$ are μ-simple [or μ-measurable]. Then $|f| \in \mathbb{R}^X$ and $gf \in E^X$ are also μ-simple [or μ-measurable]. In particular, $\mathcal{S}(X, \mu, \mathbb{K})$ and $\mathcal{L}_0(X, \mu, \mathbb{K})$ are subalgebras of $\mathbb{K}^X$.

(e) For $A \in \mathcal{A}$ and $f \in E^X$, form the restriction $\nu := \mu \,|\, (\mathcal{A} \,|\, A)$ (see Exercise IX.1.7). Then

$$\begin{aligned}
f \,|\, A \in \mathcal{S}(A, \nu, E) &\iff \chi_A f \in \mathcal{S}(X, \mu, E) , \\
f \,|\, A \in \mathcal{L}_0(A, \nu, E) &\iff \chi_A f \in \mathcal{L}_0(X, \mu, E) .
\end{aligned}$$

Proof The simple verification is left to the reader. $\blacksquare$

---

[2] We called the space of jump continuous functions $\mathcal{S}(I, E)$, but this will cause no confusion.
[3] Clearly the definition of measurability of functions is meaningful even on incomplete measure spaces.
[4] Compare the footnote to Exercise VI.6.8.

(f) Suppose $f \in \mathcal{L}_0(X, \mu, \mathbb{K})$ and $A := [f \neq 0]$. Also define $g \in \mathbb{K}^X$ through

$$g(x) := \begin{cases} 1/f(x) & \text{if } x \in A , \\ 0 & \text{if } x \notin A . \end{cases}$$

Then g is μ-measurable.

Proof The measurability of f implies the existence of a μ-null set N and a sequence (φ_j) in $\mathcal{S}(X, \mu, \mathbb{K})$ such that $\varphi_j(x) \to f(x)$ for $x \in N^c$. We set

$$\psi_j(x) := \begin{cases} 1/\varphi_j(x) & \text{if } \varphi_j(x) \neq 0 , \\ 0 & \text{if } \varphi_j(x) = 0 , \end{cases}$$

for $x \in X$ and $j \in \mathbb{N}$. By (c) and (d), $(\chi_A \psi_j)$ is a sequence in $\mathcal{S}(X, \mu, \mathbb{K})$, and one verifies easily that $(\chi_A \psi_j)(x) \to g(x)$ for every $x \in N^c$ (see Proposition II.2.6). ∎

(g) Let $e \in E \setminus \{0\}$, and suppose $\mu(X) = \infty$. Then $e\chi_X$ belongs to $\mathcal{L}_0(X, \mu, E)$ but not to $\mathcal{S}(X, \mu, E)$.

Proof It is clear that $e\chi_X$ is not μ-simple. Since X is σ-finite, there is a sequence (A_j) in $\mathcal{A}$ such that $\bigcup_j A_j = X$ and $\mu(A_j) < \infty$ for $j \in \mathbb{N}$. For $j \in \mathbb{N}$, set $X_j := \bigcup_{k=0}^{j} A_k$ and $\varphi_j := e\chi_{X_j}$. Then (φ_j) is a sequence in $\mathcal{S}(X, \mu, E)$ that converges pointwise to $e\chi_X$. ∎

A measurability criterion

A function $f \in E^X$ is said to be **$\mathcal{A}$-measurable** if the inverse images of open sets of E under f are measurable, that is, if $f^{-1}(\mathcal{T}_E) \subset \mathcal{A}$, where $\mathcal{T}_E$ is the norm topology on E. If there is a μ-null set N such that $f(N^c)$ is separable, we say f is **μ-almost separable valued.**

1.3 Remarks (a) Exercise IX.1.6 shows that the set of $\mathcal{A}$-measurable functions coincides with the set of $\mathcal{A}$-$\mathcal{B}(E)$-measurable functions.

(b) Every subspace of a separable metric space is separable.

Proof By Proposition IX.1.8, separability amounts to having a countable basis. But by restriction, a basis of a topological space yields a basis (of no greater cardinality) for any given subspace; see Proposition III.2.26. ∎

(c) Suppose E is separable and $f \in E^X$. Then f is μ-almost separable valued.
Proof This follows from (b). ∎

(d) Every finite-dimensional normed vector space is separable.[5] ∎

The next result gives a characterization of μ-measurable functions, which, besides being of theoretical significance, is very useful in practice for determining measurability.

---

[5]Compare Example V.4.3(e).

1.4 Theorem *A function in E^X is μ-measurable if and only if it is $\mathcal{A}$-measurable and μ-almost separable valued.*

Proof "$\Rightarrow$" Suppose $f \in \mathcal{L}_0(X, \mu, E)$.

(i) There exist a μ-null set N and a sequence (φ_j) in $\mathcal{S}(X, \mu, E)$ such that

$$\varphi_j(x) \to f(x) \quad (j \to \infty) \quad \text{for } x \in N^c . \tag{1.1}$$

By Proposition I.6.8, $F := \bigcup_{j=0}^{\infty} \varphi_j(X)$ is countable and therefore the closure $\overline{F}$ is separable. Because of (1.1) we have $f(N^c) \subset \overline{F}$. Remark 1.3(b) now shows that f is μ-almost separable valued.

(ii) Let O be open in E and define $O_n := \{ y \in O \;;\; \mathrm{dist}(y, O^c) > 1/n \}$ for $n \in \mathbb{N}^\times$. Then O_n is open and $\overline{O}_n \subset O$. Also let $x \in N^c$. By (1.1), $f(x)$ belongs to O if and only if there exist $n \in \mathbb{N}^\times$ and $m = m(n) \in \mathbb{N}^\times$ such that $\varphi_j(x) \in O_n$ for $j \geq m$. Therefore

$$f^{-1}(O) \cap N^c = \bigcup_{m,n \in \mathbb{N}^\times} \bigcap_{j \geq m} \varphi_j^{-1}(O_n) \cap N^c . \tag{1.2}$$

But $\varphi_j^{-1}(O_n) \in \mathcal{A}$ for $n \in \mathbb{N}^\times$ and $j \in \mathbb{N}$, because φ_j is μ-simple. Hence (1.2) says that $f^{-1}(O) \cap N^c \in \mathcal{A}$.

Furthermore, the completeness of μ shows that $f^{-1}(O) \cap N$ is a μ-null set, and altogether we obtain

$$f^{-1}(O) = \left(f^{-1}(O) \cap N \right) \cup \left(f^{-1}(O) \cap N^c \right) \in \mathcal{A} .$$

"$\Leftarrow$" Suppose f is μ-almost separable valued and $\mathcal{A}$-measurable.

(iii) We consider first the case $\mu(X) < \infty$. Take $n \in \mathbb{N}$. By assumption, there is a μ-null set N such that $f(N^c)$ is separable. If $\{ e_j \;;\; j \in \mathbb{N} \}$ is a countable dense subset of $f(N^c)$, the collection $\{ \mathbb{B}(e_j, 1/(n+1)) \;;\; j \in \mathbb{N} \}$ covers the set $f(N^c)$, and thus

$$X = N \cup \bigcup_{j \in \mathbb{N}} f^{-1}\big(\mathbb{B}(e_j, 1/(n+1))\big) .$$

Since f is $\mathcal{A}$-measurable, $X_{j,n} := f^{-1}\big(\mathbb{B}(e_j, 1/(n+1))\big)$ belongs to $\mathcal{A}$ for every $(j, n) \in \mathbb{N}^2$. The continuity of μ from below and the assumption $\mu(X) < \infty$ then imply that there are $m_n \in \mathbb{N}^\times$ and $Y_n \in \mathcal{A}$ such that

$$\bigcup_{j=0}^{m_n} X_{j,n} = Y_n^c \quad \text{and} \quad \mu(Y_n) < \frac{1}{2^{n+1}} .$$

Now define $\varphi_n \in E^X$ through

$$\varphi_n(x) := \begin{cases} e_0 & \text{if } x \in X_{0,n} , \\ e_j & \text{if } x \in X_{j,n} \backslash \bigcup_{k=0}^{j-1} X_{k,n} \text{ for } 1 \leq j \leq m_n , \\ 0 & \text{otherwise} . \end{cases}$$

Obviously $\varphi_n \in \mathcal{S}(X, \mu, E)$ for $n \in \mathbb{N}$, and

$$|\varphi_n(x) - f(x)| < 1/(n+1) \quad \text{for } x \in Y_n^c .$$

The decreasing sequence $Z_n := \bigcup_{k=0}^{\infty} Y_{n+k}$ satisfies

$$\mu(Z_n) \le \sum_{k=0}^{\infty} \mu(Y_{n+k}) \le \frac{1}{2^n} \quad \text{for } n \in \mathbb{N} .$$

It therefore follows from the continuity of μ from above that $Z := \bigcap_{n \in \mathbb{N}} Z_n$ is μ-null. We now set

$$\psi_n(x) := \begin{cases} \varphi_n(x) & \text{if } x \in Z_n^c , \\ 0 & \text{if } x \in Z_n . \end{cases}$$

Then (ψ_n) is a sequence in $\mathcal{S}(X, \mu, E)$. Also there is for every $x \in Z^c = \bigcup_n Z_n^c$ an $m \in \mathbb{N}$ such that $x \in Z_m^c$. Since $Z_m^c \subset Z_n^c$ for $n \ge m$, it follows that

$$|\psi_n(x) - f(x)| = |\varphi_n(x) - f(x)| < 1/(n+1) .$$

Altogether, $\lim \psi_n(x) = f(x)$ for every $x \in Z^c$. Therefore f is μ-measurable.

(iv) Finally, we consider the case $\mu(X) = \infty$. Remark IX.2.4(c) shows there is a disjoint sequence (X_j) in $\mathcal{A}$ such that $\bigcup_j X_j = X$ and $\mu(X_j) < \infty$. By part (iii), there exist for each $j \in \mathbb{N}$ a sequence $(\varphi_{j,k})_{k \in \mathbb{N}}$ in $\mathcal{S}(X, \mu, E)$ and a μ-null set N_j such that $\lim_k \varphi_{j,k}(x) = f(x)$ for every $x \in X_j \cap N_j^c$. With $N := \bigcup_j N_j$ and

$$\varphi_k(x) := \begin{cases} \varphi_{j,k}(x) & \text{if } x \in X_j , \quad j \in \{0, \dots, k\} , \\ 0 & \text{if } x \notin \bigcup_{j=0}^{k} X_j \end{cases}$$

for $k \in \mathbb{N}$, we have $\varphi_k \in \mathcal{S}(X, \mu, E)$ and $\lim_k \varphi_k(x) = f(x)$ for $x \in N^c$. The result follows because N is μ-null. ∎

1.5 Corollary Suppose E is separable and $f \in E^X$. The following statements are equivalent:

(i) f is μ-measurable.
(ii) f is $\mathcal{A}$-measurable.
(iii) $f^{-1}(\mathcal{S}) \subset \mathcal{A}$ for some $\mathcal{S} \subset \mathfrak{P}(E)$ such that $\mathcal{A}_\sigma(\mathcal{S}) = \mathcal{B}(E)$.
(iv) $f^{-1}(\mathcal{S}) \subset \mathcal{A}$ for any $\mathcal{S} \subset \mathfrak{P}(E)$ such that $\mathcal{A}_\sigma(\mathcal{S}) = \mathcal{B}(E)$.

Proof This follows from Theorem 1.4, Remark 1.3(c), and Exercise IX.1.6. ∎

1.6 Remark The proof of Theorem 1.4 and Remark 1.3(c) show that Corollary 1.5 remains true for incomplete measure spaces. ∎

Without much effort, we obtain from Corollary 1.5 the following properties of μ-measurable functions.

1.7 Theorem

(i) *If E and F are separable Banach spaces and if we have maps $f \in \mathcal{L}_0(X, \mu, E)$ and $g \in C\big(f(X), F\big)$, then $g \circ f$ belongs to $\mathcal{L}_0(X, \mu, F)$. In particular, $|f| \in \mathcal{L}_0(X, \mu, \mathbb{R})$.*

(ii) *A map $f = (f_1, \ldots, f_n) \colon X \to \mathbb{K}^n$ is μ-measurable if and only if each of its components f_j is.*

(iii) *Let $g, h \in \mathbb{R}^X$. Then $f = g + ih$ is μ-measurable if and only if g and h are.*

(iv) *If $f \in \mathcal{L}_0(X, \mu, E)$ and $g \in \mathcal{L}_0(X, \mu, F)$, then $(f, g) \in \mathcal{L}_0(X, \mu, E \times F)$.*

Proof (i) Let O be open in F. Since g is continuous, $g^{-1}(O)$ is open in $f(X)$. By Proposition III.2.26, there is an open subset U of E such that $g^{-1}(O) = f(X) \cap U$. Since f is Lebesgue measurable, $f^{-1}(U)$ belongs to $\mathcal{A}$ by Corollary 1.5. Because

$$(g \circ f)^{-1}(O) = f^{-1}\big(g^{-1}(O)\big) = f^{-1}\big(f(X) \cap U\big) = f^{-1}(U) ,$$

the claim follows from another application of Corollary 1.5.

(ii) The implication "$\Rightarrow$" follows from (i), because $f_j = \mathrm{pr}_j \circ f$ for $1 \le j \le n$.

"$\Leftarrow$" We consider first the case $\mathbb{K} = \mathbb{R}$. Take $I \in \mathbb{J}(n)$, and write it as $I = \prod_{j=1}^n I_j$, where $I_j \in \mathbb{J}(1)$ for $1 \le j \le n$. Because each $f_j^{-1}(I_j)$ belongs to $\mathcal{A}$, so does $f^{-1}(I) = \bigcap_{j=1}^n f_j^{-1}(I_j)$, that is, we have $f^{-1}\big(\mathbb{J}(n)\big) \subset \mathcal{A}$. Also, we know from Theorem IX.1.11 that $\mathcal{A}_\sigma\big(\mathbb{J}(n)\big) = \mathcal{B}^n$. Therefore Corollary 1.5 implies that f is μ-measurable.

Using the identification $\mathbb{C}^n = \mathbb{R}^{2n}$, the case $\mathbb{K} = \mathbb{C}$ follows immediately from what was just shown.

(iii) is a special case of (ii), and we leave (iv) as an exercise. ∎

Measurable $\overline{\mathbb{R}}$-valued functions

In the theory of integration, it is useful to consider not only real-valued functions but also maps into the extended number line $\overline{\mathbb{R}}$. Such maps are called **$\overline{\mathbb{R}}$-valued functions**. An $\overline{\mathbb{R}}$-valued function $f \colon X \to \overline{\mathbb{R}}$ is said to be **μ-measurable** if $\mathcal{A}$ contains $f^{-1}(-\infty)$, $f^{-1}(\infty)$, and $f^{-1}(O)$ for every open subset O of $\mathbb{R}$. We denote the set of all μ-measurable $\overline{\mathbb{R}}$-valued functions on X by $\mathcal{L}_0(X, \mu, \overline{\mathbb{R}})$.

1.8 Remarks (a) Any real-valued function $f \colon X \to \mathbb{R}$ can be regarded as an $\overline{\mathbb{R}}$-valued one. Thus there are in principle two notions of measurability that apply to f. But since $f^{-1}\big(\{-\infty, \infty\}\big) = \emptyset$, Corollary 1.5 implies that f is μ-measurable as a real-valued function if and only if it is μ-measurable as an $\overline{\mathbb{R}}$-valued function.

(b) Note that $\mathcal{L}_0(X, \mu, \overline{\mathbb{R}})$ is *not* a vector space. ∎

In the next result, we list simple measurability criteria for $\overline{\mathbb{R}}$-valued functions.

1.9 Proposition *For an $\overline{\mathbb{R}}$-valued function $f : X \to \overline{\mathbb{R}}$, the following statements are equivalent:*

(i) $f \in \mathcal{L}_0(X, \mu, \overline{\mathbb{R}})$.

(ii) $[f < \alpha] \in \mathcal{A}$ *for every* $\alpha \in \mathbb{Q}$ *[or* $\alpha \in \mathbb{R}$*]*.

(iii) $[f \leq \alpha] \in \mathcal{A}$ *for every* $\alpha \in \mathbb{Q}$ *[or* $\alpha \in \mathbb{R}$*]*.

(iv) $[f > \alpha] \in \mathcal{A}$ *for every* $\alpha \in \mathbb{Q}$ *[or* $\alpha \in \mathbb{R}$*]*.

(v) $[f \geq \alpha] \in \mathcal{A}$ *for every* $\alpha \in \mathbb{Q}$ *[or* $\alpha \in \mathbb{R}$*]*.

Proof "(i)$\Rightarrow$(ii)" The sets $f^{-1}(-\infty)$ and $f^{-1}((-\infty, \alpha))$ with $\alpha \in \mathbb{Q}$ [or $\alpha \in \mathbb{R}$] belong to $\mathcal{A}$. Because

$$[f < \alpha] = f^{-1}([-\infty, \alpha)) = f^{-1}(-\infty) \cup f^{-1}((-\infty, \alpha)) \ ,$$

this is also true of $[f < \alpha]$.

The implications "(ii)$\Rightarrow$(iii)$\Rightarrow$(iv)$\Rightarrow$(v)" follow from the identities

$$[f \leq \alpha] = \bigcap_{j=1}^{\infty} [f < \alpha + 1/j] \ , \quad [f > \alpha] = [f \leq \alpha]^c \ , \quad [f \geq \alpha] = \bigcap_{j=1}^{\infty} [f > \alpha - 1/j] \ .$$

"(v)$\Rightarrow$(i)" Suppose O is open in $\mathbb{R}$. By Proposition IX.5.6, there exist $(\alpha_j), (\beta_j) \in \mathbb{Q}^{\mathbb{N}}$ such that $O = \bigcup_j [\alpha_j, \beta_j)$. Therefore

$$f^{-1}(O) = \bigcup_{j \in \mathbb{N}} f^{-1}([\alpha_j, \beta_j)) = \bigcup_{j \in \mathbb{N}} ([f \geq \alpha_j] \cap [f < \beta_j]) \ ,$$

and because $[f < \alpha] = [f \geq \alpha]^c$, we conclude that $f^{-1}(O)$ belongs to $\mathcal{A}$. In addition

$$f^{-1}(-\infty) = \bigcap_{j \in \mathbb{N}} [f < -j] \quad \text{and} \quad f^{-1}(\infty) = \bigcap_{j \in \mathbb{N}} [f > j] \ .$$

Thus $f^{-1}(\pm\infty)$ also lies in $\mathcal{A}$. ∎

The lattice of measurable $\overline{\mathbb{R}}$-valued functions

An ordered set $V = (V, \leq)$ is called a **lattice** if for every pair $(a, b) \in V \times V$, the infimum $a \wedge b$ and the supremum $a \vee b$ exist in V. A subset $U \subset V$ is a **sublattice** of V if U is a lattice when given the ordering induced by V. An ordered vector space that is also a lattice is called a **vector lattice**. If a vector subspace of a vector lattice is a sublattice, we call it a **vector sublattice**.

1.10 Examples **(a)** Suppose V is a lattice [or vector lattice]. Then V^X is a lattice [or vector lattice] with respect to the pointwise ordering.

(b) $\overline{\mathbb{R}}$ is a lattice, and $\mathbb{R}$ is a vector lattice.

(c) The vector lattice $\mathbb{R}^X$ satisfies
$$f \vee g = (f + g + |f - g|)/2 , \quad f \wedge g = (f + g - |f - g|)/2 .$$

(d) $B(X, \mathbb{R})$ is a vector sublattice of $\mathbb{R}^X$.

(e) Suppose X is a topological space. Then $C(X, \mathbb{R})$ is a vector sublattice of $\mathbb{R}^X$.

Proof This follows from (c) and the fact that $|f|$ is continuous if f is. ∎

(f) $\mathcal{S}(X, \mu, \mathbb{R})$ and $\mathcal{L}_0(X, \mu, \mathbb{R})$ are vector sublattices of $\mathbb{R}^X$.

Proof The first statement is clear. The second follows from (c) and Theorem 1.7 or Remark 1.2(d). ∎

(g) Suppose V is a vector lattice and $x, y, z \in V$. Then
$$(x \vee y) + z = (x + z) \vee (y + z) ,$$
$$(-x) \vee (-y) = -(x \wedge y) ,$$
$$x + y = (x \vee y) + (x \wedge y) .$$

Proof If $u \in V$ satisfies $u \geq x$ and $u \geq y$, then clearly $u + z \geq (x + z) \vee (y + z)$. Hence
$$(x \vee y) + z \geq (x + z) \vee (y + z) .$$

Suppose $v \geq (x + z) \vee (y + z)$. Then $v - z \geq x$ and $v - z \geq y$, and hence $v \geq (x \vee y) + z$. Because this holds for every upper bound v of $\{x + z, y + z\}$, it follows that
$$(x + z) \vee (y + z) \geq (x \vee y) + z .$$

This proves the first equality. The second is none other than the trivial relation
$$\sup\{-x, -y\} = \sup(-\{x, y\}) = -\inf\{x, y\} .$$

Using this, we now find
$$x \vee y = (-y + (x + y)) \vee (-x + (x + y)) = ((-y) \vee (-x)) + (x + y)$$
$$= -(x \wedge y) + (x + y) ,$$

which proves the last claim. ∎

(h) Suppose V is a vector lattice. For $x \in V$, we set
$$x^+ := x \vee 0 , \quad x^- := (-x) \vee 0 , \quad |x| := x \vee (-x) .$$

Then[6]
$$x = x^+ - x^- , \quad |x| = x^+ + x^- , \quad x^+ \wedge x^- = 0 .$$

---

[6] See footnote 8 in Section II.8.

Proof   The first claim follows easily from (g). With this and (g), we find

$$x^+ + x^- = x + 2x^- = x + \left((-2x) \vee 0\right) = (-x) \vee x = |x| \ .$$

Analogously, we have

$$(x^+ \wedge x^-) - x^- = (x^+ - x^-) \wedge (x^- - x^-) = x \wedge 0 = -x^- \ ,$$

and therefore $x^+ \wedge x^- = 0$. ∎

If V is a vector lattice and $x \in V$, we call x^+ the **positive part** and x^- the **negative part** of x, and $|x|$ is the **modulus**[7] of x. Clearly $x^+ \geq 0$, $x^- \geq 0$, and $|x| \geq 0$.

The figures below illustrate the positive and negative parts of an element f of the vector lattice $\mathbb{R}^X$.

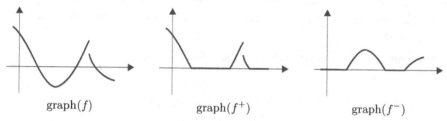

graph(f)   graph(f^+)   graph(f^-)

Suppose $f \in \overline{\mathbb{R}}^X$. Then $f^+ := f \vee 0$ is called the **positive part** of f, and $f^- := 0 \vee (-f)$ the **negative part** of f. These terms are chosen in obvious analogy to the case of the vector lattice $\mathbb{R}^X$.[8] Here too we have

$$f^+ \geq 0 \ , \quad f^- \geq 0 \ , \quad f = f^+ - f^- \ , \quad |f| = f^+ + f^- \ .$$

The next result shows that $\mathcal{L}_0(X, \mu, \overline{\mathbb{R}})$ is a sublattice of $\overline{\mathbb{R}}^X$ and that it is closed under countably many lattice operations.

1.11 Proposition   *Suppose $f \in \mathcal{L}_0(X, \mu, \overline{\mathbb{R}})$, (f_j) is a sequence in $\mathcal{L}_0(X, \mu, \overline{\mathbb{R}})$, and $k \in \mathbb{N}$. Then each of the $\overline{\mathbb{R}}$-valued functions*

$$f^+ \ , \quad f^- \ , \quad |f| \ , \quad \max_{0 \leq j \leq k} f_j \ , \quad \min_{0 \leq j \leq k} f_j \ , \quad \sup_j f_j \ , \quad \inf_j f_j \ , \quad \varlimsup_j f_j \ , \quad \varliminf_j f_j$$

belongs to $\mathcal{L}_0(X, \mu, \overline{\mathbb{R}})$.

Proof   (i) Suppose $\alpha \in \mathbb{R}$. From Proposition 1.9, we know that $[f_j > \alpha]$ belongs to $\mathcal{A}$ for $j \in \mathbb{N}$. Therefore this is also true of

$$[\sup_j f_j > \alpha] = \bigcup_j [f_j > \alpha] \ ,$$

and Proposition 1.9 implies that $\sup_j f_j$ is μ-measurable.

---

[7] This is not to be confused with the norm of the vector x if V is a normed vector space. The modulus of $x \in V$ is always a vector in V, whereas the norm is a nonnegative number.
[8] Remember that $\overline{\mathbb{R}}^X$ is a lattice but not a vector lattice.

(ii) Because f_j belongs to $\mathcal{L}_0(X, \mu, \overline{\mathbb{R}})$, so does $-f_j$. It then follows from (i) that the function $\inf_j f_j = -\sup_j(-f_j)$ is μ-measurable.

(iii) For $j \in \mathbb{N}$, set

$$g_j := \begin{cases} f_j & \text{if } 0 \leq j \leq k, \\ f_k & \text{if } \quad j > k. \end{cases}$$

Because of (i), $\sup_j g_j = \max_{0 \leq j \leq k} f_j$ belongs to $\mathcal{L}_0(X, \mu, \overline{\mathbb{R}})$. Analogously, one shows that $\min_{0 \leq j \leq k} f_j$ is μ-measurable.

(iv) From (iii), it follows that f^+, f^-, and $|f|$ belong to $\mathcal{L}_0(X, \mu, \overline{\mathbb{R}})$.

(v) We have

$$\overline{\lim_j} f_j = \inf_j \sup_{k \geq j} f_k \quad \text{and} \quad \underline{\lim_j} f_j = \sup_j \inf_{k \geq j} f_k .$$

Therefore by (i) and (ii), $\overline{\lim}_j f_j$ and $\underline{\lim}_j f_j$ also belong to $\mathcal{L}_0(X, \mu, \overline{\mathbb{R}})$. ∎

The positive cone $\mathcal{S}(X, \mu, \mathbb{R}^+)$ of $\mathcal{S}(X, \mu, \mathbb{R})$ is the set of all $f \in \mathcal{S}(X, \mu, \mathbb{R})$ such that $f(X) \subset \mathbb{R}^+$; see Remarks VI.4.7(b) and (d). Therefore it is natural to denote it by $\mathcal{S}(X, \mu, \mathbb{R}^+)$. Similarly, if $\overline{\mathbb{R}}^+ := [0, \infty]$ is the nonnegative part of the extended number line $\overline{\mathbb{R}}$, we denote by $\mathcal{L}_0(X, \mu, \overline{\mathbb{R}}^+)$ the set of all nowhere negative μ-measurable $\overline{\mathbb{R}}$-valued functions on X.

The set $\mathcal{L}_0(X, \mu, \overline{\mathbb{R}}^+)$ has an interesting characterization:

1.12 Theorem *For $f \colon X \to \overline{\mathbb{R}}^+$, the following statements are equivalent:*

(i) $f \in \mathcal{L}_0(X, \mu, \overline{\mathbb{R}}^+)$.

(ii) *There is an increasing sequence (f_j) in $\mathcal{S}(X, \mu, \mathbb{R}^+)$ such that $f_j \to f$ for $j \to \infty$.*

Proof "(i)$\Rightarrow$(ii)" By the σ-finiteness of $(\mathcal{A}, \mu)$, it suffices to consider the case $\mu(X) < \infty$ (compare part (iv) in the proof of Theorem 1.4). So for $j, k \in \mathbb{N}$, set

$$A_{j,k} := \begin{cases} [k2^{-j} \leq f < (k+1)2^{-j}] & \text{if } k = 0, \ldots, j\,2^j - 1, \\ [f \geq j] & \text{if } k = j\,2^j. \end{cases}$$

The sets $A_{j,k}$ are obviously disjoint for $k = 0, \ldots, j\,2^j$ and by Proposition 1.9 they lie in $\mathcal{A}$. Since $\mu(X) < \infty$, each $A_{j,k}$ has finite measure. By Remark 1.2(b), then,

$$f_j := \sum_{k=0}^{j\,2^j} k2^{-j} \chi_{A_{j,k}} \quad \text{for } j \in \mathbb{N}$$

belongs to $\mathcal{S}(X, \mu, \mathbb{R})$. Further one verifies that $0 \leq f_j \leq f_{j+1}$ for $j \in \mathbb{N}$.

Now suppose $x \in X$. If $f(x) = \infty$, we have $f_j(x) = j$, so $\lim_j f_j(x) = f(x)$. On the other hand, if $f(x) < \infty$, then $f_j(x) \leq f(x) < f_j(x) + 2^{-j}$ for $j > f(x)$, so $\lim_j f_j(x) = f(x)$ in this case as well. This shows (f_j) converges pointwise to f.

"(ii)⇒(i)" This follows from Proposition 1.11. ∎

Here is an illustration of the construction of the $A_{j,k}$ in the proof of Theorem 1.12:

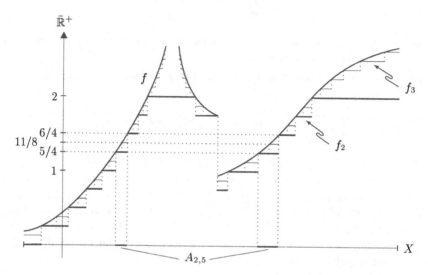

1.13 Corollary

(i) For every $f \in \mathcal{L}_0(X,\mu,\overline{\mathbb{R}})$, there is a sequence (f_j) in $\mathcal{S}(X,\mu,\mathbb{R})$ such that $f_j \to f$.

(ii) Suppose $f \in \mathcal{L}_0(X,\mu,\mathbb{R}^+)$ is bounded. Then there is an increasing sequence (f_j) in $\mathcal{S}(X,\mu,\mathbb{R}^+)$ that converges uniformly to f.

(iii) Suppose (f_j) is a sequence in $\mathcal{L}_0(X,\mu,\overline{\mathbb{R}}^+)$. Then $\sum_j f_j \in \mathcal{L}_0(X,\mu,\overline{\mathbb{R}}^+)$.

Proof (i) In view of the decomposition $f = f^+ - f^-$, this follows from Theorem 1.12 and Remark 1.2(a).

(ii) Suppose $f \in \mathcal{L}_0(X,\mu,\mathbb{R}^+)$ is bounded. For the sequence (f_j) constructed in the proof of Theorem 1.12, we have

$$f_j(x) \le f(x) < f_j(x) + 2^{-j} \quad \text{for } j > \|f\|_\infty .$$

Thus (f_j) converges uniformly to f.

(iii) By Theorem 1.12, there is for every $j \in \mathbb{N}$ an increasing sequence $(\varphi_{j,k})_{k \in \mathbb{N}}$ in $\mathcal{S}(X,\mu,\mathbb{R}^+)$ such that $\varphi_{j,k} \uparrow f_j$ for $k \to \infty$. Set $s_{k,n} := \sum_{j=0}^k \varphi_{j,n}$ for $k,n \in \mathbb{N}$. Then $(s_{k,n})_{n \in \mathbb{N}}$ is an increasing sequence in $\mathcal{S}(X,\mu,\mathbb{R}^+)$ that converges to $s_k := \sum_{j=0}^k f_j$ as $n \to \infty$. By Theorem 1.12, then, (s_k) is a sequence in $\mathcal{L}_0(X,\mu,\overline{\mathbb{R}}^+)$ such that $\lim_k s_k = \sup_k s_k = \sum_{j=0}^\infty f_j$. The claim now follows from Proposition 1.11. ∎

Pointwise limits of measurable functions

Let (f_j) be a pointwise convergent sequence in $\mathcal{L}_0(X, \mu, \mathbb{R})$. By Proposition 1.11, $f := \lim_j f_j$ is also in $\mathcal{L}_0(X, \mu, \mathbb{R})$. We will now derive an analogous statement for vector-valued sequences of functions.

1.14 Theorem *Suppose (f_j) is a sequence in $\mathcal{L}_0(X, \mu, E)$ and $f \in E^X$. If (f_j) converges μ-almost everywhere to f, then f is μ-measurable.*

Proof (i) We show first that f is μ-almost separable valued. By assumption, there is a μ-null set M such that $f_j(x) \to f(x)$ as $j \to \infty$, for any $x \in M^c$. For every $j \in \mathbb{N}$, there exists by Theorem 1.4 a μ-null set N_j such that $f_j(N_j^c)$ is separable, hence also a countable set B_j that is dense in $f_j(N_j^c)$:

$$B_j \subset f_j(N_j^c) \subset \overline{B}_j \quad \text{for } j \in \mathbb{N} .$$

With $B := \bigcup_j B_j$, we see from Corollary III.2.13(i) that $\bigcup_j \overline{B}_j \subset \overline{B}$, and we find

$$\bigcup_{j \in \mathbb{N}} f_j(N_j^c) \subset \bigcup_{j \in \mathbb{N}} \overline{B}_j \subset \overline{B} .$$

Finally let $N := M \cup \bigcup_j N_j$. Then N is a μ-null set satisfying, for any $k \in \mathbb{N}$, $N^c = M^c \cap \bigcap_j N_j^c \subset N_k^c$. Because $\lim_j f_j(x) = f(x)$ for $x \in M^c$, we thus have

$$f(N^c) \subset \overline{\bigcup_{j \in \mathbb{N}} f_j(N_j^c)} \subset \overline{\overline{B}} = \overline{B} .$$

Because B is countable, Remark 1.3(b) shows that $f(N^c)$ is separable.

(ii) Now we show that f is $\mathcal{A}$-measurable. Let O be open in E, and define $O_n := \left\{ x \in O \; ; \; \text{dist}(x, O^c) > 1/n \right\}$ for $n \in \mathbb{N}^\times$. As in (1.2), it follows that

$$f^{-1}(O) \cap M^c = \bigcup_{m,n \in \mathbb{N}^\times} \bigcap_{j \geq m} f_j^{-1}(O_n) \cap M^c .$$

By Theorem 1.4, $f_j^{-1}(O_n)$ belongs to $\mathcal{A}$ for every $j, n \in \mathbb{N}^\times$. Therefore this also applies to $f^{-1}(O) \cap M^c$. Moreover, the completeness of μ implies that $f^{-1}(O) \cap M$ is a μ-null set, and altogether we find

$$f^{-1}(O) = \left(f^{-1}(O) \cap M^c \right) \cup \left(f^{-1}(O) \cap M \right) \in \mathcal{A} .$$

The claim now follows from Theorem 1.4. ∎

1.15 Remark Theorem 1.14 generally fails for incomplete measure spaces.

Proof Let C be the Cantor set. In the proof of Corollary IX.5.29, it was shown that C contains a Borel nonmeasurable subset $N \subset C$. We take $f_j := \chi_C$ for $j \in \mathbb{N}$ and $f := \chi_N$. Remark 1.2(b) and the compactness of C imply $\chi_C \in \mathcal{S}(\mathbb{R}, \beta_1, \mathbb{R})$. Also $f_j(x) = f(x)$ for $x \in C^c \subset N^c$ and $j \in \mathbb{N}$. Because C has measure zero, (f_j) converges β_1-a.e. to f. However, because $[f > 0] = N \notin \mathcal{B}^1$, Proposition 1.9 says that f is not in $\mathcal{L}_0(\mathbb{R}, \beta_1, \mathbb{R})$. ∎

Radon measures

We conclude this section by exploring how measurability and continuity are related in vector-valued functions. Besides proving a simple measurability criterion, we prove Luzin's theorem, which exposes a surprisingly close connection between continuous and Borel measurable functions.

A metric space $X = (X, d)$ is said to be **σ-compact** if X is locally compact and there is a sequence $(X_j)_{j \in \mathbb{N}}$ of compact subsets of X such that $X = \bigcup_j X_j$.

Suppose X is a σ-compact metric space. A **Radon measure** on X is a regular, locally finite measure on a σ-algebra $\mathcal{A}$ over X such that $\mathcal{A} \supset \mathcal{B}(X)$. We say a Radon measure μ is **massive** if μ is complete and every nonempty open subset O of X satisfies $\mu(O) > 0$.

1.16 Remarks (a) Every σ-compact metric space is a σ-compact set in the sense of the definition of Section IX.5; however, a countable union of compact subsets of a metric space is not necessarily a σ-compact metric space.

Proof The first statement is clear. For the second, consider $\mathbb{Q} \subset \mathbb{R}$. ∎

(b) Every Radon measure is σ-finite.

Proof This follows from Remark IX.5.3(b). ∎

(c) Suppose X is a locally compact metric space. Then there is for every compact subset K of X a relatively compact[9] open superset of K.

Proof For every $x \in X$, we find a relatively compact open neighborhood $O(x)$ of x. Because K is compact, there are $x_0, \ldots, x_m \in K$ such that $O := \bigcup_{j=0}^{m} O(x_j)$ is an open superset of K. Corollary III.2.13(iii) implies $\overline{O} = \bigcup_{j=0}^{m} \overline{O(x_j)}$. Therefore $\overline{O}$ is compact. ∎

(d) Every open subset of $\mathbb{R}^n$ is a σ-compact metric space.

Proof Let $X \subset \mathbb{R}^n$ be open and nonempty. For every $x \in X$, there is $r > 0$ such that $\overline{\mathbb{B}}(x, r) \subset X$. Because $\overline{\mathbb{B}}(x, r)$ is compact, X is then a locally compact metric space.

For $j \in \mathbb{N}^\times$, define[10]

$$U_j := \left\{ x \in X \; ; \; \operatorname{dist}(x, U^c) > 1/j \right\} \cap \mathbb{B}(0, j) \,. \tag{1.3}$$

By Examples III.1.3(l) and III.2.22(c), the set U_j is open. Also $U_j \subset \overline{U}_j \subset U_{j+1}$, and $\bigcup_j \overline{U}_j \subset \bigcup_j U_j = X$. In particular, there exists $j_0 \in \mathbb{N}^\times$ such that $U_j \neq \emptyset$ for $j \geq j_0$. Because $\overline{U}_j$ is compact by the Heine–Borel theorem, the claim follows. ∎

(e) For a locally compact metric space X, the following statements are equivalent:

  (i) X is σ-compact.

  (ii) X is the union of a sequence $(U_j)_{j \in \mathbb{N}}$ of relatively compact open subsets with $\overline{U}_j \subset U_{j+1}$ for $j \in \mathbb{N}$.

---

[9] A subset A of a topological space is said to be **relatively compact** if $\overline{A}$ is compact.
[10] $\operatorname{dist}(x, \emptyset) := \infty$.

(iii) X is a Lindelöf space.

(iv) X satisfies the second countability axiom.

(v) X is separable.

Proof "(i)$\Rightarrow$(ii)" Let $(X_j)_{j \in \mathbb{N}}$ be a sequence of compact sets in X such that $X = \bigcup_j X_j$. By (c), there is a relatively compact open superset U_0 of X_0. Inductively choose relatively compact open subsets U_j such that $U_j \supset \overline{U}_{j-1} \cup X_j$ for $j \geq 1$. Clearly $X = \bigcup_j U_j$.

"(ii)$\Rightarrow$(iii)" Suppose $\mathcal{O} := \{ O_\alpha \; ; \; \alpha \in \mathsf{A} \}$ is an open cover of X. For every $j \in \mathbb{N}$, inductively choose $m(j) \in \mathbb{N}$ and $\alpha_0, \ldots, \alpha_{m(j)} \in \mathsf{A}$ such that $\overline{U}_j \subset \bigcup_{k=0}^{m(j)} O_{\alpha_k}$. Then $\{ O_{\alpha_k} \; ; \; k = 0, \ldots, m(j), \; j \in \mathbb{N} \}$ is a countable subcover of $\mathcal{O}$ for X.

"(iii)$\Rightarrow$(i)" By assumption, there is a sequence (x_j) in X and relatively compact open neighborhoods $O(x_j)$ of x_j $(j \in \mathbb{N})$ such that $X = \bigcup_{j \in \mathbb{N}} O(x_j)$. So $X = \bigcup_{j \in \mathbb{N}} \overline{O(x_j)}$, showing that X is σ-compact.

The remaining equivalences follow from Proposition IX.1.8. ∎

(f) Every locally finite Borel measure on a σ-compact metric space is regular and is therefore a Radon measure.

Proof This follows from (e) and Corollary VIII.1.12 in [Els99]. ∎

(g) Finite Borel measures on (nonmetrizable) compact topological spaces need not be regular.

Proof See [Flo81, Example A4.5, S. 350]. ∎

(h) Lebesgue n-measure, λ_n, is a massive Radon measure on $\mathbb{R}^n$.

Proof This follows from Theorems IX.5.1 and IX.5.4. ∎

(i) The s-dimensional Hausdorff measure $\mathcal{H}^s$ is a Radon measure on $\mathbb{R}^n$ only when $s \geq n$. It is massive if and only if $s = n$.

Proof Every Borel set is $\mathcal{H}^s$-measurable, by Example IX.4.4(c) and Theorem IX.4.3. The regularity of $\mathcal{H}^s$ for $s > 0$ follows from Corollary IX.5.22 and Theorem IX.5.4.

Suppose O is open in $\mathbb{R}^n$ and nonempty. Because O has Hausdorff dimension n (Exercise IX.3.6), it follows that

$$\mathcal{H}^s(O) = \begin{cases} 0 & \text{if } s > n \, , \\ \infty & \text{if } s < n \, . \end{cases}$$

Therefore $\mathcal{H}^s$ cannot be a Radon measure on $\mathbb{R}^n$ if $s < n$. If $s > n$, on the other hand, $\mathcal{H}^s$ is a nonmassive Radon measure.

Lemma IX.5.21 shows that $\mathcal{H}^n$ is locally finite and therefore a Radon measure on $\mathbb{R}^n$. Finally, Corollary IX.5.22 implies $\mathcal{H}^n(O) > 0$, and we are done. ∎

(j) Suppose $F : \mathbb{R} \to \mathbb{R}$ is measure generating, and denote by μ_F the Lebesgue–Stieltjes measure on $\mathbb{R}$ induced by F. Then μ_F is a Radon measure on $\mathbb{R}$, and is massive if only if F is strictly increasing.

Proof This follows from Example IX.4.4(b), Theorem IX.4.3, Exercise IX.5.19, and Proposition IX.3.5. ∎

1.17 Theorem *Suppose μ is a complete Radon measure on X. Then $C(X, E)$ is a vector subspace of $\mathcal{L}_0(X, \mu, E)$.*

Proof Take $f \in C(X, E)$ and let (X_j) be a sequence of compact sets in X such that $X = \bigcup_j X_j$. According to Exercise IX.1.6(b), f is Borel measurable and therefore $\mathcal{A}$-measurable, where $\mathcal{A}$ is the domain of μ. By Remark 1.16(e), $f(X_j)$, being a compact subset of E, is separable. Therefore $f(X) = \bigcup_j f(X_j)$ is also separable, and the claim follows from Theorem 1.4. ∎

1.18 Theorem (Luzin) *Suppose X is a σ-compact metric space, μ is a complete Radon measure on X, and $f \in \mathcal{L}_0(X, \mu, E)$. Then for every μ-measurable set A of finite measure and for every $\varepsilon > 0$, there is a compact subset K of X such that $\mu(A \backslash K) < \varepsilon$ and $f \mid K \in C(K, E)$.*

Proof (i) Because X is σ-compact, we can find a compact set $\widetilde{X}$ such that $\mu(A \backslash \widetilde{X}) < \varepsilon/2$. We set $\widetilde{f} := f \mid \widetilde{X}$ and $\widetilde{A} := A \cap \widetilde{X}$. Then $\mu(\widetilde{X}) < \infty$.

(ii) By Theorem 1.4, there is a μ-null set N of $\widetilde{X}$ such that $\widetilde{f}(N^c)$ is separable. Therefore by Proposition IX.1.8, there is a countable basis $\{ \widetilde{V}_j \; ; \; j \in \mathbb{N} \}$ of $\widetilde{f}(N^c)$, and because of Proposition III.2.26, there exist open subsets V_j in E such that $\widetilde{V}_j = V_j \cap \widetilde{f}(N^c)$.

(iii) According to Theorem 1.4, $\widetilde{f}^{-1}(V_j)$ is μ-measurable for every $j \in \mathbb{N}$. Hence it follows from the regularity of μ and the finiteness of $\mu(\widetilde{X})$ that for every $j \in \mathbb{N}$ there exist a compact set K_j and an open set U_j with $K_j \subset \widetilde{f}^{-1}(V_j) \subset U_j$ and $\mu(U_j \backslash K_j) < \varepsilon 2^{-(j+3)}$. Putting $U := \bigcup_j (U_j \backslash K_j)$, we have $\mu(U) < \varepsilon/4$.

(iv) We set $Y := (U \cup N)^c$ and show that $\widetilde{f} \mid Y$ is continuous. To verify this, let V be open in E. Then there is a subset $\{ V_{j_k} \; ; \; k \in \mathbb{N} \}$ of $\{ V_j \; ; \; j \in \mathbb{N} \}$ such that $V \cap \widetilde{f}(N^c) = \bigcup_k V_{j_k} \cap \widetilde{f}(N^c)$. This implies

$$\widetilde{f}^{-1}(V) \cap N^c = \bigcup_k \widetilde{f}^{-1}(V_{j_k}) \cap N^c .$$

Obviously $\widetilde{f}^{-1}(V_\ell) \cap Y \subset U_\ell \cap Y$ for $\ell \in \mathbb{N}$. Because

$$Y = U^c \cap N^c = \bigcap_j (U_j^c \cup K_j) \cap N^c \subset \bigcap_j (U_j^c \cup \widetilde{f}^{-1}(V_j)) \subset U_\ell^c \cup \widetilde{f}^{-1}(V_\ell) ,$$

it follows that $\widetilde{f}^{-1}(V_\ell) \cap Y = U_\ell \cap Y$, and we find

$$(\widetilde{f} \mid Y)^{-1}(V) = \widetilde{f}^{-1}(V) \cap N^c \cap U^c = \bigcup_k U_{j_k} \cap Y .$$

Because $\bigcup_k U_{j_k}$ is open in X and therefore $\bigcup_k U_{j_k} \cap Y$ is open in Y, the continuity of $\widetilde{f} \mid Y$ follows.

(v) We apply once again the regularity of μ to deduce the existence of a compact subset K of the μ-measurable set Y such that $\mu(Y \backslash K) < \varepsilon/4$. Then

$\widetilde{f} \mid K$ belongs to $C(K, E)$, and

$$\mu(\widetilde{A} \setminus K) \leq \mu(Y \setminus K) + \mu(Y^c \setminus K) \leq \mu(Y \setminus K) + \mu(U) < \varepsilon/2 .$$

Because $\mu(A \setminus K) \leq \mu(\widetilde{A} \setminus K) + \mu(A \setminus \widetilde{X}) < \varepsilon$, we are done. ∎

Exercises

1 Suppose H is a separable Hilbert space. We say $f \in H^X$ is **weakly μ-measurable** if $(f \mid e)$ belongs to $\mathcal{L}_0(X, \mu, \mathbb{K})$ for every $e \in H$. Prove:

(a) If f is weakly μ-measurable, $|f|$ is μ-measurable.

(b) f is μ-measurable if and only if f is weakly μ-measurable.

(Hints: (a) Suppose $\{e_j \; ; \; j \in \mathbb{N}\}$ is a dense subset of $\mathbb{B}_H$. Then

$$[|f| \leq \alpha] = \bigcap_j [|(f \mid e_j)| \leq \alpha] \quad \text{for } \alpha \in \mathbb{R} .$$

(b) "⇐" Using (a), we can construct as in the proof of Theorem 1.4 a sequence of a μ-simple functions that converge μ-a.e. to f.)

2 Denote by $\mathcal{S}(\mathbb{R}, E)$ the vector space of all E-valued admissible functions of $\mathbb{R}$ (see Section VI.8). Prove or disprove:

(a) $\mathcal{S}(\mathbb{R}, E) \subset \mathcal{L}_0(\mathbb{R}, \beta_1, E)$;

(b) $\mathcal{S}(\mathbb{R}, E) \supset \mathcal{L}_0(\mathbb{R}, \beta_1, E)$.

3 Prove the statement of Remark 1.2(e).

4 Show that every monotone $\overline{\mathbb{R}}$-valued function is Borel measurable.

5 Let $f, g \in \mathcal{L}_0(X, \mu, \overline{\mathbb{R}})$. Show that the sets $[f < g]$, $[f \leq g]$, $[f = g]$, and $[f \neq g]$ belong to $\mathcal{A}$.

6 Suppose (f_j) is a sequence in $\mathcal{L}_0(X, \mu, \overline{\mathbb{R}})$. Show that

$$K := \{ x \in X \; ; \; \lim_j f_j(x) \text{ exists in } \overline{\mathbb{R}} \}$$

is μ-measurable.

7 Suppose $f : X \to \overline{\mathbb{R}}$. Prove or disprove:

(a) $f \in \mathcal{L}_0(X, \mu, \overline{\mathbb{R}}) \iff f^+, f^- \in \mathcal{L}_0(X, \mu, \overline{\mathbb{R}}^+)$.

(b) $f \in \mathcal{L}_0(X, \mu, \overline{\mathbb{R}}) \iff |f| \in \mathcal{L}_0(X, \mu, \overline{\mathbb{R}}^+)$.

8 A nonempty subset B of $\overline{\mathbb{R}}^X$ is called a **Baire function space** if these statements hold:

  (i) $\alpha \in \mathbb{R}$ and $f \in B$ imply $\alpha f \in B$.

 (ii) If $f + g$ exists in $\overline{\mathbb{R}}^X$ for $f, g \in B$, then $f + g \in B$.

(iii) $\sup_j f_j$ belongs to B for every sequence (f_j) in B.

Prove:

(a) $\overline{\mathbb{R}}^X$ and $\mathcal{L}_0(X, \mu, \overline{\mathbb{R}})$ are Baire function spaces.

(b) If $\{\, B_\alpha \subset \overline{\mathbb{R}}^X \ ; \ \alpha \in A \,\}$ is a family of Baire function spaces, then $\bigcap_{\alpha \in A} B_\alpha$ is also a Baire function space.

9   For $C \subset \overline{\mathbb{R}}^X$, we call

$$\sigma(C) := \bigcap \{\, B \subset \overline{\mathbb{R}}^X \ ; \ B \supset C, \ B \text{ is a Baire function space} \,\}$$

the **Baire function space generated by** C. By Exercise 8(b), $\sigma(C)$ is a well defined Baire function space. Show that

$$\sigma\big(\mathcal{S}(X, \mu, \mathbb{R})\big) = \mathcal{L}_0(X, \mu, \overline{\mathbb{R}}) \ .$$

10   Prove that $\sigma\big(C(\mathbb{R}^n, \mathbb{R})\big) = \mathcal{L}_0(\mathbb{R}^n, \beta_n, \mathbb{R})$.

11   Show that the supremum of an uncountable family of measurable real-valued functions is generally not measurable.

12   A sequence (f_j) in E^X is said to be $\boldsymbol{\mu}$**-almost uniformly convergent** if for every $\delta > 0$ there is an $A \in \mathcal{A}$ with $\mu(A^c) < \delta$ such that the sequence $(f_j \mid A)$ converges uniformly.

(a) Suppose (f_j) is a μ-almost uniformly convergent sequence in $\mathcal{L}_0(X, \mu, E)$. Show there is an $f \in \mathcal{L}_0(X, \mu, E)$ such that $f_j \to f$ μ-a.e.

(b) Define $f_j(x) := x^j$ for $j \in \mathbb{N}$ and $x \in [0, 1]$. Verify that (f_j) converges almost uniformly (with respect to Lebesgue measure), although there is no set $N \subset [0, 1]$ of measure zero such that $(f_j \mid N^c)$ converges uniformly.

13   Suppose $(X, \mathcal{A}, \mu)$ is a finite measure space and $f_j, f \in \mathcal{L}_0(X, \mu, E)$ with $f_j \to f$ μ-a.e. Prove:

(a) For $\varepsilon > 0$ and $\delta > 0$, there exist $k \in \mathbb{N}$ and $A \in \mathcal{A}$ such that $\mu(A^c) < \delta$ and $|f_j(x) - f(x)| < \varepsilon$ for $x \in A$ and $j \geq k$.

(b) The sequence (f_j) converges μ-almost uniformly to f (**Egorov's theorem**).

(c) Part (b) is generally false if $\mu(X) = \infty$.

(Hints: (a) Consider $K := [f_j \to f]$ and $K_k := \big[\,|f_j - f| < \varepsilon \ ; \ j \geq k\,\big]$, and apply the continuity of measures from above.   (b) To obtain the A_j, choose $\varepsilon := 1/j$ and $\delta := \delta \, 2^{-j}$ in (a), and let $A := \bigcup_j A_j$.   (c) Consider the measure space $(X, \mathcal{A}, \mu) = \big(\mathbb{R}, \lambda_1, \mathcal{L}(1)\big)$ and set $f_j := \chi_{[j, j+1]}$.)

14   Suppose $(X, \mathcal{A}, \mu)$ is a measure space and $f_j, f \in \mathcal{L}_0(X, \mu, E)$. We say (f_j) **converges in measure** to f if $\lim_{j \to \infty} \mu\big(\big[\,|f_j - f| \geq \varepsilon\,\big]\big) = 0$ for every $\varepsilon > 0$.
Prove:

(a) $f_j \to f$ μ-almost uniformly $\Rightarrow$ $f_j \to f$ in measure.

(b) If (f_j) converges in measure to f and to g, then $f = g$ μ-a.e.

(c) There is a sequence of Lebesgue measurable functions on $[0, 1]$ that converges in measure, but does not converge pointwise anywhere.

(d) There is a sequence of Lebesgue measurable functions on $\mathbb{R}$ that converges pointwise but not in measure.

(Hints: (c) Set $f_j := \chi_{I_j}$, where the intervals $I_j \subset [0, 1]$ are chosen so that $\lambda_1(I_j) \to 0$ and so that the sequence $\big(f_j(x)\big)$ has two cluster points for every $x \in [0, 1]$.   (d) Consider $f_j := \chi_{[j, j+1]}$.)

15 Suppose (f_j) is a sequence in $\mathcal{L}_0(X, \mu, E)$ converging in measure to $f \in \mathcal{L}_0(X, \mu, E)$. Show that (f_j) has a subsequence that converges μ-a.e. to f.
(Hint: There is an increasing sequence $(j_k)_{k \in \mathbb{N}}$ such that

$$\mu([\,|f_m - f_n| \geq 2^{-k}]) \leq 2^{-k} \quad \text{for } m, n \geq j_k \ .$$

With the help of $B_\ell := \bigcup_{k=\ell}^\infty [\,|f_{n_{k+1}} - f_{n_k}| \geq 2^{-k}]$, conclude that $(f_{j_k})_{k \in \mathbb{N}}$ converges μ-almost uniformly. Also note Exercises 12, 14(a) and 14(b).)

16 For $x = (x_j) \in \mathbb{K}^\mathbb{N}$ and $p \in [1, \infty]$, define[11]

$$\|x\|_p := \begin{cases} \left(\sum_{j=0}^\infty |x_j|^p \right)^{1/p} & \text{if } p \in [1, \infty) \ , \\ \sup_j |x_j| & \text{if } p = \infty \ , \end{cases}$$

and

$$\ell_p := \ell_p(\mathbb{K}) := \left(\{ x \in \mathbb{K}^\mathbb{N} \ ; \ \|x\|_p < \infty \}, \ \|\cdot\|_p \right) \ .$$

Prove:

(a) If $p \in [1, \infty)$, then ℓ_p is a separable normed vector space.

(b) ℓ_∞ is not separable.

17 Suppose $x \in \overline{\mathbb{R}}$. If $x \in \mathbb{R}$, we say $U \subset \overline{\mathbb{R}}$ is a **neighborhood in $\overline{\mathbb{R}}$** of x if U contains a neighborhood *in $\mathbb{R}$* of x. For $x \in \overline{\mathbb{R}} \backslash \mathbb{R}$, neighborhoods were defined in Section II.5 as those sets that contain a semi-infinite interval of the appropriate kind. Suppose $O \subset \overline{\mathbb{R}}$. We say O is **open in $\overline{\mathbb{R}}$** if every $x \in O$ has a neighborhood U in $\overline{\mathbb{R}}$ such that $U \subset O$. Now define $\mathcal{T} := \{ O \subset \overline{\mathbb{R}} \ ; \ O \text{ is open in } \overline{\mathbb{R}} \}$. Prove:

(a) O is open in $\overline{\mathbb{R}}$ if and only if $O \cap \mathbb{R}$ is open in $\mathbb{R}$ and, in the case $\infty \in O$ [or $-\infty \in O$], there is an $a \in \mathbb{R}$ such that $(a, \infty] \subset O$ [or $[-\infty, a) \subset O$].

(b) $(\overline{\mathbb{R}}, \mathcal{T})$ is a compact topological space.

(c) $\mathcal{B}(\overline{\mathbb{R}}) = \{ B \cup F \ ; \ B \in \mathcal{B}^1, \ F \subset \{-\infty, \infty\} \}$.

(d) $\mathcal{B}(\overline{\mathbb{R}}) \,|\, \mathbb{R} = \mathcal{B}^1$.

(e) An element of $\overline{\mathbb{R}}^X$ belongs to $\mathcal{L}_0(X, \mu, \overline{\mathbb{R}})$ if and only if it is $\mathcal{A}$-$\mathcal{B}(\overline{\mathbb{R}})$-measurable.

18 Check that, if S is a separable subset of E, the closure of its span is a separable Banach space.

19 For $f \in \mathbb{K}^X$, set

$$(\operatorname{sign} f)(x) := \begin{cases} f(x)/|f(x)| & \text{if } f(x) \neq 0 \ , \\ 0 & \text{if } f(x) = 0 \ . \end{cases}$$

Demonstrate that $f \in \mathcal{L}_0(X, \mu, \mathbb{K})$ implies $\operatorname{sign} f \in \mathcal{L}_0(X, \mu, \mathbb{K})$.

---

[11]See also Proposition IV.2.17.

2 Integrable functions

In this section, we define the general Bochner–Lebesgue integral and describe its basic properties. We also prove that the vector space of integrable functions is complete with respect to the seminorm induced by the integral.

As in the previous section, suppose
- $(X, \mathcal{A}, \mu)$ is a complete σ-finite measure space;
  $E = (E, |\cdot|)$ is a Banach space.

The integral of a simple function

In Remark 1.2(c), we learned that every simple function has a unique normal form. This form will prove to be useful in the sequel, and we work with it preferentially.

> **Convention** We will always represent μ-simple functions by their normal forms, unless we say otherwise. Further, we set[1]
>
> $$\infty \cdot 0_E := -\infty \cdot 0_E := 0_E , \qquad (2.1)$$
>
> where 0_E is the zero vector of E.

For $\varphi \in \sum_{j=0}^{m} e_j \chi_{A_j} \in \mathcal{S}(X, \mu, E)$, we define **the integral of φ over X with respect to the measure μ** as the sum

$$\int_X \varphi \, d\mu := \int \varphi \, d\mu := \sum_{j=0}^{m} e_j \mu(A_j) .$$

If A is a μ-measurable set, we define the **integral of φ over A with respect to the measure μ** as

$$\int_A \varphi \, d\mu := \int_X \chi_A \varphi \, d\mu .$$

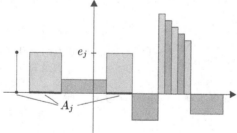

---

[1] Convention (2.1) is common in the theory of integration and makes it possible, for example, to integrate simple functions over their entire domains of definition. It is *not* to be understood in the case $E = \mathbb{R}$ as another calculation rule in $\overline{\mathbb{R}}$, but rather in the sense of "external" multiplication of $\infty, -\infty \in \overline{\mathbb{R}}$ by the zero vector of $\mathbb{R}$.

2.1 Remarks (a) For $\varphi \in \mathcal{S}(X, \mu, E)$ and $A \in \mathcal{A}$, the integral $\int_A \varphi \, d\mu$ is well defined.

Proof This follows from Remarks 1.2(c) and (d). ∎

(b) Let $\varphi = \sum_{k=0}^n f_k \chi_{B_k}$, where $f_0, \ldots, f_n \in E$ are nonzero and $B_0, \ldots, B_n \in \mathcal{A}$ are pairwise disjoint, be a μ-simple function *not necessarily in normal form*. Then

$$\int_X \varphi \, d\mu = \sum_{k=0}^n f_k \mu(B_k) \, .$$

Proof Let $\sum_{j=0}^m e_j \chi_{A_j}$ be the normal form of φ. Set

$$A_{m+1} := \bigcap_{j=0}^m A_j^c \,, \quad B_{n+1} := \bigcap_{k=0}^n B_k^c \,, \quad e_{m+1} := 0 \,, \quad f_{n+1} := 0 \,. \tag{2.2}$$

Then $X = \bigcup_{j=0}^{m+1} A_j = \bigcup_{k=0}^{n+1} B_k$, so

$$A_j = \bigcup_{k=0}^{n+1} (A_j \cap B_k) \quad \text{and} \quad B_k = \bigcup_{j=0}^{m+1} (A_j \cap B_k) \quad \text{for } j = 0, \ldots, m+1 \text{ and } k = 0, \ldots, n+1.$$

Because the sets $A_j \cap B_k$ are pairwise disjoint, we have

$$\mu(A_j) = \sum_{k=0}^{n+1} \mu(A_j \cap B_k) \quad \text{and} \quad \mu(B_k) = \sum_{j=0}^{m+1} \mu(A_j \cap B_k) \, .$$

If $A_j \cap B_k \neq \emptyset$, then $e_j = f_k$, and we find

$$\int_X \varphi \, d\mu = \sum_{j=0}^m e_j \mu(A_j) = \sum_{j=0}^{m+1} e_j \sum_{k=0}^{n+1} \mu(A_j \cap B_k) = \sum_{k=0}^{n+1} f_k \sum_{j=0}^{m+1} \mu(A_j \cap B_k)$$

$$= \sum_{k=0}^n f_k \mu(B_k) \, . \quad \blacksquare$$

(c) The integral $\int_X \cdot d\mu : \mathcal{S}(X, \mu, E) \to E$ is linear.

Proof Suppose $\varphi = \sum_{j=0}^m e_j \chi_{A_j}$ and $\psi = \sum_{k=0}^n f_k \chi_{B_k}$ are μ-simple functions and $\alpha \in \mathbb{K}$. One checks easily that $\int_X \alpha \varphi \, d\mu = \alpha \int_X \varphi \, d\mu$. With the relations (2.2), an argument like that in (b) implies

$$\chi_{A_j} = \sum_{k=0}^{n+1} \chi_{A_j \cap B_k} \quad \text{and} \quad \chi_{B_k} = \sum_{j=0}^{m+1} \chi_{A_j \cap B_k} \,,$$

and thus[2]

$$\varphi + \psi = \sum_{j=0}^{m+1} \sum_{k=0}^{n+1} (e_j + f_k) \chi_{A_j \cap B_k} \, . \tag{2.3}$$

The claim now follows from (b). ∎

---

[2] In general, (2.3) does *not* give the normal form of $\varphi + \psi$.

(d) For $A, B \in \mathcal{A}$ and $A \cap B = \emptyset$, we have

$$\int_{A \cup B} \varphi \, d\mu = \int_A \varphi \, d\mu + \int_B \varphi \, d\mu \quad \text{for } \varphi \in \mathcal{S}(X, \mu, E) \ .$$

Proof This follows from (c) and the equality $\chi_{A \cup B} \varphi = \chi_A \varphi + \chi_B \varphi$. ∎

(e) For $\varphi \in \mathcal{S}(X, \mu, E)$ and $A \in \mathcal{A}$, we have

$$\left| \int_A \varphi \, d\mu \right| \leq \int_A |\varphi| \, d\mu \leq \|\varphi\|_\infty \, \mu(A) \ .$$

Proof This follows from Remark 1.2(d) and the triangle inequality. ∎

(f) If $\varphi, \psi \in \mathcal{S}(X, \mu, \mathbb{R})$ satisfy $\varphi \leq \psi$, then $\int_A \varphi \, d\mu \leq \int_A \psi \, d\mu$.

Proof Clearly $\int_A \eta \, d\mu \geq 0$ for $\eta \in \mathcal{S}(X, \mu, \mathbb{R}^+)$. The claim now follows from (c). ∎

The $\mathcal{L}_1$-seminorm

Suppose V is a vector space over $\mathbb{K}$. A map $p \colon V \to \mathbb{R}$ is called a **seminorm** on V if it satisfies these properties:

(i) $p(v) \geq 0$ for $v \in V$;

(ii) $p(\lambda v) = |\lambda| \, p(v)$ for $v \in V$ and $\lambda \in \mathbb{K}$;

(iii) $p(v + w) \leq p(v) + p(w)$ for $v, w \in V$.

For $v \in V$ and $r > 0$, we denote by

$$\mathbb{B}_p(v, r) := \left\{ w \in V \ ; \ p(v - w) < r \right\}$$

the **open ball** in (V, p) **around** v **with radius** r. A subset O of V is said to be p-**open** if, for every $v \in O$, there is an $r > 0$ such that $\mathbb{B}_p(v, r) \subset O$.

2.2 Remarks Suppose V is a vector space and p is a seminorm on V.

(a) The seminorm p is a norm if and only if $p^{-1}(0) = \{0\}$.

(b) Suppose $K \subset \mathbb{R}^n$ is compact, $k \in \mathbb{N} \cup \{\infty\}$, and

$$p_K(f) := \max_{x \in K} |f(x)| \quad \text{for } f \in C^k(\mathbb{R}^n, E) \ .$$

Then p_K is a seminorm on $C^k(\mathbb{R}^n, E)$, but not a norm.

Proof One verifies easily that p_K is a seminorm on $C^k(\mathbb{R}^n, E)$. Let U be an open neighborhood of K. Then Exercise VII.6.7 shows that there is an $f \in C^\infty(\mathbb{R}^n, \mathbb{R})$ such that $f(x) = 1$ for $x \in K$ and $f(x) = 0$ for $x \in U^c$. For $e \in E \setminus \{0\}$, we set $g := (\chi_{\mathbb{R}^n} - f)e$. Then g belongs to $C^\infty(\mathbb{R}^n, E)$, and we have $p_K(g) = 0$, but $g \neq 0$. Therefore p_K is not a norm on $C^k(\mathbb{R}^n, E)$. ∎

(c) Let

$$\|\varphi\|_1 := \int_X |\varphi| \, d\mu \quad \text{for } \varphi \in \mathcal{S}(X, \mu, E) \,.$$

Then $\|\cdot\|_1$ is a seminorm on $\mathcal{S}(X, \mu, E)$. If there is a nonempty μ-null set in $\mathcal{A}$, then $\|\cdot\|_1$ is not a norm on $\mathcal{S}(X, \mu, E)$.

Proof It is clear that $\|\cdot\|_1$ is a seminorm on $\mathcal{S}(X, \mu, E)$. Letting N denote a nonempty μ-null set, we have $\|\chi_N\|_1 = 0$, but $\chi_N \neq 0$. ∎

(d) $\mathcal{T}_p := \{ O \subset V \ ; \ O \text{ is } p\text{-open} \}$ is a topology on V, the **topology generated by** p.

Proof One easily checks that the argument used in the proof of Proposition III.2.4 transfers to this situation. ∎

(e) The topology $\mathcal{T}_p$ is not necessarily Hausdorff. If it isn't, there is no metric on V that generates $\mathcal{T}_p$.

Proof We use the notation of (b) and set $K := \{0\}$. Further let $f \in C^k(\mathbb{R}^n, E)$ with $f(0) = 0$ and $f \neq 0$. Then $\mathbb{B}_{p_K}(f, \varepsilon) = \mathbb{B}_{p_K}(0, \varepsilon)$ for every $\varepsilon > 0$. Therefore $\mathcal{T}_{p_K}$ is not Hausdorff. The second statement follows from Proposition III.2.17. ∎

(f) A linear map $A \colon V \to E$ is said to be **(p)-bounded** if there is an $M \geq 0$ such that $|Av| \leq Mp(v)$ for $v \in V$. For a linear map $A \colon V \to E$, these statements are equivalent:

(i) A is continuous.

(ii) A is continuous at 0.

(iii) A is bounded.

Proof This follows from the proof of Theorem VI.2.5, which used only the properties of a seminorm. ∎

(g) $\int \cdot \, d\mu \colon \mathcal{S}(X, \mu, E) \to E$ is continuous.

Proof This follows from (c), (f), and Remark 2.1(c). ∎

Suppose p is a seminorm on V. We know from Remark 2.2(e) that the topology of (V, p) may not be generated by a metric on V, in which case the metric notions of Cauchy sequence and completeness are not available. Accordingly, we need new definitions: A sequence $(v_j) \in V^{\mathbb{N}}$ is called a **Cauchy sequence in** (V, p) if for every $\varepsilon > 0$ there is an $N \in \mathbb{N}$ such that $p(v_j - v_k) < \varepsilon$ for $j, k \geq N$. We call (V, p) **complete** if every Cauchy sequence in (V, p) converges.

2.3 Remarks **(a)** If (V, p) is a normed vector space, these notions agree with those of Section II.6.

(b) Suppose $(v_j) \in V^{\mathbb{N}}$ and $v \in V$. Then $v_j \to v$ if and only if $p(v - v_j) \to 0$. However, the limit of a convergent sequence is generally not uniquely determined: if p is not a norm, $v_j \to v$ implies $v_j \to w$ for every $w \in V$ such that $p(v - w) = 0$.

(c) The set of all Cauchy sequences in (V, p) forms a vector subspace of $V^{\mathbb{N}}$. ∎

In the following, we always provide the space $\mathcal{S}(X, \mu, E)$ with the topology induced by $\|\cdot\|_1$. Then we may also call a Cauchy sequence in $\mathcal{S}(X, \mu, E)$ an $\mathcal{L}_1$-**Cauchy sequence**.

A function $f \in E^X$ is called μ-**integrable** if f is a μ-a.e. limit of some $\mathcal{L}_1$-Cauchy sequence (φ_j) in $\mathcal{S}(X, \mu, E)$. We denote the set of E-valued, μ-integrable functions of X by $\mathcal{L}_1(X, \mu, E)$.

2.4 Proposition   *In the sense of vector subspaces, we have the inclusions*

$$\mathcal{S}(X, \mu, E) \subset \mathcal{L}_1(X, \mu, E) \subset \mathcal{L}_0(X, \mu, E) \ .$$

Proof   Clearly every μ-simple function is μ-integrable. That every μ-integrable function is μ-measurable follows from Remark 1.2(a) and Theorem 1.14. There remains to show that $\mathcal{L}_1(X, \mu, E)$ is a vector subspace of $\mathcal{L}_0(X, \mu, E)$. Take $f, g \in \mathcal{L}_1(X, \mu, E)$ and $\alpha \in \mathbb{K}$. There are $\mathcal{L}_1$-Cauchy sequences (φ_j) and (ψ_j) in $\mathcal{S}(X, \mu, E)$ such that $\varphi_j \to f$ and $\psi_j \to g$ μ-a.e. as $j \to \infty$. From the triangle inequality, it follows that $(\alpha\varphi_j + \psi_j)_{j \in \mathbb{N}}$ is an $\mathcal{L}_1$-Cauchy sequence in $\mathcal{S}(X, \mu, E)$ that converges μ-a.e. to $\alpha f + g$. Therefore $\alpha f + g$ is μ-integrable, as needed. ∎

The Bochner–Lebesgue integral

Let $f \in \mathcal{L}_1(X, \mu, E)$. Then there is an $\mathcal{L}_1$-Cauchy sequence (φ_j) in $\mathcal{S}(X, \mu, E)$ such that $\varphi_j \to f$ μ-a.e. We will see that the sequence $(\int_X \varphi_j \, d\mu)_{j \in \mathbb{N}}$ converges in E. It is natural, then, to define the integral of f with respect to μ as the limit of this sequence of integrals. Of course we must check that the limit is independent of the approximating sequence of simple functions; that is, we must show that $\lim_j \int \varphi_j \, d\mu = \lim_j \int \psi_j \, d\mu$ if (ψ_j) is another Cauchy sequence in $\mathcal{S}(X, \mu, E)$ such that $\psi_j \to f$ μ-a.e.

2.5 Lemma   *Suppose (φ_j) is a Cauchy sequence in $\mathcal{S}(X, \mu, E)$. Then there is a subsequence $(\varphi_{j_k})_{k \in \mathbb{N}}$ of (φ_j) and an $f \in \mathcal{L}_1(X, \mu, E)$ such that*

(i) $\varphi_{j_k} \to f$ μ-a.e. as $k \to \infty$;

(ii) *for every $\varepsilon > 0$, there exists $A_\varepsilon \in \mathcal{A}$ such that $\mu(A_\varepsilon) < \varepsilon$ and $(\varphi_{j_k})_{k \in \mathbb{N}}$ converges uniformly to f on A_ε^c.*

Proof   (α) By assumption, there exists for any $k \in \mathbb{N}$ some $j_k \in \mathbb{N}$ such that the bound $\|\varphi_\ell - \varphi_m\|_1 < 2^{-2k}$ holds for $\ell, m \geq j_k$. Without loss of generality, we can choose the sequence $(j_k)_{k \in \mathbb{N}}$ to be increasing. With $\psi_k := \varphi_{j_k}$, we then get

$$\|\psi_\ell - \psi_m\|_1 < 2^{-2\ell} \quad \text{for } m \geq \ell \geq 0 \ .$$

(β) Set $B_\ell := \big[|\psi_{\ell+1} - \psi_\ell| \geq 2^{-\ell}\big]$ for $\ell \in \mathbb{N}$. Then B_ℓ belongs to $\mathcal{A}$, and we have $\mu(B_\ell) < \infty$ for $\ell \in \mathbb{N}$, because every ψ_m is μ-simple. Therefore χ_{B_ℓ} is also μ-simple, and Remark 2.1(f) implies

$$2^{-\ell}\mu(B_\ell) = 2^{-\ell}\int_X \chi_{B_\ell}\, d\mu \leq \int_X |\psi_{\ell+1} - \psi_\ell|\, d\mu = \|\psi_{\ell+1} - \psi_\ell\|_1 < 2^{-2\ell}\,.$$

This leads to $\mu(B_\ell) < 2^{-\ell}$ for $\ell \in \mathbb{N}$.

Letting $A_n := \bigcup_{k=0}^\infty B_{n+k}$, we have $\mu(A_n) \leq 2^{-n+1}$ for $n \in \mathbb{N}$, and we see that $A := \bigcap_{n=0}^\infty A_n$ is a μ-null set.

(γ) If x lies in $A_n^c = \bigcap_{k=0}^\infty B_{n+k}^c$, then

$$|\psi_{\ell+1}(x) - \psi_\ell(x)| < 2^{-\ell} \quad \text{for } \ell \geq n\,.$$

By the Weierstrass criterion (Theorem V.1.6), the series

$$\psi_0 + \sum(\psi_{\ell+1} - \psi_\ell)$$

on A_n^c converges uniformly in E. Now we set

$$f(x) := \begin{cases} \lim_k \psi_k(x) & \text{if } x \in A^c\,, \\ 0 & \text{if } x \in A\,. \end{cases}$$

Then $\varphi_{j_k} \to f$ μ-a.e. as $k \to \infty$. Further, there is for every $\varepsilon > 0$ an $n \in \mathbb{N}$ such that $\mu(A_n) \leq 2^{-n+1} < \varepsilon$, and $(\varphi_{j_k})_{k\in\mathbb{N}}$ converges uniformly on A_n^c to f as $k \to \infty$. ∎

2.6 Lemma *Suppose (φ_j) and (ψ_j) are $\mathcal{L}_1$-Cauchy sequences in $\mathcal{S}(X, \mu, E)$ that converge μ-a.e. to the same function. Then $\lim \|\varphi_j - \psi_j\|_1 = 0$.*

Proof (i) Take $\varepsilon > 0$ and set $\eta_j := \varphi_j - \psi_j$ for $j \in \mathbb{N}$. By Remark 2.3(c), (η_j) is an $\mathcal{L}_1$-Cauchy sequence in $\mathcal{S}(X, \mu, E)$. Thus there is a natural number N such that $\|\eta_j - \eta_k\| < \varepsilon/8$ for $j, k \geq N$.

(ii) Because η_N is μ-simple, $A := [\eta_N \neq 0]$ belongs to $\mathcal{A}$ and $\mu(A) < \infty$. Moreover (η_j) converges μ-a.e. to zero. Then Lemma 2.5 says there exists $B \in \mathcal{A}$ such that $\mu(B) < \varepsilon/8(1 + \|\eta_N\|_\infty)$, and there is a subsequence $(\eta_{j_k})_{k\in\mathbb{N}}$ of (η_j) that converges uniformly to 0 on B^c. Hence there exists $K \geq N$ such that

$$|\eta_{j_K}(x)| \leq \varepsilon/8(1 + \mu(A)) \quad \text{for } x \in A \backslash B\,.$$

This implies $\int_{A\backslash B} |\eta_{j_K}|\, d\mu \leq \varepsilon/8$.

(iii) The properties of B and K imply

$$\int_B |\eta_{j_K}|\, d\mu \leq \int_B |\eta_{j_K} - \eta_N|\, d\mu + \int_B |\eta_N|\, d\mu$$
$$\leq \|\eta_{j_K} - \eta_N\|_1 + \|\eta_N\|_\infty\, \mu(B) < \varepsilon/4\,.$$

From the definition of A, we have

$$\int_{A^c} |\eta_{j_K}| \, d\mu = \int_{A^c} |\eta_{j_K} - \eta_N| \, d\mu \leq \|\eta_{j_K} - \eta_N\|_1 < \varepsilon/8 \ .$$

Altogether we obtain using Remark 2.1(d)

$$\|\eta_{j_K}\|_1 \leq \int_{A^c \cup (A \setminus B) \cup B} |\eta_{j_K}| \, d\mu < \varepsilon/2 \ ,$$

and therefore $\|\eta_j\|_1 \leq \|\eta_{j_K}\|_1 + \|\eta_j - \eta_{j_K}\|_1 < \varepsilon$ for $j \geq N$. Because $\varepsilon > 0$ was arbitrary, all is proved. ∎

2.7 Corollary Let (φ_j) and (ψ_j) be Cauchy sequences in $\mathcal{S}(X, \mu, E)$ converging μ-a.e. to the same function. The sequences $\left(\int_X \varphi_j \, d\mu \right)$ and $\left(\int_X \psi_j \, d\mu \right)$ converge in E, and

$$\lim_j \int_X \varphi_j \, d\mu = \lim_j \int_X \psi_j \, d\mu \ .$$

Proof Because

$$\left| \int_X \varphi_j \, d\mu - \int_X \varphi_k \, d\mu \right| \leq \|\varphi_j - \varphi_k\|_1 \quad \text{for } j, k \in \mathbb{N} \ ,$$

$(\int \varphi_j \, d\mu)_{j \in \mathbb{N}}$ is a Cauchy sequence in E; hence $\int \varphi_j \, d\mu \to e$ as $j \to \infty$, for some $e \in E$. Likewise there is $e' \in E$ such that $\int \psi_j \, d\mu \to e'$ for $j \to \infty$. Applying Lemma 2.6 and the continuity of the norm of E, we see that

$$|e - e'| = \lim_j \left| \int_X \varphi_j \, d\mu - \int_X \psi_j \, d\mu \right| \leq \lim_j \int_X |\varphi_j - \psi_j| \, d\mu = \lim_j \|\varphi_j - \psi_j\|_1 = 0 \ ,$$

and we are done. ∎

After these preparations, we define the integral of integrable functions in a natural way, extending the integral of simple functions. Suppose $f \in \mathcal{L}_1(X, \mu, E)$. Then there is an $\mathcal{L}_1$-Cauchy sequence (φ_j) in $\mathcal{S}(X, \mu, E)$ such that $\varphi_j \to f$ μ-a.e. According to Corollary 2.7, the quantity

$$\int_X f \, d\mu := \lim_j \int_X \varphi_j \, d\mu$$

exists in E, and is independent of the sequence (φ_j). This is called the **Bochner–Lebesgue integral of f over X with respect to the measure** μ. Other notations are often used besides $\int_X f \, d\mu$, namely,

$$\int f \, d\mu \ , \qquad \int_X f(x) \, d\mu(x) \ , \qquad \int_X f(x) \, \mu(dx) \ .$$

Clearly, in the case of simple functions, the Bochner–Lebesgue integral agrees with the integral defined at the start of this section (page 80).

The completeness of $\mathcal{L}_1$

With the help of the integral, we now define a seminorm on $\mathcal{L}_1(X, \mu, E)$ and show that $\mathcal{L}_1(X, \mu, E)$ is complete with respect to this seminorm.

2.8 Lemma If $f \in \mathcal{L}_1(X, \mu, E)$, then $|f|$ belongs to $\mathcal{L}_1(X, \mu, \mathbb{R})$. If (φ_j) is an $\mathcal{L}_1$-Cauchy sequence in $\mathcal{S}(X, \mu, E)$ such that $\varphi_j \to f$ μ-a.e., then $\int |f| \, d\mu = \lim_j \int |\varphi_j| \, d\mu$.

Proof The reverse triangle inequality (which clearly holds for seminorms too) gives

$$\big\| \, |\varphi_j| - |\varphi_k| \, \big\|_1 \leq \|\varphi_j - \varphi_k\|_1 \quad \text{and} \quad \big| \, |\varphi_j| - |\varphi_k| \, \big| \leq |\varphi_j - \varphi_k| \quad \text{for } j, k \in \mathbb{N} \,.$$

Thus $(|\varphi_j|)_{j \in \mathbb{N}}$ is an $\mathcal{L}_1$-Cauchy sequence in $\mathcal{S}(X, \mu, \mathbb{R})$ that converges μ-a.e. to $|f|$. Therefore $|f|$ belongs to $\mathcal{L}_1(X, \mu, \mathbb{R})$, and $\int |f| \, d\mu = \lim_j \int |\varphi_j| \, d\mu$. $\blacksquare$

2.9 Corollary For $f \in \mathcal{L}_1(X, \mu, E)$, let $\|f\|_1 := \int_X |f| \, d\mu$. Then $\|\cdot\|_1$ is a seminorm on $\mathcal{L}_1(X, \mu, E)$, called the $\mathcal{L}_1$-**seminorm**.

Proof Take $f, g \in \mathcal{L}_1(X, \mu, E)$ and let (φ_j) and (ψ_j) be $\mathcal{L}_1$-Cauchy sequences in $\mathcal{S}(X, \mu, E)$ such that $\varphi_j \to f$ and $\psi_j \to g$ μ-a.e. From Lemma 2.8 and Remark 2.2(c) we have

$$\|f\|_1 = \int_X |f| \, d\mu = \lim_j \int_X |\varphi_j| \, d\mu = \lim_j \|\varphi_j\|_1 \geq 0 \,,$$

as well as

$$\|f + g\|_1 = \lim_j \|\varphi_j + \psi_j\|_1 \leq \lim_j \big(\|\varphi_j\|_1 + \|\psi_j\|_1 \big) = \|f\|_1 + \|g\|_1 \,,$$

and

$$\|\alpha f\|_1 = \lim_j \|\alpha \varphi_j\| = |\alpha| \lim_j \|\varphi_j\| = |\alpha| \, \|f\|_1$$

for any $\alpha \in \mathbb{K}$. $\blacksquare$

We will always give $\mathcal{L}_1(X, \mu, E)$ the topology induced by the seminorm $\|\cdot\|_1$.

2.10 Theorem

(i) $\mathcal{S}(X, \mu, E)$ is dense in $\mathcal{L}_1(X, \mu, E)$.

(ii) The space $\mathcal{L}_1(X, \mu, E)$ is complete.

Proof (i) Suppose $f \in \mathcal{L}_1(X, \mu, E)$, and let (φ_j) denote an $\mathcal{L}_1$-Cauchy sequence of simple functions such that $\varphi_j \to f$ μ-a.e. as $j \to \infty$. Also suppose $k \in \mathbb{N}$. Then $(\varphi_j - \varphi_k)_{j \in \mathbb{N}}$ is an $\mathcal{L}_1$-Cauchy sequence in $\mathcal{S}(X, \mu, E)$ such that $(\varphi_j - \varphi_k) \to (f - \varphi_k)$

μ-a.e. for $j \to \infty$. Because of Lemma 2.8,

$$\|f - \varphi_k\|_1 = \lim_j \|\varphi_j - \varphi_k\|_1 \quad \text{for } k \in \mathbb{N} .$$

Suppose $\varepsilon > 0$. Then there is an $N \in \mathbb{N}$ such that $\|\varphi_j - \varphi_k\|_1 < \varepsilon$ for $j, k \geq N$; taking the limit $j \to \infty$ we get $\|f - \varphi_N\|_1 \leq \varepsilon$. This shows that $S(X, \mu, E)$ is dense in $\mathcal{L}_1(X, \mu, E)$.

(ii) Let (f_j) be a Cauchy sequence in $\mathcal{L}_1(X, \mu, E)$ and take $\varepsilon > 0$. Choose $M \in \mathbb{N}$ such that $\|f_j - f_k\|_1 < \varepsilon/2$ for $j, k \geq M$. We know from (i) that for any $j \in \mathbb{N}$ we can find $\varphi_j \in S(X, \mu, E)$ such that $\|f_j - \varphi_j\|_1 < 2^{-j}$. Then

$$\|\varphi_j - \varphi_k\|_1 \leq \|\varphi_j - f_j\|_1 + \|f_j - f_k\|_1 + \|f_k - \varphi_k\|_1 < 2^{-j} + 2^{-k} + \varepsilon/2$$

for $j, k \geq M$. This shows that (φ_j) is an $\mathcal{L}_1$-Cauchy sequence in $S(X, \mu, E)$. By Lemma 2.5, therefore, there is a subsequence $(\varphi_{j_k})_{k \in \mathbb{N}}$ of (φ_j) and an f in $\mathcal{L}_1(X, \mu, E)$ such that $\varphi_{j_k} \to f$ μ-a.e. as $k \to \infty$. The proof of (i) shows that there exists an $N \geq M$ such that $\|f - \varphi_{j_N}\|_1 < \varepsilon/4$, and we get

$$\|f - f_j\|_1 \leq \|f - \varphi_{j_N}\|_1 + \|\varphi_{j_N} - f_{j_N}\|_1 + \|f_{j_N} - f_j\|_1 < \varepsilon \quad \text{for } j \geq N ,$$

that is, (f_j) converges in $\mathcal{L}_1(X, \mu, E)$ to f. ∎

Elementary properties of integrals

We have seen that the integral on the space of simple functions is continuous, linear, and, for $E = \mathbb{R}$, also monotone — see Remarks 2.2(g), 2.1(c) and 2.1(f). We now show that these properties survive the extension of the integral from the space of simple functions to that of integrable functions.

2.11 Theorem

(i) $\int_X \cdot \, d\mu : \mathcal{L}_1(X, \mu, E) \to E$ is linear and continuous, and

$$\left| \int_X f \, d\mu \right| \leq \int_X |f| \, d\mu = \|f\|_1 .$$

(ii) $\int_X \cdot \, d\mu : \mathcal{L}_1(X, \mu, \mathbb{R}) \to \mathbb{R}$ is a continuous, positive linear functional.

(iii) Suppose F is a Banach space and $T \in \mathcal{L}(E, F)$. Then

$$Tf \in \mathcal{L}_1(X, \mu, F) \quad \text{and} \quad T \int_X f \, d\mu = \int_X Tf \, d\mu$$

for $f \in \mathcal{L}_1(X, \mu, E)$.

Proof (i) Proposition 2.4 showed that μ-integrable functions form a vector space. Take $f, g \in \mathcal{L}_1(X, \mu, E)$ and $\alpha \in \mathbb{K}$. Then there are $\mathcal{L}_1$-Cauchy sequences (φ_j) and (ψ_j) in $\mathcal{S}(X, \mu, E)$ such that $\varphi_j \to f$ and $\psi_j \to g$ μ-a.e. By Remark 2.1(c),

$$\int_X (\alpha \varphi_j + \psi_j) \, d\mu = \alpha \int_X \varphi_j \, d\mu + \int_X \psi_j \, d\mu \quad \text{for } j \in \mathbb{N} .$$

The linearity of the integral on $\mathcal{L}_1(X, \mu, E)$ follows by taking the limit $j \to \infty$. By Corollary 2.9, $\|\cdot\|_1$ is a seminorm on $\mathcal{L}_1(X, \mu, E)$, and Remark 2.1(e) yields

$$\left| \int_X \varphi_j \, d\mu \right| \le \int_X |\varphi_j| \, d\mu = \|\varphi_j\|_1 \quad \text{for } j \in \mathbb{N} .$$

By Lemma 2.8, we can take the limit $j \to \infty$, and we find

$$\left| \int_X f \, d\mu \right| \le \int_X |f| \, d\mu = \|f\|_1 .$$

Continuity now follows from Remark 2.2(f).

The approach just used can be adapted without difficulty to the task of proving (ii) and (iii). This is left to the reader as an exercise. ∎

2.12 Corollary

(i) *A map $f = (f_1, \ldots, f_n) \colon X \to \mathbb{K}^n$ is μ-integrable if and only if its every coordinate f_j is. In that case,*

$$\int_X f \, d\mu = \left(\int_X f_1 \, d\mu, \ldots, \int_X f_n \, d\mu \right) .$$

(ii) *Suppose $g, h \in \mathbb{R}^X$ and define $f := g + ih$. Then f is in $\mathcal{L}_1(X, \mu, \mathbb{C})$ if and only if g and h are in $\mathcal{L}_1(X, \mu, \mathbb{R})$. In that case,*

$$\int_X f \, d\mu = \int_X g \, d\mu + i \int_X h \, d\mu .$$

(iii) *A function $f \in \mathbb{R}^X$ is μ-integrable if and only if f^+ and f^- are. In that case,*

$$\int_X f \, d\mu = \int_X f^+ \, d\mu - \int_X f^- \, d\mu , \qquad \int_X |f| \, d\mu = \int_X f^+ \, d\mu + \int_X f^- \, d\mu .$$

Proof (i) "$\Rightarrow$" Take $f \in \mathcal{L}_1(X, \mu, \mathbb{K}^n)$. Since $\mathrm{pr}_j \in \mathcal{L}(\mathbb{K}^n, \mathbb{K})$ for $j = 1, \ldots, n$, it follows from Theorem 2.11(iii) that $f_j = \mathrm{pr}_j \circ f$ belongs to $\mathcal{L}_1(X, \mu, \mathbb{K})$. Moreover $\int f_j \, d\mu = \mathrm{pr}_j \int f \, d\mu$, so

$$\int_X f \, d\mu = \left(\int_X f_1 \, d\mu, \ldots, \int_X f_n \, d\mu \right) .$$

"$\Leftarrow$" For $j = 1, \ldots, n$, we consider the map

$$b_j : \mathbb{K} \to \mathbb{K}^n , \quad y \mapsto (0, \ldots, 0, y, 0, \ldots, 0) ,$$

where y is in the j-th position on the right. Then

$$b_j \in \mathcal{L}(\mathbb{K}, \mathbb{K}^n) \quad \text{and} \quad f := \sum_{j=1}^{n} b_j \circ f_j .$$

The claim now follows from Theorem 2.11(i) and (iii).

(ii) This follows from (i) and the identification of $\mathbb{C}$ with $\mathbb{R}^2$.

(iii) For $f \in \mathbb{R}^X$, we have

$$f^+ = (f + |f|)/2 , \quad f^- = (|f| - f)/2 , \quad f = f^+ - f^- , \quad |f| = f^+ + f^- .$$

Hence Theorem 2.11(i) and Lemma 2.8 imply the conclusion. ∎

2.13 Lemma   For $f \in \mathcal{L}_1(X, \mu, E)$ and $A \in \mathcal{A}$, we have $\chi_A f \in \mathcal{L}_1(X, \mu, E)$.

Proof   Suppose (φ_j) is an $\mathcal{L}_1$-Cauchy sequence in $\mathcal{S}(X, \mu, E)$ that converges μ-a.e. to f. Then $\chi_A \varphi_j$ is μ-simple by Remark 1.2(d), and $(\chi_A \varphi_j)_{j \in \mathbb{N}}$ obviously converges μ-a.e. to $\chi_A f$. By Remark 2.1(f), we have

$$\int_X |\chi_A \varphi_j - \chi_A \varphi_k| \, d\mu = \int_X \chi_A |\varphi_j - \varphi_k| \, d\mu \le \int_X |\varphi_j - \varphi_k| \, d\mu \quad \text{for } j, k \in \mathbb{N} .$$

Therefore $(\chi_A \varphi_j)_{j \in \mathbb{N}}$ is an $\mathcal{L}_1$-Cauchy sequence in $\mathcal{S}(X, \mu, E)$. This shows that $\chi_A f$ is μ-integrable. ∎

For $f \in \mathcal{L}_1(X, \mu, E)$ and $A \in \mathcal{A}$, we define the **μ-integral of f over A** by

$$\int_A f \, d\mu := \int_X \chi_A f \, d\mu .$$

This is well defined by Lemma 2.13.

2.14 Remarks   Suppose $f \in \mathcal{L}_1(X, \mu, E)$ and $A \in \mathcal{A}$.

(a) $\int_A \cdot d\mu : \mathcal{L}_1(X, \mu, E) \to E$ is linear and continuous, and

$$\left| \int_A f \, d\mu \right| \le \int_A |f| \, d\mu = \|\chi_A f\|_1 .$$

(b) Suppose $\mathcal{B} := \mathcal{A} \,|\, A$ and $\nu := \mu \,|\, \mathcal{B}$. Then $\int_A f \, d\mu = \int_A f \,|\, A \, d\nu$.

Proof   The proof is simple and is left to the reader (Exercise 1). ∎

(c) If $E = \mathbb{R}$ and $f \geq 0$, the map

$$\mathcal{A} \to [0, \infty) , \quad A \mapsto \int_A f \, d\mu$$

is a finite measure (see Exercise 11). ∎

2.15 Lemma *Suppose $f \in \mathcal{L}_1(X, \mu, E)$ and $g \in E^X$ satisfy $f = g$ μ-a.e. Then g also belongs to $\mathcal{L}_1(X, \mu, E)$, and $\int_X f \, d\mu = \int_X g \, d\mu$.*

Proof Suppose (φ_j) is an $\mathcal{L}_1$-Cauchy sequence in $\mathcal{S}(X, \mu, E)$ such that $\varphi_j \to f$ μ-a.e. Also let M and N be μ-null sets such that $\varphi_j \to f$ on M^c and $f = g$ on N^c. Then (φ_j) converges μ-a.e. to g, because $\varphi_j(x) \to g(x)$ holds for $x \in (M \cup N)^c$. Therefore g belongs to $\mathcal{L}_1(X, \mu, E)$, and $\int g \, d\mu = \lim_j \int \varphi_j \, d\mu = \int f \, d\mu$. ∎

2.16 Corollary

 (i) *Suppose $f \in E^X$ vanishes μ-a.e. Then f is μ-integrable with $\int_X f \, d\mu = 0$.*

 (ii) *Suppose $f, g \in \mathcal{L}_1(X, \mu, \mathbb{R})$ satisfy $f \leq g$ μ-a.e. Then $\int_X f \, d\mu \leq \int_X g \, d\mu$.*

Proof (i) This follows immediately from Lemma 2.15.

   (ii) Theorem 2.11(ii) and Lemma 2.15 imply $0 \leq \int_X (g - f) \, d\mu$, and therefore $\int_X f \, d\mu \leq \int_X g \, d\mu$. ∎

2.17 Proposition *For $f \in \mathcal{L}_1(X, \mu, E)$ and $\alpha > 0$, we have $\mu([\,|f| \geq \alpha]) < \infty$.*

Proof Lemma 2.5 shows there is an $\mathcal{L}_1$-Cauchy sequence (φ_j) in $\mathcal{S}(X, \mu, E)$ and a μ-measurable set A such that $\mu(A) \leq 1$ and (φ_j) converges uniformly on A^c to f. Because $|f|$ is μ-measurable, $B := A^c \cap [\,|f| \geq \alpha]$ belongs to $\mathcal{A}$. Also there is an $N \in \mathbb{N}$ such that $|\varphi_N(x) - f(x)| \leq \alpha/2$ for $x \in A^c$. Therefore

$$|\varphi_N(x)| \geq |f(x)| - |\varphi_N(x) - f(x)| \geq \alpha/2 \quad \text{for } x \in B .$$

In particular, B is contained in $[\varphi_N \neq 0]$. Thus $\mu(B) \leq \mu([\varphi_N \neq 0]) < \infty$, because φ_N is μ-simple. Since

$$[\,|f| \geq \alpha] = B \cup (A \cap [\,|f| \geq \alpha]) \subset B \cup A ,$$

it follows that $\mu([\,|f| \geq \alpha]) \leq \mu(B) + 1 < \infty$. ∎

Convergence in $\mathcal{L}_1$

For every integrable function f, there is an $\mathcal{L}_1$-Cauchy sequence of simple functions converging to it almost everywhere. We show next that *every* Cauchy sequence in $\mathcal{L}_1(X, \mu, E)$ has in fact a subsequence that converges almost everywhere to the sequence's $\mathcal{L}_1$ limit.

2.18 Theorem Let (f_j) be a sequence in $\mathcal{L}_1(X, \mu, E)$ converging to f in $\mathcal{L}_1(X, \mu, E)$.

(i) There is a subsequence $(f_{j_k})_{k \in \mathbb{N}}$ of (f_j) with the following properties:

(α) $f_{j_k} \to f$ μ-a.e. as $k \to \infty$.

(β) For every $\varepsilon > 0$, there is an $A_\varepsilon \in \mathcal{A}$ with $\mu(A_\varepsilon) < \varepsilon$ such that $(f_{j_k})_{k \in \mathbb{N}}$ converges uniformly on A_ε^c to f.

(ii) The integral $\int_X f_j \, d\mu$ converges to $\int_X f \, d\mu$ as $j \to \infty$.

Proof (i) It suffices to treat the case $f = 0$ because, if $f \neq 0$, we may consider the sequence $(f_j - f)_{j \in \mathbb{N}}$.

As in the proof of Lemma 2.5, there is a subsequence (g_k) of (f_j) such that $\|g_\ell - g_m\|_1 < 2^{-2\ell}$ for $m \geq \ell \geq 0$. The limit $m \to \infty$ gives $\|g_\ell\|_1 \leq 2^{-2\ell}$ for $\ell \in \mathbb{N}$. We set $B_\ell := [\, |g_\ell| \geq 2^{-\ell}\,]$. By Lemma 2.8, Proposition 2.4, and Proposition 1.9, B_ℓ belongs to $\mathcal{A}$, and we find

$$2^{-\ell} \mu(B_\ell) \leq \int_{B_\ell} |g_\ell| \, d\mu \leq \int_X |g_\ell| \, d\mu = \|g_\ell\|_1 \leq 2^{-2\ell} \quad \text{for } \ell \in \mathbb{N}$$

(compare Theorem 2.11(ii)). Therefore $\mu(B_\ell) \leq 2^{-\ell}$ for $\ell \in \mathbb{N}$. With $A_n := \bigcup_{k=0}^{\infty} B_{n+k}$, we have $\mu(A_n) \leq 2^{-n+1}$, and we find that $A := \bigcap_{n=0}^{\infty} A_n$ is a μ-null set. We verify easily that (g_k) converges to 0 uniformly on A_n^c and pointwise on A^c (in this connection see the proof of Lemma 2.5).

(ii) From Theorem 2.11(i) it follows that

$$\left| \int_X f_j \, d\mu - \int_X f \, d\mu \right| \leq \int_X |f_j - f| \, d\mu = \|f_j - f\|_1 \quad \text{for } j \in \mathbb{N} \,,$$

so the limit of the left-hand side as $j \to \infty$ is 0. ∎

2.19 Corollary For $f \in \mathcal{L}_1(X, \mu, E)$,

$$\|f\|_1 = 0 \Longleftrightarrow f = 0 \ \mu\text{-a.e.}$$

Proof "$\Rightarrow$" Because $\|f\|_1 = 0$, the sequence (f_j) with $f_j := 0$ for $j \in \mathbb{N}$ converges in $\mathcal{L}_1(X, \mu, E)$ to f. By Theorem 2.18 there is thus a subsequence $(f_{j_k})_{k \in \mathbb{N}}$ of (f_j) such that $f_{j_k} \to f$ μ-a.e. as $k \to \infty$. Therefore, $f = 0$ μ-a.e.

"$\Leftarrow$" By assumption, $|f| = 0$ μ-a.e.; the claim follows from Corollary 2.16(i). ∎

We conclude this section by illustrating its concepts and results in an especially simple situation.

2.20 Example (the space of summable series)   Let X denote either $\mathbb{N}$ or $\mathbb{Z}$, and let $\mathcal{H}^0$ be its 0-dimensional Hausdorff measure, or counting measure. The topology induced by $\mathbb{R}$ clearly transforms X into a σ-compact metric space in which every one-point set is open. Hence the topology of X coincides with $\mathfrak{P}(X)$: every subset of X is open. Consequently, every map of X is continuous in E, that is, $C(X, E) = E^X$.

It follows further that $\mathcal{B}(X) = \mathcal{P}(X)$, and that $\mathcal{H}^0$ is a massive Radon measure on X. Thus, by Theorem 1.17,

$$\mathcal{L}_0(X, \mathcal{H}^0, E) = C(X, E) = E^X .$$

In addition, $\mathcal{H}^0$ has no nonempty null sets. Hence $\mathcal{H}^0$-a.e. convergence is the same as pointwise convergence.

For $\varphi \in E^X$, we define the **support** of φ as the set

$$\operatorname{supp}(\varphi) := \left\{ x \in X \; ; \; \varphi(x) \neq 0 \right\} ,$$

and denote by $C_c(X, E)$ the space of continuous E-valued functions on X with compact support:

$$C_c(X, E) := \left\{ \varphi \in C(X, E) \; ; \; \operatorname{supp}(\varphi) \text{ is compact} \right\} .$$

Clearly $\varphi \in C(X, E)$ belongs to $C_c(X, E)$ if and only if $\operatorname{supp}(\varphi)$ is a finite set. Also, $C_c(X, E)$ is a vector subspace of $C(X, E)$, and we verify easily that $C_c(X, E) = \mathcal{S}(X, \mathcal{H}^0, E)$.

For $\varphi \in C_c(X)$, it follows from Remark 2.1(b) that

$$\int_X \varphi \, d\mathcal{H}^0 = \sum_{x \in \operatorname{supp}(\varphi)} \varphi(x) . \tag{2.4}$$

We now set

$$\ell_1(X, E) := \left\{ f \in E^X \; ; \; \sum_{x \in X} |f(x)| < \infty \right\} .$$

For $f \in \ell_1(X, E)$ and $n \in \mathbb{N}$, let

$$\varphi_n(x) := \begin{cases} f(x) & \text{if } |x| \leq n , \\ 0 & \text{if } |x| > n . \end{cases}$$

Then φ_n belongs to $C_c(X, E)$, and $\varphi_n \to f$ for $n \to \infty$. For $m > n$, we get from (2.4) that

$$\|\varphi_n - \varphi_m\|_1 = \sum_{n < |x| \leq m} |f(x)| .$$

Therefore (φ_n) is an $\mathcal{L}_1$-Cauchy sequence in $\mathcal{S}(X, \mathcal{H}^0, E)$, which shows that f belongs to $\mathcal{L}_1(X, \mathcal{H}^0, E)$. Therefore $\ell_1(X, E) \subset \mathcal{L}_1(X, \mathcal{H}^0, E)$, and

$$\int_X f \, d\mathcal{H}^0 = \sum_{x \in X} f(x) \quad \text{for } f \in \ell_1(X, E) . \tag{2.5}$$

Now let $f \in \mathcal{L}_1(X, \mathcal{H}^0, E)$. Then there exists an $\mathcal{L}_1$-Cauchy sequence (ψ_j) in $\mathcal{S}(X, \mu, E)$ — and therefore in $C_c(X, E)$ — that converges pointwise to f. By Lemma 2.8, $|f|$ belongs to $\mathcal{L}_1(X, \mathcal{H}^0, \mathbb{R})$, and

$$\|f\|_1 = \int_X |f| \, d\mathcal{H}^0 = \lim_{j \to \infty} \int_X |\psi_j| \, d\mathcal{H}^0 = \lim_{j \to \infty} \sum_{x \in X} |\psi_j(x)| \ .$$

Hence there is a $k \in \mathbb{N}$ such that

$$\left| \int_X |f| \, d\mathcal{H}^0 - \sum_{x \in X} |\psi_j(x)| \right| \leq 1 \quad \text{for } j \geq k \ .$$

This implies

$$\sum_{x \in X} |\psi_j(x)| \leq 1 + \int_X |f| \, d\mathcal{H}^0 =: K < \infty \quad \text{for } j \geq k \ .$$

Therefore for every $m \in \mathbb{N}$ we have

$$\sum_{|x| \leq m} |\psi_j(x)| \leq K \quad \text{for } j \geq k \ ,$$

from which, because $\psi_j \to f$ as $j \to \infty$, we obtain

$$\sum_{|x| \leq m} |f(x)| \leq K \quad \text{for } m \in \mathbb{N} \ .$$

Now Theorem II.7.7 implies that f belongs to $\ell_1(X, E)$ (and satisfies $\|f\|_1 \leq K$). Therefore we have shown that

$$\mathcal{L}_1(X, \mathcal{H}^0, E) = \ell_1(X, E) \ ,$$

whereupon we obtain from (2.5) the relation[3]

$$\|f\|_1 = \sum_{x \in X} |f(x)| \ .$$

Finally, it follows from Theorem 2.10 and Remark 2.2(a) that

$$\ell_1(X, E) := \left(\ell_1(X, E), \|\cdot\|_1 \right)$$

is a Banach space, the **space of summable** (E-valued) **series**.

If $E = \mathbb{K}$, it is customary to write $\ell_1(\mathbb{Z})$ and $\ell_1(\mathbb{N})$ for $\ell_1(X, \mathbb{K})$, and the abbreviation $\ell_1 := \ell_1(\mathbb{N})$ is also common.[4] ∎

---

[3]Note Theorem II.8.9.
[4]Compare Exercise II.8.6.

Exercises

1   Suppose $A \in \mathcal{A}$, $\mathcal{B} := \mathcal{A} \,|\, A$, and $\nu := \mu \,|\, \mathcal{B}$. Verify for $f \in E^X$ that

$$\chi_A f \in \mathcal{L}_1(X, \mu, E) \Longleftrightarrow f \,|\, A \in \mathcal{L}_1(A, \nu, E) \,.$$

For such an f, show that

$$\int_X \chi_A f \, d\mu = \int_A f \,|\, A \, d\nu \,.$$

2   Suppose (f_j) is a sequence in $\mathcal{L}_1(X, \mu, E)$ that converges uniformly to $f \in E^X$. Also suppose $\mu(X) < \infty$. Show that f belongs to $\mathcal{L}_1(X, \mu, E)$, that $f_j \to f$ in $\mathcal{L}_1(X, \mu, E)$, and that $\lim_j \int_X f_j \, d\mu = \int_X f \, d\mu$.

3   Verify that, for $f \in \mathcal{L}_1(X, \mu, \mathbb{R}^+)$,

$$\int_X f \, d\mu = \sup \left\{ \int_X \varphi \, d\mu \;;\; \varphi \in \mathcal{S}(X, \mu, \mathbb{R}^+),\ \varphi \le f \ \mu\text{-a.e.} \right\} \,.$$

4   Suppose X is an arbitrary nonempty set, $a \in X$, and δ_a is the Dirac measure with support in a. Show that $\mathcal{L}_1(X, \delta_a, \mathbb{R}) = \mathbb{R}^X$, and calculate $\int f \, d\delta_a$ for $\mathbb{R}^X$.

5   Let μ_F be the Lebesgue–Stieltjes measure of Exercise IX.4.10. Determine $\mathcal{L}_1(\mathbb{R}, \mu_F, \mathbb{K})$ and calculate $\int f \, d\mu_F$ for $f \in \mathcal{L}_1(\mathbb{R}, \mu_F, \mathbb{K})$.

6   Prove statements (ii) and (iii) of Theorem 2.11.

7   Suppose that $f \in \mathcal{L}_0(X, \mu, E)$ is bounded μ-a.e. and that $\mu(X) < \infty$. Prove or disprove that f is μ-integrable.

8   Suppose (f_j) is an *increasing* sequence in $\mathcal{L}_1(X, \mu, \mathbb{R})$ such that $f_j \ge 0$, and suppose it converges μ-a.e. to $f \in \mathcal{L}_1(X, \mu, \mathbb{R})$. Then $\int_X f_j \, d\mu \uparrow \int_X f \, d\mu$. (This is known as the **monotone convergence theorem** in $\mathcal{L}_1$).
(Hint: Show that (f_j) is a Cauchy sequence in $\mathcal{L}_1(X, \mu, \mathbb{R})$, and identify its limit.)

9   Let (f_j) be a sequence in $\mathcal{L}_1(X, \mu, \mathbb{R})$ with $f_j \ge 0$ μ-a.e. and $\sum_{j=0}^{\infty} f_j \in \mathcal{L}_1(X, \mu, \mathbb{R})$. Show that $\sum_{j=0}^{\infty} \int f_j \, d\mu = \int \left(\sum_{j=0}^{\infty} f_j \right) d\mu$. (Hint: Exercise 8.)

10   Suppose that $f \in \mathcal{L}_1(X, \mu, \mathbb{R})$ satisfies $f > 0$ μ-a.e. Show that $\int_A f \, d\mu > 0$ for every $A \in \mathcal{A}$ such that $\mu(A) > 0$.

11   Given $f \in \mathcal{L}_1(X, \mu, \mathbb{R})$ with $f \ge 0$, define $\varphi_f(A) := \int_A f \, d\mu$ for $A \in \mathcal{A}$. Prove:
(a) $(X, \varphi_f, \mathcal{A})$ is a finite measure space.
(b) $\mathcal{N}_\mu \subset \mathcal{N}_{\varphi_f}$.
(c) $\mathcal{N}_\mu = \mathcal{N}_{\varphi_f}$ if $f > 0$ μ-a.e.
In particular, show that $(X, \mathcal{A}, \varphi_f)$ is a complete finite measure space if $f > 0$ μ-a.e.
(Hints: (a) Exercise 9.   (b) Exercise 10.)

12   Suppose $f \in \mathcal{L}_1(X, \mu, \mathbb{R})$ satisfies $f > 0$ μ-a.e. and take $g \in \mathcal{L}_0(X, \mu, \mathbb{R})$. Show that g is φ_f-integrable if and only if gf is μ-integrable. In this case, show that

$$\int_X g \, d\varphi_f = \int_X fg \, d\mu \,.$$

13   For $f \in \mathcal{L}_1(X, \mu, \overline{\mathbb{R}}^+)$, prove the **Chebyshev inequality**:

$$\mu([f \geq \alpha]) \leq \frac{1}{\alpha} \int_X f \, d\mu \quad \text{for } \alpha > 0 \ .$$

14   Suppose $\mu(X) < \infty$ and let I be a perfect interval in $\mathbb{R}$. Also suppose $\varphi \in C^1(I, \mathbb{R})$ is convex. Prove **Jensen's inequality**, which says that if $f \in \mathcal{L}_1(X, \mu, \mathbb{R})$ satisfies $f(X) \subset I$ and $\varphi \circ f \in \mathcal{L}_1(X, \mu, \mathbb{R})$,

$$\varphi\left(\fint_X f \, d\mu \right) \leq \fint_X \varphi \circ f \, d\mu \ , \qquad \text{where} \qquad \fint_X f \, d\mu := \frac{1}{\mu(X)} \int_X f \, d\mu \ .$$

(Hints: Show that $\alpha := \fint f \, d\mu$ lies in I, so the bound $\varphi(y) \geq \varphi(\alpha) + \varphi'(\alpha)(y - \alpha)$ applies).

15   Suppose $f \in \mathcal{L}_1(X, \mu, E)$. Show that for every $\varepsilon > 0$ there is a $\delta > 0$ such that $\left| \int_A f \, d\mu \right| < \varepsilon$ for all $A \in \mathcal{A}$ with $\mu(A) < \delta$. (Hint: Consider Theorem 2.10.)

3 Convergence theorems

Lebesgue integration theory stands out in contrast to the Riemann theory of Chapter VI in that it contains very general and versatile criteria for the commutability of limit taking and integration. Thus the Bochner–Lebesgue integral is better suited to the needs of analysis than the (simpler) Riemann integral.

As usual, we suppose in the entire section that

- $(X, \mathcal{A}, \mu)$ is a complete σ-finite measure space;
  $E = (E, |\cdot|)$ is a Banach space.

Integration of nonnegative $\overline{\mathbb{R}}$-valued functions

In many applications of integration in mathematics, the natural sciences and other fields, real-valued functions play a prominent role. As a rule, one is interested in such cases in integrable functions, which is to say in finite integrals. However, we have already mentioned that the theory gains substantially in simplicity and elegance if it is made to encompass integrals over $\overline{\mathbb{R}}$-valued functions, ruling out infinite values neither for functions nor for integrals. As examples of results that gain from such an inclusive treatment we mention the monotone convergence theorem (Exercise 2.8 and Theorem 3.4) and the Fubini–Tonelli theorem on interchangeability of integrals (Theorem 6.11).

Because of the importance of the real-valued case, and because it offers useful additional results that rely on the total ordering of $\mathbb{R}$ and $\overline{\mathbb{R}}$, we will now develop, to complement the Bochner–Lebesgue integral, an integration theory for $\overline{\mathbb{R}}$-valued — in particular, real-valued — functions.[1]

According to Theorem 1.12, there is for every $f \in \mathcal{L}_0(X, \mu, \overline{\mathbb{R}}^+)$ an increasing sequence (f_j) in $\mathcal{S}(X, \mu, \mathbb{R}^+)$ that converges pointwise to f. It is natural to define the integral of f as the limit in $\overline{\mathbb{R}}^+$ of the increasing sequence $(\int_X f_j \, d\mu)_{j \in \mathbb{N}}$. This makes sense if we can ensure that the limit does not depend on the choice of (f_j).

3.1 Lemma *Suppose $\varphi_j, \psi \in \mathcal{S}(X, \mu, \mathbb{R}^+)$ for $j \in \mathbb{N}$. Also suppose (φ_j) is increasing and $\psi \leq \lim_j \varphi_j$. Then*

$$\int_X \psi \, d\mu \leq \lim_j \int_X \varphi_j \, d\mu \ .$$

Proof Let $\sum_{j=0}^m \alpha_j \chi_{A_j}$ be the normal form of ψ and fix $\lambda > 1$. For $k \in \mathbb{N}$, define $B_k := [\lambda \varphi_k \geq \psi]$. Because (φ_k) is increasing and $\lambda > 1$, we have $B_k \subset B_{k+1}$ for

---

[1]The theory of $\overline{\mathbb{R}}$-valued functions is the centerpiece of practically all textbooks on integration theory. It is in some ways simpler than the more general Bochner–Lebesgue theory, and suffices if one is only interested in real- and complex-valued functions, but it is inadequate for the needs of modern higher analysis, which is why we opted for a more general approach.

$k \in \mathbb{N}$ and $\bigcup_{k \in \mathbb{N}} B_k = X$. The continuity of measures from below then implies

$$\int_X \psi \, d\mu = \sum_{j=0}^{m} \alpha_j \mu(A_j) = \lim_k \sum_{j=0}^{m} \alpha_j \mu(A_j \cap B_k) = \lim_k \int_X \psi \chi_{B_k} \, d\mu \ .$$

By the definition of B_k, we have $\lambda \varphi_k \geq \psi \chi_{B_k}$, and we obtain

$$\int_X \psi \, d\mu = \lim_k \int_X \psi \chi_{B_k} \, d\mu \leq \lambda \lim_k \int_X \varphi_k \, d\mu \ .$$

Taking the limit $\lambda \downarrow 1$ now finishes the proof. ∎

3.2 Corollary *Suppose (φ_j) and (ψ_j) are increasing sequences in $\mathcal{S}(X, \mu, \mathbb{R}^+)$ such that $\lim_j \varphi_j = \lim_j \psi_j$. Then*

$$\lim_j \int_X \varphi_j \, d\mu = \lim_j \int_X \psi_j \, d\mu \qquad in \ \overline{\mathbb{R}}^+ \ .$$

Proof By assumption, $\psi_k \leq \lim_j \psi_j = \lim_j \varphi_j$ for $k \in \mathbb{N}$. By Lemma 3.1, we get

$$\int_X \psi_k \, d\mu \leq \lim_j \int_X \varphi_j \, d\mu \quad \text{for } k \in \mathbb{N} \ ,$$

and, as $k \to \infty$,

$$\lim_k \int_X \psi_k \, d\mu \leq \lim_j \int_X \varphi_j \, d\mu \ .$$

Interchanging (φ_j) and (ψ_j), we obtain the opposite inequality, and hence the desired equality. ∎

Suppose (φ_j) is an increasing sequence in $\mathcal{S}(X, \mu, \mathbb{R}^+)$ that converges pointwise to $f \in \mathcal{L}_0(X, \mu, \overline{\mathbb{R}}^+)$. We call

$$\int_X f \, d\mu := \lim_j \int_X \varphi_j \, d\mu$$

the (**Lebesgue**) **integral of f over X with respect to the measure** μ. For $A \in \mathcal{A}$,

$$\int_A f \, d\mu := \int_X \chi_A f \, d\mu$$

is the (**Lebesgue**) **integral of f over the measurable set A**.

3.3 Remarks (a) $\int_A f \, d\mu$ is well defined for all $f \in \mathcal{L}_0(X, \mu, \overline{\mathbb{R}}^+)$ and $A \in \mathcal{A}$.
Proof This follows from Theorem 1.12 and Corollary 3.2. ∎

(b) For $f, g \in \mathcal{L}_0(X, \mu, \overline{\mathbb{R}}^+)$ such that $f \leq g$ μ-a.e., we have $\int_X f \, d\mu \leq \int_X g \, d\mu$.

(c) For $f \in \mathcal{L}_0(X, \mu, \overline{\mathbb{R}}^+)$, these statements are equivalent:

(i) $\int_X f \, d\mu = 0$.

(ii) $[f > 0]$ is a μ-null set.

(iii) $f = 0$ μ-a.e.

Proof "(i)$\Rightarrow$(ii)" We set $A := [f > 0]$ and $A_j := [f > 1/j]$ for $j \in \mathbb{N}^\times$. Then (A_j) is an increasing sequence in $\mathcal{A}$ such that $A = \bigcup_j A_j$. Also $\chi_{A_j} \leq jf$. It follows that

$$0 \leq \mu(A_j) = \int_X \chi_{A_j} \, d\mu \leq j \int_X f \, d\mu = 0 \quad \text{for } j \in \mathbb{N}^\times ,$$

and continuity from below implies $\mu(A) = \lim_j \mu(A_j) = 0$.

"(ii)$\Rightarrow$(iii)" is clear.

"(iii)$\Rightarrow$(i)" Let N be a μ-null set with $f(x) = 0$ for $x \in N^c$. Then[2] $f\chi_{N^c} = 0$ and $f\chi_N \leq \infty \chi_N$. Together with the definition of the integral (see also (d) below), this yields

$$0 \leq \int_X f \, d\mu = \int_X f\chi_N \, d\mu + \int_X f\chi_{N^c} \, d\mu \leq \infty \mu(N) = 0 . \quad \blacksquare$$

(d) Suppose $f, g \in \mathcal{L}_0(X, \mu, \overline{\mathbb{R}}^+)$ and $\alpha \in [0, \infty]$. Then

$$\int_X (\alpha f + g) \, d\mu = \alpha \int_X f \, d\mu + \int_X g \, d\mu .$$

Proof We consider the case $\alpha = \infty$ and $g = 0$. Letting $\varphi_j := j\chi_{[f>0]}$ for $j \in \mathbb{N}$, we have $f_j \uparrow \infty f$, and hence

$$\int_X (\infty f) \, d\mu = \begin{cases} 0 & \text{if } \mu([f > 0]) = 0 , \\ \infty & \text{if } \mu([f > 0]) > 0 . \end{cases}$$

From (c), it now follows that $\int_X (\infty f) \, d\mu = \infty \int_X f \, d\mu$. The remaining statements follow easily from the definition of the integral and are left as exercises. $\blacksquare$

(e) (i) Suppose $f \in \mathcal{L}_0(X, \mu, \mathbb{R}^+)$ has a finite Lebesgue integral $\int_X f \, d\mu$. Then f belongs to $\mathcal{L}_1(X, \mu, \mathbb{R}^+)$ and the Lebesgue integral of f over X coincides with the Bochner–Lebesgue integral.

(ii) For $f \in \mathcal{L}_1(X, \mu, \mathbb{R}^+)$, the Lebesgue integral $\int_X f \, d\mu$ is finite and agrees with the Bochner–Lebesgue integral.

Proof (i) Theorem 1.12 guarantees the existence of a sequence (φ_j) in $\mathcal{S}(X, \mu, \mathbb{R}^+)$ such that $\varphi_j \uparrow f$. By assumption, there exists for every $\varepsilon > 0$ an $N \in \mathbb{N}$ such that $\int_X f \, d\mu - \int_X \varphi_j \, d\mu < \varepsilon$ for $j \geq N$. For $k \geq j \geq N$, the finiteness of $\int_X f \, d\mu$ now gives

$$\int_X |\varphi_k - \varphi_j| \, d\mu = \int_X (\varphi_k - \varphi_j) \, d\mu \leq \int_X (f - \varphi_j) \, d\mu = \int_X f \, d\mu - \int_X \varphi_j \, d\mu < \varepsilon .$$

Therefore (φ_j) is an $\mathcal{L}_1$-Cauchy sequence in $\mathcal{S}(X, \mu, \mathbb{R}^+)$. This shows that f belongs to $\mathcal{L}_1(X, \mu, \mathbb{R}^+)$. The second statement is a consequence of Exercise 2.8.

(ii) This follows from Theorem 1.12 and Exercise 2.8. $\blacksquare$

---

[2] We recall Convention (2.1).

(f) For every $f \in \mathcal{L}_0(X, \mu, \overline{\mathbb{R}}^+)$, we have

$$\int_X f \, d\mu = \sup\left\{ \int_X \varphi \, d\mu \; ; \; \varphi \in \mathcal{S}(X, \mu, \mathbb{R}^+) \text{ with } \varphi \le f \ \mu\text{-a.e.} \right\}. \ \blacksquare$$

The monotone convergence theorem

We now prove a significant strengthening of the monotone convergence theorem stated in Exercise 2.8 for functions in $\mathcal{L}_1(X, \mu, \mathbb{R})$. We will see that for increasing sequences in $\mathcal{L}_0(X, \mu, \overline{\mathbb{R}}^+)$, Lebesgue integration commutes with taking the limit.

3.4 Theorem (monotone convergence)   *Suppose (f_j) is an increasing sequence in $\mathcal{L}_0(X, \mu, \overline{\mathbb{R}}+)$. Then*

$$\int_X \lim_j f_j \, d\mu = \lim_j \int_X f_j \, d\mu \quad \text{in } \overline{\mathbb{R}}^+ .$$

Proof   (i) Set $f := \lim_j f_j$. By Proposition 1.11, f belongs to $\mathcal{L}_0(X, \mu, \overline{\mathbb{R}}^+)$, and $f_j \le f$ for $j \in \mathbb{N}$. By Remark 3.3(b), then, we have $\int f_j \, d\mu \le \int f \, d\mu$ for $j \in \mathbb{N}$, and hence $\lim_j \int f_j \, d\mu \le \int f \, d\mu$.

(ii) Suppose $\varphi \in \mathcal{S}(X, \mu, \mathbb{R}^+)$ with $\varphi \le f$. Take $\lambda > 1$ and set $A_j := [\lambda f_j \ge \varphi]$ for $j \in \mathbb{N}$. Then (A_j) is an increasing sequence in $\mathcal{A}$ with $\bigcup_j A_j = X$ and $\lambda f_j \ge \varphi \chi_{A_j}$. Moreover, $\varphi \chi_{A_j} \uparrow \varphi$, so

$$\int_X \varphi \, d\mu = \lim_j \int_X \varphi \chi_{A_j} \, d\mu \le \lambda \lim_j \int_X f_j \, d\mu .$$

Taking the limit $\lambda \downarrow 1$ we get $\int_X \varphi \, d\mu \le \lim_j \int_X f_j \, d\mu$ for every μ-simple function φ with $\varphi \le f$. By Remark 3.3(f), it follows that $\int_X f \, d\mu \le \lim_j \int_X f_j \, d\mu$, and we are done. $\blacksquare$

3.5 Corollary   *Suppose (f_j) is a sequence in $\mathcal{L}_0(X, \mu, \overline{\mathbb{R}}^+)$. Then*

$$\sum_{j=0}^{\infty} \int_X f_j \, d\mu = \int_X \left(\sum_{j=0}^{\infty} f_j \right) d\mu \quad \text{in } \overline{\mathbb{R}}^+ .$$

Proof   This follows from Corollary 1.13(iii) and Theorem 3.4. $\blacksquare$

3.6 Remarks   (a) The conclusion of Theorem 3.4 can fail if the sequence is not increasing.

Proof   Take $f_j := (1/j)\chi_{[0,j]}$ for $j \in \mathbb{N}^\times$. Then (f_j) is a (nonincreasing) sequence in $\mathcal{S}(\mathbb{R}, \lambda_1, \mathbb{R}^+)$ that converges uniformly to 0. But the sequence $\int f_j \, d\lambda_1$ does not converge to 0, because $\int f_j \, d\lambda_1 = 1$ for $j \in \mathbb{N}^\times$. $\blacksquare$

(b) Suppose $a_{j,k} \in \mathbb{R}^+$ for $j, k \in \mathbb{N}$. Then

$$\sum_{j=0}^{\infty} \sum_{k=0}^{\infty} a_{jk} = \sum_{k=0}^{\infty} \sum_{j=0}^{\infty} a_{jk} \ .$$

Proof We set $(X, \mu) := (\mathbb{N}, \mathcal{H}^0)$ and define $f_j : X \to \mathbb{R}^+$ by $f_j(k) := a_{jk}$ for $j, k \in \mathbb{N}$. Then (f_j) is a sequence in $\mathcal{L}_0(X, \mathcal{H}^0, \mathbb{R}^+)$ (see Example 2.20), and the claim follows from Corollary 3.5. ∎

For nonnegative double series, this result is stronger than Theorem II.8.10, because it is no longer assumed that $\sum_{jk} a_{jk}$ is summable.

Fatou's lemma

We now prove a generalization of the monotone convergence theorem for arbitrary (not necessarily increasing) sequences in $\mathcal{L}_0(X, \mu, \overline{\mathbb{R}}^+)$.

3.7 Theorem (Fatou's lemma) *For every sequence (f_j) in $\mathcal{L}_0(X, \mu, \overline{\mathbb{R}}^+)$, we have*

$$\int_X \left(\varliminf_j f_j \right) d\mu \leq \varliminf_j \int_X f_j \, d\mu \quad \text{in } \overline{\mathbb{R}}^+ \ .$$

Proof Set $g_j := \inf_{k \geq j} f_k$. By Proposition 1.11, g_j belongs to $\mathcal{L}_0(X, \mu, \overline{\mathbb{R}}^+)$, and the increasing sequence (g_j) converges to $\varliminf_j f_j$. From Theorem 3.4 we then get $\lim_j \int g_j \, d\mu = \int (\varliminf_j f_j) \, d\mu$. Also $g_j \leq f_k$, and therefore $\int g_j \, d\mu \leq \int f_k \, d\mu$ for $k \geq j$. It follows that $\int g_j \, d\mu \leq \inf_{k \geq j} \int f_k \, d\mu$, and taking the limit $j \to \infty$ finishes the proof. ∎

3.8 Corollary *Suppose (f_j) is a sequence in $\mathcal{L}_0(X, \mu, \overline{\mathbb{R}}^+)$ and $g \in \mathcal{L}_0(X, \mu, \overline{\mathbb{R}}^+)$ satisfies $\int_X g \, d\mu < \infty$ with $f_j \leq g$ μ-a.e. for $j \in \mathbb{N}$. Then*[3]

$$\varlimsup_j \int_X f_j \, d\mu \leq \int_X \left(\varlimsup_j f_j \right) d\mu \quad \text{in } \overline{\mathbb{R}}^+ \ .$$

Proof Suppose N is a μ-null set such that $f_j(x) \leq g(x)$ for $x \in N^c$ and $j \in \mathbb{N}$. Then $f_j \leq g + \infty \chi_N$ on X, and $\int_X (g + \infty \chi_N) \, d\mu = \int_X g \, d\mu$ (see Remarks 3.3(c) and (d)). Therefore we can assume without losing generality that $f_j \leq g$ for $j \in \mathbb{N}$. We set $g_j := g - f_j$ and obtain from Fatou's lemma that

$$\int_X \left(\varliminf_j g_j \right) d\mu = \int_X g \, d\mu - \int_X \left(\varlimsup_j f_j \right) d\mu \leq \varliminf_j \int_X g_j \, d\mu = \int_X g \, d\mu - \varlimsup_j \int_X f_j \, d\mu \ .$$

The claim now follows because $\int_X g \, d\mu < \infty$. ∎

---

[3]The assumption $\int_X g \, d\mu < \infty$ cannot be relaxed (see Exercise 1).

As a first application, we prove a fundamental characterization of integrable functions.

3.9 Theorem For $f \in \mathcal{L}_0(X, \mu, E)$, the following are equivalent:

(i) $f \in \mathcal{L}_1(X, \mu, E)$;

(ii) $|f| \in \mathcal{L}_1(X, \mu, \mathbb{R})$;

(iii) $\int_X |f| \, d\mu < \infty$.

If these conditions are satisfied, then $\left| \int_X f \, d\mu \right| \leq \|f\|_1 < \infty$.

Proof "(i)$\Rightarrow$(ii)" follows from Lemma 2.8, and "(ii)$\Rightarrow$(iii)" is clear. "(iii)$\Rightarrow$(ii)" was proved in Remark 3.3(e).

"(ii)$\Rightarrow$(i)" Suppose (φ_j) is a sequence in $\mathcal{S}(X, \mu, E)$ converging to f μ-a.e. Set $A_j := [\,|\varphi_j| \leq 2\,|f|\,]$ and $f_j := \varphi_j \chi_{A_j}$ for $j \in \mathbb{N}$. Theorem 1.7 and Proposition 1.9 show that A_j belongs to $\mathcal{A}$. Thus (f_j) is a sequence in $\mathcal{S}(X, \mu, E)$.

Take $N \in \mathcal{A}$ such that $\mu(N) = 0$ and $\varphi_j(x) \to f(x)$ for $x \in N^c$. If $f(x) \neq 0$ for some $x \in N^c$, there exists $k := k(x) \in \mathbb{N}$ such that $|\varphi_j(x) - f(x)| \leq 3\,|f(x)|$ for $j \geq k$. Therefore $x \in N^c \cap [\,|f| > 0\,]$ belongs to A_j for $j \geq k(x)$. This implies $f_j(x) = \varphi_j(x)$ for $j \geq k(x)$, and therefore $f_j(x) \to f(x)$ for $x \in N^c \cap [\,|f| > 0\,]$. If $f(x) = 0$ for some $x \in N^c$, then likewise $f_j(x) \to f(x)$ for $j \to \infty$. Because x belongs to A_k for some $k \in \mathbb{N}$, we find $f_k(x) = \varphi_k(x) = 0$ because $|\varphi_k(x)| \leq 2\,|f(x)| = 0$. For $x \notin A_k$, we likewise have $f_k(x) = \chi_{A_k}(x)\varphi_k(x) = 0$. This implies $|f - f_j| \to 0$ μ-a.e. Now clearly $|f - f_j| \leq 3\,|f|$ for $j \in \mathbb{N}$, so Corollary 3.8 implies

$$\overline{\lim_j} \int_X |f - f_j| \, d\mu \leq \int_X \overline{\lim_j} |f - f_j| \, d\mu = 0 \ .$$

Therefore we can find for every $\varepsilon > 0$ an $m \in \mathbb{N}$ such that $\int |f - f_j| \, d\mu < \varepsilon/2$ for $j \geq m$. It follows that, for $j, k \in \mathbb{N}$ with $j, k \geq m$,

$$\|f_j - f_k\|_1 = \int_X |f_j - f_k| \, d\mu \leq \int_X |f_j - f| \, d\mu + \int_X |f - f_k| \, d\mu < \varepsilon \ .$$

Hence (f_j) is an $\mathcal{L}_1$-Cauchy sequence in $\mathcal{S}(X, \mu, E)$, and f is μ-integrable.

The last statement follows from Theorem 2.11(i). ∎

3.10 Conclusions (a) Let $f \in \mathcal{L}_0(X, \mu, E)$, and suppose there is a sequence (f_j) in $\mathcal{L}_1(X, \mu, E)$ such that $f_j \to f$ μ-a.e. and $\underline{\lim}_j \|f_j\|_1 < \infty$. Then f belongs to $\mathcal{L}_1(X, \mu, E)$, and $\|f\|_1 \leq \underline{\lim}_j \|f_j\|_1$.

Proof By Lemma 2.15, we can assume that (f_j) converges to f on all of X. Using Fatou's lemma, we obtain

$$\int_X |f| \, d\mu = \int_X \underline{\lim_j} |f_j| \, d\mu \leq \underline{\lim_j} \int_X |f_j| \, d\mu < \infty \ ,$$

and the claim follows by Theorem 3.9. ∎

(b) Let (f_j) be a sequence in $\mathcal{L}_1(X, \mu, \mathbb{R}^+)$. Suppose there is an $f \in \mathcal{L}_1(X, \mu, \mathbb{R})$ such that

$$f_j \to f \ \mu\text{-a.e.} \quad \text{and} \quad \int_X f_j \, d\mu \to \int_X f \, d\mu \ (j \to \infty) .$$

Then[4] (f_j) converges in $\mathcal{L}_1(X, \mu, \mathbb{R})$ to f.

Proof We can assume here too that (f_j) converges to f on all of X. Then $f \geq 0$ and $|f_j - f| \leq f_j + f$. From Theorem 3.7, it follows that

$$2 \int_X f \, d\mu = \int_X \varliminf_j (f_j + f - |f_j - f|) \, d\mu \leq \varliminf_j \int_X (f_j + f - |f_j - f|) \, d\mu$$

$$= 2 \int_X f \, d\mu - \varlimsup_j \int_X |f_j - f| \, d\mu .$$

According to Theorem 3.9, $\int_X f \, d\mu$ is finite, and we find $\lim_j \int_X |f_j - f| \, d\mu = 0$. ∎

Integration of $\overline{\mathbb{R}}$-valued functions

The decomposition of an $\overline{\mathbb{R}}$-valued function into its positive and negative parts allows us also to extend the Lebesgue integral to measurable $\overline{\mathbb{R}}$-valued functions admitting negative values. We say that $f \in \mathcal{L}_0(X, \mu, \overline{\mathbb{R}})$ is **Lebesgue integrable with respect to** μ if $\int_X f^+ \, d\mu < \infty$ and $\int_X f^- \, d\mu < \infty$. In this case,

$$\int_X f \, d\mu := \int_X f^+ \, d\mu - \int_X f^- \, d\mu$$

is called the **(Lebesgue) integral over X with respect to the measure** μ.

3.11 Remarks **(a)** For $f \in \mathcal{L}_0(X, \mu, \overline{\mathbb{R}})$, these three statements are equivalent:

(i) f is Lebesgue integrable with respect to μ.

(ii) $\int_X |f| \, d\mu < \infty$;

(iii) There exists $g \in \mathcal{L}_1(X, \mu, \mathbb{R})$ such that $|f| \leq g$ μ-a.e.

Proof "(i)$\Rightarrow$(ii)" This is a consequence of $|f| = f^+ + f^-$.

"(ii)$\Rightarrow$(iii)" Theorem 3.9 says that $|f| \in \mathcal{L}_1(X, \mu, \mathbb{R})$. Hence (iii) holds with $g = |f|$.

"(iii)$\Rightarrow$(i)" This follows from $f^+ \vee f^- \leq |f| \leq g$ and Remark 3.3(b). ∎

(b) Suppose $f \in \mathcal{L}_0(X, \mu, \mathbb{R})$. Then f is Lebesgue integrable with respect to μ if and only if f is μ-integrable. In that case, the Lebesgue integral of f over X equals the Bochner–Lebesgue integral. In other words, if we consider real-valued maps, the definition of Lebesgue integrability of $\overline{\mathbb{R}}$-valued functions is consistent with the definition from Section 2.[5]

Proof This follows from (a), Theorem 3.9, and Remark 3.3(e). ∎

---

[4]Compare the statement of Theorem 2.18.

[5]See also Corollary 2.12(iii).

(c) If $f \in \mathcal{L}_0(X, \mu, \overline{\mathbb{R}})$ is Lebesgue integrable with respect to μ, then $[\,|f| = \infty]$ is a μ-null set.

Proof The assumption implies that $A := [\,|f| = \infty]$ is μ-measurable and also that $\int_X |f|\, d\mu < \infty$. Further we have $\infty \chi_A \le |f|$, and we find by Remarks 3.3(b) and (d) that

$$\infty \mu(A) = \int_X (\infty \chi_A)\, d\mu \le \int_X |f|\, d\mu < \infty .$$

Therefore $\mu(A) = 0$. ∎

Lebesgue's dominated convergence theorem

We now prove an extremely versatile and practical theorem about exchanging limits and integrals, proved by Henri Lebesgue. It is one of the cornerstones of Lebesgue integration theory and has countless applications.

3.12 Theorem (dominated convergence[6]) *Let (f_j) be a sequence in $\mathcal{L}_1(X, \mu, E)$ and suppose that there exists $g \in \mathcal{L}_1(X, \mu, \mathbb{R})$ such that*

(a) $$|f_j| \le g \quad \mu\text{-a.e. for } j \in \mathbb{N}.$$

Suppose also that, for some $f \in E^X$,

(b) $$f_j \to f \quad \mu\text{-a.e. for } j \to \infty.$$

Then f is μ-integrable, $f_j \to f$ in $\mathcal{L}_1(X, \mu, E)$, and $\int_X f_j\, d\mu \to \int_X f\, d\mu$ in E.

Proof Define

$$g_j := \sup_{k, \ell \ge j} |f_k - f_\ell|$$

for $j \in \mathbb{N}$. By Proposition 1.11, (g_j) is a sequence in $\mathcal{L}_0(X, \mu, \overline{\mathbb{R}}^+)$ that converges μ-a.e. to 0. Also $|f_k - f_\ell| \le 2g$ μ-a.e. for $k, \ell \in \mathbb{N}$, and hence $|g_j| \le 2g$ μ-a.e. for $j \in \mathbb{N}$. From Corollary 3.8 it follows that

$$0 \le \overline{\lim_j} \int_X g_j\, d\mu \le \int_X \overline{\lim_j}\, g_j\, d\mu = 0 .$$

Therefore $\left(\int_X g_j\, d\mu \right)_{j \in \mathbb{N}}$ is a (decreasing) null sequence. This means that for every $\varepsilon > 0$ there exists $N \in \mathbb{N}$ such that

$$\int_X |f_k - f_\ell|\, d\mu \le \int_X \sup_{k, \ell \ge j} |f_k - f_\ell|\, d\mu < \varepsilon \quad \text{for } k, \ell \ge j \ge N .$$

Hence (f_j) is a Cauchy sequence in $\mathcal{L}_1(X, \mu, E)$, and the claim follows from the completeness of $\mathcal{L}_1(X, \mu, E)$ and Theorem 2.18. ∎

---

[6] Also referred to as "Lebesgue's theorem".

3.13 Remark   The example in Remark 3.6(a) shows that the existence of an integrable dominating function g is essential for Theorem 3.12. ∎

As a first application of the dominated convergence theorem, we prove a simple criterion for the integrability of a measurable function.

3.14 Theorem (integrability criterion) *Suppose $f \in \mathcal{L}_0(X, \mu, E)$ and $g \in \mathcal{L}_1(X, \mu, \mathbb{R})$ satisfy $|f| \leq g$ μ-a.e. Then f belongs to $\mathcal{L}_1(X, \mu, E)$.*

Proof   Let (φ_j) be a sequence in $\mathcal{S}(X, \mu, E)$ such that $\varphi_j \to f$ μ-a.e. as $j \to \infty$. Set $A_j := [|\varphi_j| \leq 2g]$ and $f_j := \chi_{A_j} \varphi_j$ for $j \in \mathbb{N}$. Then (f_j) is a sequence in $\mathcal{S}(X, \mu, E)$ that converges μ-a.e. to f (see the proof of Theorem 3.9). Because $|f_j| \leq 2g$ for $j \in \mathbb{N}$, the claim follows from the dominated convergence theorem. ∎

3.15 Corollary

(i) *Take $f \in \mathcal{L}_1(X, \mu, E)$, $g \in \mathcal{L}_0(X, \mu, \mathbb{K})$, and $\alpha \in [0, \infty)$ with $|g| \leq \alpha$ μ-a.e. Then gf is μ-integrable, and*

$$\left| \int_X gf \, d\mu \right| \leq \alpha \|f\|_1 .$$

(ii) *Take $f \in \mathcal{L}_0(X, \mu, E)$ and $\alpha \in [0, \infty)$. If $|f| \leq \alpha$ μ-a.e. and $\mu(X) < \infty$, then f is μ-integrable with*

$$\left| \int_X f \, d\mu \right| \leq \|f\|_1 \leq \alpha \mu(X) .$$

(iii) *Let X be a σ-compact metric space and μ a complete Radon measure on X. Suppose that $f \in C(X, E)$ and that $K \subset X$ is compact. Then $\chi_K f$ belongs to $\mathcal{L}_1(X, \mu, E)$, and*

$$\left| \int_K f \, d\mu \right| \leq \|\chi_K f\|_\infty \, \mu(K) .$$

Proof   (i) By Remark 1.2(d), gf is μ-measurable. Also $|gf| \leq \alpha |f|$ μ-a.e., and $\alpha |f|$ is μ-integrable. Hence Theorem 3.14 shows that gf is μ-integrable; Theorem 2.11(i) and Corollary 2.16(ii) imply

$$\left| \int_X gf \, d\mu \right| \leq \int_X |gf| \, d\mu \leq \int_X \alpha |f| \, d\mu = \alpha \|f\|_1 .$$

(ii) Since $\mu(X)$ is finite, χ_X belongs to $\mathcal{L}_1(X, \mu, \mathbb{R})$. By Theorem 1.7(i), $|f|$ is μ-measurable. Therefore (i) shows (with $g := |f|$ and $f := \chi_X$) that $|f|$ is

μ-integrable and that

$$\int_X |f|\, d\mu \le \alpha \, \|\chi_X\|_1 = \alpha\mu(X) < \infty \ .$$

The claim now follows from Theorem 3.9.

(iii) According to Theorem 1.17, f is μ-measurable. Moreover χ_K is μ-simple, because $\mu(K)$ is finite by Remark IX.5.3(a). Therefore $\chi_K f$ is μ-measurable, and the claim follows from (ii) with $\alpha := \max_{x \in K} |f(x)|$. ∎

When dealing with a function not defined on all of X, it is occasionally useful in the theory of integration to extend its definition by setting it equal to 0 where it is not already defined. Measurability and integrability questions can then be explored with respect to the measure space $(X, \mathcal{A}, \mu)$. To that end, we set forth the following conventions.

For $f : \operatorname{dom}(f) \subset X \to E$, define the **trivial extension** $\widetilde{f} \in E^X$ of f to X by

$$\widetilde{f}(x) := \begin{cases} f(x) & \text{if } x \in \operatorname{dom}(f) \ , \\ 0 & \text{if } x \notin \operatorname{dom}(f) \ . \end{cases}$$

We say that f is **μ-measurable** or **μ-integrable** if $\widetilde{f}$ belongs to $\mathcal{L}_0(X, \mu, E)$ or $\mathcal{L}_1(X, \mu, E)$, respectively. If f is μ-integrable, we set $\int_X f\, d\mu := \int_X \widetilde{f}\, d\mu$.

3.16 Theorem (termwise integration of series) Suppose (f_j) is a sequence in $\mathcal{L}_1(X, \mu, E)$ such that $\sum_{j=0}^{\infty} \int_X |f_j|\, d\mu < \infty$. Then $\sum_j f_j$ converges absolutely μ-a.e., $\sum_j f_j$ is μ-integrable, and

$$\int_X \Big(\sum_{j=0}^{\infty} f_j \Big) d\mu = \sum_{j=0}^{\infty} \int_X f_j\, d\mu \ .$$

Proof (i) By Theorem 1.7(i) and Corollary 1.13(iii), the $\overline{\mathbb{R}}$-valued function $g := \sum_{j=0}^{\infty} |f_j|$ is μ-measurable. Corollary 3.5 implies

$$\int_X g\, d\mu = \sum_{j=0}^{\infty} \int_X |f_j|\, d\mu < \infty \ .$$

It therefore follows from Remarks 3.11(a) and (c) that $[g = \infty]$ is a μ-null set, which proves the absolute convergence of $\sum_j f_j$ for almost every $x \in X$.

(ii) Set $g_k := \sum_{j=0}^{k} f_j$ and $f(x) := \sum_{j=0}^{\infty} f_j(x)$ for $x \in [g < \infty]$. The sequence (g_k) converges μ-a.e. to f and we have the bounds $|g_k| \le \sum_{j=0}^{k} |f_j| \le g$. By the dominated convergence theorem, $\widetilde{f}$ belongs to $\mathcal{L}_1(X, \mu, E)$ and

$$\sum_{j=0}^{\infty} \int_X f_j\, d\mu = \lim_{k \to \infty} \int_X g_k\, d\mu = \int_X \lim_{k \to \infty} g_k\, d\mu = \int_X \Big(\sum_{j=0}^{\infty} f_j \Big) d\mu \ . \quad ∎$$

Parametrized integrals

As another application of the dominated convergence theorem, we investigate the continuity and differentiability of parametrized integrals.

3.17 Theorem (continuity of parametrized integrals) *Suppose M is a metric space and $f : X \times M \to E$ satisfies*

(a) $f(\cdot, m) \in \mathcal{L}_1(X, \mu, E)$ *for every* $m \in M$;

(b) $f(x, \cdot) \in C(M, E)$ *for μ-almost every* $x \in X$;

(c) *there exists* $g \in \mathcal{L}_1(X, \mu, E)$ *such that* $|f(x, m)| \leq g(x)$ *for* $(x, m) \in X \times M$.

Then

$$F : M \to E , \quad m \mapsto \int_X f(x, m) \, \mu(dx)$$

is well defined and continuous.

Proof The first statement follows immediately from (a). Suppose $m \in M$, and let (m_j) be a sequence in M converging to m. We set $f_j := f(\cdot, m_j)$ for $j \in \mathbb{N}$. From (b), it follows that $f_j \to f$ μ-a.e. Therefore by (a) and (c), we can apply the dominated convergence theorem to the sequence (f_j), and we find

$$\lim_{j \to \infty} F(m_j) = \lim_{j \to \infty} \int_X f_j \, d\mu = \int_X \lim_{j \to \infty} f_j \, d\mu = \int_X f(x, m) \, \mu(dx) = F(m) .$$

The claim now follows from Theorem III.1.4. ∎

3.18 Theorem (differentiability of parametrized integrals) *Suppose U is open in $\mathbb{R}^n$, or $U \subset \mathbb{K}$ is perfect and convex, and suppose $f : X \times U \to E$ satisfies*

(a) $f(\cdot, y) \in \mathcal{L}_1(X, \mu, E)$ *for every* $y \in U$;

(b) $f(x, \cdot) \in C^1(U, E)$ *for μ-almost every* $x \in X$;

(c) *there exists* $g \in \mathcal{L}_1(X, \mu, \mathbb{R})$ *such that*

$$\left| \frac{\partial}{\partial y^j} f(x, y) \right| \leq g(x) \quad \text{for } (x, y) \in X \times U \text{ and } 1 \leq j \leq n .$$

Then

$$F : U \to E , \quad y \mapsto \int_X f(x, y) \, \mu(dx)$$

is continuously differentiable and

$$\partial_j F(y) = \int_X \frac{\partial}{\partial y^j} f(x, y) \, \mu(dx) \quad \text{for } y \in U \text{ and } 1 \leq j \leq n .$$

Proof Take $y \in U$ and $j \in \{1, \ldots, n\}$. Let (h_k) be a null sequence in $\mathbb{K}$ such that $h_k \neq 0$ and $y + h_k e_j \in U$ for $k \in \mathbb{N}$. Finally, set

$$f_k(x) := \frac{f(x, y + h_k e_j) - f(x, y)}{h_k} \quad \text{for } x \in X \text{ and } k \in \mathbb{N},$$

The mean value theorem (Theorem VII.3.9) then gives

$$|f_k(x)| \leq \sup_{z \in U} \left| \frac{\partial}{\partial y^j} f(x, z) \right| \leq g(x) \quad \mu\text{-a.e.}$$

Because (f_k) converges μ-a.e. to $\partial f(\,\cdot\,, y)/\partial y^j$, it follows from Theorem 3.12 that

$$\lim_{k \to \infty} \frac{F(y + h_k e_j) - F(y)}{h_k} = \lim_{k \to \infty} \int_X f_k \, d\mu = \int_X \frac{\partial}{\partial y^j} f(x, y) \, \mu(dx).$$

Therefore F is partially differentiable, and $\partial_j F(y) = \int_X (\partial/\partial y^j) f(x, y) \, \mu(dx)$. The result now follows from Theorems 3.17 and VII.2.10. ∎

3.19 Corollary *Suppose U is open in $\mathbb{C}$, and $f : X \times U \to \mathbb{C}$ satisfies*
  (a) *$f(\,\cdot\,, z) \in \mathcal{L}_1(X, \mu, \mathbb{C})$ for every $z \in U$;*
  (b) *$f(x, \cdot) \in C^\omega(U, \mathbb{C})$ for μ-almost every $x \in X$;*
  (c) *there is a $g \in \mathcal{L}_1(X, \mu, \mathbb{R})$ such that $|f(x, z)| \leq g(x)$ for $(x, z) \in X \times U$.*
Then

$$F : U \to \mathbb{C}, \quad z \mapsto \int_X f(x, z) \, \mu(dx)$$

is holomorphic, and

$$F^{(n)}(z) = \int_X \frac{\partial^n}{\partial z^n} f(x, z) \, \mu(dx) \tag{3.1}$$

for every $n \in \mathbb{N}$.

Proof Take $z_0 \in U$ and $r > 0$ such that $\overline{\mathbb{D}}(z_0, r) \subset U$. Cauchy's derivative formula (Corollary VIII.5.12) gives

$$\frac{\partial}{\partial z} f(x, z) = \frac{1}{2\pi i} \int_{\partial \mathbb{D}(z, r)} \frac{f(x, \zeta)}{(\zeta - z)^2} \, d\zeta \quad \text{for } \mu\text{-almost every } x \in X \text{ and } z \in \mathbb{D}(z_0, r),$$

and we find from (c) and Proposition VIII.4.3(iv) that

$$\left| \frac{\partial}{\partial z} f(x, z) \right| \leq \frac{g(x)}{r} \quad \text{for } \mu\text{-almost every } x \in X \text{ and } z \in \mathbb{D}(z_0, r).$$

Theorem 3.18 now shows that $F \mid \mathbb{D}(z_0, r)$ belongs to $C^1(\mathbb{D}(z_0, r), \mathbb{C})$ and satisfies

$$F'(z) = \int_X \frac{\partial}{\partial z} f(x, z) \, \mu(dx) \quad \text{for } z \in \mathbb{D}(z_0, r).$$

Holomorphy is a local property, so Theorem VIII.5.11 implies that F belongs to $C^\omega(U, \mathbb{C})$. The validity of (3.1) now follows from a simple induction argument. ∎

Exercises

1 Find a measure space $(X, \mathcal{A}, \mu)$, a sequence (f_j) in $\mathcal{L}_0(X, \mu, \mathbb{R}^+)$, and a function g in $\mathcal{L}_0(X, \mu, \overline{\mathbb{R}}^+)$ such that

$$f_j \leq g \text{ for } j \in \mathbb{N} \quad \text{and} \quad \varlimsup_j \int_X f_j \, d\mu > \int_X \left(\varlimsup_j f_j \right) d\mu .$$

2 Suppose $f \in \mathcal{L}_1(X, \mu, E)$ and $\varepsilon > 0$. Show that there exists $A \in \mathcal{A}$ such that

$$\mu(A) < \infty \quad \text{and} \quad \left| \int_X f \, d\mu - \int_B f \, d\mu \right| < \varepsilon$$

for every $B \in \mathcal{A}$ with $B \supset A$.

3 Suppose (f_j) is a sequence in $\mathcal{L}_1(X, \mu, E)$ converging in measure to $f \in \mathcal{L}_0(X, \mu, E)$. Also suppose there is $g \in \mathcal{L}_1(X, \mu, \mathbb{R})$ such that $|f_j| \leq g$ μ-a.e. for all $j \in \mathbb{N}$. Then f belongs to $\mathcal{L}_1(X, \mu, E)$,

$$f_j \to f \text{ in } \mathcal{L}_1(X, \mu, E) , \quad \text{and} \quad \int_X f_j \, d\mu \to \int_X f \, d\mu \text{ in } E .$$

(Hint: If $\left(\int_X f_j \, d\mu \right)$ does not converge to $\int_X f \, d\mu$, there is a subsequence $(f_{j_k})_{k \in \mathbb{N}}$ and a $\delta > 0$ such that

$$\| f_{j_k} - f \|_1 \geq \delta \quad \text{for } k \in \mathbb{N} . \tag{3.2}$$

Use Exercise 1.15 and Theorem 3.12 to derive a contradiction from (3.2).

4 Let $f, g \in \mathcal{L}_0(X, \mu, \overline{\mathbb{R}})$ be Lebesgue integrable functions. Prove:

(i) If $f \leq g$ μ-a.e., then $\int_X f \, d\mu \leq \int_X g \, d\mu$.

(ii) $\left| \int_X f \, d\mu \right| \leq \int_X |f| \, d\mu$.

(iii) $f \wedge g$ and $f \vee g$ are Lebesgue integrable, and

$$- \int_X (|f| + |g|) \, d\mu \leq \int_X (f \wedge g) \, d\mu \leq \int_X (f \vee g) \, d\mu \leq \int_X (|f| + |g|) \, d\mu .$$

5 Suppose the sequence (f_j) in $\mathcal{L}_0(X, \mu, \overline{\mathbb{R}}^+)$ converges in measure to $f \in \mathcal{L}_0(X, \mu, \overline{\mathbb{R}}^+)$. Prove that

$$\int_X f \, d\mu \leq \varliminf_j \int_X f_j \, d\mu .$$

6 For $x \in \mathbb{R}^n \setminus \{0\}$, define

$$k_n(x) := \begin{cases} x^+ & \text{if } n = 1 , \\ \log |x| & \text{if } n = 2 , \\ |x|^{2-n} & \text{if } n \geq 3 . \end{cases}$$

Further suppose $U \subset \mathbb{R}^n$ is open and nonempty, that $A \in \mathcal{L}(n)$ satisfies $A \subset U^c$, and that $f \in C_c(\mathbb{R}^n)$.

(a) The map $A \to \mathbb{R}$, $x \mapsto f(x) k_n(|y - x|)$ is λ_n-integrable for every $y \in U$.

(b) The map $U \to \mathbb{R}$, $y \mapsto \int_A f(x) k_n(|y - x|) \, \lambda_n(dx)$ is smooth and harmonic.

7 Verify that

(i) $\mathcal{L}_1(\mathbb{R}^n, \lambda_n, E) \cap BC(\mathbb{R}^n, E) \subsetneq C_0(\mathbb{R}^n, E)$;

(ii) $\mathcal{L}_1(\mathbb{R}^n, \lambda_n, E) \cap BUC(\mathbb{R}^n, E) \subseteq C_0(\mathbb{R}^n, E)$.

4 Lebesgue spaces

We saw in Corollary VI.7.4 that the space of continuous $\mathbb{K}$-valued functions over a compact interval I is not complete with respect to the L_2 norm. The framework of Lebesgue integration theory now gives us the means to complete the inner product space $\left(C(I,\mathbb{K}), (\cdot\,|\,\cdot)_2\right)$: we will construct a vector space L_2 and an extension of $(\cdot\,|\,\cdot)_2$ onto $L_2 \times L_2$ — also denoted by $(\cdot\,|\,\cdot)_2$ — such that $\left(L_2, (\cdot\,|\,\cdot)_2\right)$ is a Hilbert space containing $C(I,\mathbb{K})$ as a dense subspace.

This construction can be generalized in a natural way, leading to a new family of Banach spaces, the Lebesgue L_p-spaces. These are of great importance in many areas of mathematics.

In the following, we suppose that

- $(X, \mathcal{A}, \mu)$ is a complete σ-finite measure space;
  $E = (E, |\cdot|)$ is a Banach space.

Essentially bounded functions

We say that a function $f \in \mathcal{L}_0(X, \mu, E)$ is **μ-essentially bounded** if there exists $\alpha \geq 0$ such that $\mu([\,|f| > \alpha]) = 0$. The **$\mu$-essential supremum** of f is then[1]

$$\|f\|_\infty := \underset{x \in X}{\text{ess-sup}}\, |f(x)| := \inf\left\{ \alpha \geq 0 \,;\, \mu([\,|f| > \alpha]) = 0 \right\}.$$

4.1 Remarks **(a)** Let $f \in \mathcal{L}_0(X, \mu, E)$. There is equivalence between:

(i) f is μ-essentially bounded;

(ii) $\|f\|_\infty < \infty$;

(iii) f is bounded μ-a.e.

Proof "(i)$\Rightarrow$(ii)$\Rightarrow$(iii)" is clear.

"(iii)$\Rightarrow$(i)" Suppose N is a μ-null set and take $\alpha \geq 0$ such that $|f(x)| \leq \alpha$ for $x \in N^c$. Then $[\,|f| > \alpha] \subset N$, and the completeness of μ implies that $\mu([\,|f| > \alpha]) = 0$. $\blacksquare$

(b) Suppose $f \in \mathcal{L}_0(X, \mu, E)$. Then $|f| \leq \|f\|_\infty$ μ-a.e.

Proof The case $\|f\|_\infty = \infty$ is clear. If $\|f\|_\infty < \infty$, then $[\,|f| > \|f\|_\infty + 2^{-j}]$ is a μ-null set for every $j \in \mathbb{N}$, and hence so is the set $[\,|f| > \|f\|_\infty] = \bigcup_{j \in \mathbb{N}}[\,|f| > \|f\|_\infty + 2^{-j}]$. $\blacksquare$

(c) Suppose f and g are μ-essentially bounded and $\alpha \in \mathbb{K}$. Then $\alpha f + g$ is also μ-essentially bounded, and

$$\|\alpha f + g\|_\infty \leq |\alpha|\,\|f\|_\infty + \|g\|_\infty.$$

---

[1]Note that now $\|\cdot\|_\infty$ has two meanings, namely, the essential supremum of a measurable function and the supremum norm of a bounded function. The two values need *not* be the same; see (d) and (e) in Remark 4.1. When necessary we denote the supremum norm by $\|\cdot\|_{B(X,E)}$.

Proof By (a) and (b), there exist μ-null sets M and N such that $|f(x)| \leq \|f\|_\infty$ for $x \in M^c$ and $|g(x)| \leq \|g\|_\infty$ for $x \in N^c$. Therefore

$$|\alpha f(x) + g(x)| \leq |\alpha| \, \|f\|_\infty + \|g\|_\infty \quad \text{for } x \in (M \cup N)^c = M^c \cap N^c \; .$$

Hence $\alpha f + g$ is μ-essentially bounded and $\|\alpha f + g\|_\infty \leq |\alpha| \, \|f\|_\infty + \|g\|_\infty$. ∎

(d) Suppose $f \in \mathcal{L}_0(X, \mu, E)$ is bounded. Then $\|f\|_\infty \leq \|f\|_{B(X,E)}$ (supremum norm). If N is a nonempty μ-null set, then $\|\chi_N\|_\infty = 0$ and $\|\chi_N\|_{B(X,E)} = 1$.

(e) Suppose X is σ-compact metric space and μ is a massive Radon measure on X. Then

$$\|f\|_\infty = \|f\|_{B(X,E)} \quad \text{for } f \in BC(X, E) \; .$$

Proof By Theorem 1.17, any $f \in BC(X, E)$ is μ-measurable, and by (d) we just have to show that $\|f\|_{B(X,E)} \leq \|f\|_\infty$. Assume otherwise. Then there exists $x \in X$ such that

$$\|f\|_\infty < |f(x)| \leq \|f\|_{B(X,E)} \; ,$$

and in view of the continuity of f there is an open neighborhood O of x in X such that $\|f\|_\infty < |f(y)|$ for $y \in O$. From (b) it follows that $\mu(O) = 0$, contradicting the assumption that μ is massive. ∎

The Hölder and Minkowski inequalities

Suppose $f \in \mathcal{L}_0(X, \mu, E)$. For $p \in (0, \infty)$, we set

$$\|f\|_p := \left(\int_X |f|^p \, d\mu \right)^{1/p}$$

with the convention that $\infty^{1/p} := \infty$. We define the **Lebesgue space over X with respect to the measure μ** as[2]

$$\mathcal{L}_p(X, \mu, E) := \big\{ f \in \mathcal{L}_0(X, \mu, E) \; ; \; \|f\|_p < \infty \big\} \quad \text{for } p \in (0, \infty] \; .$$

For $p \in [1, \infty]$, we define the **dual exponent to p** as

$$p' := \begin{cases} \infty & \text{if } p = 1 \; , \\ p/(p-1) & \text{if } p \in (1, \infty) \; , \\ 1 & \text{if } p = \infty \; . \end{cases}$$

With this assignment, we obviously have

$$\frac{1}{p} + \frac{1}{p'} = 1 \quad \text{for } p \in [1, \infty] \; .$$

We are now in a position to state and prove two important inequalities.

---

[2]Theorem 3.9 shows that the notation $\mathcal{L}_p(X, \mu, E)$ is consistent in the case $p = 1$ with that of Section 2. In the following, we concentrate on the Lebesgue spaces $\mathcal{L}_p$ with $p \in [1, \infty]$. The case $p \in (0, 1)$ will be treated in Exercise 13.

4.2 Theorem *Suppose $p \in [1, \infty]$.*

(i) *For $f \in \mathcal{L}_p(X, \mu, \mathbb{K})$ and $g \in \mathcal{L}_{p'}(X, \mu, \mathbb{K})$, we have $fg \in \mathcal{L}_1(X, \mu, \mathbb{K})$, and*

$$\left| \int_X fg \, d\mu \right| \leq \int_X |fg| \, d\mu \leq \|f\|_p \|g\|_{p'} \qquad \textbf{(Hölder's}^3 \textbf{ inequality)}.$$

(ii) *Suppose $f, g \in \mathcal{L}_p(X, \mu, E)$. Then $f + g \in \mathcal{L}_p(X, \mu, E)$, and*

$$\|f + g\|_p \leq \|f\|_p + \|g\|_p \qquad \textbf{(Minkowski's inequality)}.$$

Proof (i) We consider first the case $p = 1$. By Remark 4.1(b), there is a μ-null set N such that $|g(x)| \leq \|g\|_\infty$ for $x \in N^c$. It then follows from Remarks 1.2(d) and 3.3(b) and Lemma 2.15 that

$$\int_{N^c} |fg| \, d\mu \leq \|g\|_\infty \int_{N^c} |f| \, d\mu = \|f\|_1 \|g\|_\infty < \infty .$$

Hence Remark 3.11(a), Theorem 3.9 and Lemma 2.15 result in fg being integrable, and Theorem 2.11(i) implies

$$\left| \int_X fg \, d\mu \right| \leq \int_X |fg| \, d\mu = \int_{N^c} |fg| \, d\mu \leq \|f\|_1 \|g\|_\infty .$$

Suppose now $p \in (1, \infty)$. If

$$f = 0 \ \mu\text{-a.e.} \quad \text{or} \quad g = 0 \ \mu\text{-a.e.} , \tag{4.1}$$

then fg also vanishes μ-a.e., and the claim follows from Corollary 2.16. On the other hand, if (4.1) does not apply, Corollary 2.19 gives $\|f\|_p > 0$ and $\|g\|_{p'} > 0$. We then set $\xi := |f|/\|f\|_p$, $\eta := |g|/\|g\|_{p'}$, and obtain from Young's inequality (Theorem IV.2.15) that

$$\frac{|fg|}{\|f\|_p \|g\|_{p'}} \leq \frac{1}{p} \frac{|f|^p}{\|f\|_p^p} + \frac{1}{p'} \frac{|g|^{p'}}{\|g\|_{p'}^{p'}} .$$

It follows that

$$\int_X |fg| \, d\mu \leq \frac{1}{p} \|f\|_p^{1-p} \|g\|_{p'} \int_X |f|^p \, d\mu + \frac{1}{p'} \|f\|_p \|g\|_{p'}^{1-p'} \int_X |g|^{p'} \, d\mu$$
$$= \|f\|_p \|g\|_{p'} ,$$

and we conclude using Theorem 3.9 that fg belongs to $\mathcal{L}_1(X, \mu, E)$. Therefore

$$\left| \int_X fg \, d\mu \right| \leq \|fg\|_1 \leq \|f\|_p \|g\|_{p'} .$$

The case $p = \infty$ is treated analogously to the case $p = 1$.

---

^{3}For $p = 2$, this is the **Cauchy–Schwarz inequality**.

(ii) Because of Corollary 2.9 and Remark 4.1(c), it suffices to consider the case $p \in (1, \infty)$. In addition, we can assume without loss of generality that $\|f+g\|_p > 0$. We will first prove that $f + g$ belongs to $\mathcal{L}_p(X, \mu, E)$. Noting the inequality

$$|a + b|^p \le (2(|a| \vee |b|))^p \le 2^p(|a|^p + |b|^p) \quad \text{for } a, b \in E , \tag{4.2}$$

we obtain

$$\int_X |f + g|^p \, d\mu \le 2^p \left(\int_X |f|^p \, d\mu + \int_X |g|^p \, d\mu \right) < \infty$$

because $f, g \in \mathcal{L}_p(X, \mu, E)$. Therefore $\|f + g\|_p < \infty$. Due to the equivalence

$$|f + g|^{p-1} \in \mathcal{L}_{p'}(X, \mu, \mathbb{R}) \iff |f + g| \in \mathcal{L}_p(X, \mu, \mathbb{R}) ,$$

it follows from Hölder's inequality that

$$\int_X |h| \, |f + g|^{p-1} \, d\mu \le \|h\|_p \big\| |f + g|^{p-1} \big\|_{p'} = \|h\|_p \, \|f + g\|_p^{p/p'}$$

for $h \in \mathcal{L}_p(X, \mu, E)$, and we find

$$\int_X |f + g|^p \, d\mu \le \int_X |f| \, |f + g|^{p-1} \, d\mu + \int_X |g| \, |f + g|^{p-1} \, d\mu \tag{4.3}$$
$$\le (\|f\|_p + \|g\|_p) \, \|f + g\|_p^{p/p'} .$$

The claim follows, because $\|f + g\|_p < \infty$ and $p/p' = p - 1$. ∎

4.3 Corollary *Suppose $p \in [1, \infty]$. Then $\mathcal{L}_p(X, \mu, E)$ is a vector subspace of $\mathcal{L}_0(X, \mu, E)$, and $\|\cdot\|_p$ is a seminorm on $\mathcal{L}_p(X, \mu, E)$.*

4.4 Remarks (a) Set $\mathcal{N} := \{ f \in \mathcal{L}_0(X, \mu, E) \,;\, f = 0 \ \mu\text{-a.e.} \}$. For $f \in \mathcal{L}_0(X, \mu, E)$ the following statements are equivalent:

(i) $\|f\|_p = 0$ for all $p \in [1, \infty]$.

(ii) $\|f\|_p = 0$ for some $p \in [1, \infty]$.

(iii) $f \in \mathcal{N}$.

Proof "(i)⇒(ii)" is trivial. "(ii)⇒(iii)" follows from Corollary 2.19 and Remark 4.1(b).

"(iii)⇒(i)" For $p \in [1, \infty)$, use Lemma 2.15. The case $p = \infty$ is clear. ∎

(b) $\mathcal{N}$ is a vector subspace of $\mathcal{L}_p(X, \mu, E)$ for every $p \in [1, \infty] \cup \{0\}$.

Proof The case $p = 0$ is clear; in particular, $\mathcal{N}$ is a vector space. For $p \in [1, \infty]$, the claim then follows from (a), "(iii)⇒(i)". ∎

(c) For $p \in [1, \infty]$, we have these inclusions of vector subspaces:

$$\mathcal{S}(X, \mu, E) \subset \mathcal{L}_p(X, \mu, E) \subset \mathcal{L}_0(X, \mu, E) .$$

Proof It is clear that every μ-simple function is μ-essentially bounded. Take $p \in [1, \infty)$ and let $\varphi \in \mathcal{S}(X, \mu, E)$ have normal form $\sum_{j=0}^m e_j \chi_{A_j}$. Then $|\varphi|^p \le \sum_{j=0}^m |e_j|^p \chi_{A_j}$, so $\|\varphi\|_p < \infty$. The claim follows by Remark 1.2(a) and Corollary 4.3. ∎

Lebesgue spaces are complete

We now generalize Theorem 2.10(ii), proving that all Lebesgue spaces $\mathcal{L}_p(X,\mu,E)$ with $p \in [1,\infty]$ are complete. For $p \in (1,\infty)$, this depends on the following lemma.

4.5 Lemma *Suppose V is a vector space and q is a seminorm on V. The following statements are equivalent:*

(i) *(V,q) is complete.*

(ii) *For every sequence $(v_j) \in V^{\mathbb{N}}$ such that $\sum_{j=0}^{\infty} q(v_j) < \infty$, the series $\sum_j v_j$ converges in V.*

Proof "(i)$\Rightarrow$(ii)" Suppose $(v_j) \in V^{\mathbb{N}}$ and $\sum_{j=0}^{\infty} q(v_j) < \infty$. For every $\varepsilon > 0$ there exists $K \in \mathbb{N}$ such that $\sum_{j=\ell+1}^{\infty} q(v_j) < \varepsilon$ for $\ell \geq K$ (see Exercise II.7.4). We set $w_k := \sum_{j=0}^{k} v_j$ for $k \in \mathbb{N}$ and get

$$q(w_m - w_\ell) = q\Big(\sum_{j=\ell+1}^{m} v_j \Big) \leq \sum_{j=\ell+1}^{m} q(v_j) \leq \sum_{j=\ell+1}^{\infty} q(v_j) < \varepsilon \quad \text{for } m > \ell \geq K .$$

Therefore (w_k) is a Cauchy sequence in V, and so converges to some $v \in V$ by the completeness of V. Hence the series $\sum_j v_j$ converges.

"(ii)$\Rightarrow$(i)" Let (v_j) be a Cauchy sequence in V. For $k \in \mathbb{N}$, take $j_k \in \mathbb{N}$ such that $q(v_{j_{k+1}} - v_{j_k}) < 2^{-(k+1)}$. Setting $w_k := v_{j_{k+1}} - v_{j_k}$, we have $\sum_{k=0}^{\infty} q(w_k) \leq 1$, and we can find by assumption a $v \in V$ such that $q\big(v - \sum_{k=0}^{\ell} w_k\big) \to 0$ as $\ell \to \infty$. Let $\varepsilon > 0$ and $L \in \mathbb{N}$ be such that $q\big(v - \sum_{k=0}^{\ell} w_k\big) < \varepsilon/2$ for $\ell \geq L$. Because (v_j) is a Cauchy sequence in V, there exists $K \geq L$ such that $q(v_{j_{\ell+1}} - v_k) < \varepsilon/2$ for $k,\ell \geq K$. Finally setting $\widetilde{v} := v + v_{j_0}$, we have for $k \geq K$ that

$$q(\widetilde{v} - v_k) = q(v + v_{j_0} - v_{j_{K+1}} + v_{j_{K+1}} - v_k)$$
$$\leq q\Big(v - \sum_{k=0}^{K} w_k\Big) + q(v_{j_{K+1}} - v_k) < \varepsilon .$$

This shows that (v_k) converges to $\widetilde{v}$. ∎

4.6 Theorem *For $p \in [1,\infty]$, $\mathcal{L}_p(X,\mu,E)$ is complete.*

Proof (i) Consider first the case $p \in (1,\infty)$. Let (f_j) be a sequence in $\mathcal{L}_p(X,\mu,E)$ such that $\sum_{j=0}^{\infty} \|f_j\|_p < \infty$. Set $g_k := \sum_{j=0}^{k} |f_j|$ for $k \in \mathbb{N}$ and $g := \sum_{j=0}^{\infty} |f_j|$. By Corollary 1.13(iii), g belongs to $\mathcal{L}_0(X,\mu,\mathbb{R}^+)$, and we have $|g_k|^p \to |g|^p$. Because

$$\|g_k\|_p \leq \sum_{j=0}^{k} \|f_j\|_p \leq \sum_{j=0}^{\infty} \|f_j\|_p < \infty ,$$

Conclusion 3.10(a) tells us that $g \in \mathcal{L}_p(X,\mu,\mathbb{R})$. By Remark 3.11(c), then, there is a μ-null set N such that $g(x) < \infty$ for $x \in N^c$. Therefore $f(x) := \sum_{j=0}^{\infty} f_j(x)$ is well defined for every $x \in N^c$ by the Weierstrass criterion (Theorem V.1.6). Also, since $|f|^p \le g^p$ μ-a.e. and $g \in \mathcal{L}_p(X, \mu, \mathbb{R})$, Theorem 3.14 implies that $\tilde{f}$ belongs to $\mathcal{L}_p(X, \mu, E)$. Finally Fatou's lemma shows that

$$\left\| \tilde{f} - \sum_{j=0}^{k} f_j \right\|_p^p = \int_X \left| \lim_{\ell \to \infty} \sum_{j=k+1}^{\ell} f_j \right|^p d\mu \le \lim_{\ell \to \infty} \int_X \left| \sum_{j=k+1}^{\ell} f_j \right|^p d\mu = \lim_{\ell \to \infty} \left\| \sum_{j=k+1}^{\ell} f_j \right\|_p^p ,$$

and we find

$$\left\| \tilde{f} - \sum_{j=0}^{k} f_j \right\|_p \le \lim_{\ell \to \infty} \sum_{j=k+1}^{\ell} \|f_j\|_p = \sum_{j=k+1}^{\infty} \|f_j\|_p \quad \text{for } k \in \mathbb{N} .$$

Because $\sum_{j=0}^{\infty} \|f_j\|_p < \infty$, the sequence $\left(\sum_{j=k+1}^{\infty} \|f_j\|_p \right)_{k \in \mathbb{N}}$ converges to zero. Therefore so does $\left(\left\| \tilde{f} - \sum_{j=0}^{k} f_j \right\|_p \right)_{k \in \mathbb{N}}$. Now it follows from Lemma 4.5 that $\mathcal{L}_p(X, \mu, E)$ is complete.

(ii) Now suppose (f_j) is a Cauchy sequence in $\mathcal{L}_\infty(X, \mu, E)$. We set

$$A_j := [|f_j| > \|f_j\|_\infty], \quad B_{k,\ell} := [|f_k - f_\ell| > \|f_k - f_\ell\|_\infty] \quad \text{for } j, k, \ell \in \mathbb{N}$$

and $N := \bigcup_j A_j \cup \bigcup_{k,\ell} B_{k,\ell}$. By Remarks 4.1(b) and IX.2.5(b), N is a null set and

$$|f_j(x)| \le \|f_j\|_\infty , \quad |f_k(x) - f_\ell(x)| \le \|f_k - f_\ell\|_\infty \quad \text{for } j, k, \ell \in \mathbb{N} , \ x \in N^c .$$

Therefore $(f_j \,|\, N^c)$ is a Cauchy sequence in the Banach space $B(N^c, E)$, and we can find an $f \in B(N^c, E)$ such that $(f_j \,|\, N^c)$ converges uniformly to f. Thus (f_j) converges μ-a.e. to $\tilde{f}$. We know the function $\tilde{f}$ is μ-essentially bounded because $\big[|wtf| > \|f\|_{B(N^c,E)} \big] = \emptyset$, and we have

$$|\tilde{f}(x) - f_j(x)| \le \|f - f_j \,|\, N^c\|_{B(N^c,E)} \quad \text{for } x \in N^c \text{ and } j \in \mathbb{N} .$$

Hence (f_j) converges in $\mathcal{L}_\infty(X, \mu, E)$ to $\tilde{f}$.

(iii) The case $p = 1$ was dealt with in Theorem 2.10(ii). $\blacksquare$

4.7 Corollary Let $p \in [1, \infty]$, and suppose $f_j, f \in \mathcal{L}_p(X, \mu, E)$ satisfy $f_j \to f$ in $\mathcal{L}_p(X, \mu, E)$.

(i) If $p = \infty$, then (f_j) converges μ-a.e. to f.

(ii) If $p \in [1, \infty)$, there is a subsequence $(f_{j_k})_{k \in \mathbb{N}}$ of (f_j) converging μ-a.e. to f.

Proof Because (f_j) converges in $\mathcal{L}_p(X, \mu, E)$ to f, we know (f_j) is a Cauchy sequence in $\mathcal{L}_p(X, \mu, E)$. Statement (i) now follows immediately from the proof of Theorem 4.6.

If $p \in (1, \infty)$, choose a subsequence $(f_{j_k})_{k \in \mathbb{N}}$ of (f_j) such that $\|f_{j_{k+1}} - f_{j_k}\|_p < 2^{-(k+1)}$. Then the proof of Theorem 4.6 shows that there is a $g \in \mathcal{L}_p(X, \mu, E)$ such that $(f_{j_k} - f_{j_0}) \to g$ in $\mathcal{L}_p(X, \mu, E)$ and $(f_{j_k} - f_{j_0}) \to g$ μ-a.e. as $k \to \infty$. Because (f_j) converges in $\mathcal{L}_p(X, \mu, E)$ to f, we have $\|f - (g + f_{j_0})\|_p = 0$. Remark 4.4(a) implies $f = g + f_{j_0}$ μ-a.e., from which the claim follows.

The case $p = 1$ was treated in Theorem 2.18. $\blacksquare$

4.8 Proposition $\mathcal{S}(X, \mu, E)$ *is dense in* $\mathcal{L}_p(X, \mu, E)$ *for* $p \in [1, \infty)$.[4]

Proof   Suppose $f \in \mathcal{L}_p(X, \mu, E)$. Then f is μ-measurable by Remark 4.4(c). Thus there is a sequence (φ_j) in $\mathcal{S}(X, \mu, E)$ such that $\varphi_j \to f$ μ-a.e. as $j \to \infty$. We set $A_j := [\,|\varphi_j| \le 2\,|f|\,]$ and $\psi_j := \chi_{A_j} \varphi_j$. Then (ψ_j) is a sequence in $\mathcal{S}(X, \mu, E)$ that converges μ-a.e. to f. Moreover,

$$|\psi_j - f|^p \le (|\psi_j| + |f|)^p \le 3^p\,|f|^p \quad \text{for } j \in \mathbb{N} \ .$$

Because $3^p\,|f|^p$ belongs to $\mathcal{L}_1(X, \mu, \mathbb{R})$, we can apply the dominated convergence theorem, and we find

$$\|\psi_j - f\|_p^p = \int_X |\psi_j - f|^p\, d\mu \to 0 \quad \text{as } j \to \infty \ ,$$

from which the claim follows. ∎

L_p-spaces

We proved in Remark 4.4(b) that

$$\mathcal{N} := \big\{\, f \in \mathcal{L}_0(X, \mu, E) \ ; \ f = 0 \ \mu\text{-a.e.} \,\big\}$$

is a vector subspace of $\mathcal{L}_p(X, \mu, E)$ for $p \in \{0\} \cup [1, \infty]$. Hence the quotient spaces

$$L_p(X, \mu, E) := \mathcal{L}_p(X, \mu, E)/\mathcal{N} \quad \text{for } p \in \{0\} \cup [1, \infty]$$

are well defined vector spaces over $\mathbb{K}$, by Example I.12.3(i). By Remark 4.4(c), we also have

$$L_p(X, \mu, E) \subset L_0(X, \mu, E) \quad \text{for } p \in [1, \infty] \ ,$$

in the sense of vector subspaces. Suppose $[f] \in L_0(X, \mu, E)$, and let g be a representative of $[f]$. Then $f - g \in \mathcal{N}$, that is, f and g agree μ-a.e. By Remark 4.4(a), the map

$$\||\cdot\|| : L_0(X, \mu, E) \to \overline{\mathbb{R}}^+ \ , \quad [f] \mapsto \|f\|_p$$

is well defined for every $p \in [1, \infty]$, and for $[f] \in L_p(X, \mu, E)$, we have

$$\|| [f] \||_p = \|f\|_p = 0 \iff f = 0 \ \mu\text{-a.e.} \iff [f] = 0 \ . \tag{4.4}$$

Since $\||\cdot\||_p$ obviously inherits the properties of the seminorm $\|\cdot\|_p$, (4.4) shows that $\||\cdot\||_p$ is a norm on $L_p(X, \mu, E)$. Therefore $L_p(X, \mu, E)$ is a normed vector space, whereas the space we constructed it from, $\mathcal{L}_p(X, \mu, E)$, is only seminormed. So limits in $\mathcal{L}_p(X, \mu, E)$ are generally not unique, but limits in $L_p(X, \mu, E)$ are.[5] The price we pay for the better topological structure of $L_p(X, \mu, E)$ is that its elements are not functions on X but rather cosets of the vector subspace $\mathcal{N}$ of $\mathcal{L}_p(X, \mu, E)$. In other words, we identify functions that coincide μ-a.e. Experience shows that the following simplified notation does not lead to misunderstandings.

---

[4]The statement can fail if $p = \infty$; see Exercise 8 (but also Exercise 9).
[5]See Remark 2.3(b).

Convention   Suppose $p \in \{0\} \cup [1, \infty]$. Then we write the coset $[f] = f + \mathcal{N}$ in $L_p(X, \mu, E)$ as f and identify with each other functions that agree μ-a.e. Further, if $p \in [1, \infty]$, we denote the norm in $L_p(X, \mu, E)$ by $\|\cdot\|_p$ and set

$$L_p(X, \mu, E) := \big(L_p(X, \mu, E), \|\cdot\|_p \big) \quad \text{for } p \in [1, \infty] .$$

4.9 Remarks   (a) For $f \in L_0(X, \mu, E)$ and $x \in X$, $f(x)$ is *undefined* if μ has nonempty null sets. That is, elements of $L_0(X, \mu, E)$ cannot be "evaluated pointwise". (Of course, if one chooses a representative $\overset{*}{f}$ of f, then $\overset{*}{f}(x)$ is defined.)

(b) For $p \in [1, \infty]$,

$$L_p(X, \mu, E) = \big\{ f \in L_0(X, \mu, E) \ ; \ \|f\|_p < \infty \big\} .$$

Proof   "$\subseteq$" Let $f \in L_p(X, \mu, E)$. Any representative $\overset{*}{f}$ of f lies in $\mathcal{L}_p(X, \mu, E)$, that is, it is μ-measurable and satisfies $\|\overset{*}{f}\|_p < \infty$. Hence f belongs to $L_0(X, \mu, E)$, and $\|f\|_p < \infty$.

"$\supseteq$" Consider $f \in L_0(X, \mu, E)$ with $\|f\|_p < \infty$. Every representative $\overset{*}{f}$ of f is μ-measurable, with $\|f\|_p = \|\overset{*}{f}\|_p < \infty$. Thus $\overset{*}{f}$ belongs to $\mathcal{L}_p(X, \mu, E)$, and so f belongs to $L_p(X, \mu, E)$. ∎

(c) Suppose $f, g \in L_0(X, \mu, \mathbb{R})$, and let $\overset{*}{f}, \overset{*}{g}$ be representatives of f, g. If we write

$$f \leq g \ :\Longleftrightarrow \ \overset{*}{f} \leq \overset{*}{g} \ \mu\text{-a.e.} ,$$

we obtain a well defined ordering $\leq$ on $L_0(X, \mu, \mathbb{R})$, which makes this space into a vector lattice.

Proof   We leave the simple proof as an exercise. ∎

(d) Suppose $(F, \leq)$ is a vector lattice and $(F, \|\cdot\|)$ is a Banach space. If $|x| \leq |y|$ implies $\|x\| \leq \|y\|$, we call $(F, \leq, \|\cdot\|)$ a **Banach lattice**.

(e) $\big(L_p(X, \mu, \mathbb{R}), \leq, \|\cdot\|_p \big)$ is a Banach lattice for every $p \in [1, \infty]$.

Proof   It is clear that $L_p(X, \mu, \mathbb{R})$ is a vector sublattice of $L_0(X, \mu, \mathbb{R})$. Also it follows immediately from the monotony of integrals and of the map $t \mapsto t^p$ that $L_p(X, \mu, \mathbb{R})$ is a Banach lattice in the case $p \in [1, \infty)$.

Suppose $f, g \in L_\infty(X, \mu, \mathbb{R})$ with $|f| \leq |g|$, and let $\overset{*}{f}, \overset{*}{g}$ be representatives thereof. Then $|\overset{*}{f}| \leq |\overset{*}{g}|$ μ-a.e. In addition, Remark 4.1(b) shows that $|\overset{*}{g}| \leq \|g\|_\infty$ μ-a.e. Therefore $|\overset{*}{f}| \leq \|g\|_\infty$ μ-a.e., and hence $\|f\|_\infty \leq \|g\|_\infty$. ∎

4.10 Theorem

 (i) $L_p(X, \mu, E)$ is a Banach space for every $p \in [1, \infty]$.

 (ii) If H is a Hilbert space, then so is $L_2(X, \mu, H)$ with respect to the scalar product

$$(\cdot \mid \cdot)_2 : L_2(X, \mu, H) \times L_2(X, \mu, H) \to \mathbb{K} , \quad (f, g) \mapsto \int_X (f \mid g)_H \, d\mu .$$

Proof (i) Suppose $p \in [1, \infty]$. We already know that $L_p(X, \mu, E)$ is a normed vector space. Let (f_j) be a Cauchy sequence in $L_p(X, \mu, E)$, and $(\overset{*}{f}_j)$ a corresponding sequence of representatives. Then $(\overset{*}{f}_j)$ is a Cauchy sequence in $\mathcal{L}_p(X, \mu, E)$. By Theorem 4.6, there exists $\overset{*}{f} \in \mathcal{L}_p(X, \mu, E)$ such that $\|\overset{*}{f}_j - \overset{*}{f}\|_p \to 0$ as $j \to \infty$. Letting $f := \overset{*}{f} + \mathcal{N}$, we have $f \in L_p(X, \mu, E)$ and $\|f_j - f\|_p = \|\overset{*}{f}_j - \overset{*}{f}\|_p \to 0$. Therefore $L_p(X, \mu, E)$ is complete.

(ii) Using statements (i) and (iv) of Theorem 1.7 and Hölder's inequality, we easily prove that $(\cdot \mid \cdot)_2$ is a scalar product on $L_2(X, \mu, H)$ satisfying $|(f \mid f)_2| = \|f\|_2^2$ for $f \in L_2(X, \mu, H)$. The claim then follows from (i). ∎

4.11 Corollary $L_2(X, \mu, \mathbb{K})$ is a Hilbert space with respect to the scalar product

$$(f \mid g)_2 = \int_X f \bar{g} \, d\mu \quad \text{for } f, g \in L_2(X, \mu, \mathbb{K}) \ .$$

Continuous functions with compact support

Let Y be a topological space. For $f \in E^Y$, we call

$$\operatorname{supp}(f) := \overline{\{ x \in Y \; ; \; f(x) \neq 0 \}}$$

the **support** of f. Here, as usual, the bar denotes the closure (in Y). Continuous functions with compact support are particularly significant. We therefore define

$$C_c(Y, E) := \{ f \in C(Y, E) \; ; \; \operatorname{supp}(f) \text{ is compact} \} \ .$$

4.12 Examples (a) For the Dirichlet function $\chi_{\mathbb{Q}} \in \mathbb{R}^{\mathbb{R}}$ of Example III.1.3(c), we have

$$\operatorname{supp}(\chi_{\mathbb{Q}}) = \operatorname{supp}(\chi_{\mathbb{R}-\mathbb{Q}}) = \mathbb{R} \ .$$

Proof This follows from Propositions I.10.8 and I.10.11. ∎

(b) Suppose $X = \mathbb{Z}$ or $X = \mathbb{N}$, and provide X with the metric induced from $\mathbb{R}$. Let $\mathcal{H}^0$ be the counting measure on $\mathfrak{P}(X)$. Then[6]

$$C_c(X, E) = \mathcal{S}(X, \mathcal{H}^0, E) = \{ \varphi \in E^X \; ; \; \operatorname{Num}[\varphi \neq 0] < \infty \} \ .$$

(c) Suppose X is a metric space. Then $C_c(X, E)$ is a vector subspace of $BC(X, E)$. If X is compact, then $C_c(X, E) = C(X, E) = BC(X, E)$.

Proof The first statement follows from Corollary III.3.7. The second is a consequence of Exercise III.3.2 and Corollary III.3.7. ∎

---

[6] Compare Example 2.20.

4.13 Proposition *Suppose X is a metric space and A and B are closed, disjoint nonempty subsets of X. There exists $\varphi \in C(X)$ such that $0 \leq \varphi \leq 1$, $\varphi \mid A = 1$, and $\varphi \mid B = 0$. Such a function is called a* **Urysohn function**.

Proof If $D \subset X$ is nonempty, Example III.1.3(l) shows that the distance function $d(\cdot, D)$ belongs to $C(X)$. If D is also closed, we have $d(x, D) = 0$ if and only if $x \in D$. Using these properties, we easily prove that the function defined by

$$\varphi(x) := \frac{d(x, B)}{d(x, A) + d(x, B)} \quad \text{for } x \in X ,$$

has the stated properties. ∎

With help from Urysohn functions, we can now prove an important approximation theorem.

4.14 Theorem *Suppose X is a σ-compact metric space and μ is a Radon measure on X. Then $C_c(X, E)$ is a dense vector subspace of $\mathcal{L}_p(X, \mu, E)$ for $p \in [1, \infty)$.*

Proof Suppose $\varepsilon > 0$. According to Proposition 4.8, $\mathcal{S}(X, \mu, E)$ is dense in $\mathcal{L}_p(X, \mu, E)$. Thus, because of Theorem 1.17 and Minkowski's inequality (that is, the triangle inequality), it suffices to verify that for every μ-measurable set A of finite measure and every $e \in E \backslash \{0\}$, there exists $f \in C_c(X, E)$ such that $\|f - \chi_A e\|_p < \varepsilon$.

Suppose then that $A \in \mathcal{A}$ with $\mu(A) < \infty$. Because μ is regular, we can find a compact subset K and an open subset U of X such that $K \subset A \subset U$ and

$$\mu(U \backslash K) = \mu(U) - \mu(K) < (\varepsilon/|e|)^p .$$

Proposition 4.13 secures the existence of a Urysohn function φ on X with $\varphi \mid K = 1$ and $\varphi \mid U^c = 0$. Setting $f := \varphi e$, we get, as needed,

$$\|\chi_A e - f\|_p^p \leq |e|^p \int_X \chi_{U \backslash K} \, d\mu \leq |e|^p \, \mu(U \backslash K) < \varepsilon^p . \quad ∎$$

Embeddings

Suppose X and Y are topological spaces, and X is a subset of Y. Denoting by $j : X \to Y$, $x \mapsto x$ the inclusion[7] of X in Y, we say X is **continuously embedded** in Y if j is continuous.[8] In this case, we write $X \hookrightarrow Y$. We write $X \overset{d}{\hookrightarrow} Y$ if X is also a dense subset of Y. If X and Y are vector spaces, the notation $X \hookrightarrow Y$ (and the term "continuously embedded") will always mean in addition that X is a vector subspace of Y, not just any odd subset.

---

[7]See Example I.3.2(b).
[8]These notions become important when X is *not* provided with the topology induced by Y; see Remark 4.15(a).

4.15 Remarks (a) Suppose V and W are normed vector spaces. V is continuously embedded in W if and only if V is a vector subspace of W and there is an $\alpha > 0$ such that $\|v\|_W \le \alpha \|v\|_V$ for $v \in V$, that is, if the norm of V is stronger than the norm induced from W on V.

If V carries the norm induced by W, then $V \hookrightarrow W$ always.

(b) Suppose X is open in $\mathbb{R}^n$. Then

$$BUC^k(X, E) \hookrightarrow BUC^\ell(X, E) \quad \text{for } k \ge \ell .$$

If X is bounded as well, then

$$BUC^k(X, \mathbb{K}) \overset{d}{\hookrightarrow} BUC(X, \mathbb{K}) \quad \text{for } k \in \mathbb{N} .$$

Proof The first statement is clear. The second follows from the Stone–Weierstrass approximation theorem (Corollary V.4.8) and then Application VI.2.2. ∎

Simple examples (see Exercise 5.1) show that Lebesgue spaces are generally not contained in one another. Under suitable extra assumptions on the measure space $(X, \mathcal{A}, \mu)$, continuous embeddings exist for Lebesgue spaces. For example, if $\mathcal{H}^0$ is the counting measure on $\mathfrak{P}(\mathbb{N})$, the spaces ℓ_p introduced in Exercise 1.16 coincide with $\mathcal{L}_p(\mathbb{N}, \mathcal{H}^0, \mathbb{K})$ for $1 \le p \le \infty$, and we have the embeddings

$$\ell_1 \hookrightarrow \ell_p \hookrightarrow \ell_q \hookrightarrow \ell_\infty \quad \text{for } 1 \le p \le q \le \infty ,$$

(see Exercise 11).

Finite measure spaces present an altogether different situation:

4.16 Theorem Let $(X, \mathcal{A}, \mu)$ be a finite *complete* measure space. Then

$$L_q(X, \mu, E) \overset{d}{\hookrightarrow} L_p(X, \mu, E) \quad \text{for } 1 \le p < q \le \infty$$

and

$$\|f\|_p \le \mu(X)^{1/p - 1/q} \|f\|_q \quad \text{for } f \in L_q(X, \mu, E) . \tag{4.5}$$

Proof (i) Take $f \in L_q(X, \mu, E)$ and set $r := q/p$. Let $g \in \mathcal{L}_q(X, \mu, E)$ be a representative of f. Then $|g|^p$ belongs to $\mathcal{L}_r(X, \mu, \mathbb{R})$, and $1/r' = (q - p)/q$. Further, χ_X belongs to $\mathcal{L}_{r'}(X, \mu, \mathbb{R})$, because μ is a finite measure. Thus in the case $q < \infty$ Hölder's inequality gives

$$\|g\|_p^p = \int_X \chi_X |g|^p \, d\mu \le \left(\int_X \chi_X^{r'} \, d\mu \right)^{1/r'} \left(\int_X |g|^{pr} \, d\mu \right)^{1/r} = \mu(X)^{(q-p)/q} \|g\|_q^p ,$$

and we find $\|g\|_p \le \mu(X)^{1/p - 1/q} \|g\|_q$; this clearly also holds in the case $q = \infty$. Because g is an arbitrary representative of f, we see that f belongs to $L_p(X, \mu, E)$ and (4.5) holds. By Remark 4.15(a), it follows that $L_q(X, \mu, E) \hookrightarrow L_p(X, \mu, E)$.

(ii) $M := \{\, [\varphi] \in L_0(X, \mu, E) \; ; \; \varphi \in \mathcal{S}(X, \mu, E) \,\}$ satisfies $M \subset L_q(X, \mu, E)$, and, because $p < \infty$, Proposition 4.8 implies that M is dense in $L_p(X, \mu, E)$. Therefore $L_q(X, \mu, E)$ is also dense in $L_p(X, \mu, E)$. ∎

The next theorem shows that, in the case of a massive Radon measure μ, an element of $L_0(X, \mu, E)$ has at most one continuous representative. In this case, then, we can identify each function in $C(X, E)$ with the equivalence class it generates in $L_0(X, \mu, E)$, and regard $C(X, E)$ as a vector subspace of $L_0(X, \mu, E)$.

4.17 Proposition *Suppose μ is a massive Radon measure on a σ-compact space X. Then the map*

$$C(X, E) \to L_0(X, \mu, E) , \quad f \mapsto [f] \tag{4.6}$$

is linear and injective.

Proof Theorem 1.17 shows that the map (4.6) is well defined and linear.

Take $f, g \in C(X, E)$ with $[f] = [g]$. There exists $h \in \mathcal{N}$ such that $f - g = h$, that is, $f - g = 0$ μ-a.e. Assume for a contradiction that $f(x) \neq g(x)$ for some $x \in X$. By continuity, $(f - g)(y) \neq 0$ for all y in some open neighborhood U of x. But $\mu(U) > 0$, contrary to the assumption that $f - g = 0$ μ-a.e. Therefore $f = g$, which proves the asserted injectivity. ∎

> **Convention** Let μ be a massive Radon measure on a σ-compact space X. We identify $C(X, E)$ with its image in $L_0(X, \mu, E)$ under the injection (4.6) and so regard $C(X, E)$ as a vector subspace of $L_0(X, \mu, E)$. Then
>
> $$\|f\|_{B(X,E)} = \|f\|_\infty \quad \text{for } f \in BC(X, E) .$$

The following result is a simple consequence of this convention.

4.18 Theorem *Let μ be a massive Radon measure on a σ-compact metric space X.*

  (i) *$C_c(X, E)$ is a dense vector subspace of $L_p(X, \mu, E)$ for every $p \in [1, \infty)$.*

  (ii) *$BC(X, E)$ is a closed vector subspace of $L_\infty(X, \mu, E)$.*

Proof The first statement follows from Theorem 4.14. The second is obvious. ∎

Continuous linear functionals on L_p

For the rest of this section, we use for $p \in [1, \infty]$ the abbreviations

$$L_p(X) := L_p(X, \mu, \mathbb{K}) \quad \text{and} \quad L_p'(X) := \big(L_p(X)\big)' ,$$

the prime on the right indicating the dual space (Remark VII.2.13(a)). From Hölder's inequality, it follows that, for every $f \in L_{p'}(X)$, the map

$$T_f : L_p(X) \to \mathbb{K} , \quad g \mapsto \int_X f g \, d\mu$$

is a continuous linear functional on $L_p(X)$, that is, an element of $L'_p(X)$; it satisfies

$$\|T_f\|_{L'_p(X)} \le \|f\|_{p'} . \tag{4.7}$$

In fact (4.7) holds with equality:

4.19 Proposition   *The map*

$$T \colon L_{p'}(X) \to L'_p(X) , \quad f \mapsto T_f$$

is a linear isometry for every $p \in [1, \infty]$.

Proof   (i) Clearly T is linear. Also, in view of (4.7), we need only show that for every $f \in L_{p'}(X)$ satisfying $f \neq 0$ and every $\varepsilon > 0$, there is $g \in L_p(X)$ such that

$$\|g\|_p = 1 \quad \text{and} \quad \|f\|_{p'} < \left| \int_X fg \, d\mu \right| + \varepsilon .$$

(ii) First assume $p \in (1, \infty)$, so $p' \in (1, \infty)$. Therefore

$$g := \overline{\operatorname{sign} f} \, \|f\|_{p'}^{1-p'} \, |f|^{p'-1}$$

is well defined and μ-measurable (see Exercise 1.19 and Theorem 1.7(i)). Also

$$\int_X |g|^p \, d\mu = \|f\|_{p'}^{p(1-p')} \int_X |f|^{p(p'-1)} \, d\mu = \|f\|_{p'}^{-p'} \|f\|_{p'}^{p'} = 1$$

and $fg = \|f\|_{p'}^{1-p'} |f|^{p'}$. Therefore $\|f\|_{p'} = \int_X fg \, d\mu$.

For $p = \infty$, we set $g := \overline{\operatorname{sign} f}$. Then

$$\|g\|_\infty = 1 \quad \text{and} \quad \|f\|_1 = \int_X fg \, d\mu .$$

(iii) Now suppose that $p = 1$. Suppose $0 < \varepsilon < \|f\|_\infty$ and set $\alpha := \|f\|_\infty - \varepsilon$. Because $[\,|f| > \alpha\,]$ has positive measure and μ is σ-finite, we can find $A \in \mathcal{A}$ such that $A \subset [\,|f| > \alpha\,]$ and $\mu(A) \in (0, \infty)$. Therefore $g := \overline{\operatorname{sign} f} \, (1/\mu(A)) \chi_A$ is well defined and μ-measurable. Clearly $\|g\|_1 = 1$ and

$$\int_X fg \, d\mu = \frac{1}{\mu(A)} \int_A |f| \, d\mu \ge \alpha = \|f\|_\infty - \varepsilon .$$

This concludes the proof. ∎

4.20 Remarks (a) One can show that the map T of Proposition 4.19 is surjective for every $p \in [1, \infty)$, that is, every continuous linear functional on $L_p(X)$ can be represented is of the form T_f for an appropriate $f \in L_{p'}(X)$; see [Rud83, Theorem 6.1.6], for example. Consequently $T : L_{p'}(X) \to L'_p(X)$ is an isometric isomorphism for every $p \in [1, \infty)$. *This isomorphism allows us to identify* $L_{p'}(X)$ *with* $L'_p(X)$ *for* $p \in [1, \infty)$. The dual pairing $\langle \cdot, \cdot \rangle_{L_p} : L'_p(X) \times L_p(X) \to \mathbb{K}$ satisfies

$$\langle g, f \rangle_{L_p} = \int_X fg \, d\mu \quad \text{for } (g, f) \in L_{p'}(X) \times L_p(X) .$$

(b) In the case $p = \infty$, the map $T : L_1(X) \to L'_\infty(X)$ is generally not surjective; see [Fol99, S. 191].

(c) Denote by $\langle \cdot, \cdot \rangle_E : E' \times E \to \mathbb{K}$ the duality pairing between E and E'. Then the map

$$\kappa : E \to [E']' , \quad e \mapsto \langle \cdot, e \rangle_E$$

is linear and bounded. Its norm is at most 1.

Proof Clearly κ is linear. Suppose $e \in E$ with $\|e\|_E \leq 1$. Then

$$\left| \langle \kappa(e), e' \rangle_{E'} \right| = |\langle e', e \rangle_E| \leq \|e'\|_{E'} \quad \text{for } e' \in E' ,$$

and we find $\|\kappa(e)\|_{(E')'} \leq 1$, from which the claim follows. ∎

(d) With tools from functional analysis, one can show that κ is an isometry and therefore injective. We call κ the **canonical injection** of E into the **double dual space** $E'' := (E')'$ of E. If κ is surjective as well, and hence an isometric isomorphism, we say E is **reflexive**. In this case, the canonical isomorphism κ allows us to identify E with its double dual E''.

(e) $L_p(X)$ reflexive for $p \in (1, \infty)$.

Proof This follows from (a). ∎

(f) The spaces $L_1(X)$ and $L_\infty(X)$ are generally not reflexive; see, for instance, [Ada75, Theorem 2.35]. ∎

Exercises

1 Let $S(X, \mu, E) := \{ [f] \in L_0(X, \mu, E) ; [f] \cap S(X, \mu, E) \neq \emptyset \}$. Prove that $S(X, \mu, E)$ a dense vector subspace of $L_p(X, \mu, E)$ for $1 \leq p < \infty$.

2 For $a \in \mathbb{R}^n$, we define $\tau_a : E^{(\mathbb{R}^n)} \to E^{(\mathbb{R}^n)}$, the **right translation by** a, by

$$(\tau_a \varphi)(x) := \varphi(x - a) \quad \text{for } x \in \mathbb{R}^n , \quad \varphi \in E^{(\mathbb{R}^n)} .$$

Set $\tau_a[f] := [\tau_a f]$ for $[f] \in L_p$. Prove:

(i) $(\mathbb{R}^n, +) \to \left(\mathcal{L}\mathrm{aut}(L_p(\mathbb{R}^n, \lambda_n, E)), \circ \right)$, $a \mapsto \tau_a$ is a group homomorphism with $\|\tau_a\|_{\mathcal{L}(L_p)} = 1$ for every $p \in [1, \infty]$.

(ii) For $p \in [1, \infty)$ and $f \in L_p(\mathbb{R}^n, \lambda_n, E)$, we have $\lim_{a \to 0} \|\tau_a f - f\|_p = 0$.

(iii) If $\lim_{a \to 0} \|\tau_a f - f\|_\infty = 0$, there exists $g \in BUC(\mathbb{R}^n, E)$ such that $f = g$ μ-a.e.

3   Suppose μ is a complete Radon measure on a σ-compact space X, and let $(X_j)_{j \in \mathbb{N}}$ be a sequence of relatively compact open subsets of X covering X. For $p \in [1, \infty]$, set

$$q_{j,p}(f) := \|\chi_{X_j} f\|_p \quad \text{for } j \in \mathbb{N}, \quad f \in L_0(X, \mu, E),$$

$$L_{p,\mathrm{loc}}(X, \mu, E) := \big\{ f \in L_0(X, \mu, E) ; \ q_{j,p}(f) < \infty, \ j \in \mathbb{N} \big\}.$$

Finally, define

$$d_p(f, g) := \sum_{j=0}^{\infty} \frac{2^{-j} q_{j,p}(f - g)}{1 + q_{j,p}(f - g)} \quad \text{for } f, g \in L_{p,\mathrm{loc}}(X, \mu, E).$$

(i) $L_{p,\mathrm{loc}}(X, \mu, E)$ is well defined, that is, independent of the particular sequence (X_j).

(ii) $\big(L_{p,\mathrm{loc}}(X, \mu, E), d_p\big)$ is a complete metric space.

(iii) $L_p(X, \mu, E) \xhookrightarrow{d} L_{p,\mathrm{loc}}(X, \mu, E) \xhookrightarrow{d} L_{1,\mathrm{loc}}(X, \mu, E)$.

(iv) The topology generated by d_p is independent of the sequence (X_j).

4   Suppose $p, q \in [1, \infty]$ and define

$$L_p \cap L_q := (L_p \cap L_q)(X, \mu, E) := L_p(X, \mu, E) \cap L_q(X, \mu, E),$$

$$L_p + L_q := (L_p + L_q)(X, \mu, E) := L_p(X, \mu, E) + L_q(X, \mu, E).$$

Also set $\|f\|_{L_p \cap L_q} := \|f\|_p + \|f\|_q$ for $f \in L_p \cap L_q$, and put

$$\|f\|_{L_p + L_q} := \inf\big\{ \|g\|_p + \|h\|_q ; \ g \in L_p(X, \mu, E), \ h \in L_q(X, \mu, E) \text{ with } f = g + h \big\}$$

for $f \in L_p + L_q$.

(i) Check that the **interpolation inequality**

$$\|f\|_r \leq \|f\|_p^{1-\theta} \|f\|_q^\theta, \qquad \text{where} \quad \frac{1}{r} := \frac{1-\theta}{p} + \frac{\theta}{q},$$

holds for $f \in L_p \cap L_q$ and $\theta \in [0, 1]$.

(ii) $(L_p \cap L_q, \|\cdot\|_{L_p \cap L_q})$ and $(L_p + L_q, \|\cdot\|_{L_p + L_q})$ are Banach spaces with

$$(L_p \cap L_q)(X, \mu, E) \hookrightarrow L_r(X, \mu, E) \hookrightarrow (L_p + L_q)(X, \mu, E) \hookrightarrow L_{1,\mathrm{loc}}(X, \mu, E)$$

for $1 \leq p \leq r \leq q \leq \infty$.

(Hints: (i) Hölder's inequality.   (ii) Take $f \in L_p + L_q$ with $\|f\|_{L_p + L_q} = 0$. To show it vanishes, note that $L_r \hookrightarrow L_{1,\mathrm{loc}}$ for $r \in [1, \infty]$ (see Exercise 3). To prove the completeness of $L_p + L_q$ apply Lemma 4.5. The embedding $L_p \cap L_q \hookrightarrow L_r$ follows from (a).)

5   Suppose $p \in [1, \infty)$ and $f \in (L_p \cap L_\infty)(X, \mu, E)$. Prove that $\lim_{q \to \infty} \|f\|_q = \|f\|_\infty$.

6   Prove that the map

$$L_\infty(X, \mu, \mathbb{K}) \times L_p(X, \mu, E) \to L_p(X, \mu, E), \quad ([\varphi], [f]) \mapsto [\varphi f]$$

is bilinear and continuous and has norm at most 1.

7 Suppose $\mu(X) < \infty$, and for $f, g \in L_0(X, \mu, E)$ put

$$d_0(f, g) := \int_X \frac{|f - g|}{1 + |f - g|} \, d\mu \ .$$

(i) $\big(L_0(X, \mu, E), d_0\big)$ is a metric space.

(ii) (f_j) converges to 0 in $\big(L_0(X, \mu, E), d_0\big)$ if and only if it converges to 0 in measure.

8 Let μ be a Radon measure on a σ-compact space X and let E be separable. Prove:

(i) $C_c(X, \mathbb{K})$ is separable.

(ii) $C_c(X, E)$ is separable.

(iii) $L_p(X, \mu, E)$ is separable for $p \in [1, \infty)$.

(iv) $L_\infty(X, \mu, E)$ is generally not separable.

(v) $\mathcal{S}(X, \mu, E)$ is generally not dense in $\mathcal{L}_\infty(X, \mu, E)$.

(Hints: (i) Corollary V.4.8 and Remark 1.16(e).   (ii) Take $A \subset C_c(X, \mathbb{K})$ and let B be countable and dense in E. For $a \in A$ and $b \in B$, set $(a \otimes b)(x) := a(x)b$ for $x \in X$ and consider

$$\big\{ \textstyle\sum_{j=0}^m a_j \otimes b_j \ ; \ m \in \mathbb{N}, \ (a_j, b_j) \in A \times B, \ j = 0, \dots, m \big\} \ .$$

(iii) Theorem 4.14.   (iv) Find an uncountable subset A of L_∞ such that $\|f - g\|_\infty \geq 1$ for all distinct $f, g \in A$.)

9 If μ finite and E is finite-dimensional, show that $\mathcal{S}(X, \mu, E)$ is dense in $\mathcal{L}_\infty(X, \mu, E)$.

10 Prove the statement of Remark 4.9(c).

11 Prove:

(i) $\ell_p = \mathcal{L}_p(\mathbb{N}, \mathcal{H}^0, \mathbb{K})$ for $1 \leq p \leq \infty$.

(ii) $\ell_p \hookrightarrow \ell_q$ with $\|\cdot\|_q \leq \|\cdot\|_p$ if $1 \leq p \leq q \leq \infty$.

(iii) $\ell_p \overset{d}{\hookrightarrow} \ell_q \overset{d}{\hookrightarrow} c_0 \hookrightarrow \ell_\infty$ if $1 \leq p \leq q < \infty$ (see Section II.2).

12 Suppose $p, q \in [1, \infty]$ with $1 \leq p \leq q \leq \infty$. Prove:

(i) $L_\infty(X, \mu, E) \subset L_1(X, \mu, E) \Rightarrow L_q(X, \mu, E) \hookrightarrow L_p(X, \mu, E)$.

(ii) $L_1(X, \mu, E) \subset L_\infty(X, \mu, E) \Rightarrow L_p(X, \mu, E) \hookrightarrow L_q(X, \mu, E)$.

(iii) There exists a complete σ-finite measure space $(X, \mathcal{A}, \mu)$ [or $(Y, \mathcal{B}, \nu)$] realizing the embedding $L_\infty(X, \mu, \mathbb{R}) \hookrightarrow L_1(X, \mu, \mathbb{R})$ [or $L_1(Y, \nu, \mathbb{R}) \hookrightarrow L_\infty(Y, \nu, \mathbb{R})$].

(Hints: (i) Hölder's inequality.   (ii) Show that $L_p \hookrightarrow L_\infty$ and apply Exercise 4(i).)

13 For $p \in (0, 1)$, prove:

(i) $\|f + g\|_p^p \leq \|f\|_p^p + \|g\|_p^p$ for $f, g \in \mathcal{L}_0(X, \mu, E)$.

(ii) $\|f + g\|_p \leq 2^{1/p - 1}(\|f\|_p + \|g\|_p)$ for $f, g \in \mathcal{L}_0(X, \mu, E)$.

(iii) $\mathcal{L}_p(X, \mu, E)$ is a vector subspace of $\mathcal{L}_0(X, \mu, E)$.

(iv) $\mathcal{N} := \big\{ f \in \mathcal{L}_0(X, \mu, E) \ ; \ f = 0 \ \mu\text{-a.e.} \big\}$ is a vector subspace of $\mathcal{L}_p(X, \mu, E)$, and

$$\mathcal{N} = \big\{ f \in \mathcal{L}_p(X, \mu, E) \ ; \ \|f\|_p = 0 \big\} \ .$$

(v) Putting $\rho(f, g) := \|f - g\|_p^p$ induces a metric on

$$L_p(X, \mu, E) := \mathcal{L}_p(X, \mu, E)/\mathcal{N} \ .$$

(vi) $\big(L_p(X, \mu, E), \rho\big)$ is complete.

(vii) For $f, g \in \mathcal{L}_p(X, \mu, \mathbb{R})$ with $f \geq 0$ and $g \geq 0$, we have $\|f + g\|_p \geq \|f\|_p + \|g\|_p$.

(viii) The map

$$L_p(X, \mu, \mathbb{R}) \to \mathbb{R}^+ , \quad [f] \mapsto \|f\|_p$$

is *not* a norm.

(Hints: (i) For $a > 0$, the map $\left[t \mapsto a^p + t^p - (a + t)^p \right]$ is increasing on $\mathbb{R}^+$. (ii) For $a > 0$, examine $\left[t \mapsto (a^{1/p} + t^{1/p})/(a + t)^{1/p} \right]$. (vi) Adapt the proof of Lemma 4.5 and Theorem 4.6. (vii) Theorem 4.2.)

14 Suppose $p_j \in [1, \infty]$ for $j = 1, \ldots, m$; let $1/r := \sum_{j=1}^m 1/p_j$. For $f_j \in L_{p_j}(X, \mu, \mathbb{K})$, show that $\prod_{j=1}^m f_j$ belongs to $L_r(X, \mu, \mathbb{K})$ and that

$$\left\| \prod_{j=1}^m f_j \right\|_r \leq \prod_{j=1}^m \|f_j\|_{p_j} .$$

(Hint: Hölder's inequality.)

15 Suppose X is a metric space. The function $f \in E^X$ **vanishes at infinity** if for every $\varepsilon > 0$ there is a compact subset K of X such that $|f(x)| < \varepsilon$ for all $x \in K^c$. Verify that

$$C_0(X, E) := \left\{ f \in C(X, E) ; \ f \text{ vanishes at infinity} \right\}$$

is the closure of $C_c(X, E)$ in $BUC(X, E)$.

16 For $f \in \mathcal{L}_0(X, \mu, E)$, set

$$\lambda_f(t) := \mu([\, |f| > t \,]) \quad \text{and} \quad f^*(t) := \inf\{ s \geq 0 ; \ \lambda_f(s) \leq t \} \quad \text{for } t \in [0, \infty) .$$

We call $f^* : [0, \infty) \to [0, \infty]$ the **decreasing rearrangement** of f. Prove:

(i) λ_f and f^* are decreasing, continuous from the right, and Lebesgue measurable.

(ii) If $|f| \leq |g|$ for $g \in \mathcal{L}_0(X, \mu, E)$, then $\lambda_f \leq \lambda_g$ and $f^* \leq g^*$.

(iii) If (f_j) is an increasing sequence such that $|f_j| \uparrow |f|$, then $\lambda_{f_j} \uparrow \lambda_f$ and $f_j^* \uparrow f^*$.

(iv) For $p \in (0, \infty)$,

$$\int_X |f|^p \, d\mu = p \int_{\mathbb{R}^+} t^{p-1} \lambda_f(t) \, \lambda_1(dt) = \int_{\mathbb{R}^+} (f^*)^p \, d\lambda_1 .$$

(v) $\|f\|_\infty = f^*(0)$.

(vi) $\lambda_f = \lambda_{f^*}$.

(Hint for (iv): Consider first simple functions and then apply (iii) together with Theorems 1.12 and 3.4.)

17 For $j \in \mathbb{N}$, let $I_{j,k} := \left[k2^{-j}, (k+1)2^{-j} \right]$ for $k = 0, \ldots, 2^{j-1}$. Further let $\{ J_n ; n \in \mathbb{N} \}$ be a relabeling of $\{ I_{j,k} ; j \in \mathbb{N}, k = 0, \ldots, 2^{j-1} \}$ and set $f_n := \chi_{J_n}$ for $j \in \mathbb{N}$. Prove that (f_n) is a null sequence in $L_p([0, 1])$ for every $p \in [1, \infty)$, even though $(f_n(x))$ diverges for every $x \in [0, 1]$.

18 Suppose (f_k) is a sequence in $L_p(X)$, where $1 \leq p < \infty$. We say that (f_k) **converges weakly** in $L_p(X)$ to $f \in L_p(X)$ if

$$\int_X f_k \varphi \, dx \to \int_X f\varphi \, dx \quad \text{for } \varphi \in L_{p'}(X) .$$

In this case, f is called a **weak limit** of (f_k) in $L_p(X)$.

Prove:

(i) Weak limits in $L_p(X)$ are unique.

(ii) Every convergent sequence in $L_p(X)$ converges weakly in $L_p(X)$.

(iii) If (f_k) converges weakly in $L_p(X)$ to f and converges μ-a.e. to $g \in L_p(X)$, then $f = g$.

(iv) If (f_k) converges weakly in $L_2(X)$ to f and $\|f_k\|_2 \to \|f\|_2$, then (f_k) converges in $L_2(X)$ to f.

(v) Let $e_k(t) := (2\pi)^{-1/2} e^{ikt}$ for $0 < t < 2\pi$ and $k \in \mathbb{N}$. Then the sequence (e_k) converges weakly to 0 in $L_2((0, 2\pi))$, even though it diverges in $L_2((0, 2\pi))$.

(Hints: (i) For $f \in L_p(X)$ consider $\varphi(x) := \overline{f(x)} \, |f(x)|^{p/p'-1}$.   (ii) Hölder's inequality. (iii) Show that $g \in L_p(X)$, so $[\,|g| = \infty\,]$ is a μ-null set. If $X_n := [\sup_{k \geq n} |f_k(x)| \geq n]$ then $\bigcap X_n$ is also a μ-null set. Now consider $\lim \int_{X_n^c} f_n \varphi \, dx$ for $\varphi \in L_{p'}(X)$.   (iv) Apply the parallelogram identity in $L_2(X)$.   (v) The first statement follows from Bessel's inequality, the second from (ii).)

5 The n-dimensional Bochner–Lebesgue integral

In this short section, we discuss the relationship between the Bochner–Lebesgue integral and the Cauchy–Riemann integral defined in Chapter VI. We show that every jump continuous function is Lebesgue measurable and that the corresponding integrals are equal. This connection will allow us to bring into Lebesgue integration theory the methods we developed for the Cauchy–Riemann integral.

We also show that a bounded scalar-valued function on a compact interval is Riemann integrable if and only if the set of its discontinuities has measure zero. From this it follows that there are Lebesgue integrable functions that are not Riemann integrable. Thus the Lebesgue integral is a proper extension of the Riemann integral—and therefore also of the Cauchy–Riemann integral.

In this entire section, suppose

- $X \subset \mathbb{R}^n$ is a λ_n-measurable set of positive measure;
  $E = (E, |\cdot|)$ is a Banach space.

Lebesgue measure spaces

From Exercise IX.1.7, we know that $\mathcal{L}_X := \mathcal{L}(n) \,|\, X$ is a σ-algebra over X. Thus the restriction $\lambda_n \,|\, X := \lambda_n \,|\, \mathcal{L}_X$ is a measure on X, called n-**dimensional Lebesgue measure** (or **Lebesgue n-measure**) on X. We denote this restriction by λ_n as well. We check easily that $(X, \mathcal{L}_X, \lambda_n)$ is a complete σ-finite measure space. If there is no danger of misunderstanding, we drop the qualifier "Lebesgue" (or "λ_n") from the words measurable, measure, integrable and so on.

If $f \in E^X$ is integrable, we call

$$\int_X f \, d\lambda_n := \int_X f \, d(\lambda_n \,|\, X) = \int_{\mathbb{R}^n} \tilde{f} \chi_X \, d\lambda_n$$

the (n-**dimensional**) (**Bochner–Lebesgue**) **integral** of f over X. The notations

$$\int_X f(x) \, d\lambda_n(x) \quad \text{and} \quad \int_X f(x) \, \lambda_n(dx)$$

are also common.

For short, we set

$$\mathcal{L}_p(X, E) := \mathcal{L}_p(X, \lambda_n, E) \text{ and } L_p(X, E) := L_p(X, \lambda_n, E) \ .$$

We also set $\mathcal{L}_p(X) := \mathcal{L}_p(X, \mathbb{K})$ and $L_p(X) := L_p(X, \mathbb{K})$ for $p \in [1, \infty] \cup \{0\}$.

The next theorem lists important properties of n-dimensional integrals.

5.1 Theorem *Suppose X is open in $\mathbb{R}^n$ or, in the case $n = 1$, a perfect interval. Then:*

(i) *λ_n is a massive Radon measure on X.*

(ii) *$C(X, E)$ is a vector subspace of $L_0(X, E)$.*

(iii) *$BC(X, E)$ is a closed vector subspace of $L_\infty(X, E)$.*

(iv) *$C_c(X, E)$ is a dense vector subspace of $L_p(X, E)$ for $p \in [1, \infty)$. If K is a compact subset of X, then*

$$\|f\|_p \le \lambda_n(K)^{1/p}\|f\|_\infty \quad \text{for } f \in C_c(X, E) \text{ such that } \operatorname{supp}(f) \subset K .$$

(v) *If X has finite measure and $1 \le p < q \le \infty$, then*

$$L_q(X, E) \xrightarrow{d} L_p(X, E)$$

and

$$\|f\|_p \le \lambda_n(X)^{1/p-1/q}\|f\|_q \quad \text{for } f \in L_q(X, E) .$$

Proof (i) X is a σ-compact metric space — by Remark 1.16(e) if X is open, and for obvious reasons if X is an interval. Now the claim follows from Remark 1.16(h) and Exercise IX.5.21.

(ii) and (iii) are covered respectively by Proposition 4.17 and Theorem 4.18(ii).

(iv) The first statement is a consequence of Theorem 4.18(i), and the second is obvious.

(v) is a special case of Theorem 4.16. ∎

5.2 Remark Suppose X is measurable and its boundary ∂X is a λ_n-null set. Then the Borel set $\mathring{X}$ belongs to $\mathcal{L}(n)$, and we have $\lambda_n(\mathring{X}) = \lambda_n(X)$. Further, one checks easily that the map

$$L_p(X, E) \to L_p(\mathring{X}, E) , \quad [f] \mapsto [f \mid \mathring{X}]$$

is a vector space isomorphism for $p \in [1, \infty] \cup \{0\}$. If $p \in [1, \infty]$, it is an isometry. Thus we can identify $L_p(X, E)$ and $L_p(\mathring{X}, E)$ for $p \in [1, \infty] \cup \{0\}$. In particular, for an interval X in $\mathbb{R}$ with endpoints $a := \inf X$ and $b := \sup X$, we have

$$L_p(X, E) = L_p([a, b], E) = L_p([a, b), E) = L_p((a, b], E) = L_p((a, b), E)$$

for $p \in [1, \infty] \cup \{0\}$.

The Lebesgue integral of absolutely integrable functions

We now show that every absolutely integrable function is Lebesgue integrable, and its integral in the sense of Section VI.8 equals the Lebesgue integral.

5.3 Theorem *Suppose* $f : (a, b) \to E$ *is absolutely integrable, where* $a, b \in \overline{\mathbb{R}}$ *and* $a < b$. *Then* f *belongs to* $\mathcal{L}_1((a, b), E)$, *and*

$$\int_{(a,b)} f \, d\lambda_1 = \int_a^b f \, .$$

Proof (i) Suppose $a < \alpha < \beta < b$. If $g : [\alpha, \beta] \to E$ is a staircase function, then g is obviously λ_1-simple and

$$\int_{(\alpha,\beta)} g \, d\lambda_1 = \int_\alpha^\beta g \, . \tag{5.1}$$

Now suppose $g : [\alpha, \beta] \to E$ is jump continuous. Then there is a sequence (g_j) of staircase functions that converges uniformly to g. Therefore g is measurable, and Remark VI.1.1(d) and Corollary 3.15(ii) show that g belongs to $\mathcal{L}_1((\alpha, \beta), E)$. Because g is bounded and the sequence (g_j) converges uniformly, there is an $M \geq 0$ such that $|g_j| \leq M$ for all $j \in \mathbb{N}$. Therefore it follows from Lebesgue's dominated convergence theorem that

$$\lim_{j \to \infty} \int_{(\alpha,\beta)} g_j \, d\lambda_1 = \int_{(\alpha,\beta)} g \, d\lambda_1$$

in E, and we conclude using (5.1) and the definition of the Cauchy–Riemann integral that

$$\int_\alpha^\beta g = \lim_{j \to \infty} \int_\alpha^\beta g_j = \lim_{j \to \infty} \int_{(\alpha,\beta)} g_j \, d\lambda_1 = \int_{(\alpha,\beta)} g \, d\lambda_1 \, .$$

(ii) We fix $c \in (a, b)$ and choose a sequence (β_j) in (c, b) such that $\beta_j \uparrow b$. We also set[1]

$$g := \chi_{[c,b)} f \quad , \quad g_j := \chi_{[c,\beta_j]} f \quad \text{for } j \in \mathbb{N} \, .$$

By (i), (g_j) is a sequence in $\mathcal{L}_1(\mathbb{R}, E)$. Obviously (g_j) converges pointwise to g and $(|g_j|)$ is an increasing sequence converging to $|g|$. Therefore g is measurable. From (i), it follows that

$$\int_{\mathbb{R}} |g_j| \, d\lambda_1 = \int_{(c,\beta_j)} |f| \, d\lambda_1 = \int_c^{\beta_j} |f| \, ,$$

and the absolute convergence of $\int_c^b f$ implies

$$\lim_{j \to \infty} \int_{\mathbb{R}} |g_j| \, d\lambda_1 = \lim_{j \to \infty} \int_c^{\beta_j} |f| = \int_c^b |f| \, . \tag{5.2}$$

---

[1]Here and in similar situations, we regard $\chi_{[c,b)} f$ as a function on $\mathbb{R}$. Writing $\chi_{[c,b)} \widetilde{f}$ would be more precise but cumbersome.

On the other hand, the monotone convergence theorem shows that

$$\int_{\mathbb{R}} |g| \, d\lambda_1 = \lim_{j \to \infty} \int_{\mathbb{R}} |g_j| \, d\lambda_1 \; ,$$

and we see from (5.2) that g belongs to $\mathcal{L}_1(\mathbb{R}, E)$. Therefore we can apply the dominated convergence theorem to the sequence (g_j), to get

$$\lim_{j \to \infty} \int_{\mathbb{R}} g_j \, d\lambda_1 = \int_{\mathbb{R}} g \, d\lambda_1 = \int_{[c,b)} f \, d\lambda_1$$

in E. Further, it follows from (i) that

$$\int_{\mathbb{R}} g_j \, d\lambda_1 = \int_{[c,\beta_j)} f \, d\lambda_1 = \int_c^{\beta_j} f \; ,$$

and hence, by Proposition VI.8.7,

$$\lim_{j \to \infty} \int_{\mathbb{R}} g_j \, d\lambda_1 = \lim_{j \to \infty} \int_c^{\beta_j} f = \int_c^b f$$

in E. Thus the limits $\int_{[c,b)} f \, d\lambda_1$ and $\int_c^b f$ are equal. In similar fashion, we show that $\chi_{(a,c]} f$ belongs to $\mathcal{L}_1(\mathbb{R}, E)$ and that $\int_{(a,c]} f \, d\lambda_1 = \int_a^c f$. This shows that f is Lebesgue integrable with $\int_{(a,b)} f \, d\lambda_1 = \int_a^b f$. ∎

5.4 Corollary For $-\infty < a < b < \infty$, we have $\mathcal{S}([a,b], E) \hookrightarrow \mathcal{L}_1([a,b], E)$ and

$$\int_{[a,b]} f \, d\lambda_1 = \int_a^b f \quad \text{for } f \in \mathcal{S}([a,b], E) \; .$$

Proof This follows from Theorem 5.3 and Proposition VI.8.3. ∎

5.5 Remarks Fix $a, b \in \overline{\mathbb{R}}$ with $a < b$.

(a) Suppose $f : (a,b) \to E$ is admissible and $\int_a^b f$ exists as an improper integral. Then f need not belong to $\mathcal{L}_1((a,b), E)$.

Proof We define $f : \mathbb{R} \to \mathbb{R}$ by

$$f(x) := \begin{cases} 0 & \text{if } x \in (-\infty, 0) \; , \\ (-1)^j/j & \text{if } x \in [j-1, j), \text{ where } j \in \mathbb{N}^{\times} \; . \end{cases}$$

Obviously f is admissible, and $\int_{-\infty}^{\infty} f$ exists in $\mathbb{R}$, since

$$\int_{-\infty}^{\infty} f = \sum_{j=1}^{\infty} (-1)^j/j \; .$$

If f belonged to $\mathcal{L}_1(\mathbb{R})$, we would have $\int_{\mathbb{R}} |f|\, d\lambda_1 < \infty$, contradicting the monotone convergence theorem, which gives

$$\int_{\mathbb{R}} |f|\, d\lambda_1 = \lim_{k\to\infty} \int_{\mathbb{R}} \chi_{[0,k]} |f|\, d\lambda_1 = \lim_{k\to\infty} \sum_{j=1}^{k} 1/j = \infty. \quad \blacksquare$$

(b) Suppose $f : (a,b) \to E$ is admissible and f belongs to $\mathcal{L}_1\big((a,b), E\big)$. Then f is absolutely integrable, and

$$\int_{(a,b)} f\, d\lambda_1 = \int_a^b f \qquad \text{in } E.$$

Proof Take $c \in (a,b)$ and let (α_j) be a sequence in (a,c) with $\alpha_j \to a$. Also let $f_j := \chi_{[\alpha_j,c]} f$. Then (f_j) converges pointwise to $\chi_{(a,c]} f$, and we have $|f_j| \le |f|$ for $j \in \mathbb{N}$. Because f is admissible, Proposition VI.4.3 shows that $|f|\,\big|\,[\alpha_j,c]$ belongs to $\mathcal{S}\big([\alpha_j,c],\mathbb{R}\big)$. Thus it follows from Corollary 5.4 and the dominated convergence theorem that

$$\int_{\alpha_j}^c |f| = \int_{[\alpha_j,c]} |f|\, d\lambda_1 \to \int_{(a,c]} |f|\, d\lambda_1.$$

Therefore $\int_a^c |f|$ exists. Analogously, we show the existence of $\int_c^b |f|$ and thus the absolute convergence of $\int_a^b f$. The second statement now follows from Theorem 5.3. $\blacksquare$

Suppose $f \in \mathcal{L}_1\big((a,b), E\big)$. Remark 5.5(b) shows that no misunderstanding should arise in this case if we denote $\int_{(a,b)} f\, d\lambda_1$ by $\int_a^b f$ or $\int_a^b f(x)\, dx$. From now on, we will usually write in the n-dimensional case

$$\int_X f\, dx := \int_X f\, d\lambda_n.$$

Theorem 5.3 and its corollary allow us to transfer the integration methods developed in Volume II to the framework of Lebesgue theory. In combination with the integrability criterion of Theorem 3.14 and the dominated convergence theorem, these provide very effective tools for proving the existence of integrals. This will be made clear in the remaining sections of this chapter, when we develop procedures for the concrete evaluation of "multidimensional" integrals.

A characterization of Riemann integrable functions

Theorem 5.3 showed that the Lebesgue integral is an extension of the Cauchy–Riemann integral. We now characterize Riemann integrable functions and show that this extension is proper.

5.6 Theorem *Let I be a compact interval, and let $f : I \to \mathbb{K}$ be bounded. Then f is Riemann integrable if and only if it is continuous λ_1-a.e. In this case, f is Lebesgue integrable, and the Riemann and Lebesgue integrals are equal.*

Proof (i) We can take without loss of generality the case $\mathbb{K} = \mathbb{R}$ and $I := [0,1]$. For $k \in \mathbb{N}$, let $\mathfrak{Z}_k := (\xi_{0,k}, \ldots, \xi_{2^k,k})$ be the partition of $[0,1]$ with $\xi_{j,k} := j \, 2^{-k}$ for $j = 0, \ldots, 2^k$. Also suppose

$$I_{0,k} := [\xi_{0,k}, \xi_{1,k}] \quad , \quad I_{j,k} := (\xi_{j,k}, \xi_{j+1,k}] \quad \text{for } j = 1, \ldots, 2^k - 1 \ .$$

Finally, set $\alpha_{j,k} := \inf_{x \in \bar{I}_{j,k}} f(x)$, $\beta_{j,k} := \sup_{x \in \bar{I}_{j,k}} f(x)$, and

$$g_k := \sum_{j=0}^{2^k - 1} \alpha_{j,k} \chi_{I_{j,k}} \ , \quad h_k := \sum_{j=0}^{2^k - 1} \beta_{j,k} \chi_{I_{j,k}} \quad \text{for } k \in \mathbb{N} \ .$$

Then (g_k) is an increasing and (h_k) a decreasing sequence of λ_1-simple functions. Therefore their pointwise limits $g := \lim_k g_k$ and $h := \lim_k h_k$ are defined and λ_1-measurable, and $g \leq f \leq h$. Furthermore, we have

$$\int_{[0,1]} g_k \, d\lambda_1 = \underline{S}(f,k) \quad \text{and} \quad \int_{[0,1]} h_k \, d\lambda_1 = \overline{S}(f,k) \ ,$$

where $\underline{S}(f,k)$ and $\overline{S}(f,k)$ stand for the lower and upper sums of f on $[0,1]$ with respect to the partition $\mathfrak{Z}_k$ (see Exercise VI.3.7). Denoting by $\underline{\int} f$ and $\overline{\int} f$ the lower and upper Riemann integrals of f on $[0,1]$, we find from the monotone convergence theorem that

$$\int_{[0,1]} (h - g) \, d\lambda_1 = \overline{\int} f - \underline{\int} f \ . \tag{5.3}$$

(ii) Let $R := \bigcup_{k \in \mathbb{N}} \{\xi_{0,k}, \ldots, \xi_{2^k,k}\}$ be the set of endpoints of the intervals $I_{j,k}$. Let C be the set of continuous points of f. Then

$$[g = h] \cap R^c \subset C \subset [g = h] \ . \tag{5.4}$$

To see this, take $\varepsilon > 0$. Suppose first that $x_0 \in R^c$ and $g(x_0) = h(x_0)$. We can find a $k \in \mathbb{N}$ such that $h_k(x_0) - g_k(x_0) < \varepsilon$ and a $j \in \{0, \ldots, 2^k - 1\}$ such that x_0 lies in the interval $(\xi_{j,k}, \xi_{j+1,k})$. For $x \in I_{j,k}$, we thus have

$$|f(x) - f(x_0)| \leq \sup_{y \in \bar{I}_{j,k}} f(y) - \inf_{y \in \bar{I}_{j,k}} f(y) = h_k(x_0) - g_k(x_0) < \varepsilon \ ,$$

which proves the continuity of f at x_0.

Now suppose $x_0 \in C$. Take $\delta > 0$ such that $|f(x) - f(x_0)| < \varepsilon/2$ for $x \in [x_0 - \delta, x_0 + \delta] \cap [0,1]$. Choose $k_0 \in \mathbb{N}$ with $2^{-k_0} \leq \delta$ and take for every $k \geq k_0$ a $j \in \{0, \ldots, 2^k - 1\}$ such that $x_0 \in I_{j,k} \subset [x_0 - \delta, x_0 + \delta]$. Then

$$0 \leq h_k(x_0) - g_k(x_0) = \sup_{x \in \bar{I}_{j,k}} \big(f(x) - f(x_0)\big) - \inf_{x \in \bar{I}_{j,k}} \big(f(x) - f(x_0)\big) < \varepsilon \ .$$

It follows that $h(x_0) - g(x_0) = \lim_k \big(h_k(x_0) - g_k(x_0)\big) = 0$. This proves (5.4).

(iii) If f is a Riemann integrable function, then $\underline{\int} f = \overline{\int} f = \int f$ (Exercise VI.3.10). Therefore (5.3) shows that

$$h = g = f \qquad \lambda_1\text{-a.e.} \,, \tag{5.5}$$

which implies the λ_1-measurability of f. Since f is bounded, $f \in \mathcal{L}_1([0,1])$. We also have $|g_k| \leq \|f\|_\infty$ λ_1-a.e. for $k \in \mathbb{N}$. Then Lebesgue's dominated convergence theorem results in

$$\int_{[0,1]} g \, d\lambda_1 = \lim_k \int_{[0,1]} g_k \, d\lambda_1 = \lim_k \underline{S}(f,k) = \int_0^1 f \,,$$

where, in the last equality, we have once more used Exercise VI.3.10. From (5.5) and Lemma 2.15, it follows that $\int_{[0,1]} f \, d\lambda_1 = \int_0^1 f$. Finally (5.4), (5.5), and the countability of R imply that the discontinuous points of f form a set of Lebesgue measure zero.

(iv) Suppose conversely that C^c has measure zero. By (5.4), so does $[g \neq h]$, and the Riemann integrability of f follows from (5.3). This finishes the proof. ∎

5.7 Corollary *Some Lebesgue integrable functions are not Riemann integrable. Thus the Lebesgue integral is a proper extension of the Riemann integral.*

Proof Consider the Dirichlet function

$$f : [0,1] \to \mathbb{R} \,, \quad f(x) := \begin{cases} 1 & \text{if } x \in \mathbb{Q} \,, \\ 0 & \text{if } x \notin \mathbb{Q} \,, \end{cases}$$

on $[0,1]$. By Lemma 2.15, f belongs to $\mathcal{L}_1([0,1])$, since f vanishes almost everywhere. But we know from Example III.1.3(c) that f is nowhere continuous, hence not Riemann integrable by Theorem 5.6. ∎

The equivalence class of maps that agree a.e. with the Dirichlet function contains Riemann integrable functions — for example, the null function. So this example is uninteresting from the viewpoint of L_1-spaces. However, in Exercise 13, it will be shown that there exists $f \in \mathcal{L}_1([0,1], \mathbb{R})$ such that no $g \in [f]$ is Riemann integrable. This implies that the Riemann integral is inadequate for the theory of L_p-spaces.

Exercises

1   For $p, q \in [1, \infty]$ with $p \neq q$, show that $L_p(\mathbb{R}, E) \not\subset L_q(\mathbb{R}, E)$.

2   Suppose J is an open interval and $f \in C^1(J, E)$ has compact support. Then $\int_J f' = 0$.

3   Suppose $f \in \mathcal{L}_0([0,1], \mathbb{R}^+)$ is bounded. Show that

$$\underline{\int} f \leq \int_{[0,1]} f \, d\lambda_1 \leq \overline{\int} f \,.$$

4 Suppose I is a compact interval, and define the **space of functions of bounded variation on I** by

$$BV(I, E) := \{ f : I \to E \ ; \ \mathrm{Var}(f, I) < \infty \} .$$

(a) In the sense of vector subspaces, we have the inclusions

$$C^{1-}(I, E) \subset BV(I, E) \subset B(I, E) .$$

(b) Let $\alpha := \inf I$ and $f \in \mathcal{L}_1(I, E)$. Then $F : I \to E$, $x \mapsto \int_\alpha^x f(t)\, dt$ belongs to $BV(I, E)$, and $\mathrm{Var}(F, I) \leq \|f\|_1$.

(c) For every $f \in BV(I, \mathbb{R})$, there are increasing maps $s^\pm : I \to \mathbb{R}$ such that $f = s^+ - s^-$.

(d) $BV(I, \mathbb{R})$ is a vector subspace of the space $\mathcal{S}(I, \mathbb{R})$ of jump continuous functions $I \to \mathbb{R}$.

(e) Every monotone function belongs to $BV(I, \mathbb{R})$.

(Hint for (c): For $\alpha := \inf I$, consider the functions $s^+ := \left(x \mapsto \mathrm{Var}(f^+, [\alpha, x]) \right)$ and $s^- := s - f$.)

5 Suppose H is a separable Hilbert space. Show[2] that $BV\big([a, b], H\big)$ is a vector subspace of $\mathcal{L}_\infty\big([a, b], H\big)$ and that

$$\int_a^{b-h} \|f(t+h) - f(t)\|\, dt \leq h \, \mathrm{Var}(f, [a, b]) \quad \text{for } 0 < h < b - a .$$

(Hints: Note Exercises 1.1 and 4(d). For $0 < h < b - a$ and $t \in [a, b - h]$, show that $\|f(t+h) - f(t)\| \leq \mathrm{Var}(f, [a, t+h]) - \mathrm{Var}(f, [a, t])$.)

6 Suppose $J \subset \mathbb{R}$ is a perfect interval. A function $f : J \to E$ is **absolutely continuous** if for every $\varepsilon > 0$ there is $\delta > 0$ such that

$$\sum_{k=0}^m |f(\beta_k) - f(\alpha_k)| < \varepsilon$$

for every finite family $\{ (\alpha_k, \beta_k) \ ; \ k = 0, \ldots, m \}$ of pairwise disjoint subintervals of J with $\sum_{k=0}^m (\beta_k - \alpha_k) < \delta$. We denote by $W_1^1(J, E)$ the set of all absolutely continuous functions in E^J. Prove:

(a) In the sense of vector subspaces, we have the inclusions

$$BC^1(J, E) \subset W_1^1(J, E) \subset C(J, E) .$$

(b) If J compact, then $W_1^1(J, E) \subset BV(J, E)$.

(c) The Cantor function (Exercise III.3.8) is continuous but not absolutely continuous.

(d) Set $\alpha := \inf J$ and take $f \in \mathcal{L}_1(J, E)$. Then $F : J \to E$, $x \mapsto \int_\alpha^x f(t)\, dt$ is absolutely continuous.

---

[2]One can show that the statement of Exercise 5 remains true if H is replaced by an arbitrary Banach space.

7 For $j = 1, 2$, define $f_j : [0, 1] \to \mathbb{R}$ by

$$f_j(x) := \begin{cases} x^2 \sin(1/x^j) & \text{if } x \in (0, 1] , \\ 0 & \text{if } x = 0 ; \end{cases}$$

compare Exercise IV.1.2. Prove:

(a) $f_1 \in BV([0, 1], \mathbb{R})$.

(b) $f_2 \notin BV([0, 1], \mathbb{R})$.

8 Let μ and ν be measures on a measurable space $(X, \mathcal{A})$. We say ν is **μ-absolutely continuous** if every μ-null set is also a ν-null set. In this case, we write $\nu \ll \mu$.

(a) Let $(X, \mathcal{A}, \mu)$ be a σ-finite complete measure space. For $f \in \mathcal{L}_0(X, \mu, \overline{\mathbb{R}}^+)$, define

$$f \cdot \mu : \mathcal{A} \to [0, \infty] , \quad A \mapsto \int_A f \, d\mu .$$

Show that $f \cdot \mu$ is a complete measure on $(X, \mathcal{A})$ with $f \cdot \mu \ll \mu$.

(b) Let $\mathcal{A} := \mathcal{L}_{[0,1]}$, $\nu := \lambda_1$, and $\mu := \mathcal{H}^0$. Check:

  (i) $\nu \ll \mu$.

  (ii) there is no $f \in \mathcal{L}_0([0, 1], \mathcal{A}, \mu)$ such that $\nu = f \cdot \mu$.

9 Suppose $(X, \mathcal{A}, \nu)$ is a finite measure space and μ is measure on $(X, \mathcal{A})$. The following statements are equivalent:

  (i) $\nu \ll \mu$.

  (ii) For every $\varepsilon > 0$ there is $\delta > 0$ such that $\nu(A) < \varepsilon$ for all $A \in \mathcal{A}$ with $\mu(A) < \delta$.

10 For $f \in \mathcal{L}_0(\mathbb{R}, \lambda_1, \overline{\mathbb{R}}^+)$, let $F(x) := \int_{-\infty}^x f(t) \, dt$ for $x \in \mathbb{R}$, and denote by μ_F the Lebesgue–Stieltjes measure on $\mathbb{R}$ generated by F. Prove:

(a) $F \in W_1^1(\mathbb{R}, \mathbb{R})$ implies $\mu_F \ll \lambda_1$.

(b) $\mu_F \ll \beta_1$ implies $F \in W_1^1(\mathbb{R}, \mathbb{R})$ if μ_F is finite.

11 Let I is an interval and take $f \in \mathcal{L}_1(I, \mathbb{R}^n)$. For a fixed $a \in I$, suppose $\int_a^x f(t) \, dt = 0$ for $x \in I$. Show that $f(x) = 0$ for almost every $x \in I$.

12 Let $0 \le a < b < \infty$ and $I := (-b, -a) \cup (a, b)$, and suppose $f \in \mathcal{L}_1(I, E)$. Show that $\int_I f \, dx = 0$ if f is odd, and $\int_I f \, dx = 2 \int_a^b f \, dx$ if f is even.

13 Define
$$K_0 := [0, 1] ,$$
$$K_1 := K_0 \backslash (3/8, 5/8) ,$$
$$K_2 := K_1 \backslash \left((5/32, 7/32) \cup (25/32, 27/32) \right) , \ \ldots$$

Generally, K_{n+1} is derived from K_n by the removal of open "middle fourths" of length $(1/4)^{n+1}$, rather than middle thirds as in the construction of the traditional Cantor set (Exercise III.3.8). Set $K := \bigcap K_n$ and $f := \chi_K$. Show that f belongs to $\mathcal{L}_1([0, 1])$ and that no $g \in [f]$ is Riemann integrable.

6 Fubini's theorem

The heart of this section is the proof that the Lebesgue integral of functions of multiple variables can be calculated iteratively and that this sequence of one-dimensional integrations can be performed in any order. Therefore multivariable integration reduces to integrating functions of only one variable. With the results of the previous section and the procedures developed in Volume II, multidimensional integrals can be calculated explicitly in many cases.

The method of iterative evaluation of integrals has wide-reaching theoretical applications, a few of which we will present.

Throughout this section, we suppose

- m, n are positive integers and E is a Banach space.

In addition, we will generally identify $\mathbb{R}^{m+n}$ with $\mathbb{R}^m \times \mathbb{R}^n$.

Maps defined almost everywhere

Suppose $(X, \mathcal{A}, \mu)$ is a measure space. We will often consider nonnegative $\overline{\mathbb{R}}$-valued functions that are only defined μ-a.e. For these, we shall simply write $x \mapsto f(x)$, without specifying the precise domain of definition. We say such a function $x \mapsto f(x)$ is **measurable** if there is a μ-null set N such that $f \mid N^c \colon N^c \to \overline{\mathbb{R}}^+$ is defined and μ-measurable. Therefore $\int_{N^c} f \, d\mu$ is defined. If M is another μ-null set such that $f \mid M^c \colon M^c \to \overline{\mathbb{R}}^+$ is defined and μ-measurable, the equalities $\mu(N) = \mu(M) = \mu(M \cup N) = 0$ and Remarks 3.3(a) and (b) imply that

$$\int_{N^c} f \, d\mu = \int_{M^c \cap N^c} f \, d\mu = \int_{M^c} f \, d\mu \ .$$

Therefore

$$\int_X f \, d\mu := \int_{N^c} f \, d\mu \tag{6.1}$$

is well defined and independent of the chosen null set N.

For an E-valued function $x \mapsto f(x)$ defined μ-a.e., we define measurability just as above. We say such an f is **integrable** if $f \mid N^c$ belongs to $\mathcal{L}_1(N^c, \mu, E)$. In this case, $\int_X f \, d\mu$ is also defined through (6.1), and Lemma 2.15 shows this definition is meaningful.

Consider for example $A \in \mathcal{L}(m+n)$, and assume that the cross section $A_{[x]}$ is λ_n-measurable for λ_m-almost every $x \in \mathbb{R}^m$. Then $x \mapsto \lambda_n(A_{[x]})$ is a nonnegative $\overline{\mathbb{R}}$-valued function defined λ_m-a.e. If $x \mapsto \lambda_n(A_{[x]})$ is measurable, the integral $\int_{\mathbb{R}^m} \lambda_n(A_{[x]}) \, dx$ is well defined.

Cavalieri's principle

We denote by $\mathcal{C}(m, n)$ the set of all $A \in \mathcal{L}(m + n)$ for which

  (i) $A_{[x]} \in \mathcal{L}(n)$ for λ_m-almost every $x \in \mathbb{R}^m$;

  (ii) $x \mapsto \lambda_n(A_{[x]})$ is λ_m-measurable;

(iii) $\lambda_{m+n}(A) = \int_{\mathbb{R}^m} \lambda_n(A_{[x]}) \, dx$.

We want to show that $\mathcal{C}(m, n)$ agrees with $\mathcal{L}(m + n)$, but we need some preliminaries.

6.1 Remarks (a) Suppose $A \in \mathcal{C}(1, n)$ is bounded and $\mathrm{pr}_1(A)$ is an interval with endpoints a and b. Then

$$\lambda_{n+1}(A) = \int_a^b \lambda_n(A_{[x]}) \, dx \ .$$

This statement is called **Cavalieri's principle** and makes precise the geometric idea that the measure (volume) of A can be determined by partitioning A into thin parallel slices and continuously summing (integrating) the volumes of these slices.

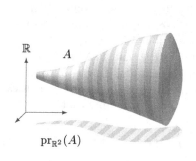
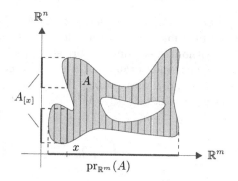

(b) $\mathcal{L}(m) \boxtimes \mathcal{L}(n) \subset \mathcal{C}(m, n)$.

(c) For every ascending sequence (A_j) in $\mathcal{C}(m, n)$, the union $\bigcup_j A_j$ belongs to $\mathcal{C}(m, n)$.

Proof (i) For $j \in \mathbb{N}$, let M_j be a λ_m-null set such that $A_{j,[x]} := (A_j)_{[x]} \in \mathcal{L}(n)$ for $x \in M_j^c$. Letting $A := \bigcup_j A_j$ and $M := \bigcup_j M_j$, we then have $A_{[x]} = \bigcup_j A_{j,[x]} \in \mathcal{L}(n)$ for $x \in M^c$. The continuity of λ_n from below implies $\lambda_n(A_{[x]}) = \lim_j \lambda_n(A_{j,[x]})$ for $x \in M^c$, and we conclude with the help of Proposition 1.11 that $x \mapsto \lambda_n(A_{[x]})$ is λ_m-measurable.

  (ii) Because $A_j \in \mathcal{C}(m, n)$, we have

$$\int_{\mathbb{R}^m} \lambda_n(A_{j,[x]}) \, dx = \lambda_{m+n}(A_j) \quad \text{for } j \in \mathbb{N} \ ,$$

and from the monotone convergence theorem, it follows that

$$\lim_j \int_{\mathbb{R}^m} \lambda_n(A_{j,[x]}) \, dx = \int_{\mathbb{R}^m} \lambda_n(A_{[x]}) \, dx \ . \tag{6.2}$$

The continuity of λ_{m+n} from below thus shows that

$$\lambda_{m+n}(A) = \lim_j \lambda_{m+n}(A_j) = \lim_j \int_{\mathbb{R}^m} \lambda_n(A_{j,[x]}) \, dx = \int_{\mathbb{R}^m} \lambda_n(A_{[x]}) \, dx \; .$$

Therefore A belongs to $\mathcal{C}(m,n)$. ∎

(d) Suppose (A_j) is a descending sequence in $\mathcal{C}(m,n)$ and there is a $k \in \mathbb{N}$ such that $\lambda_{m+n}(A_k) < \infty$. Then $\bigcap_j A_j$ belongs to $\mathcal{C}(m,n)$.

Proof We set $A := \bigcap_j A_j$. The measurability of λ_m-almost all cross sections $A_{[x]}$ and of $x \mapsto \lambda_n(A_{[x]})$ follow as in (c). Next, Lebesgue's dominated convergence theorem shows that (6.2) is true in this case. The claim now follows as in (c). ∎

(e) Suppose (A_j) is a disjoint sequence in $\mathcal{C}(m,n)$. Then $\bigcup_j A_j$ also belongs to $\mathcal{C}(m,n)$.

Proof Because of (c), it suffices to prove the statement for finite disjoint sequences. We leave this to the reader as an exercise. ∎

(f) Every open set in $\mathbb{R}^{m+n}$ belongs to $\mathcal{C}(m,n)$.

Proof This follows from Proposition IX.5.6, (e) and (b). ∎

(g) Every bounded G_δ-set in $\mathbb{R}^{m+n}$ belongs to $\mathcal{C}(m,n)$.

Proof This follows from (f) and (d). ∎

(h) Suppose A is a λ_{m+n}-null set. Then A belongs to $\mathcal{C}(m,n)$, and there is a λ_m-null set M such that $A_{[x]}$ is a λ_n-null set for every $x \in M^c$.

Proof It suffices to verify there is a λ_m-null set M such that $\lambda_n(A_{[x]}) = 0$ for $x \in M^c$. So let $A_j := A \cap (j\mathbb{B}^{m+n})$ for $j \in \mathbb{N}$. Then (A_j) is an ascending sequence of bounded λ_{m+n}-null sets with $\bigcup_j A_j = A$. By Corollary IX.5.5, there is a sequence (G_j) of bounded G_δ-sets with $G_j \supset A_j$ and $\lambda_{m+n}(G_j) = 0$ for $j \in \mathbb{N}$. From (g), it therefore follows that

$$0 = \lambda_{m+n}(G_j) = \int_{\mathbb{R}^m} \lambda_n(G_{j,[x]}) \, dx \; .$$

Hence, there is for every $j \in \mathbb{N}$ a λ_m-null set M_j such that $\lambda_n(G_{j,[x]}) = 0$ for $x \in M_j^c$ (see Remark 3.3(c)). Because

$$\bigcup_j G_{j,[x]} \supset \bigcup_j A_{j,[x]} = \Big(\bigcup_j A_j \Big)_{[x]} = A_{[x]} \quad \text{for } x \in \mathbb{R}^m \; ,$$

$M := \bigcup_j M_j$ has the desired property. ∎

After these remarks, we can now show the equality of $\mathcal{C}(m,n)$ and $\mathcal{L}(m+n)$.

6.2 Proposition $\mathcal{C}(m,n) = \mathcal{L}(m+n)$.

Proof We need only check the inclusion $\mathcal{L}(m+n) \subset \mathcal{C}(m,n)$.

(i) Suppose $A \in \mathcal{L}(m+n)$ is bounded. By Corollary IX.5.5, there is a bounded G_δ-set G such that $G \supset A$ and $\lambda_{m+n}(G) = \lambda_{m+n}(A)$. Because A has finite measure, $G \setminus A$ is a bounded λ_{m+n}-null set by Proposition IX.2.3(ii), and we conclude using Remark 6.1(h) that $(G \setminus A)_{[x]} = G_{[x]} \setminus A_{[x]}$ is a λ_n-null set for λ_m-almost every $x \in \mathbb{R}^m$. By Remark 6.1(g), $G_{[x]}$ belongs to $\mathcal{L}(n)$ for λ_m-almost every $x \in \mathbb{R}^m$. Because

$$A_{[x]} = G_{[x]} \cap (G_{[x]} \setminus A_{[x]})^c \quad \text{for } x \in \mathbb{R}^m ,$$

this is also true of λ_m-almost every intersection $A_{[x]}$. In addition, $\lambda_n(A_{[x]}) = \lambda_n(G_{[x]})$ for λ_m-almost every $x \in \mathbb{R}^m$. We know by Remark 6.1(g) that G belongs to $\mathcal{C}(m,n)$. Therefore $x \mapsto \lambda_n(A_{[x]})$ is measurable, and

$$\lambda_{m+n}(G) = \int_{\mathbb{R}^m} \lambda_n(G_{[x]}) \, dx = \int_{\mathbb{R}^m} \lambda_n(A_{[x]}) \, dx .$$

Therefore A belongs to $\mathcal{C}(m,n)$.

(ii) If A is not bounded, we set $A_j := A \cap (j\mathbb{B}^{m+n})$ for $j \in \mathbb{N}$. Then (A_j) is an ascending sequence in $\mathcal{L}(m+n)$ with $\bigcup_j A_j = A$. The claim now follows from (i) and Remark 6.1(c). ∎

6.3 Corollary If $A \in \mathcal{L}(m+n)$ has finite measure, then $\lambda_n(A_{[x]}) < \infty$ for λ_m-a.e. $x \in \mathbb{R}^m$.

Proof Because Proposition 6.2 implies

$$\int_{\mathbb{R}^m} \lambda_n(A_{[x]}) \, dx = \lambda_{m+n}(A) < \infty ,$$

the claim follows from Remark 3.11(c). ∎

For $A \in \mathcal{L}(m+n)$ and $x \in \mathbb{R}^m$, we have $\chi_A(x, \cdot) = \chi_{A_{[x]}}$, so Proposition 6.2 can also be formulated in terms of characteristic functions. It is then easy to apply the statement to linear combinations of characteristic functions and therefore to simple functions.

6.4 Lemma Suppose $f \in \mathcal{S}(\mathbb{R}^{m+n}, E)$.
(i) $f(x, \cdot) \in \mathcal{S}(\mathbb{R}^n, E)$ for λ_m-almost every $x \in \mathbb{R}^m$.
(ii) the E-valued function $x \mapsto \int_{\mathbb{R}^n} f(x,y) \, dy$ is λ_m-integrable.
(iii) $\int_{\mathbb{R}^{m+n}} f \, d(x,y) = \int_{\mathbb{R}^m} \left[\int_{\mathbb{R}^n} f(x,y) \, dy \right] dx$.

Proof (i) With $f = \sum_{j=0}^k e_j \chi_{A_j}$, we have $f(x, \cdot) = \sum_{j=0}^k e_j \chi_{A_{j,[x]}}$ for $x \in \mathbb{R}^m$. Then it follows easily from Proposition 6.2 and Corollary 6.3 that there is a λ_m-null set M such that $f(x, \cdot)$ belongs to $\mathcal{S}(\mathbb{R}^n, E)$ for every $x \in M^c$.

(ii) We set

$$g(x) := \int_{\mathbb{R}^n} f(x,y) \, dy = \sum_{j=0}^k e_j \lambda_n(A_{j,[x]}) \quad \text{for } x \in M^c . \tag{6.3}$$

Then Proposition 6.2 and Remark 1.2(d) show that $x \mapsto g(x)$ is λ_m-measurable. In addition, we have

$$
\int_{\mathbb{R}^m} |g| \, dx \leq \sum_{j=0}^{k} |e_j| \int_{\mathbb{R}^m} \lambda_n(A_{j,[x]}) \, dx = \sum_{j=0}^{k} |e_j| \lambda_{m+n}(A_j) < \infty .
$$

Therefore $x \mapsto g(x)$ is λ_m-integrable.

(iii) Finally, it follows from Proposition 6.2 and (6.3) that

$$
\int_{\mathbb{R}^{m+n}} f \, d(x,y) = \sum_{j=0}^{k} e_j \lambda_{m+n}(A_j) = \sum_{j=0}^{k} e_j \int_{\mathbb{R}^m} \lambda_n(A_{j,[x]}) \, dx = \int_{\mathbb{R}^m} g \, dx
$$

$$
= \int_{\mathbb{R}^m} \left[\int_{\mathbb{R}^n} f(x,y) \, dy \right] dx ,
$$

which completes the proof. ∎

6.5 Remark In the definition of the set $\mathcal{C}(m,n)$, we chose to single out the first m coordinates of $\mathbb{R}^{m+n}$. We could just as well have chosen the last n coordinates and made the same argument not with $\lambda_n(A_{[x]})$ but with $\lambda_m(A^{[y]})$ for λ_n-almost every $y \in \mathbb{R}^n$. With this definition of $\mathcal{C}(m,n)$, we would obviously have found that $\mathcal{C}(m,n) = \mathcal{L}(m+n)$. Thus the roles of x and y in Lemma 6.4 can be exchanged, and we conclude that, for $f \in \mathcal{S}(\mathbb{R}^{m+n}, E)$,

(i) $f(\cdot, y) \in \mathcal{S}(\mathbb{R}^m, E)$ for λ_n-almost every $y \in \mathbb{R}^n$;

(ii) the E-valued function $y \mapsto \int_{\mathbb{R}^m} f(x,y) \, dx$ is λ_n-integrable;

(iii) $\int_{\mathbb{R}^{m+n}} f \, d(x,y) = \int_{\mathbb{R}^n} \left[\int_{\mathbb{R}^m} f(x,y) \, dx \right] dy$.

In particular, we find

$$
\int_{\mathbb{R}^m} \left[\int_{\mathbb{R}^n} f(x,y) \, dy \right] dx = \int_{\mathbb{R}^n} \left[\int_{\mathbb{R}^m} f(x,y) \, dx \right] dy
$$

for $f \in \mathcal{S}(\mathbb{R}^{m+n}, E)$. In other words, the integral $\int_{\mathbb{R}^{m+n}} f \, d(x,y)$ can be calculated iteratively in the case of simple functions, and the order in which the integrals are performed is irrelevant. ∎

Applications of Cavalieri's principle

The main result of this section is that the statement of Remark 6.5 about the iterative calculation of integrals remains true for arbitrary integrable functions f. Before we prove this theorem, we first give a few applications of Cavalieri's principle, meaning that we are working in the case $f = \chi_A$.

6.6 Examples   **(a)** (geometric interpretation of the integral)   For $M \in \mathcal{L}(m)$ and $f \in \mathcal{L}_0(M, \mathbb{R}^+)$, the set

$$S_f := S_{f,M} := \left\{ (x,y) \in \mathbb{R}^m \times \mathbb{R} \; ; \; 0 \le y \le f(x), \; x \in M \right\}$$

belongs to $\mathcal{L}(m+1)$, and

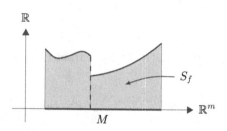

$$\int_M f \, dx = \lambda_{m+1}(S_f) \, ,$$

that is, the integral $\int_M f \, dx$ equals the $(m+1)$-dimensional Lebesgue measure of the set of points under the graph of f.[1]

Proof   Set $f_1 := \mathrm{pr}_{\mathbb{R}}$ and $f_2 := f \circ \mathrm{pr}_{\mathbb{R}^m}$. Then f_1 and f_2 belong to $\mathcal{L}_0(M \times \mathbb{R}, \overline{\mathbb{R}}^+)$, and $S_f = [0 \le f_1 \le f_2]$. Therefore Proposition 1.9 implies the λ_{m+1}-measurability of S_f. Because $(S_f)_{[x]} = \big[0, f(x)\big]$ for $x \in M$, it follows that $\lambda_1\big((S_f)_{[x]}\big) = f(x)$, and hence

$$\lambda_{m+1}(S_f) = \int_{\mathbb{R}^m} \lambda_1\big((S_f)_{[x]}\big) \, dx = \int_M f \, dx \, ,$$

by Proposition 6.2. ∎

(b) (substitution rule for linear maps)   Suppose $T \in \mathcal{L}(\mathbb{R}^m)$, $a \in \mathbb{R}^m$ and $M \in \mathcal{L}(m)$. Also let $\varphi(x) := a + Tx$ for $x \in \mathbb{R}^m$ and $f \in \mathcal{L}_1(\varphi(M))$. Then $f \circ \varphi$ belongs to $\mathcal{L}_1(M)$, and

$$\int_{\varphi(M)} f \, dy = |\det T| \int_M (f \circ \varphi) \, dx \, . \tag{6.4}$$

In particular, the Lebesgue integral is **affine isometry invariant**, that is, for every affine isometry φ of $\mathbb{R}^m$, we have

$$\int_{\mathbb{R}^m} f = \int_{\mathbb{R}^m} f \circ \varphi \quad \text{for } f \in \mathcal{L}_1(\mathbb{R}^m) \, .$$

Proof   (i) By Theorem IX.5.12, φ maps the σ-algebra $\mathcal{L}(m)$ into itself. Therefore $\varphi(M)$ belongs to $\mathcal{L}(m)$, and Theorem 1.4 implies that $f \circ \varphi$ lies in $\mathcal{L}_0(M)$. The decomposition $f = f_1 - f_2 + i(f_3 - f_4)$ with $f_j \in \mathcal{L}_1(\varphi(M), \mathbb{R}^+)$ shows that we can limit ourselves to the case of $f \in \mathcal{L}_1(\varphi(M), \mathbb{R}^+)$. Then (a) says that

$$\int_{\varphi(M)} f = \lambda_{m+1}(S_{f, \varphi(M)}) \, , \quad \int_M f \circ \varphi = \lambda_{m+1}(S_{f \circ \varphi, M}) \, . \tag{6.5}$$

(ii) We set $\widehat{a} := (a, 0) \in \mathbb{R}^m \times \mathbb{R}$ and $\widehat{T}(x,t) := (Tx, t)$ for $(x,t) \in \mathbb{R}^m \times \mathbb{R}$. Then $\widehat{a} + \widehat{T}(S_{f \circ \varphi}) = S_f$ and $\det T = \det \widehat{T}$, because the representation matrix $\widehat{T}$ has the block

---

[1]Compare the introductory remarks to Section VI.3.

structure

$$[\widehat{T}] = \begin{bmatrix} [T] & 0 \\ 0 & 1 \end{bmatrix}.$$

Corollary IX.5.23 and Theorem IX.5.25 therefore imply

$$\lambda_{m+1}(S_f) = \lambda_{m+1}\big(\widehat{T}(S_{f\circ\varphi})\big) = |\det T|\,\lambda_{m+1}(S_{f\circ\varphi})\,,$$

which, due to (6.5), proves (6.4). The integrability of $f\circ\varphi$ follows from Remark 3.11(a). ∎

(c) (the volume of the unit ball in $\mathbb{R}^m$) For $m \in \mathbb{N}^\times$, we have

$$\lambda_m(\mathbb{B}^m) = \frac{\pi^{m/2}}{\Gamma(1+m/2)}\,;$$

in particular, $\lambda_1(\mathbb{B}^1) = 2$, $\lambda_2(\mathbb{B}^2) = \pi$, and $\lambda_3(\mathbb{B}^3) = 4\pi/3$.

Proof Setting $\omega_m := \lambda_m(\mathbb{B}^m)$, we obtain from Cavalieri's principle and Remarks IX.5.26(b) and 6.5 that

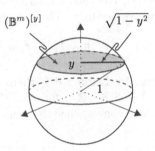

$$\omega_m = \int_{-1}^{1} \lambda_{m-1}\big((\mathbb{B}^m)^{[y]}\big)\,dy$$

$$= \int_{-1}^{1} \lambda_{m-1}\big(\sqrt{1-y^2}\,\mathbb{B}^{m-1}\big)\,dy$$

$$= \omega_{m-1}\int_{-1}^{1} \big(\sqrt{1-y^2}\,\big)^{m-1}\,dy\,.$$

To calculate the integral

$$B_m := \int_{-1}^{1} (1-y^2)^{(m-1)/2}\,dy = 2\int_{0}^{1}(1-y^2)^{(m-1)/2}\,dy \quad \text{for } m \in \mathbb{N}^\times\,,$$

we let $y = -\cos x$, so that $dy = \sin x\,dx$. This gives $B_m = 2\int_0^{\pi/2}\sin^m x\,dx$. It follows from the proof of Example VI.5.5(d) that

$$B_{2m} = \frac{(2m-1)(2m-3)\cdot\,\cdots\,\cdot 1}{2m(2m-2)\cdot\,\cdots\,\cdot 2}\cdot\pi\,,\quad B_{2m+1} = \frac{2m(2m-2)\cdot\,\cdots\,\cdot 2}{(2m+1)(2m-1)\cdot\,\cdots\,\cdot 1}\cdot 2\,.$$

Thus we find $B_m B_{m-1} = 2\pi/m$ and

$$\omega_m = B_m\omega_{m-1} = B_m B_{m-1}\omega_{m-2} = \frac{2\pi}{m}\omega_{m-2}\,. \tag{6.6}$$

Since $\omega_1 = 2$, we obtain $\omega_2 = B_2\omega_1 = 2B_2 = \pi$ and therefore, with (6.6),

$$\omega_{2m} = \frac{\pi^m}{m!}\,,\quad \omega_{2m+1} = \frac{(2\pi)^m}{1\cdot 3\cdot 5\cdot\,\cdots\,\cdot(2m+1)}\cdot 2\,.$$

These two expressions can be unified with the help of the Gamma function, because

$$\Gamma(m+1) = m!\,,\quad \Gamma\Big(m+\frac{3}{2}\Big) = \frac{\sqrt{\pi}}{2^{m+1}}\cdot 1\cdot 3\cdot\,\cdots\,\cdot(2m+1)$$

(see Theorem VI.9.2 and Exercise VI.9.1). ∎

Tonelli's theorem

We now prove the advertised theorem that justifies the iterative calculation of integrals of nonnegative $\overline{\mathbb{R}}$-valued functions. This version, Tonelli's theorem, will give us an important integrability criterion in the case of E-valued functions.

6.7 Theorem (Tonelli) For $f \in \mathcal{L}_0(\mathbb{R}^{m+n}, \overline{\mathbb{R}}^+)$,

   (i) $f(x, \cdot) \in \mathcal{L}_0(\mathbb{R}^n, \overline{\mathbb{R}}^+)$ for λ_m-a.a. $x \in \mathbb{R}^m$,
   $f(\cdot, y) \in \mathcal{L}_0(\mathbb{R}^m, \overline{\mathbb{R}}^+)$ for λ_n-a.a. $y \in \mathbb{R}^n$;

   (ii) $x \mapsto \int_{\mathbb{R}^n} f(x, y) \, dy$ is λ_m-measurable,
   $y \mapsto \int_{\mathbb{R}^m} f(x, y) \, dx$ is λ_n-measurable;

   (iii) $\int_{\mathbb{R}^{m+n}} f \, d(x, y) = \int_{\mathbb{R}^m} \left[\int_{\mathbb{R}^n} f(x, y) \, dy \right] dx = \int_{\mathbb{R}^n} \left[\int_{\mathbb{R}^m} f(x, y) \, dx \right] dy$.

Proof   (i) By Theorem 1.12, there is an increasing sequence (f_j) in $\mathcal{S}(\mathbb{R}^{m+n}, \mathbb{R}^+)$ that converges to f. The monotone convergence theorem then gives

$$\lim_j \int_{\mathbb{R}^{m+n}} f_j \, d(x, y) = \int_{\mathbb{R}^{m+n}} f \, d(x, y) \qquad \text{in } \overline{\mathbb{R}}^+ . \tag{6.7}$$

Further, by Lemma 6.4, there is for every $j \in \mathbb{N}$ a λ_m-null set M_j such that $f_j(x, \cdot) \in \mathcal{S}(\mathbb{R}^n, \mathbb{R}^+)$ for $x \in M_j^c$. If we set $M := \bigcup_j M_j$, we then see from the monotone convergence theorem that

$$\int_{\mathbb{R}^n} f_j(x, y) \, dy \uparrow \int_{\mathbb{R}^n} f(x, y) \, dy \quad \text{for } x \in M^c . \tag{6.8}$$

Lemma 6.4(ii), Proposition 1.11, the fact that M has measure zero, and (6.8) imply that the $\overline{\mathbb{R}}$-valued function $x \mapsto \int_{\mathbb{R}^n} f(x, y) \, dy$ is λ_m-measurable. Next, it follows from (6.7), Lemma 6.4(iii), (6.8), and the monotone convergence theorem that

$$\int_{\mathbb{R}^{m+n}} f \, d(x, y) = \lim_j \int_{\mathbb{R}^{m+n}} f_j \, d(x, y) = \lim_j \int_{\mathbb{R}^m} \left[\int_{\mathbb{R}^n} f_j(x, y) \, dy \right] dx$$

$$= \int_{\mathbb{R}^m} \left[\int_{\mathbb{R}^n} f(x, y) \, dy \right] dx .$$

The remaining statements are proved analogously (paying heed to Remark 6.5). ∎

6.8 Corollary   For $f \in \mathcal{L}_0(\mathbb{R}^{m+n}, E)$, suppose $f = 0$ λ_{m+n}-a.e. Then there is a λ_m-null set M such that $f(x, \cdot)$ vanishes λ_n-a.e. for every $x \in M^c$, and a λ_n-null set N such that $f(\cdot, y) = 0$ λ_m-a.e. for every $y \in N^c$.

Proof   Clearly, it suffices to prove the existence of M (compare Remark 6.5). Tonelli's theorem gives

$$\int_{\mathbb{R}^m} \left[\int_{\mathbb{R}^n} |f(x, y)| \, dy \right] dx = \int_{\mathbb{R}^{m+n}} |f| \, d(x, y) = 0 .$$

Thus according to Remark 3.3(c) there is a λ_m-null set M such that

$$\int_{\mathbb{R}^n} |f(x,y)|\, dy = 0 \quad \text{for } x \in M^c ,$$

from which the claim follows again by Remark 3.3(c). ∎

Fubini's theorem for scalar functions

It is now easy to extend Tonelli's to the case of integrable $\mathbb{K}$-valued functions, which is of particular interest for applications.

6.9 Theorem (Fubini) For $f \in \mathcal{L}_1(\mathbb{R}^{m+n})$,

(i) $f(x,\cdot) \in \mathcal{L}_1(\mathbb{R}^n)$ for λ_m-almost every $x \in \mathbb{R}^m$,
$f(\cdot,y) \in \mathcal{L}_1(\mathbb{R}^m)$ for λ_n-almost every $y \in \mathbb{R}^n$;

(ii) $x \mapsto \int_{\mathbb{R}^n} f(x,y)\, dy$ is λ_m-integrable,
$y \mapsto \int_{\mathbb{R}^m} f(x,y)\, dx$ is λ_n-integrable;

(iii) $\int_{\mathbb{R}^{m+n}} f\, d(x,y) = \int_{\mathbb{R}^m} \left[\int_{\mathbb{R}^n} f(x,y)\, dy\right] dx = \int_{\mathbb{R}^n} \left[\int_{\mathbb{R}^m} f(x,y)\, dx\right] dy.$

Proof (a) For $f \in \mathcal{L}_1(\mathbb{R}^{m+n}, \mathbb{R}^+)$, the claim follows from Tonelli's theorem and Remark 3.3(e).

(b) Given the representation $f = f_1 - f_2 + i\,(f_3 - f_4)$, with $f_j \in \mathcal{L}_1(\mathbb{R}^{m+n}, \mathbb{R}^+)$, the general case now follows by Corollary 2.12 and the linearity of the integral. ∎

6.10 Corollary Suppose $A \in \mathcal{L}(m)$ and $f \in \mathcal{L}_1(A)$. Let $(j_1, \ldots, j_m)$ denote a permutation of $(1, \ldots, m)$. Then

$$\int_A f\, dx = \int_{\mathbb{R}} \left(\int_{\mathbb{R}} \cdots \left(\int_{\mathbb{R}} \widetilde{f}(x_1 \ldots, x_m)\, dx_{j_1} \right) \cdots dx_{j_{m-1}} \right) dx_{j_m} .$$

Fubini's theorem guarantees that *integrable* functions can be integrated in any order. In combination with Tonelli's theorem, we obtain a simple, versatile, and extraordinarily important criterion for the integrability of functions of multiple variables, as well as a method for explicitly calculating integrals.

6.11 Theorem (Fubini–Tonelli) Suppose $A \in \mathcal{L}(m+n)$ and $f \in \mathcal{L}_0(A)$.

(i) *If one of the integrals*

$$\int_{\mathbb{R}^m} \left[\int_{\mathbb{R}^n} |\widetilde{f}(x,y)|\, dy\right] dx , \quad \int_{\mathbb{R}^n} \left[\int_{\mathbb{R}^m} |\widetilde{f}(x,y)|\, dx\right] dy , \quad \int_A |f|\, d(x,y)$$

is finite, then so is each of the others, and they are all equal. In that case, f is integrable, and the statement of Theorem 6.9 holds for $\widetilde{f}$.

(ii) If $\mathrm{pr}_{\mathbb{R}^m}(A)$ is measurable[2] and f is integrable, then

$$\int_A f \, d(x,y) = \int_{\mathrm{pr}_{\mathbb{R}^m}(A)} \left[\int_{A_{[x]}} f(x,y) \, dy \right] dx \ .$$

Proof Because $\tilde{f}$ belongs to $\mathcal{L}_0(\mathbb{R}^{m+n})$, the first statement follows immediately from Tonelli's theorem. Then by Theorem 3.9, $\tilde{f}$ is integrable, and hence so is f. The claim is now clear. ∎

6.12 Remarks **(a)** We have lost no generality by choosing the first m coordinates, because by Corollary 6.10, this order can always be achieved by a permutation.

(b) Typically we omit the brackets in $\int_{\mathbb{R}^n} \left[\int_{\mathbb{R}^m} f(x,y) \, dx \right] dy$ and instead write, say,

$$\int_{\mathbb{R}^n} \int_{\mathbb{R}^m} f(x,y) \, dx \, dy \ . \tag{6.9}$$

In this notation, it is understood that the integrals are to be evaluated from the inside to the outside.[3] The **iterated integral** (6.9) is to be distinguished from the $(m+n)$-dimensional integral

$$\int_{\mathbb{R}^{m+n}} f \, d(x,y) = \int_{\mathbb{R}^{m+n}} f \, d\lambda_{m+n} \ .$$

(c) There exists $f \in \mathcal{L}_0(\mathbb{R}^2) \setminus \mathcal{L}_1(\mathbb{R}^2)$ such that

$$\int_{\mathbb{R}} \int_{\mathbb{R}} f(x,y) \, dx \, dy = \int_{\mathbb{R}} \int_{\mathbb{R}} f(x,y) \, dy \, dx = 0 \ .$$

Therefore the existence and equality of the iterated integral does not imply that f is integrable.

Proof Define $f : \mathbb{R}^2 \to \mathbb{R}$ by

$$f(x,y) := \begin{cases} \dfrac{xy}{(x^2+y^2)^2} & \text{if } (x,y) \neq (0,0) \ , \\ 0 & \text{if } (x,y) = (0,0) \ . \end{cases} \tag{6.10}$$

Then f is λ_2-measurable. For every $y \in \mathbb{R}$, the improper Riemann integral $\int_{\mathbb{R}} f(x,y) \, dx$ converges absolutely. Also $f(\cdot,y)$ is odd. Hence $\int_{\mathbb{R}} f(x,y) \, dx = 0$ for every[4] $y \in \mathbb{R}$, and therefore, because $f(x,y) = f(y,x)$, we have

$$\int_{\mathbb{R}} \int_{\mathbb{R}} f(x,y) \, dx \, dy = \int_{\mathbb{R}} \int_{\mathbb{R}} f(x,y) \, dy \, dx = 0 \ .$$

---

[2] As Remark IX.5.14(b) shows, this is not generally the case.

[3] That is, the integral $\int_{\mathbb{R}^m} f(x,y) \, dx$ is calculated for fixed y and the result is then integrated over y in $\mathbb{R}^n$.

[4] The case $y = 0$ is covered in the given argument, although it follows more simply from the fact that $f(\cdot,0) = 0$.

Now suppose f were integrable. Then, by Fubini's theorem, $x \mapsto \int_{\mathbb{R}} |f(x,y)| \, dy$ would also be integrable, which, because

$$\int_{\mathbb{R}} \frac{|xy|}{(x^2 + y^2)^2} \, dy = \frac{1}{|x|} \quad \text{for } x \neq 0 ,$$

cannot be true. ∎

(d) There exists $g \in \mathcal{L}_0(\mathbb{R}^2) \setminus \mathcal{L}_1(\mathbb{R}^2)$ such that

$$0 < \left| \int_{\mathbb{R}} \int_{\mathbb{R}} g(x,y) \, dx \, dy \right| = \left| \int_{\mathbb{R}} \int_{\mathbb{R}} g(x,y) \, dy \, dx \right| < \infty .$$

Proof Let f be the function from (6.10) and take $h \in \mathcal{L}_1(\mathbb{R}^2)$ with $\int h \, d(x,y) > 0$. Then $g := f + h$ has the stated properties. ∎

6.13 Examples (a) (multidimensional Gaussian integrals) For $n \in \mathbb{N}^\times$, we have

$$\int_{\mathbb{R}^n} e^{-|x|^2} \, dx = \pi^{n/2} .$$

Proof Using $|x|^2 = x_1^2 + \cdots + x_n^2$ and the properties of the exponential function, it follows from Tonelli's theorem that

$$\int_{\mathbb{R}^n} e^{-|x|^2} \, dx = \int_{\mathbb{R}} \cdots \cdots \int_{\mathbb{R}} e^{-x_1^2} e^{-x_2^2} \cdot \cdots \cdot e^{-x_n^2} \, dx_1 \cdots \cdot dx_n$$

$$= \prod_{j=1}^{n} \int_{\mathbb{R}} e^{-x_j^2} \, dx_j = \left(\int_{\mathbb{R}} e^{-t^2} \, dt \right)^n .$$

Now the claim follows from Application VI.9.7. ∎

(b) (a representation of the beta function[5]) For $v, w \in [\operatorname{Re} z > 0]$,

$$\mathsf{B}(v, w) = \frac{\Gamma(v) \Gamma(w)}{\Gamma(v + w)} .$$

Proof Set $A := \{ (s,t) \in \mathbb{R}^2 \; ; \; 0 < t < s \}$ and define $\gamma_{v,w} : A \to \mathbb{C}$ by $\gamma_{v,w}(s,t) := t^{v-1}(s-t)^{w-1} e^{-s}$ for $v, w \in [\operatorname{Re} z > 0]$. Setting $\gamma_v(t) := t^{v-1} e^{-t}$ for $t > 0$, we find from Tonelli's theorem that

$$\int_A |\gamma_{v,w}(s,t)| \, d(s,t) = \int_0^\infty \int_t^\infty |\gamma_{v,w}(s,t)| \, ds \, dt$$

$$= \left(\int_0^\infty \gamma_{\operatorname{Re} v}(t) \, dt \right) \left(\int_0^\infty \gamma_{\operatorname{Re} w}(s) \, ds \right) = \Gamma(\operatorname{Re} v) \Gamma(\operatorname{Re} w) < \infty .$$

Therefore $\gamma_{v,w}$ is integrable, and Fubini's theorem analogously gives

$$\int_A \gamma_{v,w}(s,t) \, d(s,t) = \int_0^\infty \int_t^\infty \gamma_{v,w}(s,t) \, ds \, dt = \Gamma(v) \Gamma(w) . \tag{6.11}$$

---

[5] Compare Remark VI.9.12(a).

Since $\mathrm{pr}_1(A) = \mathbb{R}^+$ and $A_{[s]} = [0, s]$ for $s > 0$, we obtain from (6.11) and Theorem 6.11(ii)

$$\Gamma(v)\Gamma(w) = \int_0^\infty \left(\int_0^s t^{v-1}(s-t)^{w-1}\, dt \right) e^{-s}\, ds \ .$$

The substitution $r = t/s$ in the inner integral and the definition of the beta function give

$$\Gamma(v)\Gamma(w) = \int_0^\infty \left(\int_0^1 r^{v-1}(1-r)^{w-1}\, dr \right) s^{v+w-1} e^{-s}\, ds = \mathsf{B}(v,w)\,\Gamma(v+w) \ , \qquad (6.12)$$

which completes the proof. ∎

Example (b) shows that complicated integrals can be simplified by a deft choice of integration order.

Fubini's theorem for vector-valued functions[6]

We now want to show that Fubini's theorem also holds for E-valued functions, and offer some applications. A few preliminary remarks will prove helpful.

Suppose $A \in \mathcal{L}(m+n)$ has finite measure. By Proposition 6.2 and Corollary 6.3, there is a λ_m-null set M such that $A_{[x]} \in \mathcal{L}(n)$ and $\lambda_n(A_{[x]}) < \infty$ for $x \in M^c$. We fix $q \in [1, \infty)$. Because $|\chi_{A_{[x]}}|^q = \chi_{A_{[x]}}$, we have

$$\int_{\mathbb{R}^n} |\chi_{A_{[x]}}(y)|^q\, dy = \int_{\mathbb{R}^n} \chi_{A_{[x]}}(y)\, dy = \lambda_n(A_{[x]}) < \infty \ .$$

If, as agreed to in Section 4, we identify $\chi_{A_{[x]}}$ with the equivalence class of all functions that coincide λ_n-a.e. with $y \mapsto \chi_{A_{[x]}}(y)$, we obtain the map

$$M^c \to F := L_q(\mathbb{R}^n) \ , \qquad x \mapsto \chi_{A_{[x]}} \ .$$

Because F is a Banach space, we can study its measurability and integrability properties.

6.14 Lemma *Suppose $A \in \mathcal{L}(m+n)$ has finite measure. Then the F-valued map $x \mapsto \chi_{A_{[x]}}$, which is defined λ_m-everywhere, is λ_m-measurable.*

Proof We denote by $\psi_A : \mathbb{R}^m \to F$ the trivial extension of $x \mapsto \chi_{A_{[x]}}$.

(i) Suppose A is a λ_{m+n}-null set. By Remark 6.1(h), there is a λ_m-null set M such that $A_{[x]}$ is a λ_n-null set for $x \in M^c$. Therefore $\psi_A(x) = 0$ in F for $x \in M^c$. The claim follows.

(ii) Now suppose A is an interval of the form $[a, b)$ with $a, b \in \mathbb{R}^{m+n}$. Set $J_1 := \prod_{j=1}^m [a_j, b_j)$ and $J_2 := \prod_{j=m+1}^{m+n} [a_j, b_j)$. Because $A = J_1 \times J_2$, we have

$$\chi_{A_{[x]}} = \chi_{J_1}(x)\chi_{J_2} \quad \text{for } x \in \mathbb{R}^m \ ,$$

and we see that in this case ψ_A belongs to $\mathcal{S}(\mathbb{R}^m, F)$.

---

[6]This section may be skipped on first reading.

(iii) Suppose $A \subset \mathbb{R}^{m+n}$ is open and (I_j) is a disjoint sequence of intervals of the form $[a, b)$ with $A = \bigcup_j I_j$ (see Proposition IX.5.6). We set

$$f_k := \sum_{j=0}^{k} \psi_{I_j} \quad \text{for } k \in \mathbb{N} .$$

By (ii) and Remark 1.2(a), (f_k) is a sequence in $\mathcal{S}(\mathbb{R}^m, F)$. Also, there is a set M of Lebesgue measure zero such that

$$\|\psi_A(x) - f_k(x)\|_F^q = \int_{\mathbb{R}^n} \left| \chi_{A_{[x]}}(y) - \Big(\sum_{j=0}^{k} \chi_{(I_j)_{[x]}}(y) \Big) \right|^q dy$$

$$= \lambda_n \Big(\bigcup_{j=k+1}^{\infty} (I_j)_{[x]} \Big) = \sum_{j=k+1}^{\infty} \lambda_n \big((I_j)_{[x]} \big)$$

for $x \in M^c$. In addition, $A_{[x]}$ has finite measure by Corollary 6.3, and $\lambda_n(A_{[x]}) = \sum_{j=0}^{\infty} \lambda_n((I_j)_{[x]})$ for $x \in M^c$. Therefore (f_k) converges λ_m-a.e. to ψ_A in F, and we see that ψ_A belongs to $\mathcal{L}_0(\mathbb{R}^m, F)$.

(iv) Suppose A is a G_δ-set. The proof of Corollary IX.5.5 shows that there is a sequence (O_j) of open sets such that $\lambda_{m+n}(O_j) < \infty$ and $A = \bigcap O_j$. Set

$$f_k := \psi_{\bigcap_{j=0}^k O_j} , \qquad R_k := \bigcap_{j=0}^{k} O_j \backslash A \quad \text{for } k \in \mathbb{N} .$$

Then (f_k) is a sequence in $\mathcal{L}_0(\mathbb{R}^m, F)$ by (iii), and (R_k) is a descending sequence with $\bigcap_{k=0}^{\infty} R_k = \emptyset$ and $\lambda_{m+n}(R_0) < \infty$. Also, we have

$$\|f_k(x) - \psi_A(x)\|_F^q = \int_{\mathbb{R}^n} \left| \chi_{(\bigcap_{j=0}^k O_j)_{[x]}}(y) - \chi_{A_{[x]}}(y) \right|^q dy = \lambda_n \big((R_k)_{[x]} \big)$$

for λ_m-almost every $x \in \mathbb{R}^m$. The continuity of λ_n from above therefore implies that (f_k) converges λ_m-a.e. to ψ_A. From Theorem 1.14, it now follows that ψ_A belongs to $\mathcal{L}_0(\mathbb{R}^m, F)$.

(v) To conclude, consider $A \in \mathcal{L}(m + n)$ such that $\lambda_{m+n}(A)$ is finite. By Corollary IX.5.5, there is a G_δ-set G containing A and having the same measure as A. By Proposition IX.2.3(ii), $N := G \backslash A$ is a λ_{m+n}-null set with $\psi_A = \psi_G - \psi_N$ λ_m-a.e. Now the claim follows from (i) and (iv). ∎

6.15 Corollary Let $p, q \in [1, \infty)$, and suppose $\varphi \in \mathcal{S}(\mathbb{R}^{m+n}, E)$ has compact support. Then the $L_q(\mathbb{R}^n, E)$-valued function $x \mapsto [\varphi(x, \cdot)]$ is defined λ_m-a.e. and is L_p-integrable, that is,

$$\int_{\mathbb{R}^m} \|\varphi(x, \cdot)\|_{L_q(\mathbb{R}^n, E)}^p \, dx < \infty .$$

If $p = q$, this holds for every $\varphi \in \mathcal{S}(\mathbb{R}^{m+n}, E)$.

Proof By Minkowski's inequality, it suffices to prove this for $\varphi := e\chi_A$ with $e \in E$ and $A \in \mathcal{L}(m+n)$, where A has finite measure if $p = q$, and A is bounded if $p \neq q$.

By Lemma 6.14, there is a λ_m-null set M such that the function

$$M^c \to L_q(\mathbb{R}^n) , \quad x \mapsto \chi_{A_{[x]}}$$

is λ_m-measurable. Because $\varphi(x,\cdot) = e\chi_{A_{[x]}}$, $(x \mapsto \varphi(x,\cdot)) \in \mathcal{L}_0(M^c, L_q(\mathbb{R}^n, E))$. From

$$\|\varphi(x,\cdot)\|_{L_q(\mathbb{R}^n,E)} = \left(\int_{\mathbb{R}^n} |e|^q \chi_{A_{[x]}}(y)\,dy \right)^{1/q} = |e| \left[\lambda_n(A_{[x]}) \right]^{1/q} \quad \text{for } x \in M^c ,$$

we obtain

$$\int_{\mathbb{R}^m} \|\varphi(x,\cdot)\|_{L_q(\mathbb{R}^n,E)}^p\,dx = |e|^p \int_{\mathbb{R}^m} \lambda_n(A_{[x]})^{p/q}\,dx .$$

In the case $p = q$, Proposition 6.2 implies

$$\int_{\mathbb{R}^n} \lambda_n(A_{[x]})\,dx = \lambda_{m+n}(A) < \infty .$$

Suppose therefore $p \neq q$. Because φ has compact support, there are compact subsets $K \subset \mathbb{R}^m$ and $L \subset \mathbb{R}^n$ such that $A \subset K \times L$. Thus $A_{[x]} \subset L$, which implies $\lambda_n(A_{[x]}) \leq \lambda_n(L)$ for λ_m-almost every $x \in \mathbb{R}^m$. From this we deduce

$$\int_{\mathbb{R}^m} \lambda_n(A_{[x]})^{p/q}\,dx = \int_K \lambda_n(A_{[x]})^{p/q}\,dx \leq \lambda_n(L)^{p/q} \lambda_m(K) < \infty . \quad \blacksquare$$

These preparations are more general than necessary for our current purpose, but will prove useful for further applications. We are ready to prove Fubini's theorem in the E-valued case.

6.16 Theorem (Fubini) For $f \in \mathcal{L}_1(\mathbb{R}^{m+n}, E)$,
  (i) $f(x,\cdot) \in \mathcal{L}_1(\mathbb{R}^n, E)$ for λ_m-almost every $x \in \mathbb{R}^m$;
      $f(\cdot,y) \in \mathcal{L}_1(\mathbb{R}^m, E)$ for λ_n-almost every $y \in \mathbb{R}^n$;
  (ii) $x \mapsto \int_{\mathbb{R}^n} f(x,y)\,dy$ is λ_m-integrable;
       $y \mapsto \int_{\mathbb{R}^m} f(x,y)\,dx$ is λ_n-integrable;
  (iii) $\int_{\mathbb{R}^{m+n}} f\,d(x,y) = \int_{\mathbb{R}^m} \left[\int_{\mathbb{R}^n} f(x,y)\,dy \right] dx = \int_{\mathbb{R}^n} \left[\int_{\mathbb{R}^m} f(x,y)\,dx \right] dy.$

Proof (a) Let $f \in \mathcal{L}_1(\mathbb{R}^{m+n}, E)$. Then there is an $\mathcal{L}_1$-Cauchy sequence (f_j) in $\mathcal{S}(\mathbb{R}^{m+n}, E)$ and a λ_{m+n}-null set L such that $f_j(x,y) \to f(x,y)$ for $(x,y) \in L^c$. By Remark 6.1(h), there is a set M_1 of measure zero such that

$$f_j(x,\cdot) \to f(x,\cdot) \quad \lambda_n\text{-a.e.} , \tag{6.13}$$

for $x \in M_1^c$. We set $F := L_1(\mathbb{R}^n, E)$ and denote by φ_j the trivial extension of $x \mapsto f_j(x,\cdot)$. According to Corollary 6.15, (φ_j) is a sequence in $\mathcal{L}_1(\mathbb{R}^m, F)$ for which

$$\|\varphi_j - \varphi_k\|_1 = \int_{\mathbb{R}^m} \|\varphi_j(x) - \varphi_k(x)\|_F \, dx = \int_{\mathbb{R}^m} \int_{\mathbb{R}^n} |f_j(x,y) - f_k(x,y)| \, dy \, dx \ .$$

Further, Lemma 6.4 shows

$$\int_{\mathbb{R}^m} \int_{\mathbb{R}^n} |f_j(x,y) - f_k(x,y)| \, dy \, dx = \int_{\mathbb{R}^{m+n}} |f_j - f_k| \, d(x,y) = \|f_j - f_k\|_1 \ ,$$

and we see that (φ_j) is a Cauchy sequence in $\mathcal{L}_1(\mathbb{R}^m, F)$. By Theorems 2.10 and 2.18, there is thus a $\widehat{g} \in \mathcal{L}_1(\mathbb{R}^m, F)$, a λ_m-null set M_2, and a subsequence of (φ_j), which, for simplicity, we also denote by (φ_j), such that

$$\lim_{j \to \infty} \varphi_j(x) = \widehat{g}(x) \quad \text{for } x \in M_2^c \tag{6.14}$$

in F and $\varphi_j \to \widehat{g}$ in $\mathcal{L}_1(\mathbb{R}^m, F)$. For $x \in M_2^c$, let $g(x) \in \mathcal{L}_1(\mathbb{R}^n, E)$ be a representative of $\widehat{g}(x)$. Then there is a set $N(x)$ of Lebesgue measure zero and a subsequence of $(\varphi_j(x))$, which we also write as $(\varphi_j(x))$, such that, in E,

$$\lim_{j \to \infty} f_j(x,y) = \lim_{j \to \infty} \varphi_j(x)(y) = g(x)(y) \quad \text{for } x \in M_2^c \text{ and } y \in (N(x))^c \ .$$

Hence (6.13) implies that for every $x \in M_1^c \cap M_2^c$ the maps $f(x, \cdot), g(x) \colon \mathbb{R}^n \to E$ are equal λ_n-a.e. Lemma 2.15 now shows that $f(x, \cdot)$ belongs to $\mathcal{L}_1(\mathbb{R}^n, E)$ and that

$$\int_{\mathbb{R}^n} g(x)(y) \, dy = \int_{\mathbb{R}^n} f(x,y) \, dy \quad \text{for } x \in M_1^c \cap M_2^c \ . \tag{6.15}$$

Furthermore, it follows from (6.13), (6.14), and Theorem 2.18(ii) that

$$\int_{\mathbb{R}^n} f_j(x,y) \, dy = \int_{\mathbb{R}^n} \varphi_j(x)(y) \, dy \to \int_{\mathbb{R}^n} g(x)(y) \, dy = \int_{\mathbb{R}^n} f(x,y) \, dy \tag{6.16}$$

for $x \in M_1^c \cap M_2^c$.

(b) For $\varphi \in F = L_1(\mathbb{R}^n, E)$, let $A\varphi := \int_{\mathbb{R}^n} \varphi \, dy$. By Theorem 2.11(i), A belongs to $\mathcal{L}(F, E)$. Theorem 2.11(iii) implies that $g_j := A\varphi_j$ defines a sequence in $\mathcal{L}_1(\mathbb{R}^m, E)$.

Because

$$g_j(x) = \int_{\mathbb{R}^n} \varphi_j(x)(y) \, dy = \int_{\mathbb{R}^n} f_j(x,y) \, dy \ , \tag{6.17}$$

we know from Theorem 2.11(i) that

$$|g_j(x) - g_k(x)| = \left| \int_{\mathbb{R}^n} (f_j(x,y) - f_k(x,y)) \, dy \right| \le \int_{\mathbb{R}^n} |f_j(x,y) - f_k(x,y)| \, dy \ .$$

Therefore Theorem 2.11(ii) gives

$$\int_{\mathbb{R}^m} |g_j - g_k| \, dx \le \int_{\mathbb{R}^m} \int_{\mathbb{R}^n} |(f_j(x,y) - f_k(x,y))| \, dy \, dx = \|f_j - f_k\|_1 \ ,$$

where the last equality follows from Tonelli's theorem. Therefore (g_j) is a Cauchy sequence in $\mathcal{L}_1(\mathbb{R}^m, E)$, and by completeness there is some $h \in \mathcal{L}_1(\mathbb{R}^m, E)$ such that $g_j \to h$ in $\mathcal{L}_1(\mathbb{R}^m, E)$. Hence we can find a λ_m-null set M_3 and a subsequence of (g_j), which we also denote by (g_j), such that $g_j(x) \to h(x)$ for $x \in M_3^c$ and $j \to \infty$. In view of (6.17), it follows from (6.16) that

$$h(x) = \int_{\mathbb{R}^n} f(x, y) \, dy \quad \text{for } x \in M_1^c \cap M_2^c \cap M_3^c , \tag{6.18}$$

which proves the first statement of (ii).

(c) Since $g_j \to h$ in $\mathcal{L}_1(\mathbb{R}^m, E)$ and because of (6.17) and (6.18), Theorem 2.18(ii) implies

$$\int_{\mathbb{R}^m} \int_{\mathbb{R}^n} f_j(x, y) \, dy \, dx \to \int_{\mathbb{R}^m} \int_{\mathbb{R}^n} f(x, y) \, dy \, dx .$$

Finally, it follows from Lemma 6.4 that

$$\int_{\mathbb{R}^m} \int_{\mathbb{R}^n} f_j(x, y) \, dy \, dx = \int_{\mathbb{R}^{m+n}} f_j \, d(x, y) ,$$

and with $\int_{\mathbb{R}^{m+n}} f \, d(x, y) = \lim_j \int_{\mathbb{R}^{m+n}} f_j \, d(x, y)$, we have

$$\int_{\mathbb{R}^m} \int_{\mathbb{R}^n} f(x, y) \, dy \, dx = \int_{\mathbb{R}^{m+n}} f \, d(x, y) .$$

We have proved the first part of each of the statements (i) and (ii), and the first equality in (iii). The remaining claims follow by exchanging the roles of x and y. ∎

6.17 Remark   The analogues of the Fubini–Tonelli theorem and Corollary 6.10 clearly also hold in the E-valued case. ∎

Minkowski's inequality for integrals

As an application we now prove a continuous version of Minkowski's inequality.

Fix $p, q \in [1, \infty)$. For $f \in \mathcal{L}_0(\mathbb{R}^{m+n}, E)$, Theorem 1.7(i) shows that $|f|^q$ belongs to $\mathcal{L}_0(\mathbb{R}^{m+n}, \mathbb{R}^+)$. Hence Tonelli's theorem implies that $|f(x, \cdot)|^q$ lies in $\mathcal{L}_0(\mathbb{R}^n, \mathbb{R}^+)$ for λ_m-almost every $x \in \mathbb{R}^m$ and that the $\overline{\mathbb{R}}^+$-valued function

$$x \mapsto \int_{\mathbb{R}^n} |f(x, y)|^q \, dy ,$$

which is defined λ_m-a.e., is λ_m-measurable. Therefore

$$\|f\|_{(p,q)} := \left(\int_{\mathbb{R}^m} \left[\int_{\mathbb{R}^n} |f(x, y)|^q \, dy \right]^{p/q} dx \right)^{1/p}$$

is defined in $\overline{\mathbb{R}}^+$. We easily check that

$$\mathcal{L}_{(p,q)}(\mathbb{R}^{m+n}, E) := \left\{ f \in \mathcal{L}_0(\mathbb{R}^{m+n}, E) \; ; \; \|f\|_{(p,q)} < \infty \right\}$$

is a vector subspace of $\mathcal{L}_0(\mathbb{R}^{m+n}, E)$ and that $\|\cdot\|_{(p,q)}$ defines a seminorm on $\mathcal{L}_{(p,q)}(\mathbb{R}^{m+n}, E)$. Finally, we set

$$\mathcal{S}_c(\mathbb{R}^{m+n}, E) := \left\{ f \in \mathcal{S}(\mathbb{R}^{m+n}, E) \; ; \; \mathrm{supp}(f) \text{ is compact} \right\} .$$

6.18 Lemma $\mathcal{S}_c(\mathbb{R}^{m+n}, E)$ *is a dense vector subspace of* $\mathcal{L}_{(p,q)}(\mathbb{R}^{m+n}, E)$.

Proof (i) Take $f \in \mathcal{L}_{(p,q)}(\mathbb{R}^{m+n}, E)$ and let (g_k) be a sequence in $\mathcal{S}(\mathbb{R}^{m+n}, E)$ such that $g_k \to f$ a.e. Set $A_k := \left[|g_k| \leq 2\,|f| \right] \cap k\mathbb{B}^{m+n}$ and $f_k := \chi_{A_k} g_k$. Then (f_k) is a sequence in $\mathcal{S}_c(\mathbb{R}^{m+n}, E)$, and there is a λ_{m+n}-null set L such that

$$f_k(x,y) \to f(x,y) \quad \text{for } (x,y) \in L^c . \tag{6.19}$$

Moreover

$$|f_k - f| \leq |f_k| + |f| \leq 3\,|f| \quad \text{for } k \in \mathbb{N} . \tag{6.20}$$

(ii) By (6.20), it follows from Tonelli's theorem and Theorem 3.9 that there is a λ_m-null set M_0 such that

$$|f(x,\cdot) - f_k(x,\cdot)|^q, \, |f(x,\cdot)|^q \in \mathcal{L}_1(\mathbb{R}^n) \quad \text{for } x \in M_0^c \text{ and } k \in \mathbb{N} . \tag{6.21}$$

Remark 6.1(h) says there is a λ_m-null set M_1 such that $L_{[x]}$ is a λ_n-null set for every $x \in M_1^c$. Set $M := M_0 \cup M_1$ and choose $x \in M^c$. From (6.19), we read off that $f_k(x,y) \to f(x,y)$ for $y \in (L_{[x]})^c$. By (6.20) and (6.21), we can apply Lebesgue's dominated convergence theorem to the sequence $\left(|f(x,\cdot) - f_k(x,\cdot)|^p \right)_{k \in \mathbb{N}}$, and we find

$$\lim_{k \to \infty} \int_{\mathbb{R}^n} |f(x,y) - f_k(x,y)|^q \, dy = 0 \quad \text{for } x \in M^c .$$

Now define

$$\varphi_k := \left(x \mapsto \left(\int_{\mathbb{R}^n} |f(x,y) - f_k(x,y)|^q \, dy \right)^{p/q} \right)^{\sim} \quad \text{for } k \in \mathbb{N} .$$

Then the sequence (φ_k) converges λ_m-a.e. to 0.

(iii) Finally, set

$$\varphi := \left(x \mapsto 3^p \left(\int_{\mathbb{R}^n} |f(x,y)|^q \, dy \right)^{p/q} \right)^{\sim} .$$

Because $f \in \mathcal{L}_{(p,q)}(\mathbb{R}^{m+n}, E)$, we know φ belongs to $\mathcal{L}_1(\mathbb{R}^m)$, and (6.20) implies $0 \leq \varphi_k \leq \varphi$ λ_m-a.e. for $k \in \mathbb{N}$. Hence we can apply dominated convergence theorem to (φ_k) to see that $(\int_{\mathbb{R}^m} \varphi_k)_{k \in \mathbb{N}}$ is a null sequence in $\mathbb{R}^+$. The claim now follows because

$$\int_{\mathbb{R}^m} \varphi_k = \int_{\mathbb{R}^m} \left[\int_{\mathbb{R}^n} |f(x,y) - f_k(x,y)|^q \, dy \right]^{p/q} dx = \|f - f_k\|_{(p,q)}^p . \quad \blacksquare$$

One easily checks that $\mathcal{N} := \{ f \in \mathcal{L}_0(\mathbb{R}^{m+n}, E) \; ; \; f = 0 \text{ a.e.} \}$ is a vector subspace of $\mathcal{L}_{(p,q)}(\mathbb{R}^{m+n}, E)$ and that f belongs to $\mathcal{N}$ if and only if $\|f\|_{(p,q)} = 0$. Therefore

$$L_{(p,q)}(\mathbb{R}^{m+n}, E) := \mathcal{L}_{(p,q)}(\mathbb{R}^{m+n}, E)/\mathcal{N}$$

is a well defined vector space, and the assignment $[f] \mapsto \|f\|_{(p,q)}$ defines a norm on $L_{(p,q)}(\mathbb{R}^{m+n}, E)$, which we again denote by $\|\cdot\|_{(p,q)}$. In what follows, we always provide the space $L_{(p,q)}(\mathbb{R}^{m+n}, E)$ with the topology induced by $\|\cdot\|_{(p,q)}$.

We set

$$S_c(\mathbb{R}^{m+n}, E) := \{ [f] \in L_0(\mathbb{R}^{m+n}, E) \; ; \; [f] \cap \mathcal{S}_c(\mathbb{R}^{m+n}, E) \neq \emptyset \} \; .$$

6.19 Remarks   (a) $S_c(\mathbb{R}^{m+n}, E)$ is a dense vector subspace of $L_{(p,q)}(\mathbb{R}^{m+n}, E)$.

Proof   This follows from Lemma 6.18. ∎

(b) Let $f \in \mathcal{L}_0(\mathbb{R}^{m+n}, E)$. If $f(x, \cdot)$ belongs to $\mathcal{L}_q(\mathbb{R}^n, E)$ for almost every $x \in \mathbb{R}^m$ and

$$\left[x \mapsto \left(\int_{\mathbb{R}^n} |f(x,y)|^q \, dx \right)^{1/q} \right]^{\sim} \in \mathcal{L}_p(\mathbb{R}^n) \; ,$$

then $[f]$ belongs to $L_{(p,q)}(\mathbb{R}^{m+n}, E)$.

(c) $L_{(p,p)}(\mathbb{R}^{m+n}, E) = L_p(\mathbb{R}^{m+n}, E)$.

Proof   This follows from Remark 4.9(b) and the Fubini–Tonelli theorem. ∎

(d) $S_c(\mathbb{R}^n, E)$ is a dense vector subspace of $L_p(\mathbb{R}^n, E)$.

Proof   This is a consequence of (a) and (c). ∎

Consider $g \in \mathcal{S}_c(\mathbb{R}^{m+n}, E)$. By Corollary 6.15, $T_0g := (x \mapsto [g(x, \cdot)])^{\sim}$ belongs to $\mathcal{L}_p(\mathbb{R}^m, L_q(\mathbb{R}^n, E))$. Denoting by $[T_0g]$ the equivalence class of T_0g with respect to the vector subspace of all elements of $\mathcal{L}_0(\mathbb{R}^m, L_q(\mathbb{R}^n, E))$ that vanish λ_m-a.e., we have $[T_0g] \in L_p(\mathbb{R}^m, L_q(\mathbb{R}^n, E))$. Further, it follows from Corollary 6.8 that $[T_0g] = [T_0h]$ if $g, h \in \mathcal{S}_c(\mathbb{R}^{m+n}, E)$ coincide λ_{m+n}-a.e. Thus

$$T \colon S_c(\mathbb{R}^{m+n}, E) \to L_p(\mathbb{R}^m, L_q(\mathbb{R}^n, E)) \; , \quad [g] \mapsto [T_0g]$$

is a well defined linear map.

6.20 Lemma   *There is a unique extension*

$$\overline{T} \in \mathcal{L}(L_{(p,q)}(\mathbb{R}^{m+n}, E), L_p(\mathbb{R}^m, L_q(\mathbb{R}^n, E)))$$

of T, and $\overline{T}$ is an isometry with a dense image.

Proof (i) For $f \in \mathcal{S}_c(\mathbb{R}^{m+n}, E)$, let $g \in f \cap \mathcal{S}_c(\mathbb{R}^{m+n}, E)$. Then

$$\int_{\mathbb{R}^m} \|Tf\|_{L_q(\mathbb{R}^n, E)}^p \, dx = \int_{\mathbb{R}^m} \left(\int_{\mathbb{R}^n} |g(x,y)|^q \, dy \right)^{p/q} dx = \|g\|_{(p,q)}^p = \|f\|_{(p,q)}^p \, .$$

Therefore $T \in \mathcal{L}\big(\mathcal{S}_c(\mathbb{R}^{m+n}, E), L_p(\mathbb{R}^m, L_q(\mathbb{R}^n, E))\big)$ is an isometry. Now it follows from Theorem VI.2.6 and Remark 6.19(a) that there is a uniquely determined isometric extension $\overline{T}$ of T.

(ii) We set $F := L_q(\mathbb{R}^n, E)$ and choose $w \in L_p(\mathbb{R}^m, F)$ and $\varepsilon > 0$. It follows from Remark 6.19(d) that there is a $\varphi \in \mathcal{S}_c(\mathbb{R}^m, F)$ such that $\|w - \varphi\|_p < \varepsilon/2$. Let $\sum_{j=0}^r \chi_{A_j} \hat{f}_j$ be the normal form of φ. Then $\bigcup_{j=0}^r A_j$ is bounded in $\mathbb{R}^m$, and $\alpha := \sum_{j=0}^r \lambda_m(A_j)$ is finite. In the case $\alpha = 0$, we have

$$\|w\|_p = \|w - T0\|_p < \varepsilon/2 \, .$$

In the case $\alpha > 0$, we choose for every $j \in \{0, \ldots, r\}$ a representative f_j of $\hat{f}_j$ and a $\psi_j \in \mathcal{S}_c(\mathbb{R}^n, E)$ such that

$$\|\psi_j - f_j\|_q < \alpha^{-1/p}(r+1)^{-1/q'} \varepsilon \, .$$

Also let

$$h(x,y) := \sum_{j=0}^r \chi_{A_j}(x)\psi_j(y) \quad \text{for } (x,y) \in \mathbb{R}^{m+n} \, .$$

With $\psi_j = \sum_{k_j=0}^{s_j} \chi_{B_{k_j}} e_{k_j}$ for $j \in \{0, \ldots, r\}$, we then have

$$h = \sum_{j=0}^r \sum_{k_j=0}^{s_j} \chi_{A_j} \chi_{B_{k_j}} e_{k_j} = \sum_{j=0}^r \sum_{k_j=0}^{s_j} \chi_{A_j \times B_{k_j}} e_{k_j} \, ,$$

and we see that h belongs to $\mathcal{S}_c(\mathbb{R}^{m+n}, E)$. Finally, let g be the equivalence class of h in $L_0(\mathbb{R}^{m+n}, E)$. Then g belongs to $\mathcal{S}_c(\mathbb{R}^{m+n}, E)$, and $Tg = \sum_{j=0}^r [\chi_{A_j} \psi_j]$. From Hölder's inequality (for sums) and the equality $\chi_A^2 = \chi_A$, it follows that

$$\int_{\mathbb{R}^m} \|Tg - \varphi\|_F^p = \int_{\mathbb{R}^m} \left[\int_{\mathbb{R}^n} \left| \sum_{j=0}^r \chi_{A_j}(x)(\psi_j(y) - f_j(y)) \right|^q dy \right]^{p/q} dx$$

$$\leq (r+1)^{p/q'} \int_{\mathbb{R}^m} \left[\int_{\mathbb{R}^n} \sum_{j=0}^r \chi_{A_j}(x) |\psi_j(y) - f_j(y)|^q \, dy \right]^{p/q} dx$$

$$= (r+1)^{p/q'} \int_{\mathbb{R}^m} \left[\sum_{j=0}^r \chi_{A_j}(x) \|\psi_j - f_j\|_F^q \right]^{p/q} dx$$

$$\leq (r+1)^{p/q'} \sum_{j=0}^r \lambda_m(A_j) \|\psi_j - f_j\|_F^p \, .$$

Therefore,

$$\|Tg - \varphi\|_p \le \alpha^{1/p} (r+1)^{1/q'} \max_j \|\psi_j - f_j\|_F < \varepsilon/2 \ ,$$

and consequently $\|Tg - w\|_p < \varepsilon$. Because this holds for every choice of w and ε, we see that the image of T, and *a fortiori* that of $\overline{T}$, is dense. ∎

As usual, we lighten the notation by writing T for $\overline{T}$. In addition, as stated in Section 4, our notation for elements of Lebesgue spaces does not distinguish between cosets and their representatives. This means that for $f \in L_{(p,q)}(\mathbb{R}^{m+n}, E)$ we may write $Tf(x)$ as $f(x, \cdot)$. With these conventions, Lemma 6.20 says that

$$T : L_{(p,q)}(\mathbb{R}^{m+n}, E) \to L_p(\mathbb{R}^m, L_q(\mathbb{R}^n, E)) \ , \quad f \mapsto (x \mapsto f(x, \cdot)) \qquad (6.22)$$

is a linear isometry whose image is dense.

Now it is easy to prove our continuous Minkowski's inequality.

6.21 Proposition (Minkowski's inequality for integrals) For $1 \le q < \infty$, we have:

(i) $\left(\int_{\mathbb{R}^n} \left[\int_{\mathbb{R}^m} |f(x,y)| \, dx \right]^q dy \right)^{1/q} \le \int_{\mathbb{R}^m} \left[\int_{\mathbb{R}^n} |f(x,y)|^q dy \right]^{1/q} dx$

$$\text{for } f \in \mathcal{L}_0(\mathbb{R}^{m+n}, E).$$

(ii) $\left(\int_{\mathbb{R}^n} \left| \int_{\mathbb{R}^m} f(x,y) \, dx \right|^q dy \right)^{1/q} \le \int_{\mathbb{R}^m} \left[\int_{\mathbb{R}^n} |f(x,y)|^q \, dy \right]^{1/q} dx < \infty$

$$\text{for } f \in \mathcal{L}_{(1,q)}(\mathbb{R}^{m+n}, E).$$

Proof In case (i), we can assume without loss of generality that

$$\int_{\mathbb{R}^m} \left[\int_{\mathbb{R}^n} |f(x,y)|^q \, dy \right]^{1/q} dx < \infty \ .$$

Then $|f|$ belongs to $\mathcal{L}_{(1,q)}(\mathbb{R}^{m+n}, \mathbb{R})$, and the claim is a special case of (ii), with f replaced by $|f|$ and E by $\mathbb{R}$. Suppose therefore that $f \in \mathcal{L}_{(1,q)}(\mathbb{R}^{m+n}, E)$. It follows from Lemma 6.20 and Theorem 2.11(i) (with E replaced by $L_q(\mathbb{R}^n, E)$) that

$$\int_{\mathbb{R}^m} Tf \, dx = \int_{\mathbb{R}^m} f(x, \cdot) \, dx \in L_q(\mathbb{R}^n, E)$$

and

$$\left(\int_{\mathbb{R}^n} \left| \int_{\mathbb{R}^m} f(x,y) \, dx \right|^q dy \right)^{1/q} = \left\| \int_{\mathbb{R}^m} Tf \, dx \right\|_{L_q(\mathbb{R}^n, E)} \le \int_{\mathbb{R}^m} \|Tf\|_{L_q(\mathbb{R}^n, E)} \, dx$$

$$= \int_{\mathbb{R}^m} \left(\int_{\mathbb{R}^n} |f(x,y)|^q \, dy \right)^{1/q} dx \ .$$

This completes the proof. ∎

A characterization of $L_p(\mathbb{R}^{m+n}, E)$

As another consequence of Lemma 6.20, we obtain an often useful generalization and sharpening of Fubini's theorem.

6.22 Theorem   For $1 \le p < \infty$,
$$L_p(\mathbb{R}^{m+n}, E) \to L_p(\mathbb{R}^m, L_p(\mathbb{R}^n, E)) , \quad f \mapsto (x \mapsto f(x, \cdot))$$
is an isometric isomorphism.

Proof   Suppose $v \in L_p(\mathbb{R}^m, L_p(\mathbb{R}^n, E))$. By Lemma 6.20, there is a sequence (f_j) in $L_p(\mathbb{R}^{m+n}, E)$ such that $\lim_j T f_j = v$ in $L_p(\mathbb{R}^m, L_p(\mathbb{R}^n, E))$. Because T is a linear isometry, it follows easily that (f_j) is a Cauchy sequence in $L_p(\mathbb{R}^{m+n}, E)$. Denoting by f its limit in $L_p(\mathbb{R}^{m+n}, E)$, we have $Tf = v$. Therefore T is surjective. This proves the claim. ∎

By means of this isometric isomorphism, we can identify the Banach spaces $L_p(\mathbb{R}^{m+n}, E)$ and $L_p(\mathbb{R}^m, L_p(\mathbb{R}^n, E))$:
$$L_p(\mathbb{R}^{m+n}, E) = L_p(\mathbb{R}^m, L_p(\mathbb{R}^n, E)) .$$

6.23 Remarks   (a) The statement of Theorem 6.22 is false for $p = \infty$, that is
$$L_\infty(\mathbb{R}^{m+n}, E) \ne L_\infty(\mathbb{R}^m, L_\infty(\mathbb{R}^n, E)) .$$

Proof   Take $A := \{ (x,y) \in \mathbb{R}^2 ; 0 \le y \le x \le 1 \}$ and $f := \chi_A$. Because A is Lebesgue measurable, f belongs to $L_\infty(\mathbb{R}^2)$. If we set
$$g(x) := f(x, \cdot) = \begin{cases} \chi_{[0,x]} & \text{if } 0 \le x \le 1 , \\ 0 & \text{otherwise} , \end{cases}$$
then $g(x)$ belongs to $L_\infty(\mathbb{R})$, and $\|g(x)\|_\infty \le 1$ for $x \in \mathbb{R}$. But g nevertheless does not belong to $L_\infty(\mathbb{R}, L_\infty(\mathbb{R}))$, because the map $g \colon \mathbb{R} \to L_\infty(\mathbb{R})$ is not λ_1-measurable. To see this, it suffices by Theorem 1.4 to show that g is not λ_1-almost separable-valued. To check this, note that
$$\|g(x) - g(r)\|_{L_\infty(\mathbb{R})} = 1 \quad \text{for } r \in \mathbb{R} \backslash \{x\} \tag{6.23}$$
for $x \in (0,1]$. Were g λ_1-almost separable valued, there would be a λ_1-null set $N \subset \mathbb{R}$ and a sequence (r_j) in $\mathbb{R}$ such that
$$\inf_{j \in \mathbb{N}} \|g(x) - g(r_j)\|_\infty < 1/2 \quad \text{for } x \in N^c . \tag{6.24}$$
Because $\lambda_1((0,1] \backslash N) = 1$, the set $(0,1] \backslash N$ is uncountable. Hence it follows from (6.23) that (6.24) cannot hold, and g is not λ_1-measurable. ∎

(b) Generalizing Theorem 6.22, one can show that for any $p, q \in [1, \infty)$, the map
$$L_{(p,q)}(\mathbb{R}^{m+n}, E) \to L_p(\mathbb{R}^m, L_q(\mathbb{R}^n, E)) , \quad f \mapsto (x \mapsto f(x, \cdot))$$
is an isometric isomorphism. Therefore $L_{(p,q)}(\mathbb{R}^{m+n}, E)$ is complete. ∎

A trace theorem

From Example IX.5.2 and the invariance of the Lebesgue measure under isometries, it follows that every hyperplane Γ in $\mathbb{R}^n$ is a λ_n-null set. Hence for $u \in L_p(\mathbb{R}^n)$, the restriction $u \,|\, \Gamma$, or *trace* of u on Γ, is not defined, because u can be "arbitrarily changed" on Γ. As another application of Fubini–Tonelli, we now show that one can nevertheless define such a trace on Γ for elements of certain vector subspaces of $L_p(\mathbb{R}^n)$. Of course, this is trivially the case for the vector subspace $C_c^1(\mathbb{R}^n)$. The significance of what follows is that this space is given not the supremum norm, but rather the L_p norm, with derivatives thrown into the mix. In the next section, we will understand better the significance of these subspaces of $L_p(\mathbb{R}^n)$.

Consider the coordinate hyperplane $\Gamma := \mathbb{R}^{n-1} \times \{0\}$, which we identify with $\mathbb{R}^{n-1}$. For $u \in C(\mathbb{R}^n)$, we let $\gamma u := u \,|\, \Gamma$ be the **trace** of u on Γ:

$$(\gamma u)(x) := u(x, 0) \quad \text{for } x \in \mathbb{R}^{n-1} .$$

Then $\gamma \colon C_c^1(\mathbb{R}^n) \to C_c(\mathbb{R}^{n-1})$, $u \mapsto \gamma u$ is a well defined linear map.

Now take $1 \leq p < \infty$, and give $C_c^1(\mathbb{R}^n)$ the norm

$$\|u\|_{1,p} := \left(\|u\|_p^p + \sum_{j=1}^n \|\partial_j u\|_p^p \right)^{1/p} .$$

Further, set

$$\widehat{H}_p^1(\mathbb{R}^n) := \left(C_c^1(\mathbb{R}^n), \| \cdot \|_{1,p} \right) .$$

Since $C_c(\mathbb{R}^{n-1})$ is a vector subspace of $L_p(\mathbb{R}^{n-1})$,

$$\gamma \colon \widehat{H}_p^1(\mathbb{R}^n) \to L_p(\mathbb{R}^{n-1}) , \quad u \mapsto \gamma u$$

is a well defined linear map, the **trace operator** with respect to $\Gamma = \mathbb{R}^{n-1}$. The following trace theorem shows that γ is continuous.

6.24 Proposition $\gamma \in \mathcal{L}(\widehat{H}_p^1(\mathbb{R}^n), L_p(\mathbb{R}^{n-1}))$ for $1 \leq p < \infty$.

Proof Define $h \in C^1(\mathbb{R})$ by $h(t) := |t|^{p-1} t$. For $v \in C_c^1(\mathbb{R}^n)$, it follows from the chain rule that $\partial_n h(v) = h'(v) \partial_n v$. Since v has compact support, the fundamental theorem of calculus then implies that

$$-h(v(x, 0)) = \int_0^\infty \partial_n h(v)(x, y) \, dy = \int_0^\infty h'(v(x, y)) \partial_n v(x, y) \, dy \quad \text{for } x \in \mathbb{R}^{n-1} .$$

Because $h'(t) = p \, |t|^{p-1}$, we find

$$|v(x, 0)|^p = \big| h(v(x, 0)) \big| \leq \int_0^\infty \big| h'(v(x, y)) \big| \, |\partial_n v(x, y)| \, dy$$

$$= p \int_0^\infty |v(x, y)|^{p-1} \, |\partial_n v(x, y)| \, dy .$$

Also, Young's inequality gives $\xi^{p-1}\eta \leq \dfrac{p-1}{p}\xi^p + \dfrac{1}{p}\eta^p$ for $\xi, \eta \in [0,\infty)$, so

$$|v(x,0)|^p \leq (p-1)\int_0^\infty |v(x,y)|^p\,dy + \int_0^\infty |\partial_n v(x,y)|^p\,dy\ .$$

With $c_p := \max\{p-1,1\}$, it now follows from Fubini–Tonelli that

$$\int_{\mathbb{R}^{n-1}} |v(x,0)|^p\,dx$$
$$\leq c_p\left(\int_{\mathbb{R}^{n-1}\times\mathbb{R}} |v(x,y)|^p\,d(x,y) + \int_{\mathbb{R}^{n-1}\times\mathbb{R}} |\partial_n v(x,y)|^p\,d(x,y)\right)\ . \tag{6.25}$$

Therefore

$$\|\gamma v\|_{L_p(\mathbb{R}^{n-1})} \leq c\,\|v\|_{\widehat{H}_p^1(\mathbb{R}^n)} \quad \text{for } v \in \widehat{H}_p^1(\mathbb{R}^n)\ ,$$

where $c := c_p^{1/p}$. This proves the theorem. ∎

6.25 Remark Denote by $\mathbb{H}^n$ the upper half-space of $\mathbb{R}^n$:

$$\mathbb{H}^n := \mathbb{R}^{n-1}\times(0,\infty) = \left\{\,(x,y)\in\mathbb{R}^{n-1}\times\mathbb{R}\ ;\ y > 0\,\right\}\ .$$

Then $\Gamma = \mathbb{R}^{n-1}\times\{0\} = \partial\mathbb{H}^n$. If we set

$$\widehat{H}_p^1(\mathbb{H}^n) := \left(\{\,u\,|\,\mathbb{H}^n\ ;\ u\in C_c^1(\mathbb{R}^n)\,\}, \ \|\cdot\|_{1,p}\right)\ ,$$

then $\widehat{H}_p^1(\mathbb{H}^n)$ is a vector subspace of $L_p(\mathbb{H}^n)$, and from a statement analogous to (6.25), it follows that

$$\gamma \in \mathcal{L}(\widehat{H}_p^1(\mathbb{H}^n), L_p(\mathbb{R}^{n-1}))\ .$$

In this case, γu for $u\in\widehat{H}_p^1(\mathbb{R}^n)$ is the **trace** of u on **the boundary** $\partial\mathbb{H}^n$. ∎

Exercises

1 Suppose $B \in \mathcal{L}(n)$ and $a \in \mathbb{R}^{n+1}$. Denote by

$$Z_a(B) := \left\{\,(x,0) + ta \in \mathbb{R}^{n+1}\ ;\ x\in B,\ t\in[0,1]\,\right\}$$

the cylinder with base B and edge a, and let

$$K_a(B) := \left\{\,(1-t)(x,0) + ta \in \mathbb{R}^{n+1}\ ;\ x\in B,\ t\in[0,1]\,\right\}$$

be the cone with base B and tip a. Prove:

(a) $\lambda_{n+1}\big(Z_a(B)\big) = |a_{n+1}|\lambda_n(B)$;

(b) $\lambda_{n+1}\big(K_a(B)\big) = |a_{n+1}|\lambda_n(B)/(n+1)$.

If one interprets $|a_{n+1}|$ as the height of the cylinder $Z_a(B)$ or the cone $K_a(B)$, then (b) says the volume of an n-dimensional cone is equal to the total volume of n cylinders with the same base and height.

2 For $0 < r < a$, let $V_{a,r}$ be the region in $\mathbb{R}^3$ enclosed by the 2-torus $T_{a,r}^2$. Show that $V_{a,r} = 2\pi^2 ar^2$.

3 Suppose $J \subset \mathbb{R}$ is an interval with endpoints $a := \inf J$ and $b := \sup J$. Also let $f \in \mathcal{L}_0(J, \mathbb{R}^+)$, and denote by

$$R_f := \big\{ (x,t) \in \mathbb{R}^n \times J \; ; \; |x| \le f(t) \big\}$$

the **solid of revolution** arising by rotation of the graph of f around the t-axis. Prove that

$$\lambda_{n+1}(R_f) = \omega_n \int_a^b (f(t))^n \, dt \; ,$$

where ω_n is the volume of $\mathbb{B}^n$. Interpret this formula geometrically in the case $n = 2$.

4 Suppose K is compact in $\mathbb{R}^n$ and $\rho_K := \int_K \rho(x) \, dx > 0$ for $\rho \in \mathcal{L}_1(K, \mathbb{R}^+)$. Then

$$S(K, \rho) := \frac{1}{\rho_K} \int_K x \rho(x) \, dx \in \mathbb{R}^n$$

is the **centroid of K with respect to the density ρ**. We set $S(K) := S(K, 1)$. Now suppose $J := [a, b]$ is a perfect, compact interval in $\mathbb{R}$, and let $f \in \mathcal{L}_0(J, \mathbb{R}^+)$. Also put

$$A_f := \big\{ (x,y) \in \mathbb{R}^2 \; ; \; 0 \le y \le f(x), \; x \in J \big\} \; ,$$

and denote by R_f the solid of revolution in $\mathbb{R}^3$ generated by f (by rotating about the x-axis). Prove:

(a) For $f \in \mathcal{L}_1(J, \mathbb{R}^+)$,

$$S(A_f) = \big(S_1(A_f), S_2(A_f)\big) = \frac{1}{\|f\|_1} \left(\int_a^b x f(x) \, dx, \frac{1}{2} \int_a^b (f(x))^2 \, dx \right) .$$

(b) For $f \in \mathcal{L}_2(J, \mathbb{R}^+)$,

$$S(R_f) = \left(\frac{1}{\|f\|_2^2} \int_a^b t(f(t))^2 \, dt, 0, 0 \right) .$$

(c) For $f \in \mathcal{L}_1(J, \mathbb{R}^+)$, we have **Guldin's first rule**

$$\lambda_3(R_f) = \pi \int_a^b (f(x))^2 \, dx = 2\pi S_2(A_f) \lambda_2(A_f) \; .$$

In words, *the volume of a solid of revolution is equal to the area of a meridional slice[7] times the circumference of the circle drawn by the centroid of that slice during a full revolution.*[8]

5 (a) For $\alpha \in [0, \pi/2)$, let $a := (\cos\alpha, 0, \sin\alpha)$. Determine the centroid of the cylinder $Z_a(\mathbb{B}_2)$ and the cone $K_a(\mathbb{B}_2)$ with respect to the density **1**.

(b) Let $A_\lambda := \big\{ (x,y) \in \mathbb{R}^2 \; ; \; 0 \le y \le e^{-\lambda x}, \; x \ge 0 \big\}$ for $\lambda > 0$. Show that $S(A_\lambda) \in A_\lambda$.

(c) Give an example where $S(A_f) \notin A_f$.

---

[7] That is, the intersection with a plane containing the rotation axis.

[8] Guldin's first rule also holds for solids of revolution not arising from the rotation of a graph; see Exercise XII.1.11.

6 Let $K \subset \mathbb{R}^n$ be convex and compact. Check that $S(K, \rho) \in K$ for $\rho \in \mathcal{L}_1(K, \mathbb{R}^+)$.

7 Denote by $\Delta_n := \{ x \in \mathbb{R}^n \; ; \; x_j \geq 0, \; \sum_{j=1}^n x_j \leq 1 \}$ the **standard simplex** in $\mathbb{R}^n$. Prove:

(a) $\lambda_n(\Delta_n) = 1/n!$.

(b) $S(\Delta_n) = (1/(n+1), 1/(n+1), \dots, 1/(n+1))$.

8 Given $f \in \mathcal{L}_1(\mathbb{R}^m, \mathbb{K})$, $g \in \mathcal{L}_1(\mathbb{R}^n, E)$, define $F(x, y) := f(x)g(y)$ for $(x, y) \in \mathbb{R}^m \times \mathbb{R}^n$. Show that F belongs to $\mathcal{L}_1(\mathbb{R}^{m+n}, E)$ and that

$$\int_{\mathbb{R}^{m+n}} F(x, y) \, d(x, y) = \int_{\mathbb{R}^m} f(x) \, dx \int_{\mathbb{R}^n} g(y) \, dy \ .$$

9 For $D := \{ (x, y) \in \mathbb{R}^2 \; ; \; x, y \geq 0, \; x + y \leq 1 \}$, show that

$$\int_D x^m y^n \, d(x, y) = \frac{1}{n+1} B(m+1, n+2) \quad \text{for } m, n \in \mathbb{N} \ .$$

10 Show that $\int_{[0,1] \times [0,1]} y/\sqrt{x} \, d(x, y) = 1$.

11 Show that $\int_{\mathbb{R}^n} \partial_j \varphi \, dx = 0$ for $\varphi \in C_c^1(\mathbb{R}^n, E)$ and $j \in \{1, \dots, n\}$.

12 For each of the following maps $f : (0, 1) \times (0, 1) \to \mathbb{R}$, calculate

$$\int_0^1 \int_0^1 f(x, y) \, dx \, dy \ , \quad \int_0^1 \int_0^1 f(x, y) \, dy \, dx \ , \quad \int_0^1 \int_0^1 |f(x, y)| \, dx \, dy \ , \quad \int_0^1 \int_0^1 |f(x, y)| \, dy \, dx \ .$$

(a) $f(x, y) := (x - y)/(x^2 + y^2)^{3/2}$.

(b) $f(x, y) := 1/(1 - xy)^\alpha$ for $\alpha > 0$.

13 Let $p, q \in [1, \infty]$. Prove:

(a) $L_p(\mathbb{R}^n) \not\subset L_q(\mathbb{R}^n)$ if $p \neq q$.

(b) if $X \subset \mathbb{R}^n$ is open and bounded, then $L_p(X) \subsetneq L_q(X)$ if $p > q$.

7 The convolution

In this section we use the translation invariance of the Lebesgue measure to introduce a new product on $L_1(\mathbb{R}^n)$, the convolution, which rests on the Lebesgue integral. We show that this operation is defined not only on $L_1(\mathbb{R}^n)$ but also on other function spaces, and that it has important smoothing properties. Among its applications are certain approximation theorems which we prove here for their great usefulness in later constructions.

We will consider mainly spaces of $\mathbb{K}$-valued functions defined on all of $\mathbb{R}^n$. For such spaces we omit the domain and image from the notation. In other words, if $\mathfrak{F}(\mathbb{R}^n) = \mathfrak{F}(\mathbb{R}^n, \mathbb{K})$ is a vector space of $\mathbb{K}$-valued functions on $\mathbb{R}^n$, we write simply $\mathfrak{F}$ if there is no risk of confusion. Thus L_p stands for $L_p(\mathbb{R}^n) = L_p(\mathbb{R}^n, \mathbb{K})$, and so on. Also $\int f\, dx$ will always mean $\int_{\mathbb{R}^n} f\, dx$.

Defining the convolution

Let F be a $\mathbb{K}$-vector space. For $f \in \mathrm{Funct}(\mathbb{R}^n, F)$, we define another function $\check{f} \in \mathrm{Funct}(\mathbb{R}^n, F)$ by $\check{f}(x) := f(-x)$, where $x \in \mathbb{R}^n$. The map $f \mapsto \check{f}$ is called **inversion** (about the origin).

Recall from IX.5.15 the definition of the translation group $\mathfrak{T} := \{\, \tau_a \; ; \; a \in \mathbb{R}^n \,\}$. Now we define an action[1] of this group on $\mathrm{Funct}(\mathbb{R}^n, F)$ by

$$\mathfrak{T} \times \mathrm{Funct}(\mathbb{R}^n, F) \to \mathrm{Funct}(\mathbb{R}^n, F) , \quad (\tau_a, f) \mapsto \tau_a f , \tag{7.1}$$

where

$$\tau_a f(x) := f(x - a) \quad \text{for } a, x \in \mathbb{R}^n . \tag{7.2}$$

Therefore

$$\tau_a f = f \circ \tau_{-a} = (\tau_{-a})^* f ,$$

where $(\tau_{-a})^*$ is the pull back defined in Section VIII.3.

7.1 Remarks (a) For $f \in \mathrm{Funct} := \mathrm{Funct}(\mathbb{R}^n, \mathbb{K})$, we have $\check{f} = (-\mathrm{id}_{\mathbb{R}^n})^* f$.

(b) Inversion is an involutive[2] vector space isomorphism on Funct and on $\mathcal{L}_p$ for $p \in [1, \infty] \cup \{0\}$.

(c) Suppose $E \in \{\, BC^k, BUC^k, C_0 \; ; \; k \in \mathbb{N} \,\}$. Then inversion belongs to $\mathcal{L}\mathrm{aut}(E)$.

(d) For $f \in \mathrm{Funct}$ and $x \in \mathbb{R}^n$, we have

$$(\tau_{-x} f)\check{\;}(y) = \tau_x \check{f}(y) = f(x - y) \quad \text{for } y \in \mathbb{R}^n .$$

---

[1] See Exercise I.7.6.
[2] A map $f \in X^X$ is said to be **involutive** if $f \circ f = \mathrm{id}_X$.

(e) Suppose $n = 1$ and $a > 0$. Then $\tau_a : \mathbb{R} \to \mathbb{R}$, $x \mapsto x + a$ is the **right translation** on $\mathbb{R}$ by a. Definition (7.2) means that τ_a also translates the graph of f to the right by a.

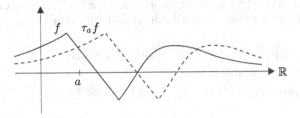

Therefore $\mathfrak{T}$ acts as a **right translation** on $\mathrm{Funct}(\mathbb{R}, F)$, which clarifies defining $\tau_a f$ as the pull back of the left translation τ_{-a} of $\mathbb{R}$. ∎

Take $f, g \in \mathcal{L}_1$ and $x \in \mathbb{R}^n$, and let O be open in $\mathbb{K}$.

$$(\tau_{-x} f)^{-1}(O) = (f \circ \tau_x)^{-1}(O) = \tau_{-x}(f^{-1}(O)) .$$

Therefore it follows from Corollary 1.5 and Lemma IX.5.16 that $(\tau_{-x} f)^{-1}(O)$ is measurable. Hence, again by Corollary 1.5, $\tau_{-x} f$ belongs to $\mathcal{L}_0$. Now we deduce from Remark 1.2(d) and parts (b) and (d) of Remark 7.1 that $y \mapsto f(x - y)g(y)$ belongs to $\mathcal{L}_0$ for every $x \in \mathbb{R}^n$. If this function is integrable, we define the **convolution of f with g at x** by

$$f * g(x) := \int f(x - y)g(y)\, dy .$$

We say f and g are **convolvable** if $f * g(x)$ is defined for almost every $x \in \mathbb{R}^n$. In this case the a.e.-defined function

$$f * g := \big(x \mapsto f * g(x)\big)$$

is called the **convolution of f with g**. If f and g are convolvable and $(f * g)^p$ is integrable (or $f * g$ is essentially bounded for $p = \infty$), we write $f * g \in \mathcal{L}_p$, in a slight abuse of notation.[3])

We now show that every pair $(f, g) \in \mathcal{L}_p \times \mathcal{L}_1$ with $p \in [1, \infty]$ is convolvable. The following observation will be helpful.

7.2 Lemma   For $f \in \mathcal{L}_0$ and $(x, y) \in \mathbb{R}^n \times \mathbb{R}^n = \mathbb{R}^{2n}$, let

$$F_1(x, y) := f(x) \quad \text{and} \quad F_2(x, y) := f(x - y) .$$

Then F_1 and F_2 belong to $\mathcal{L}_0(\mathbb{R}^{2n})$.

---

[3] We literally mean that the trivial extension of $f * g$ belongs to $\mathcal{L}_p$.

Proof   (i) Suppose O is open in $\mathbb{K}$ and $A := f^{-1}(O)$. Then A belongs to $\mathcal{L}(n)$. Therefore Remark 6.1(b) and Proposition 6.2 show that $F_1^{-1}(O) = A \times \mathbb{R}^n$ is λ_{2n}-measurable. Now the claim for F_1 follows from Corollary 1.5.

(ii) Set $\varphi(x, y) := (x - y, y)$ for $(x, y) \in \mathbb{R}^n \times \mathbb{R}^n$. Then $\varphi \in \mathcal{L}\mathrm{aut}(\mathbb{R}^{2n})$ and $F_2 = F_1 \circ \varphi$. The claim then follows from (i) and Theorem IX.5.12. ∎

7.3 Theorem   *Suppose $p \in [1, \infty]$ and $(f, g) \in \mathcal{L}_p \times \mathcal{L}_1$.*

(i) *f and g are convolvable.*

(ii) *(Young's inequality) $f * g \in \mathcal{L}_p$ and $\|f * g\|_p \le \|f\|_p \|g\|_1$.*

Proof   (a) Suppose first that $p \in [1, \infty)$. By Lemma 7.2 and Remark 1.2(d), the map $(x, y) \mapsto f(x - y)g(y)$ belongs to $\mathcal{L}_0(\mathbb{R}^{2n})$. Using Hölder's inequality, we deduce that

$$
\int |f(x - y)g(y)| \, dy = \int |f(x - y)| \, |g(y)|^{1/p} \, |g(y)|^{1/p'} \, dy
$$
$$
\le \left(\int |f(x - y)|^p \, |g(y)| \, dy \right)^{1/p} \left(\int |g(y)| \, dy \right)^{1/p'} .
$$

From this and Tonelli's theorem, we get

$$
\int \left(\int |f(x - y)g(y)| \, dy \right)^p dx \le \|g\|_1^{p/p'} \int\!\!\int |f(x - y)|^p \, |g(y)| \, dy \, dx
$$
$$
= \|g\|_1^{p/p'} \int\!\!\int |f(x - y)|^p \, dx \, |g(y)| \, dy
$$
$$
= \|g\|_1^{1 + p/p'} \|f\|_p^p < \infty ,
$$

where in the last step we once more used the translation invariance of the Lebesgue integral. Thus we find[4]

$$
\left(\int \left[\int |f(x - y)g(y)| \, dy \right]^p dx \right)^{1/p} \le \|f\|_p \|g\|_1 < \infty . \tag{7.3}
$$

Now from Remark 3.11(c), we conclude that $\int |f(x - y)g(y)| \, dy < \infty$ for almost every $x \in \mathbb{R}^n$; by Remark 3.11(a), this suffices to show that f and g are convolvable. Part (ii) of the theorem now follows from (7.3).

(b) In the case $p = \infty$, we have

$$
\int |f(x - y)g(y)| \, dy \le \|f\|_\infty \|g\|_1 < \infty \quad \text{for almost every } x \in \mathbb{R}^n ,
$$

which immediately implies (i) and (ii). ∎

---

[4]Those readers who worked through the last part of the previous section will recognize that this bound can also be easily derived from Minkowski's inequality for integrals.

7.4 Corollary Let $([f], [g]) \in L_p \times L_1$ with $p \in [1, \infty]$. Then

$$f * g = \overset{*}{f} * \overset{*}{g} \qquad a.e. \text{ in } \mathbb{R}^n$$

for $(\overset{*}{f}, \overset{*}{g}) \in ([f], [g])$.

Proof By Theorem 7.3, $f * g$, $\overset{*}{f} * \overset{*}{g}$, and $f * \overset{*}{g}$ are defined a.e. and belong to $\mathcal{L}_p$. Because

$$f * g - \overset{*}{f} * \overset{*}{g} = f * (g - \overset{*}{g}) + (f - \overset{*}{f}) * \overset{*}{g} ,$$

we obtain from Young's inequality that

$$\left\| f * g - \overset{*}{f} * \overset{*}{g} \right\|_p \leq \|f\|_p \left\| g - \overset{*}{g} \right\|_1 + \left\| f - \overset{*}{f} \right\|_p \left\| \overset{*}{g} \right\|_1 = 0 ,$$

from which the claim follows. ∎

We can now define the convolution for elements of $L_p \times L_1$ with $p \in [1, \infty]$: indeed, Corollary 7.4 guarantees that the map

$$* : L_p \times L_1 \to L_p , \qquad ([f], [g]) \mapsto [f * g]$$

is well defined. We call this the $*$ the **convolution product** on $L_p \times L_1$, and $[f] * [g] := [f * g]$ **convolution of** $[f]$ with $[g]$. It is clear that the convolution can also be defined on $L_1 \times L_p$, and we use the symbol $*$ for this as well.

The translation group

To be able to better explore further properties of the convolution, we first gather some important definitions and facts about the representation of the translation group $(\mathbb{R}^n, +)$ on function spaces.

Let F be a $\mathbb{K}$-vector space and let V be a vector subspace of $\text{Funct}(\mathbb{R}^n, F)$ that is **invariant** under the action (7.1) of the translation group $\mathfrak{T}$ of $\mathbb{R}^n$, meaning that $\tau_a(V) \subset V$ for all $a \in \mathbb{R}^n$. By **restriction**, (7.1) induces an action

$$\mathfrak{T} \times V \to V , \qquad (\tau_a, v) \mapsto \tau_a v$$

of the translation group $\mathfrak{T}$ on V. For every $a \in \mathbb{R}^n$, the map $T_a := (v \mapsto \tau_a v)$ is a linear map from V into itself. Because

$$\tau_a \tau_b v = \tau_{a+b} v \quad \text{and} \quad \tau_0 v = v ,$$

T_a is a vector space automorphism of V and $(T_a)^{-1} = T_{-a}$. Hence[5]

$$(\mathbb{R}^n, +) \to \text{Aut}(V) , \qquad a \mapsto T_a$$

---

[5]See Remarks I.12.2(d) and I.7.6(e).

is a group homomorphism, a **linear representation** of the group $(\mathbb{R}^n, +)$ on V. In particular,

$$\mathfrak{T}_V := \{ T_a \in \mathrm{Aut}(V) \; ; \; a \in \mathbb{R}^n \}$$

is a subgroup of $\mathrm{Aut}(V)$, called the **group of translations on** V. Instead of T_a, we tend to use the same symbol τ_a if there is no fear of misunderstanding. The invariance of V under (7.1) is also expressed by saying that $(\mathbb{R}^n, +)$ is **linearly representable** on V.

If V is a (semi)normed vector space, the group $\mathfrak{T}_V$ is said to be **strongly continuous** if $\lim_{a \to 0} \tau_a v = v$ for every $v \in V$.

7.5 Remarks   (a) $(\mathbb{R}^n, +)$ is linearly representable on Funct and on $B := B(\mathbb{R}^n)$.

(b) $(\mathbb{R}^n, +)$ is linearly representable on $\mathcal{L}_\infty$, and $\|\tau_a f\|_\infty = \|f\|_\infty$ for $f \in \mathcal{L}_\infty$.

Proof   Take $f \in \mathcal{L}_\infty$. For every $\alpha > \|f\|_\infty$ there is a set N of Lebesgue measure zero such that $|f(x)| \le \alpha$ for $x \in N^c$. By translation invariance (Theorem IX.5.17), $N_a := \tau_a(N)$ also has measure zero and

$$|\tau_a f(x)| = |f(x - a)| \le \alpha \quad \text{for } x \in N_a^c .$$

Therefore $\tau_a f$ is essentially bounded, and $\|\tau_a f\|_\infty \le \|f\|_\infty$. The claim follows since

$$\|f\|_\infty = \|\tau_{-a}(\tau_a f)\|_\infty \le \|\tau_a f\|_\infty . \ \blacksquare$$

(c) The translation groups $\mathfrak{T}_B$ and $\mathfrak{T}_{\mathcal{L}_\infty}$ are not strongly continuous.

Proof $\|\tau_a \chi_{\mathbb{B}^n} - \chi_{\mathbb{B}^n}\|_\infty = 1$ for $a \in \mathbb{R}^n \setminus \{0\}$. $\blacksquare$

(d) If $\mathfrak{T}_V$ is strongly continuous, then

$$(a \mapsto \tau_a f) \in C(\mathbb{R}^n, V) \quad \text{for } f \in V .$$

Proof   This follows from $\tau_a f - \tau_b f = \tau_{a-b}(\tau_b f) - \tau_b f$ for $f \in V$ and $a, b \in \mathbb{R}^n$. $\blacksquare$

7.6 Theorem   *Suppose $V = \mathcal{L}_p$ with $p \in [1, \infty)$ or $V = BUC^k$ with $k \in \mathbb{N}$. Then $(\mathbb{R}^n, +)$ is linearly representable on V, and the translation group $\mathfrak{T}_V$ is strongly continuous. Also $\|\tau_a f\|_V = \|f\|_V$ for $a \in \mathbb{R}^n$ and $f \in V$.*

Proof   (i) We consider first the case $V = BUC^k$. Take $f \in BUC^k$, $a \in \mathbb{R}^n$, and $\varepsilon > 0$. Then there is $\delta > 0$ such that $|f(x) - f(y)| < \varepsilon$ for all $x, y \in \mathbb{R}^n$ satisfying $|x - y| < \delta$. It follows that

$$|\tau_a f(x) - \tau_a f(y)| = |f(x - a) - f(y - a)| < \varepsilon \tag{7.4}$$

for $x, y \in \mathbb{R}^n$ such that $|x - y| < \delta$. Therefore $\tau_a f$ belongs to BUC, and because

$$\partial^\alpha \tau_a f = \tau_a \partial^\alpha f \quad \text{for } \alpha \in \mathbb{N}^n \text{ and } |\alpha| \le k , \tag{7.5}$$

we obtain $\tau_a f \in BUC^k$. Consequently $(\mathbb{R}^n, +)$ is linearly representable on BUC^k. From Remark 7.5(b) and (7.5), we find $\|\tau_a f\|_{BC^k} = \|f\|_{BC^k}$.

Now take $x \in \mathbb{R}^n$. If $|a| < \delta$, we can set $y = x + a$ in (7.4), and we get

$$|\tau_a f(x) - f(x)| < \varepsilon \quad \text{for } x \in \mathbb{R}^n ,$$

that is, $\|\tau_a f - f\|_\infty < \varepsilon$ for $a \in \delta \mathbb{B}^n$. Analogously, we can show with (7.5) that there is a $\delta_1 > 0$ such that $\|\tau_a f - f\|_{BC^k} < \varepsilon$ for $a \in \delta_1 \mathbb{B}^n$. Therefore $\mathfrak{T}_{BUC^k}$ is strongly continuous.

(ii) Let $p \in [1, \infty)$ and $f \in \mathcal{L}_p$. The equality $\|\tau_a f\|_p = \|f\|_p$ follow from the translation invariance of the Lebesgue integral.

Now take $\varepsilon > 0$. By Theorem 4.14, there is a $g \in C_c$ such that $\|f - g\|_p < \varepsilon/3$. Because g has compact support, there is a compact subset K of $\mathbb{R}^n$ such that $\operatorname{supp}(\tau_a g) \subset K$ for $|a| \leq 1$. Also, since g is uniformly continuous, there exists $\delta \in (0, 1]$ such that

$$\|\tau_a g - g\|_\infty < \varepsilon/3\lambda_n(K)^{1/p} \quad \text{for } a \in \delta \mathbb{B}^n .$$

Suppose $a \in \delta \mathbb{B}^n$. Because $\operatorname{supp}(\tau_a g - g) \subset K$, Theorem 5.1(iv) implies that

$$\|\tau_a g - g\|_p < \varepsilon/3 \quad \text{for } a \in \delta \mathbb{B}^n .$$

Since

$$\|\tau_a f - f\|_p \leq \|\tau_a f - \tau_a g\|_p + \|\tau_a g - g\|_p + \|g - f\|_p$$

and $\|\tau_a f - \tau_a g\|_p = \|\tau_a(f - g)\|_p = \|f - g\|_p$, we get $\|\tau_a f - f\|_p < \varepsilon$ for $a \in \delta \mathbb{B}^n$, and we are done. ∎

We now define an action of $\mathfrak{T}$ on L_p for $p \in [1, \infty]$. By Remark 7.5(b) and Theorem 7.6, τ_a is an isometry of $\mathcal{L}_p$ for every $a \in \mathbb{R}^n$. Therefore the map

$$L_p \to L_p , \quad [f] \mapsto [\tau_a f]$$

is well defined for every $a \in \mathbb{R}^n$. We denote it by τ_a also, that is, we set

$$\tau_a[f] := [\tau_a f] \quad \text{for } f \in L_p \text{ and } a \in \mathbb{R}^n .$$

Then

$$\|\tau_a[f]\|_p = \|[\tau_a f]\|_p = \|\tau_a f\|_p = \|f\|_p = \|[f]\|_p . \tag{7.6}$$

Clearly

$$\mathfrak{T} \times L_p \to L_p , \quad (\tau_a, f) \mapsto \tau_a f$$

is an action of the translation group $\mathfrak{T}$ of $\mathbb{R}^n$ on L_p. By Remark 7.5(b) and Theorem 7.6, $T_a := (f \mapsto \tau_a f)$ is a linear isometry on L_p for every $a \in \mathbb{R}^n$. Again writing T_a as τ_a, we conclude that

$$(\mathbb{R}^n, +) \to \mathcal{L}\mathrm{aut}(L_p) , \quad a \mapsto \tau_a$$

is a representation of the additive group of $\mathbb{R}^n$ by linear isometries on L_p. In particular, the **translation group on L_p**, namely

$$\mathfrak{T}_{L_p} := \left\{ \tau_a \in \mathcal{L}\mathrm{aut}(L_p) ; \ a \in \mathbb{R}^n \right\} ,$$

is a subgroup of $\mathcal{L}\mathrm{aut}(L_p)$ consisting of isometries.

7.7 Corollary *The translation group on L_p is strongly continuous for $1 \leq p < \infty$.*

Proof This is an immediate consequence of Theorem 7.6 and (7.6). ∎

Elementary properties of the convolution

After this digression about the translation group, we return to the convolution and derive its chief properties.

7.8 Theorem *Consider $(f, g) \in L_p \times L_1$ with $p \in [1, \infty]$.*

(i) *The convolution $f * g$ belongs to L_p, and satisfies **Young's inequality***

$$\|f * g\|_p \leq \|f\|_p \|g\|_1 .$$

(ii) $f * g = g * f$.

(iii) *If $p = \infty$, the convolution $f * g$ belongs to[6] BUC.*

(iv) *For $\varphi \in BC^k$, we have $\varphi * g \in BUC^k$,*

$$\partial^\alpha(\varphi * g) = \partial^\alpha \varphi * g \quad \text{for } \alpha \in \mathbb{N}^n , \quad |\alpha| \leq k ,$$

and $\|\varphi * g\|_{BC^k} \leq \|\varphi\|_{BC^k} \|g\|_1$.

Proof (i) follows from Theorem 7.3(ii) and Corollary 7.4.

(ii) Take $x \in \mathbb{R}$ and let $\overset{*}{f}$ and $\overset{*}{g}$ be representatives of f and g. Also set $\psi(y) := x - y$ for $y \in \mathbb{R}^n$. Then ψ is an involutive isometry of $\mathbb{R}^n$. It follows from Theorem 7.3(i) and Example 6.6(b) that

$$\overset{*}{f} * \overset{*}{g}(x) = \int \overset{*}{f}(x - y)\overset{*}{g}(y) \, dy = \int \left(\overset{*}{f} \circ \psi\right)\left(\left(\overset{*}{g} \circ \psi\right) \circ \psi\right) dy$$

$$= \int \left(\overset{*}{g} \circ \psi\right)\overset{*}{f} \, dy = \int \overset{*}{g}(x - y)\overset{*}{f}(y) \, dy = \overset{*}{g} * \overset{*}{f}(x) .$$

Therefore $f * g = g * f$ by Corollary 7.4.

(iii) The motion invariance of the Lebesgue integral yields $\|\breve{g}\|_1 = \|g\|_1$. From part (ii), and because the elements of $\mathfrak{T}_{L_1}$ are isometries, we then get

$$\left|\overset{*}{f} * \overset{*}{g}(x) - \overset{*}{f} * \overset{*}{g}(y)\right| \leq \int \left|\overset{*}{f}(z)\left(\overset{*}{g}(x - z) - \overset{*}{g}(y - z)\right)\right| dz \leq \|\overset{*}{f}\|_\infty \|\tau_x \breve{g} - \tau_y \breve{g}\|_1$$

$$= \|f\|_\infty \|\tau_y(\tau_{x-y}\breve{g} - \breve{g})\|_1 = \|f\|_\infty \|\tau_{x-y}\breve{g} - \breve{g}\|_1$$

for $x, y \in \mathbb{R}^n$. Because $\breve{g} \in L_1$, the strong continuity of $\mathfrak{T}_{L_1}$ together with part (i) implies that $\overset{*}{f} * \overset{*}{g} \in BUC$. The claim follows.

---

[6] See Theorem 4.18.

(iv) In view of (iii), it suffices to consider the case $k \geq 1$. To this end we define $h(x, y) := \varphi(x - y)g(y)$ for $(x, y) \in \mathbb{R}^n \times \mathbb{R}^n$. Then h satisfies the assumptions of Theorem 3.18, and it follows that $\partial_j(\varphi * g) = \partial_j\varphi * g$ for $j \in \{1, \ldots, n\}$. By (iii) and Theorem VII.2.10, we have $\varphi * g \in BUC^1$. We now see inductively that $\varphi * g$ belongs to BUC^k and satisfies $\partial^\alpha(\varphi * g) = \partial^\alpha\varphi * g$ for every $\alpha \in \mathbb{N}^n$ with $|\alpha| \leq k$. Finally, by (i), we have

$$\|\varphi * g\|_{BC^k} = \max_{|\alpha| \leq k} \|\partial^\alpha(\varphi * g)\|_\infty = \max_{|\alpha| \leq k} \|(\partial^\alpha\varphi) * g\|_\infty \leq \Big(\max_{|\alpha| \leq k} \|\partial^\alpha\varphi\|_\infty\Big) \|g\|_1$$

$$= \|\varphi\|_{BC^k} \|g\|_1 . \blacksquare$$

7.9 Corollary

(i) *Let* $p \in (1, \infty)$ *and* $k \in \mathbb{N}$. *The convolution satisfies*

$$* \in \begin{cases} \mathcal{L}^2_{\text{sym}}(L_1, L_1) , \\ \mathcal{L}(L_p, L_1; L_p) , \\ \mathcal{L}(L_\infty, L_1; BUC) , \\ \mathcal{L}(BC^k, L_1; BUC^k) , \end{cases}$$

and all these maps have norm at most 1.

(ii) $(L_1, +, *)$ *is a commutative Banach algebra without a multiplicative identity.*

Proof (i) and the first statement of (ii) follow immediately from Theorem 7.8. We now assume there is $e \in L_1$ such that $e * f = f$ for every $f \in L_1$. We choose a representative $\overset{*}{e}$ of e and then find by Exercise 2.15 a $\delta > 0$ such that

$$\left| \int_{\delta\mathbb{B}^n} \overset{*}{e}(x - y) \, dy \right| = \left| \int_{\mathbb{B}^n(x, \delta)} \overset{*}{e}(z) \, dz \right| < 1 \quad \text{for } x \in \mathbb{R}^n .$$

Furthermore, there is a set N of Lebesgue measure zero such that $\chi_{\delta\mathbb{B}^n}(x) = \overset{*}{e} * \chi_{\delta\mathbb{B}^n}(x)$ for $x \in N^c$. However, for $x \in \delta\mathbb{B}^n \cap N^c$, we have

$$1 = \chi_{\delta\mathbb{B}^n}(x) = \overset{*}{e} * \chi_{\delta\mathbb{B}^n}(x) = \int_{\mathbb{R}^n} \overset{*}{e}(x - y)\chi_{\delta\mathbb{B}^n}(y) \, dy = \int_{\delta\mathbb{B}^n} \overset{*}{e}(x - y) \, dy < 1 ,$$

which is not possible. $\blacksquare$

7.10 Theorem (additivity of supports)   *Suppose* $f, g \in \mathcal{L}_0$ *are convolvable and* f *has compact support. Then*

$$\text{supp}(f * g) \subset \text{supp}(f) + \text{supp}(g) .$$

Proof (i) We can assume $f * g \neq 0$. For $x \in [f * g \neq 0]$, there is a $y \in \mathbb{R}^n$ such that $f(x - y)g(y) \neq 0$. It follows that $y \in \text{supp}(g)$ and $x \in y + \text{supp}(f)$, and thus x belongs to $\text{supp}(f) + \text{supp}(g)$. Hence $[f * g \neq 0] \subset \text{supp}(f) + \text{supp}(g)$.

(ii) We show that $\operatorname{supp}(f) + \operatorname{supp}(g)$ is closed. Let (x_k) be a sequence in $\operatorname{supp}(f) + \operatorname{supp}(g)$ such that $x_k \to x$ for some $x \in \mathbb{R}^n$. Then there are sequences (a_k) in $\operatorname{supp}(f)$ and (b_k) in $\operatorname{supp}(g)$ such that $x_k = a_k + b_k$ for $k \in \mathbb{N}$. Because $\operatorname{supp}(f)$ is compact, there is a subsequence $(a_{k_\ell})_{\ell \in \mathbb{N}}$ of (a_k) and an $a \in \operatorname{supp}(f)$ such that $a_{k_\ell} \to a$ as $\ell \to \infty$. Thus $b_{k_\ell} = x_{k_\ell} - a_{k_\ell} \to x - a$ as $k \to \infty$. Because $\operatorname{supp}(g)$ is closed, we know $x-a$ belongs to $\operatorname{supp}(g)$. Hence there exists $b \in \operatorname{supp}(g)$ such that $x = a+b$. This shows that $\operatorname{supp}(f)+\operatorname{supp}(g)$ is closed. The claim follows from Corollary III.2.13. $\blacksquare$

Approximations to the identity

We saw in Corollary 7.9 that the convolution algebra L_1 has no multiplicative identity. However, the next theorem secures the existence of "approximations to the identity", elements $\varphi \in L_1$ that satisfy $\|\varphi * f - f\|_1 < \varepsilon$ for every $f \in L_1$ (for a given $\varepsilon > 0$).

7.11 Theorem (approximation theorem)   Given $E \in \{L_p \; ; \; 1 \leq p < \infty\}$ or $E \in \{BUC^k \; ; \; k \in \mathbb{N}\}$, set $\varphi \in \mathcal{L}_1$ and

$$a := \int \varphi \, dx \, , \quad \varphi_\varepsilon(x) := \varepsilon^{-n} \varphi(x/\varepsilon) \quad \text{for } x \in \mathbb{R}^n \, , \quad \varepsilon > 0 \, .$$

Then $\lim_{\varepsilon \to 0} \varphi_\varepsilon * f = af$ in E for $f \in E$.

Proof (i) Fix $\varepsilon > 0$. By the substitution rule — Example 6.6(b) — we know that $\varphi_\varepsilon \in L_1$ and $\int \varphi_\varepsilon \, dx = a$. Thus Theorem 7.8 shows that $\varphi_\varepsilon * f \in E$ for $f \in E$.

(ii) To prove the limit as $\varepsilon \to 0$, consider first the case $E = L_p$. Take $f \in \mathcal{L}_p$ and $\varepsilon > 0$. By Theorem 7.3(i) and the proof of Theorem 7.8(ii), and using the transformation $y \mapsto y/\varepsilon$ in Example 6.6(b), we obtain

$$\varphi_\varepsilon * f(x) - af(x) = f * \varphi_\varepsilon(x) - af(x) = \int [f(x-y) - f(x)] \varphi_\varepsilon(y) \, dy$$

$$= \int [f(x - \varepsilon z) - f(x)] \varphi(z) \, dz = \int [\tau_{\varepsilon z} f(x) - f(x)] \varphi(z) \, dz \tag{7.7}$$

for almost every $x \in \mathbb{R}^n$. Corollary 7.7 and Remark 7.5(d) imply that

$$\left(z \mapsto (\tau_{\varepsilon z} f - f) \right) \in C(\mathbb{R}^n, E) \quad \text{for } \varepsilon > 0 \, , \tag{7.8}$$

and

$$\lim_{\varepsilon \to 0} \|\tau_{\varepsilon z} f - f\|_E = 0 \quad \text{for } z \in \mathbb{R}^n \, . \tag{7.9}$$

Now set

$$g^\varepsilon(z) := (\tau_{\varepsilon z} f - f) \varphi(z) \quad \text{for } z \in \mathbb{R}^n \text{ and } \varepsilon > 0 \, .$$

Then it follows from (7.8), Theorem 1.17, and Remark 1.2(d) that g^ε belongs to $\mathcal{L}_0(\mathbb{R}^n, E)$ for every $\varepsilon > 0$. Because $\|\tau_{\varepsilon z} f\|_E = \|f\|_E$, we also derive from the triangle inequality that

$$\|g^\varepsilon(z)\|_E \leq 2\,\|f\|_E\,|\varphi(z)| \quad \text{for } z \in \mathbb{R}^n \text{ and } \varepsilon > 0 \ .$$

Because $\varphi \in \mathcal{L}_1(\mathbb{R}^n)$, we therefore conclude that $g^\varepsilon \in \mathcal{L}_1(\mathbb{R}^n, E)$. Then (7.7) and Theorem 2.11(i) imply the bound[7]

$$\|\varphi_\varepsilon * f - af\|_E = \left\|\int g^\varepsilon(z)\,dz\right\|_E \leq \int \|g^\varepsilon(z)\|_E\,dz \ .$$

Now the dominated convergence theorem shows that $\varphi_\varepsilon * f$ converges in E to af as $\varepsilon \to 0$, because, by (7.9), we have $\lim_{\varepsilon \to 0} \|g^\varepsilon(z)\|_E = 0$ for almost every $z \in \mathbb{R}^n$.

(iii) Now suppose $f \in BUC^k$. If $\varphi = 0$ λ_n-a.e., the claim is obviously true. So suppose $m := \int |\varphi|\,dx > 0$. From Theorem 7.8(ii) and (iv), it follows that

$$\partial^\alpha(\varphi_\varepsilon * f - af) = \varphi_\varepsilon * \partial^\alpha f - a\partial^\alpha f \quad \text{for } \alpha \in \mathbb{N}^n \text{ and } |\alpha| \leq k \ .$$

Therefore it suffices to consider the case $k = 0$.

Let $\eta > 0$. Then there is a $\delta > 0$ such that

$$|f(x - y) - f(x)| \leq \eta/2m \quad \text{for } x, y \in \mathbb{R}^n \ , \quad |y| < \delta \ ,$$

and we obtain

$$
\begin{aligned}
|\varphi_\varepsilon * f(x) - af(x)| &\leq \int |f(x-y) - f(x)|\,|\varphi_\varepsilon(y)|\,dy \\
&\leq \frac{\eta}{2m} \int_{[|y|<\delta]} |\varphi_\varepsilon(y)|\,dy + 2\,\|f\|_\infty \int_{[|y|\geq\delta]} |\varphi_\varepsilon(y)|\,dy \quad (7.10) \\
&\leq \frac{\eta}{2} + 2\,\|f\|_\infty \int_{[|y|\geq\delta]} |\varphi_\varepsilon(y)|\,dy
\end{aligned}
$$

for $x \in \mathbb{R}^n$. The substitution rule then gives

$$\int_{[|y|\geq\delta]} |\varphi_\varepsilon(y)|\,dy = \varepsilon^{-n} \int_{[|y|\geq\delta]} |\varphi(y/\varepsilon)|\,dy = \int_{[|z|\geq\delta/\varepsilon]} |\varphi(z)|\,dz \ .$$

By the dominated convergence theorem, then, there exists $\varepsilon_0 > 0$ such that

$$\int_{[|y|\geq\delta]} |\varphi_\varepsilon(y)|\,dy \leq \frac{\eta}{4\,\|f\|_\infty} \quad \text{for } \varepsilon \in (0, \varepsilon_0] \ .$$

Now the claim follows from (7.10). ∎

---

[7]This also follows from Minkowski's inequality for integrals.

Suppose $\varphi \in \mathcal{L}_1$ satisfies $\int \varphi \, dx = 1$ and set

$$\varphi_\varepsilon(x) := \varepsilon^{-n} \varphi(x/\varepsilon) \quad \text{for } x \in \mathbb{R}^n \text{ and } \varepsilon > 0 . \tag{7.11}$$

The family $\{ \varphi_\varepsilon \; ; \; \varepsilon > 0 \}$ is called an **approximating kernel** or an **approximation to the identity**. If

$$\varphi \in C^\infty(\mathbb{R}^n, \mathbb{R}) , \quad \check{\varphi} = \varphi , \quad \varphi \geq 0 , \quad \operatorname{supp}(\varphi) \subset \overline{\mathbb{B}}^n , \quad \int \varphi \, dx = 1 ,$$

we call $\{ \varphi_\varepsilon \; ; \; \varepsilon > 0 \}$ a **mollifier** or **smoothing kernel**. Every smoothing kernel obviously satisfies

$$\operatorname{supp}(\varphi_\varepsilon) \subset \varepsilon \overline{\mathbb{B}}^n \quad \text{for } \|\varphi_\varepsilon\|_1 = 1 \text{ and } \varepsilon > 0 .$$

7.12 Examples[8] (a) The **Gaussian kernel** is the family $\{ k_\varepsilon \; ; \; \varepsilon > 0 \}$ defined by

$$k(x) := (4\pi)^{-n/2} \, e^{-|x|^2/4} \quad \text{for } x \in \mathbb{R}^n .$$

It is an approximating kernel.

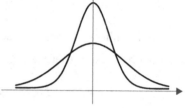

Proof From Example 6.13(a), we know that

$$\int g(x) \, dx = 1$$

for $g(x) := \pi^{-n/2} e^{-|x|^2}$. Since $k(x) = 2^{-n} g(x/2)$ for $x \in \mathbb{R}^n$, it follows from the substitution rule that $\int k(x) \, dx = 1$. $\blacksquare$

(b) Let

$$\varphi(x) := \begin{cases} c \, e^{1/(|x|^2 - 1)} & \text{if } |x| < 1 , \\ 0 & \text{if } |x| \geq 1 , \end{cases}$$

where $c := \left(\int_{\mathbb{B}^n} e^{1/(|x|^2 - 1)} \, dx \right)^{-1}$ is chosen so that ϕ integrates to 1. Then the family $\{ \varphi_\varepsilon \; ; \; \varepsilon > 0 \}$ is a smoothing kernel.

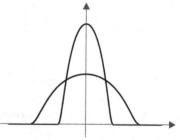

Proof Because $x \mapsto |x|^2 - 1$ is smooth on $\mathbb{R}^n$, Example IV.1.17 shows that φ belongs to $C^\infty(\mathbb{R}^n, \mathbb{R})$ (see Exercise VII.5.16). The claim follows easily. $\blacksquare$

Test functions

Let X be a metric space, and let A and B be subsets of X. We say A is **compactly contained in** B (in symbols: $A \subset\subset B$) if $\overline{A}$ is compact and is contained in the interior of B.

---

[8]In both examples, the area under the graphs is always 1, so smaller values of ε give correspondingly higher maxima.

If X is open in $\mathbb{R}^n$ and E is a normed vector space, we call

$$\mathcal{D}(X, E) := \{\, \varphi \in C^\infty(X, E) \;;\; \operatorname{supp}(\varphi) \subset\subset X \,\}$$

the **space of (E-valued) test functions** on X. When $E = \mathbb{K}$, we write $\mathcal{D}(X) := \mathcal{D}(X, \mathbb{K})$, as usual. Clearly $\mathcal{D}(X, E)$ is a vector subspace of $C^\infty(X, E)$ and of $C_c(X, E)$, and $\mathcal{D}(X, E) = C^\infty(X, E) \cap C_c(X, E)$. Because the map

$$j : C_c(X, E) \to C_c(\mathbb{R}^n, E) , \quad g \mapsto \widetilde{g} ,$$

is linear and injective, we can identify $C_c(X, E)$ with a vector subspace of $C_c(\mathbb{R}^n, E)$ and regard (as needed) each element of the former as an element of the latter. Likewise, we identify $\mathcal{D}(X, E)$ with a vector subspace of $\mathcal{D}(\mathbb{R}^n, E)$. With these notations, we have the following inclusions of vector subspaces for every $p \in [1, \infty]$:

$$\mathcal{D}(X, E) \subset \mathcal{D}(\mathbb{R}^n, E) \subset C_c(\mathbb{R}^n, E) \subset L_p(\mathbb{R}^n, E) .$$

7.13 Theorem *Suppose X is open in $\mathbb{R}^n$ and $p \in [1, \infty)$. Then $\mathcal{D}(X)$ is a dense vector subspace of $L_p(X)$ and of $C_0(X)$.*

Proof (i) Take $g \in C_c(X)$ and $\eta > 0$. Also let $\{\varphi_\varepsilon \;;\; \varepsilon > 0\}$ be a smoothing kernel. By Theorem 7.8, $\varphi_\varepsilon * g$ belongs to BUC^k and therefore to BUC^∞ for every $k \in \mathbb{N}$. Because g has compact support, there is $\varepsilon_0 > 0$ such that[9] $\operatorname{dist}(\operatorname{supp}(g), X^c) \geq \varepsilon_0$. From Theorem 7.10, it follows that

$$\operatorname{supp}(\varphi_\varepsilon * g) \subset \operatorname{supp}(\varphi_\varepsilon) + \operatorname{supp}(g) \subset \operatorname{supp}(g) + \varepsilon \overline{\mathbb{B}}^n \quad \text{for } \varepsilon > 0 .$$

Then $\varphi_\varepsilon * g$ belongs to $\mathcal{D}(X)$ for $\varepsilon \in (0, \varepsilon_0)$. Finally by Theorem 7.11 we can find for every $q \in [1, \infty]$ some $\varepsilon_1 \in (0, \varepsilon_0)$ such that $\|\varphi_{\varepsilon_1} * g - g\|_q < \eta/2$.

(ii) Now suppose $f \in L_p(X)$. By Theorem 5.1, we can find $g \in C_c(X)$ such that $\|f - g\|_p < \eta/2$. By (i), there is $h \in \mathcal{D}(X)$ such that $\|f - h\|_p < \eta$.

(iii) For $f \in C_0(X)$, let K be a compact subset of X such that $|f(x)| < \eta/2$ for $x \in X \backslash K$. By Proposition 4.13, we can choose a $\varphi \in C_c(X)$ such that $0 \leq \varphi \leq 1$ and $\varphi \,|\, K = 1$. We set $g := \varphi f$. Because $f(x) = g(x)$ for $x \in K$, it follows that

$$|f(x) - g(x)| = |f(x)| \, |1 - \varphi(x)| < \eta/2 \quad \text{for } x \in X .$$

Therefore $\|f - g\|_\infty \leq \eta/2$. The claim then follows from (i). $\blacksquare$

Smooth partitions of unity

In Section 4, we proved the existence of continuous Urysohn functions in general metric spaces. This result can be distinctly improved in the special case of $\mathbb{R}^n$, where we can use mollifiers to actually construct smooth cutoff functions.

---

[9]$\operatorname{dist}(\operatorname{supp}(g), \emptyset) := \infty$.

7.14 Proposition (smooth cutoff functions) *Suppose $K \subset \mathbb{R}^n$ is compact, and set*

$$K_\rho := \left\{ x \in \mathbb{R}^n \; ; \; \mathrm{dist}(x, K) < \rho \right\} \quad \text{for } \rho > 0 .$$

Then for every $\alpha \in \mathbb{N}^n$ and every $\rho > 0$ there exist a positive constant $c(\alpha)$ and a map $\varphi \in \mathcal{D}(K_\rho)$ such that $0 \le \varphi \le 1$, $\varphi | K = 1$, and $\|\partial^\alpha \varphi\|_\infty \le c(\alpha) \rho^{-|\alpha|}$.

Proof Set $\{ \psi_\varepsilon \; ; \; \varepsilon > 0 \}$ be a smoothing kernel. Let $\delta := \rho/3$ and $\varphi := \psi_\delta * \chi_{K_\delta}$. Then φ belongs to BUC^∞, and it follows from Theorem 7.10 that

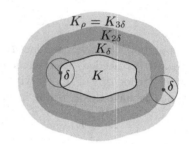

$$\mathrm{supp}(\varphi) \subset \mathrm{supp}(\psi_\delta) + \overline{K}_\delta \subset \delta \overline{\mathbb{B}}^n + \overline{K}_\delta$$
$$\subset \overline{K}_{2\delta} \subset K_{3\delta} = K_\rho .$$

Therefore φ belongs to $\mathcal{D}(K_\rho)$. Moreover

$$\varphi(x) = \int \psi_\delta(x - y) \chi_{K_\delta}(y) \, dy \le \int \psi_\delta(x - y) \, dy = 1$$

for $x \in \mathbb{R}^n$, and hence $0 \le \varphi \le 1$. If x lies in K, then

$$\varphi(x) = \int \psi_\delta(y) \chi_{K_\delta}(x - y) \, dy = \int \psi_\delta(y) \, dy = 1 ,$$

and therefore $\varphi | K = 1$. Finally, since $\partial^\alpha \psi_\delta = \delta^{-|\alpha|}(\partial^\alpha \psi_1)_\delta$ for $\alpha \in \mathbb{N}^n$, we have from Theorem 7.8(iv) that

$$\partial^\alpha \varphi = \partial^\alpha (\psi_\delta * \chi_{K_\delta}) = \partial^\alpha \psi_\delta * \chi_{K_\delta} = \delta^{-|\alpha|}(\partial^\alpha \psi_1)_\delta * \chi_{K_\delta} .$$

Now $c(\alpha) := 3^{|\alpha|} \|\partial^\alpha \psi_1\|_1$ is independent of $\delta > 0$, and so it follows from Young's inequality that $\|\partial^\alpha \varphi\|_\infty \le c(\alpha) \rho^{-|\alpha|}$. ∎

Let $K \subset \mathbb{R}^n$ be compact and denote by $\{ X_j \; ; \; 0 \le j \le m \}$ a finite open cover of K. If for every $j \in \{0, \ldots, m\}$, there is a $\varphi_j \in C^\infty(\mathbb{R}^n)$ such that

(i) $0 \le \varphi_j \le 1$,

(ii) $\mathrm{supp}(\varphi_j) \subset X_j$, and

(iii) $\sum_{j=0}^m \varphi_j(x) = 1$ for $x \in K$,

then $\{ \varphi_j \; ; \; 0 \le j \le m \}$ is called a **smooth partition of unity on K** subordinate to the cover $\{ X_j \; ; \; 0 \le j \le m \}$.

If X_0 is open in $\mathbb{R}^n$ and $K \subset X_0$, then $\mathrm{dist}(K, X_0^c) > 0$, and Proposition 7.14 (with $\rho := \mathrm{dist}(K, X_0^c)$) secures the existence of a smooth partition of unity on K subordinate to the one-element cover $\{X_0\}$ of K. To treat the general case of a finite cover, we need a technical result:

7.15 Lemma (shrinking lemma) *Let $\{ X_j \; ; \; 0 \leq j \leq m \}$ be a finite open cover of a compact subset K of $\mathbb{R}^n$. Then there is an open cover $\{ U_j \; ; \; 0 \leq j \leq m \}$ of K such that $U_j \subset\subset X_j$ for $j \in \{0, \ldots, m\}$.*

Proof Given $x \in K$, choose $j \in \{0, \ldots, m\}$ such that $x \in X_j$ and $r_x > 0$ such that $V_x := \mathbb{B}^n(x, r_x)$ is compact and contained in X_j. Then $\{ V_x \; ; \; x \in K \}$ is an open cover of K, and there exist $k \in \mathbb{N}$ and $\{x_0, \ldots, x_k\} \subset K$ with $K \subset \bigcup_{i=0}^{k} V_{x_i}$. With $U_j := \bigcup\{ V_{x_i} \; ; \; V_{x_i} \subset X_j \}$ for $j \in \{0, \ldots, m\}$, we have a family $\{ U_j \; ; \; 0 \leq j \leq m \}$ having the desired properties. ∎

7.16 Theorem (smooth partitions of unity) *If K is a compact subset of $\mathbb{R}^n$, every finite open cover of K has a smooth partition of unity subordinate to it.*

Proof Suppose $\{ X_j \; ; \; 0 \leq j \leq m \}$ is a finite open cover of K. By Lemma 7.15, there is an open cover $\{ U_j \; ; \; 0 \leq j \leq m \}$ such that $U_j \subset\subset X_j$ for $j \in \{0, \ldots, m\}$. We define $K_j := \overline{U}_j$. Then K_j is compact, and $\mathrm{dist}(K_j, X_j^c)$ is positive for every $j \in \{0, \ldots, m\}$. Proposition 7.14 now shows there is a $\psi_j \in \mathcal{D}(X_j)$ such that $0 \leq \psi_j \leq 1$ and $\psi_j \,|\, K_j = 1$. Defining

$$\varphi_0 := \psi_0 \quad \text{and} \quad \varphi_k := \psi_k \prod_{j=0}^{k-1}(1 - \psi_j) \quad \text{for } 1 \leq k \leq m ,$$

it is easy to check by induction that $\sum_{j=0}^{m} \varphi_j = 1 - \prod_{j=0}^{m}(1 - \psi_j)$. The claim now follows because $K \subset \bigcup_{j=0}^{m} K_j$. ∎

We next present some simple applications of Theorem 7.16. Additional, more complicated situations will be described in succeeding chapters.

7.17 Applications (a) Suppose X is open in $\mathbb{R}^n$. Then for $f \in L_0(X)$ the following statements are equivalent:

  (i) $f \in L_{1,\mathrm{loc}}(X)$;

  (ii) $\varphi f \in L_1(X)$ for every $\varphi \in \mathcal{D}(X)$;

  (iii) $\widetilde{f \,|\, K} \in L_1(X)$ for every $K = \overline{K} \subset\subset X$.

Proof Let $(U_j)_{j \in \mathbb{N}}$ be an ascending sequence of relatively compact open subsets of X with $X = \bigcup_j U_j$ (see Remarks 1.16(d) and (e)). Then (see Exercise 4.3)

$$L_{1,\mathrm{loc}}(X) = \big\{ f \in L_0(X) \; ; \; \chi_{U_j} f \in L_1(X), \, j \in \mathbb{N} \big\} .$$

"(i)⟹(ii)" Let $\varphi \in \mathcal{D}(X)$. Since $K := \mathrm{supp}(\varphi)$ is compact and $(U_j)_{j \in \mathbb{N}}$ is ascending, there is a $k \in \mathbb{N}$ such that $K \subset U_k$. By virtue of Proposition 7.14, we find a $\psi \in \mathcal{D}(U_k)$ such that $0 \leq \psi \leq 1$ and $\psi \,|\, K = 1$. Then

$$\int_X |\varphi f| \, dx = \int_X |\varphi \psi f| \, dx \leq \|\varphi\|_\infty \int_X |\psi f| \, dx \leq \|\varphi\|_\infty \, \|\chi_{U_k} f\|_1 < \infty .$$

Therefore φf belongs to $L_1(X)$.

"(ii)$\Rightarrow$(iii)" Take $K = \overline{K} \subset\subset X$ and $\varphi \in \mathcal{D}(X)$ with $\varphi \,|\, K = 1$. Then

$$\int_K |f| = \int_K |\varphi f| \le \|\varphi f\|_1 < \infty \, ,$$

and therefore $\widetilde{f \,|\, K} \in L_1(X)$.

"(iii)$\Rightarrow$(i)" This implication is clear because every $\overline{U}_j$ is compact. ∎

(b) Suppose X is open in $\mathbb{R}^n$. Then $C(X) \subset L_{1,\mathrm{loc}}(X)$.

Proof Take $f \in C(X)$ and $\varphi \in \mathcal{D}(X)$. Then φf belongs to $C_c(X)$. By Theorem 5.1, we have $\varphi f \in L_1(X)$, and the claim follows from (a). ∎

(c) The linear representation of the group $(\mathbb{R}^n, +)$ in BUC^k is injective, hence a group isomorphism onto its image $\mathfrak{T}_{BUC^k}$.

Proof For $a \in \mathbb{R}^n$, suppose $\tau_a = \mathrm{id}_{BUC^k}$. We choose $r > |a|$ and a cutoff function $\varphi \in \mathcal{D}(\mathbb{R}^n)$ for $r\overline{\mathbb{B}}^n$. Then $f_j := \varphi \,\mathrm{pr}_j$ belongs to BUC^k, and we find

$$-a_j = \tau_a f_j(0) = f_j(0) = 0 \quad \text{for } j \in \{1, \ldots, n\} \, .$$

Therefore $a = 0$. This implies the injectivity of the representation $a \mapsto \tau_a$. ∎

(d) Suppose X is open in $\mathbb{R}^n$ and bounded. Also let $\{\, X_j \;;\; 0 \le j \le m \,\}$ be a finite open cover of $\overline{X}$, and let $\{\, \varphi_j \;;\; 0 \le j \le m \,\}$ be a smooth partition of unity subordinate to it. Finally let $k \in \mathbb{N}$ and

$$\|\|u\|\|_{BC^k} := \sum_{j=0}^m \|\varphi_j u\|_{BC^k} \quad \text{for } u \in BC^k(X) \, .$$

Then $\|\|\cdot\|\|_{BC^k}$ is an equivalent norm on $BC^k(X)$.

Proof Take $u \in BC^k(X)$. Obviously

$$\|u\|_{BC^k} = \Big\| \sum_{j=0}^m \varphi_j u \Big\|_{BC^k} \le \sum_{j=0}^m \|\varphi_j u\|_{BC^k} = \|\|u\|\|_{BC^k} \, .$$

From Leibniz's rule (see Exercise VII.5.21), we obtain

$$\|\|u\|\|_{BC^k} = \sum_{j=0}^m \max_{|\alpha| \le k} \|\partial^\alpha (\varphi_j u)\|_\infty = \sum_{j=0}^m \max_{|\alpha| \le k} \Big\| \sum_{\beta \le \alpha} \binom{\alpha}{\beta} \partial^\beta \varphi_j \, \partial^{\alpha - \beta} u \Big\|_\infty$$

$$\le \sum_{j=0}^m c_k \|\varphi_j\|_{BC^k} \|u\|_{BC^k} \le C \|u\|_{BC^k} \, ,$$

where we have set $c_k := \max_{|\alpha| \le k} \sum_{\beta \le \alpha} \binom{\alpha}{\beta}$ and $C := c_k \sum_{j=0}^m \|\varphi_j\|_{BC^k}$. ∎

Convolutions of E-valued functions

A look back at preceding proofs shows that the convolution $f * g$ can also be defined when one of the two functions takes values in a Banach space F and the other is scalar-valued. All proofs carry through without change[10] so long as the substitution rule for isometries still holds for F-valued functions. This is indeed the case, as we shall show in the next section. In particular, the key approximation result in Theorem 7.11 remains true for the spaces $L_p(\mathbb{R}^n, F)$ and $BUC^k(\mathbb{R}^n, F)$ with $1 \le p < \infty$ and $k \in \mathbb{N}$. One consequence of this is an analogue of Theorem 7.13 to the effect that $\mathcal{D}(X, F)$ is a dense vector subspace of $C_0(X, F)$ and of $L_p(X, F)$ for $1 \le p < \infty$.

Distributions[11]

Suppose X is a nonempty open subset of $\mathbb{R}^n$. A scalar function on X, as is well known, is a rule for assigning a real or complex number to every point in X. But this definition is just an abstraction, since the individual points of X cannot in practice be discerned. If, for example, we want to determine the temperature distribution of some medium that occupies the set X — we must rely on an experimental probe. However, such a probe, being of nonzero size, can only determine values of f in an extended region; whatever value it assigns to $f(x_0)$ represents not the actual value at x_0 (if indeed such a thing is physically meaningful) but rather some kind of average around x_0: mathematically, an integral $\int_X \varphi f \, dx$, where φ is a "test function" that depends on the probe. Of course, the measurement will better approximate the exact value $f(x_0)$ the more the test function φ is concentrated about x_0, that is, the less the probe smears the data.

To claim complete knowledge of $f(x_0)$, one might imagine bringing to bear all conceivable probes, or in other words, determining the averages $\int_X \varphi f \, dx$ over all possible test functions φ. In mathematical terms, we'd be replacing the pointwise function $f: X \to \mathbb{K}$ by a functional defined on the space of all test functions, namely, the map

$$T_f : \mathcal{D}(X) \to \mathbb{K} , \quad \varphi \mapsto \int_X \varphi f \, dx . \tag{7.12}$$

Our choice of $\mathcal{D}(X)$ as the space of test functions is to a large extent arbitrary. For conceptual simplicity, we might want to consider $C_c(X)$ instead of $\mathcal{D}(X)$. At the same time, we would like to avoid performing more "measurements" than necessary; this warrants choosing a test space that is small in some sense. But the space must be large enough that the averages $\int_X \varphi f \, dx$ do determine f. That is, we want the equality of $\int_X \varphi f \, dx$ and $\int_X \varphi g \, dx$ for all test functions φ to imply $f = g$.

---

[10]Naturally, the commutativity formula $f * g = g * f$ must be interpreted correctly.

[11]The rest of this section is meant to provide glimpses of applications and more advanced theories; it can be skipped over on first reading.

The next theorem shows that this is indeed the case if we choose $\mathcal{D}(X)$ as the test space and work with "functions" in $L_{1,\text{loc}}(X)$. (By Application 7.17(a), $L_{1,\text{loc}}(X)$ is the largest vector subspace E of $L_0(X)$ such that $\int_X \varphi f \, dx$ is well defined for all $f \in E$ and all $\varphi \in \mathcal{D}(X)$.)

7.18 Theorem *Suppose $f \in L_{1,\text{loc}}(X)$. If*

$$\int_X \varphi f \, dx = 0 \quad \text{for } \varphi \in \mathcal{D}(X) , \tag{7.13}$$

then $f = 0$.

Proof Suppose $f \neq 0$, and let $\overset{*}{f} \in \mathcal{L}_{1,\text{loc}}(X)$ be a representative of f. By regularity, there is a compact subset K of X of positive measure such that $\overset{*}{f}(x) \neq 0$ for $x \in K$. Take $\eta \in \mathcal{D}(X)$ with $\eta \mid K = 1$, and let $g := \eta \overset{*}{f}$. By Application 7.17(a), g belongs to $\mathcal{L}_1$. Also $g(x) \neq 0$ for $x \in K$. Let $\{ \varphi_\varepsilon \; ; \; \varepsilon > 0 \}$ be a smoothing kernel. Then $\lim_{\varepsilon \to 0} \varphi_\varepsilon * g = g$ in $\mathcal{L}_1$. By Corollary 4.7, there is a null sequence (ε_j) and a set N of Lebesgue measure zero such that

$$\lim_{j \to \infty} \varphi_{\varepsilon_j} * g(x) = g(x) \quad \text{for } x \in N^c . \tag{7.14}$$

Given $x_0 \in K \cap N^c$, set $\psi_j := \eta \tau_{x_0} \varphi_{\varepsilon_j} \in \mathcal{D}(X)$ for $j \in \mathbb{N}$. Since $\breve{\varphi}_\varepsilon = \varphi_\varepsilon$ by Remark 7.1(d), equality (7.13) gives

$$\varphi_{\varepsilon_j} * g(x_0) = \int g(y) \, \varphi_{\varepsilon_j}(x_0 - y) \, dy = \int_X (\eta \overset{*}{f})(y) \, \varphi_{\varepsilon_j}(x_0 - y) \, dy$$

$$= \int_X \overset{*}{f}(y) \, \psi_j(y) \, dy = 0 .$$

However, because of (7.14) this contradicts $g(x_0) \neq 0$. The claim follows because the representative $\overset{*}{f}$ of f was chosen arbitrarily. ∎

Clearly the map T_f is a linear functional on $\mathcal{D}(X)$. For the interpretation of $T_f \varphi = \int_X \varphi f \, dx$ as a measurement value to be meaningful, $T_f \varphi$ must "depend continuously" on the measuring device; that is, small perturbations in the probe, hence in the test function φ, should cause only small changes in the measured value. Mathematically speaking, this means that T_f must be a continuous linear functional on $\mathcal{D}(X)$. So we must introduce a topology on $\mathcal{D}(X)$.

Since our treatment here is introductory, we will limit ourselves to stating what it means for a sequence to converge in $\mathcal{D}(X)$. This convergence should be compatible with the vector structure on $\mathcal{D}(X)$, so it suffices to consider the case where the limit is 0.

We say that a sequence (φ_j) **converges to 0** (or **is a null sequence**) **in $\mathcal{D}(X)$** if the following conditions are satisfied:

$(\mathcal{D}_1)$ There exists $K \subset\subset X$ such that $\text{supp}(\varphi_j) \subset K$ for $j \in \mathbb{N}$.

$(\mathcal{D}_2)$ $\varphi_j \to 0$ in $BC^k(X)$ for every $k \in \mathbb{N}$.

Obviously $(\mathcal{D}_2)$ is equivalent to:

The sequence $(\partial^\alpha \varphi_j)_{j \in \mathbb{N}}$ converges uniformly to 0 for every $\alpha \in \mathbb{N}^n$. (7.15)

So for φ_j to converge in $\mathcal{D}(X)$ to 0, not only must (7.15) hold, but the supports of the functions φ_j must all be contained in a fixed compact subset of X.

A linear functional $T : \mathcal{D}(X) \to \mathbb{K}$ is **continuous** if $T\varphi_j \to 0$ for every null sequence (φ_j) in $\mathcal{D}(X)$. A continuous linear functional on $\mathcal{D}(X)$ is also called a **Schwartz distribution**, or simply **distribution**, on X. The set of distributions on X is denoted by $\mathcal{D}'(X)$; it is clearly a vector subspace of $\mathrm{Hom}(\mathcal{D}(X), \mathbb{K})$.

(In functional analysis — more precisely, in the theory of topological vector spaces — one shows that there is exactly one Hausdorff topology on $\mathcal{D}(X)$ that is locally convex,[12] compatible with the vector structure, and such that sequences converge to 0 in the sense above if and only if they converge to 0 in the topology. With respect to this topology, $\mathcal{D}'(X)$ is the dual of $\mathcal{D}(X)$, that is, the space of all continuous linear functionals on $\mathcal{D}(X)$. See, for example, [Sch66] or [Yos65].)

7.19 Examples (a) For every $f \in L_{1,\mathrm{loc}}(X)$, the linear functional T_f defined by (7.12) is a distribution on X.

Proof Let (φ_j) be a sequence in $\mathcal{D}(X)$ such that $\varphi_j \to 0$ in $\mathcal{D}(X)$. Then there is a compact subset K of X such that $\mathrm{supp}(\varphi_j) \subset X$ for $j \in \mathbb{N}$. It follows that

$$|T_f \varphi_j| = \left| \int_X \varphi_j f \, dx \right| \le \int_K |\varphi_j| \, |f| \, dx \le \|f\|_{L_1(K)} \|\varphi_j\|_\infty$$

for $j \in \mathbb{N}$. Because $\|f\|_{L_1(K)} < \infty$, we find that $T_f \varphi_j \to 0$ in $\mathbb{K}$, because $(\mathcal{D}_2)$ implies that $\|\varphi_j\|_\infty \to 0$. ∎

(b) Let μ be a Radon measure on X. Then

$$\mathcal{D}(X) \to \mathbb{K} , \quad \varphi \mapsto \int_X \varphi \, d\mu$$

defines a distribution on X.

Proof Suppose (φ_j) is a sequence in $\mathcal{D}(X)$ such that $\varphi_j \to 0$ in $\mathcal{D}(X)$. Also suppose $K = \overline{K} \subset\subset X$ contains $\mathrm{supp}(\varphi_j)$ for all $j \in \mathbb{N}$. Then

$$\left| \int_X \varphi_j \, d\mu \right| \le \int_K |\varphi_j| \, d\mu \le \mu(K) \|\varphi_j\|_\infty \quad \text{for } j \in \mathbb{N} .$$

As in the proof of (a), this implies that φ is a distribution on X. ∎

(c) Let δ be the Dirac measure on $\mathbb{R}^n$ with support at 0. Then

$$\varphi \mapsto \langle \delta, \varphi \rangle := \int_X \varphi \, d\delta = \varphi(0) \quad \text{for } \varphi \in \mathcal{D}(\mathbb{R}^n) ,$$

---

[12]This means the origin has an open neighborhood basis of convex sets.

is a distribution on $\mathbb{R}^n$, the **Dirac distribution**

$$\delta : \mathcal{D}(\mathbb{R}^n) \to \mathbb{K} , \quad \varphi \mapsto \varphi(0) .$$

There is no $u \in L_{1,\mathrm{loc}}(\mathbb{R}^n)$ such that $T_u = \delta$.

Proof   The first statement is a special case of (b).

Suppose now that $u \in L_{1,\mathrm{loc}}(\mathbb{R}^n)$ with $T_u = \delta$, that is,

$$\int_{\mathbb{R}^n} \varphi u \, dx = \varphi(0) \quad \text{for } \varphi \in \mathcal{D}(\mathbb{R}^n) . \tag{7.16}$$

Choosing only such $\varphi \in \mathcal{D}(\mathbb{R}^n)$ that $\mathrm{supp}(\varphi) \subset\subset X := \mathbb{R}^n \setminus \{0\}$, we have $\varphi(0) = 0$, and from Theorem 7.18, it follows that $u \,|\, X = 0$ in $L_{1,\mathrm{loc}}(X)$. But X and $\mathbb{R}^n$ differ only on a set of measure zero (a single point!), so $u = 0$ in $L_{1,\mathrm{loc}}(\mathbb{R}^n)$, contradicting (7.16). ∎

(d) Let $\alpha \in \mathbb{N}^n$. Then

$$S_\alpha : \mathcal{D}(\mathbb{R}^n) \to \mathbb{K} , \quad \varphi \mapsto \partial^\alpha \varphi(0)$$

defines a distribution. There is no $u \in L_{1,\mathrm{loc}}(\mathbb{R}^n)$ such that $T_u = S_\alpha$.

Proof   Let (φ_j) be a sequence in $\mathcal{D}(\mathbb{R}^n)$ such that $\varphi_j \to 0$ in $\mathcal{D}(\mathbb{R}^n)$, and suppose $K = \overline{K} \subset\subset \mathbb{R}^n$ with $\mathrm{supp}(\varphi_j) \subset K$ for $j \in \mathbb{N}$. We can assume that 0 lies in K. Then we have the estimate

$$|\partial^\alpha \varphi_j(0)| \le \max_{x \in K} |\partial^\alpha \varphi_j(x)| \le \|\varphi_j\|_{BC^{|\alpha|}} \quad \text{for } j \in \mathbb{N} .$$

Thus $(\mathcal{D}_2)$ implies that $\partial^\alpha \varphi_j(0) \to 0$ in $\mathbb{K}$, which shows that S_α is a distribution. The second statement is proved as in (c). ∎

The following key result is now a simple consequence of Theorem 7.18.

7.20 Theorem   *The map*

$$L_{1,\mathrm{loc}}(X) \to \mathcal{D}'(X) , \quad f \mapsto T_f$$

is linear and injective.

Proof   Example 7.19(a) shows the map is well defined. It is linear because integration is. It is injective by Theorem 7.18. ∎

By Theorem 7.20, we can identify $L_{1,\mathrm{loc}}(X)$ with its image in $\mathcal{D}'(X)$. In other words, we can regard $L_{1,\mathrm{loc}}(X)$ as a vector subspace of the space of all Schwartz distributions, by identifying a function $f \in L_{1,\mathrm{loc}}(X)$ with the distribution

$$T_f = \left(\varphi \mapsto \int_X \varphi f \, dx \right) \in \mathcal{D}'(X) .$$

In this sense, every $f \in L_{1,\mathrm{loc}}(X)$ is a distribution. The elements of $L_{1,\mathrm{loc}}(X)$ are called **regular distributions**. All other distributions are **singular**. Examples 7.19(c) and (d) illustrate singular distributions.

The theory of distributions plays an important role in higher analysis, especially in the study of partial differential equations, and in theoretical physics. We cannot elaborate here, but see for example [Sch65], [RS72].

Linear differential operators

Let X be open in $\mathbb{R}^n$. Given functions $a_\alpha \in C^\infty(X)$, for each $\alpha \in \mathbb{N}^n$ such that $|\alpha| \leq m \in \mathbb{N}$, we set

$$A(\partial)u := \sum_{|\alpha| \leq m} a_\alpha \partial^\alpha u \quad \text{for } u \in \mathcal{D}(X) .$$

Obviously, $A(\partial)$ is a linear map of $\mathcal{D}(X)$ onto itself; we say it is a **linear differential operator on X** of order $\leq m$ (with smooth coefficients). It has **order m** if

$$\sum_{|\alpha|=m} \|a_\alpha\|_\infty \neq 0 ,$$

that is, if at least one coefficient a_α of the **leading part** $\sum_{|\alpha|=m} a_\alpha \partial^\alpha$ of $A(\partial)$ does not vanish identically. We denote by $\mathcal{D}\mathrm{iffop}(X)$ the set of all linear differential operators on X; those of order $\leq m$ are denoted by $\mathcal{D}\mathrm{iffop}_m(X)$.

A linear map $T : \mathcal{D}(X) \to \mathcal{D}(X)$ is said to be **continuous**[13] if $T\varphi_j \to 0$ in $\mathcal{D}(X)$ for every sequence (φ_j) in $\mathcal{D}(X)$ such that $\varphi_j \to 0$ in $\mathcal{D}(X)$. The set of all continuous endomorphisms of $\mathcal{D}(X)$ is a vector subspace of $\mathrm{End}(\mathcal{D}(X))$, which we denote by $\mathcal{L}(\mathcal{D}(X))$.[13]

7.21 Proposition $\mathcal{D}\mathrm{iffop}(X)$ *is a vector subspace of* $\mathcal{L}(\mathcal{D}(X))$, *and* $\mathcal{D}\mathrm{iffop}_m(X)$ *is a vector subspace of* $\mathcal{D}\mathrm{iffop}(X)$.

Proof Let $m \in \mathbb{N}$, and take $A(\partial) := \sum_{|\alpha| \leq m} a_\alpha \partial^\alpha \in \mathcal{D}\mathrm{iffop}_m(X)$. Let (φ_j) be a null sequence in $\mathcal{D}(X)$, and let $K = \overline{K} \subset\subset X$ contain $\mathrm{supp}(\varphi_j)$ for all $j \in \mathbb{N}$. Then $\mathrm{supp}(A(\partial)\varphi_j) \subset K$ for $j \in \mathbb{N}$. For $\beta \in \mathbb{N}^n$, the Leibniz rule gives

$$\|\partial^\beta(a_\alpha \partial^\alpha \varphi_j)\|_{C(K)} = \left\| \sum_{\gamma \leq \beta} \binom{\beta}{\gamma} \partial^\gamma a_\alpha \partial^{\beta-\gamma+\alpha}\varphi_j \right\|_{C(K)}$$

$$\leq c(\alpha, \beta) \max_{\gamma \leq \beta} \|\partial^\gamma a_\alpha\|_{C(K)} \|\partial^{\beta-\gamma+\alpha}\varphi_j\|_\infty .$$

From this we derive for $k \in \mathbb{N}$ the inequality

$$\|A(\partial)\varphi_j\|_{BC^k(X)} \leq c(k) \sum_{|\alpha| \leq m} \|a_\alpha\|_{BC^k(K)} \|\varphi_j\|_{BC^{k+m}(X)} \quad \text{for } j \in \mathbb{N} ,$$

---

[13]It is shown in functional analysis that these definitions are consistent with our previous definitions for continuity and $\mathcal{L}(E)$.

where the constant $c(k)$ is independent of j. Now $\mathcal{A}(\partial)\varphi_j \to 0$ in $BC^k(X)$ follows from $(\mathcal{D}_2)$. Because this is true for every $k \in \mathbb{N}$, we see that $\mathcal{A}(\partial)\varphi_j \to 0$ in $\mathcal{D}(X)$. This proves that $\mathrm{Diffop}_m(X) \subset \mathcal{L}(\mathcal{D}(X))$. The other statements are clear. ∎

Let $(\cdot \mid \cdot)$ denote the inner product in $L_2(X)$, and suppose $\mathcal{A}(\partial)$ belongs to $\mathrm{Diffop}(X)$. If there is a differential operator $\mathcal{A}^\sharp(\partial) \in \mathrm{Diffop}(X)$ such that

$$\bigl(\mathcal{A}(\partial)u \mid v\bigr) = \bigl(u \mid \mathcal{A}^\sharp(\partial)v\bigr) \quad \text{for } u, v \in \mathcal{D}(X) \,,$$

we say $\mathcal{A}^\sharp(\partial)$ is the **formal adjoint** of $\mathcal{A}(\partial)$. Because

$$\bigl(u \mid \mathcal{A}^\sharp(\partial)v\bigr) = \int_X u\overline{\mathcal{A}^\sharp(\partial)v}\, dx$$

and $\overline{\mathcal{A}^\sharp(\partial)v} \in \mathcal{D}(X) \subset L_{1,\mathrm{loc}}(X)$ for $v \in \mathcal{D}(X)$, it follows easily from Theorem 7.18 that $\mathcal{A}(\partial)$ has at most one formal adjoint. If $\mathcal{A}(\partial)$ has a formal adjoint $\mathcal{A}^\sharp(\partial)$ that coincides with $\mathcal{A}(\partial)$, then $\mathcal{A}(\partial)$ is **formally self-adjoint**.

We will now show that every $\mathcal{A}(\partial) \in \mathrm{Diffop}(X)$ has a differential operator formally adjoint to it, and we derive an explicit form for $\mathcal{A}^\sharp(\partial)$. First we need this:

7.22 Proposition (integration by parts) For $f \in C^1(X)$ and $g \in C_c^1(X)$,

$$\int_X (\partial_j f)g\, dx = -\int_X f\partial_j g\, dx \quad \text{for } j \in \{1, \dots, n\} \,.$$

Proof We need only consider the case $j = 1$; the general case will follow by permutation of coordinates, in view of Corollary 6.10. So write $x = (x_1, x') \in \mathbb{R} \times \mathbb{R}^{n-1}$. Since fg has compact support, it follows from integrating by parts that

$$\int_{-\infty}^{\infty} [\partial_1 f(x_1, x')]g(x_1, x')\, dx_1 = -\int_{-\infty}^{\infty} f(x_1, x')\partial_1 g(x_1, x')\, dx_1$$

for every $x' \in \mathbb{R}^{n-1}$. From Fubini's theorem, we now get

$$\int_X (\partial_1 f)g\, dx = \int_{\mathbb{R}^n} (\partial_1 f)g\, dx$$

$$= \int_{\mathbb{R}^{n-1}} \left(\int_{-\infty}^{\infty} \partial_1 f(x_1, x')g(x_1, x')\, dx_1 \right) dx'$$

$$= -\int_{\mathbb{R}^{n-1}} \left(\int_{-\infty}^{\infty} f(x_1, x')\partial_1 g(x_1, x')\, dx_1 \right) dx'$$

$$= -\int_{\mathbb{R}^n} f\partial_1 g\, dx = -\int_X f\partial_1 g\, dx \,. \quad ∎$$

7.23 Corollary *Suppose* $f \in C^k(X)$ *and* $g \in C_c^k(X)$. *Then*

$$\int_X (\partial^\alpha f) g \, dx = (-1)^{|\alpha|} \int_X f \partial^\alpha g \, dx$$

for $\alpha \in \mathbb{N}^n$ *such that* $|\alpha| \leq k$.

Integration by parts is also the core of the proof of the next result.

7.24 Proposition *Every differential operator*

$$\mathcal{A}(\partial) = \sum_{|\alpha| \leq m} a_\alpha \partial^\alpha \in \mathcal{D}\text{iffop}(X)$$

has a unique formal adjoint, which is explicitly given by

$$\mathcal{A}^\sharp(\partial) v = \sum_{|\alpha| \leq m} (-1)^{|\alpha|} \partial^\alpha (\bar{a}_\alpha v) \quad \text{for } v \in \mathcal{D}(X) . \tag{7.17}$$

If $\mathcal{A}(\partial)$ *has order* m, *then* $\mathcal{A}^\sharp(\partial)$ *is also an* m-*th order differential operator.*

Proof We already know that there is at most one formal adjoint, so we need only prove existence and the validity of (7.17).

Take $u, v \in \mathcal{D}(X)$. Integrating by parts, we find

$$\left(\mathcal{A}(\partial) u \mid v \right) = \int_X (\mathcal{A}(\partial) u) \bar{v} \, dx = \sum_{|\alpha| \leq m} \int_X (a_\alpha \partial^\alpha u) \bar{v} \, dx$$

$$= \sum_{|\alpha| \leq m} (-1)^{|\alpha|} \int_X u \partial^\alpha (a_\alpha \bar{v}) \, dx = \int_X u \overline{\sum_{|\alpha| \leq m} (-1)^{|\alpha|} \partial^\alpha (\bar{a}_\alpha v)} \, dx .$$

Therefore

$$\left(\mathcal{A}(\partial) u \mid v \right) = \left(u \mid \mathcal{A}^\sharp(\partial) v \right) \quad \text{for } u, v \in \mathcal{D}(X)$$

if $\mathcal{A}^\sharp(\partial) v$ is as in (7.17). By Leibniz's rule, there exist $b_\alpha \in C^\infty(X)$ for $\alpha \in \mathbb{N}^n$ with $|\alpha| \leq m - 1$, such that

$$\mathcal{A}^\sharp(\partial) = (-1)^m \sum_{|\alpha| = m} \bar{a}_\alpha \partial^\alpha + \sum_{|\alpha| \leq m-1} b_\alpha \partial^\alpha .$$

Therefore $\mathcal{A}^\sharp(\partial)$ belongs to $\mathcal{D}\text{iffop}(X)$. The claim is now clear. ■

For differential operators that describe the time evolution of systems, it is usual to treat time as a distinguished variable. We recall, for instance, the wave operator $\partial_t^2 - \Delta_x$ and the heat operator $\partial_t - \Delta_x$ in the variables $(t, x) \in \mathbb{R} \times \mathbb{R}^n$ (see Exercise VII.5.10). Another example is the **Schrödinger operator** $(1/i)\partial_t - \Delta_x$. All three operators are second order differential operators.

7.25 Examples (a) The wave operator and the Schrödinger operator are formally self-adjoint.[15]

(b) The heat operator has $(\partial_t - \Delta_x)^\sharp = -\partial_t - \Delta_x$ as its adjoint. It is therefore *not* formally self-adjoint.

(c) For $\mathcal{A}(\partial) := \partial_t - \sum_{j=1}^n \partial_j$, we have $\mathcal{A}^\sharp(\partial) = -\mathcal{A}(\partial)$.

(d) Suppose $a_{jk}, a_j, a_0 \in C^\infty(X, \mathbb{R})$ with

$$\sum_{j,k=1}^n \|a_{jk}\|_\infty \neq 0 \ , \quad a_{jk} = a_{kj} \quad \text{for } j, k \in \{1, \ldots, n\} \ .$$

Also define $\mathcal{A}(\partial) \in \mathcal{D}\text{iffop}_2(X)$ by

$$\mathcal{A}(\partial)u := \sum_{j,k=1}^n \partial_j(a_{jk}\partial_k u) + \sum_{j=1}^n a_j \partial_j u + a_0 u \quad \text{for } u \in \mathcal{D}(X) \ .$$

Then we say $\mathcal{A}(\partial)$ is a **divergence form** operator.[16] In this case, we have

$$\mathcal{A}^\sharp(\partial)v = \sum_{j,k=1}^n \partial_j(a_{jk}\partial_k v) - \sum_{j=1}^n a_j \partial_j v + \left(a_0 - \sum_{j=1}^n \partial_j a_j\right)v \quad \text{for } v \in \mathcal{D}(X) \ .$$

Therefore the formal adjoint is also of divergence form, and $\mathcal{A}(\partial)$ is formally self-adjoint if and only if $a_j = 0$ for $j = 1, \ldots, n$.

Proof This follows easily from Proposition 7.22. ∎

(e) The Laplace operator Δ is a formally self-adjoint second-order differential operator of divergence form.

Proof This follows from (d) by taking $a_{jk} = \delta_{jk}$ (the Kronecker delta). ∎

Weak derivatives

We now explain briefly how the concept of derivative can be generalized so functions that are not differentiable in the classical sense can be assigned a generalized derivative.

Suppose X is open in $\mathbb{R}^n$. We say $u \in L_{1,\mathrm{loc}}(X)$ is **weakly differentiable** if there exists $u_j \in L_{1,\mathrm{loc}}(X)$ such that

$$\int_X (\partial_j \varphi)u \, dx = -\int_X \varphi u_j \, dx \quad \text{for } \varphi \in \mathcal{D}(X) \text{ and } 1 \leq j \leq n \ . \tag{7.18}$$

---

[15]These facts are of particular importance in mathematical physics.
[16]The reason for this language will be clarified in Section XI.6.

More generally, if $m \geq 2$ is an integer, we say $u \in L_{1,\mathrm{loc}}(X)$ is **m-times weakly differentiable on X** if there exists $u_\alpha \in L_{1,\mathrm{loc}}(X)$ such that

$$\int_X (\partial^\alpha \varphi) u \, dx = (-1)^{|\alpha|} \int_X \varphi u_\alpha \, dx \quad \text{for } \varphi \in \mathcal{D}(X) , \tag{7.19}$$

for all $\alpha \in \mathbb{N}^n$ with $|\alpha| \leq m$. If this is the case, then it immediately follows from Theorem 7.18 that $u_\alpha \in L_{1,\mathrm{loc}}(X)$ is uniquely determined by u (and α). We call u_α the **α-th weak partial derivative** and set $\partial^\alpha u := u_\alpha$. In the case $m = 1$, we set $\partial_j u := u_j$. These notations are justified by the first of the following remarks.

7.26 Remarks (a) Suppose $m \in \mathbb{N}^\times$. Then every $u \in C^m(X)$ is m-times weakly differentiable, and the weak derivatives agree with the classical, or usual, partial derivative.

Proof This follows from Corollary 7.23. ∎

(b) Let $W_{1,\mathrm{loc}}^m(X)$ be the set of all m-times weakly differentiable functions on X. Then $W_{1,\mathrm{loc}}^m(X)$ is a vector subspace of $L_{1,\mathrm{loc}}(X)$, and for every $\alpha \in \mathbb{N}^n$ with $|\alpha| \leq m$, the map

$$W_{1,\mathrm{loc}}^m(X) \to W_{1,\mathrm{loc}}^{m-|\alpha|}(X) , \quad u \mapsto \partial^\alpha u$$

is well defined and linear.

Proof We leave the simple proof to the reader as an exercise. ∎

(c) For $u \in W_{1,\mathrm{loc}}^m(X)$ and $\alpha, \beta \in \mathbb{N}^n$ with $|\alpha| + |\beta| \leq m$, we have $\partial^\alpha \partial^\beta u = \partial^\beta \partial^\alpha u$.

Proof This follows immediately from the defining equations (7.19) and the properties of smooth functions. ∎

(d) Suppose $u \in L_{1,\mathrm{loc}}(\mathbb{R})$ is defined by $u(x) := |x|$ for $x \in \mathbb{R}$. Then u is weakly differentiable, and $\partial u = \mathrm{sign}$.

Proof First, the absolute value function $|\cdot|$ is smooth on $\mathbb{R}^\times$, and its derivative is $\mathrm{sign} \,|\, \mathbb{R}^\times$ there. Now suppose $\varphi \in \mathcal{D}(\mathbb{R})$. Integration by parts gives

$$\int_\mathbb{R} \varphi' u \, dx = \int_0^\infty \varphi' u \, dx + \int_{-\infty}^0 \varphi' u \, dx$$

$$= \varphi(x) x \big|_0^\infty - \int_0^\infty \varphi(x) \, dx - \varphi(x) x \big|_{-\infty}^0 + \int_{-\infty}^0 \varphi(x) \, dx$$

$$= -\int_\mathbb{R} \varphi(x) \, \mathrm{sign}(x) \, dx .$$

The claim follows since sign belongs to $L_{1,\mathrm{loc}}(\mathbb{R})$. ∎

(e) The function sign belongs to $L_{1,\mathrm{loc}}(\mathbb{R})$ and is smooth on $\mathbb{R}^\times$. Nevertheless it is not weakly differentiable. Thus the absolute value function of item (d) is not twice weakly differentiable.

Proof   For $\varphi \in \mathcal{D}(\mathbb{R})$, we have

$$\int_{\mathbb{R}} \varphi' \operatorname{sign} dx = \int_0^\infty \varphi'(x)\,dx - \int_{-\infty}^0 \varphi'(x)\,dx = -2\varphi(0) \ . \tag{7.20}$$

Were sign weakly differentiable, then there would be a $v \in L_{1,\mathrm{loc}}(\mathbb{R})$ such that

$$\int_{\mathbb{R}} \varphi v \, dx = 2\varphi(0) \quad \text{for } \varphi \in \mathcal{D}(\mathbb{R}) \ ,$$

which is false: see Example 7.19(c). $\blacksquare$

In terms of the Dirac distribution δ, (7.20) assumes the form

$$\int_{\mathbb{R}} \varphi' \operatorname{sign} dx = -2\delta(\varphi) \quad \text{for } \varphi \in \mathcal{D}(X) \ .$$

Denoting the duality pairing as usual by

$$\langle \cdot, \cdot \rangle : \mathcal{D}'(X) \times \mathcal{D}(X) \to \mathbb{K} \ ,$$

so $\langle T, \varphi \rangle$ is the value of the continuous linear functional T on the element φ, we have

$$\langle \operatorname{sign}, \varphi' \rangle = -2\langle \delta, \varphi \rangle \quad \text{for } \varphi \in \mathcal{D}(\mathbb{R}) \ , \tag{7.21}$$

where we have identified $\operatorname{sign} \in L_{1,\mathrm{loc}}(\mathbb{R})$ with the regular distribution $T_{\operatorname{sign}} \in \mathcal{D}'(X)$, as discussed right after the proof of Theorem 7.20. A comparison of (7.19) and (7.21) suggests the following definition: Let $S, T \in \mathcal{D}'(X)$ and $\alpha \in \mathbb{N}^n$. Then S is called the α-th **distributional derivative** of T if

$$\langle T, \partial^\alpha \varphi \rangle = (-1)^{|\alpha|} \langle S, \varphi \rangle \quad \text{for } \varphi \in \mathcal{D}(X) \ .$$

In this case, S is clearly defined by T (and α), so we can set $\partial^\alpha T := S$. We see easily that every distribution has distributional derivatives of every order and that for every $\alpha \in \mathbb{N}^n$ the distributional derivative

$$\partial^\alpha : \mathcal{D}'(X) \to \mathcal{D}'(X) \ , \quad T \mapsto \partial^\alpha T$$

is a linear map.[17] In particular, (7.21) shows that, in the sense of distributions,

$$\partial(\operatorname{sign}) = 2\delta \ .$$

We cannot go any further here into the theory of distributions, but we want to briefly introduce Sobolev spaces. Suppose $m \in \mathbb{N}$ and $1 \le p \le \infty$. Because

---

[17]See Exercise 13.

$L_p(X) \subset L_{1,\text{loc}}(X)$, every $u \in L_p(X)$ has distributional derivatives of all orders. We set[18]

$$W_p^m(X) := \{ u \in L_p(X) ; \partial^\alpha u \in L_p(X), |\alpha| \leq m \},$$

where ∂^α denotes the α-th distributional derivative. Also let

$$\|u\|_{m,p} := \begin{cases} \left(\sum_{|\alpha| \leq m} \|\partial^\alpha u\|_p^p \right)^{1/p}, & 1 \leq p < \infty, \\ \max_{|\alpha| \leq m} \|\partial^\alpha u\|_\infty, & p = \infty. \end{cases} \tag{7.22}$$

We verify easily that

$$W_p^m(X) := \left(W_p^m(X), \|\cdot\|_{m,p} \right)$$

is a normed vector space, called the **Sobolev space of order** m. In particular, $W_p^0(X) = L_p(X)$.

7.27 Theorem

(i) $W_p^m(X)$ is continuously embedded in $L_p(X)$, and $u \in L_p(X)$ belongs to $W_p^m(X)$ if and only if u is m-times weakly differentiable and all weak derivatives of order $\leq m$ belong to $L_p(X)$.

(ii) $W_p^m(X)$ is a Banach space.

Proof (i) This is obvious.

(ii) Let (u_j) be a Cauchy sequence in $W_p^m(X)$. It follows immediately from (7.22) that $(\partial^\alpha u_j)_{j \in \mathbb{N}}$ is a Cauchy sequence in $L_p(X)$ for every $\alpha \in \mathbb{N}^n$ such that $|\alpha| \leq m$. Because $L_p(X)$ is complete, there exists a unique $u_\alpha \in L_p(X)$ such that $\partial^\alpha u_j \to u_\alpha$ in $L_p(X)$ for $j \to \infty$ and $|\alpha| \leq m$. We set $u := u_0$. Then it follows from (7.19) that, for all $j \in \mathbb{N}$,

$$\int_X (\partial^\alpha \varphi) u_j \, dx = (-1)^{|\alpha|} \int_X \varphi \partial^\alpha u_j \, dx \quad \text{for } \varphi \in \mathcal{D}(X) \text{ and } |\alpha| \leq m. \tag{7.23}$$

From Hölder's inequality, we deduce

$$\left| \int_X (\partial^\alpha \varphi) u_j \, dx - \int_X (\partial^\alpha \varphi) u \, dx \right| = \left| \int_X \partial^\alpha \varphi (u_j - u) \, dx \right| \leq \|\partial^\alpha \varphi\|_{p'} \|u_j - u\|_p,$$

which shows that

$$\int_X (\partial^\alpha \varphi) u_j \, dx \to \int_X (\partial^\alpha \varphi) u \, dx \quad \text{for } \varphi \in \mathcal{D}(X).$$

Analogously, we find that

$$\int_X \varphi \partial^\alpha u_j \, dx \to \int_X \varphi u_\alpha \, dx \quad \text{for } \varphi \in \mathcal{D}(X).$$

---

[18] If X is an interval in $\mathbb{R}$, then one can show that $W_1^1(X)$ coincides with the space introduced in Exercise 5.6.

Thus it follows from (7.23) that

$$\int_X (\partial^\alpha \varphi) u \, dx = (-1)^{|\alpha|} \int_X \varphi u_\alpha \, dx \quad \text{for } \varphi \in \mathcal{D}(X) .$$

This shows that u_α is the α-th weak derivative of u, and we see that u is m-times weakly differentiable. Because $u_\alpha \in L_p(X)$ for $|\alpha| \leq m$, we also have $u \in W_p^m(X)$, and it is clear that $u_j \to u$ in $W_p^m(X)$. Therefore $W_p^m(X)$ is complete. ∎

7.28 Corollary $W_2^m(X)$ *is a Hilbert space with the inner product*

$$(u \mid v)_m := \sum_{|\alpha| \leq m} (\partial^\alpha u \mid \partial^\alpha v) \quad \text{for } u, v \in W_2^m(X) .$$

We will conclude this section by proving the so-called trace theorem for Sobolev spaces. For $m \in \mathbb{N}$ and $1 \leq p < \infty$, set

$$\widehat{H}_p^m(X) := \left(\{ u \mid X ; \ u \in C_c^m(\mathbb{R}^n) \}, \ \|\cdot\|_{m,p} \right) .$$

Clearly $\widehat{H}_p^m(X)$ is a vector subspace of $W_p^m(X)$. If the boundary ∂X of X is sufficiently nice (for example, if $\overline{X} \subset \mathbb{R}^n$ is an n-dimensional submanifold with boundary,[19]) one can show that $\widehat{H}_p^m(X)$ is dense in $W_p^m(X)$. In particular, this is the case for $X := \mathbb{R}^n$ or $X := \mathbb{H}^n$.

7.29 Theorem (trace theorem) *Let $1 \leq p < \infty$ and $X = \mathbb{R}^n$ or $X = \mathbb{H}^n$. Then there is a unique* **trace operator** *$\gamma \in \mathcal{L}\big(W_p^1(X), L_p(\mathbb{R}^{n-1})\big)$ such that $\gamma u = u \mid \mathbb{R}^{n-1}$ for $u \in \mathcal{D}(\mathbb{R}^n)$ (more precisely, for $u \in \widehat{H}_p^1(X)$). Here $\mathbb{R}^{n-1}$ is identified with $\mathbb{R}^{n-1} \times \{0\} \subset \mathbb{R}^n$.*

Proof Since $\widehat{H}_p^1(X)$ is dense in $W_p^1(X)$, the claim follows from Proposition 6.24, Remark 6.25, and Theorem VI.2.6. ∎

This theorem says in particular that every element $u \in W_p^1(\mathbb{H}^n)$ has boundary values $\gamma u \in L_p(\partial \mathbb{H}^n)$. Because u is generally not continuous on $\overline{\mathbb{H}}^n$, γu cannot be simply determined by restriction.

The existence of a trace is the foundation for the treatment of boundary value problems in partial differential equations by the methods of functional analysis.

Exercises

1 For $a > 0$, calculate $\chi_{[-a,a]} * \chi_{[-a,a]}$ and $\chi_{[-a,a]} * \chi_{[-a,a]} * \chi_{[-a,a]}$.

2 Let $p, p' \in (1, \infty)$ satisfy $1/p + 1/p' = 1$. Prove:

---

[19]See Section XI.1.

(a) $f * g$ belongs to C_0 for $(f, g) \in L_p \times L_{p'}$, and $\|f * g\|_\infty \le \|f\|_p \|g\|_{p'}$.

(b) The convolution is a well defined, bilinear, continuous map from $L_p \times L_{p'}$ into C_0.

3 Let $p, q, r \in [1, \infty]$ with $1/p + 1/q = 1 + 1/r$. Verify that

$$* : L_p \times L_q \to L_r , \quad (f, g) \mapsto f * g$$

is well defined, bilinear, and continuous. Also verify the **generalized Young inequality**

$$\|f * g\|_r \le \|f\|_p \|g\|_q \quad \text{for } (f, g) \in L_p \times L_q .$$

(Hint: The cases $r = 1$ and $r = \infty$ are covered by Theorem 7.3 and Exercise 2, respectively. For $r \in (1, \infty)$, consider

$$|f(x - y)g(y)| = |f(x - y)|^{1 - p/r} \left(|f(x - y)|^p |g(y)|^q \right)^{1/r} |g(y)|^{1 - q/r}$$

and apply Hölder's inequality.)

4 Show that $f * g$ belongs to C^k for $(f, g) \in C_c^k \times L_{1,\mathrm{loc}}$.

5 Suppose $f \in L_{1,\mathrm{loc}}$ satisfies $\partial^\alpha f \in L_{1,\mathrm{loc}}$ for a given $\alpha \in \mathbb{N}^n$. Verify that

$$\partial^\alpha (f * \varphi) = (\partial^\alpha f) * \varphi = f * \partial^\alpha \varphi \quad \text{for } \varphi \in BC^\infty .$$

6 Exhibit a vector subspace of Funct in which $(\mathbb{R}, +)$ is not linearly representable.

7 Given $p \in [1, \infty)$, suppose $K \subset L_p$ is compact. Prove that for every $\varepsilon > 0$ there is a $\delta > 0$ such that $\|\tau_a f - f\|_p < \varepsilon$ for all $f \in K$ and all $a \in \mathbb{R}^n$ with $|a| < \delta$. (Hint: Recall Theorem III.3.10 and Theorem 5.1(iv).)

8 Show that every nontrivial ideal of $(L_1, *)$ is dense in L_1.

9 Let $p \in [1, \infty]$, and denote by k the Gaussian kernel of Example 7.12(a). Prove:

(a) $\partial^\alpha k \in L_p$ for $\alpha \in \mathbb{N}^n$.

(b) $k * u \in BUC^\infty$ for $u \in L_p$.

10 Let $f \in L_1$, and suppose $\partial^\alpha f \in L_1$ for some $\alpha \in \mathbb{N}^n$. Show that

$$\int (\partial^\alpha f) \varphi \, dx = (-1)^{|\alpha|} \int f \partial^\alpha \varphi \, dx \quad \text{for } \varphi \in BC^\infty .$$

11 Let $V \in \{\, \text{Funct}, B, L_p \,;\, 1 \le p \le \infty \,\}$. Show that the linear representation of $(\mathbb{R}^n, +)$ on V by translations is a group isomorphism.

12 For $f, g, h \in L_0$, suppose f is convolvable with g and g with h. If $f * g$ is convolvable with h and f with $g * h$, show that $(f * g) * h = f * (g * h)$. Thus convolution is associative on L_1.

13 Show that the distributional derivative

$$\partial^\alpha : \mathcal{D}'(X) \to \mathcal{D}'(X) , \quad T \mapsto \partial^\alpha T$$

is a well defined linear map for every $\alpha \in \mathbb{N}^n$.

14 Show that $(f \mapsto fu) \in \mathcal{L}(BC^m(X), W_p^m(X))$ for $u \in W_p^m(X)$ with $1 \le p \le \infty$ and $m \in \mathbb{N}$.

15 Suppose (T_j) is a sequence in $\mathcal{D}'(X)$ and that $T \in \mathcal{D}'(X)$. We say (T_j) **converges in** $\mathcal{D}'(X)$ to T if

$$\lim_j \langle T_j, \varphi \rangle = \langle T, \varphi \rangle \quad \text{for } \varphi \in \mathcal{D}(X) .$$

Let $\{ \varphi_\varepsilon \; ; \; \varepsilon > 0 \}$ be an approximation to the identity, and let (ε_j) be a null sequence. Show that $(\varphi_{\varepsilon_j})$ converges in $\mathcal{D}'(\mathbb{R}^n)$ to δ.

8 The substitution rule

In our treatment of the Cauchy–Riemann integral, we encountered the substitution rule of Theorem VI.5.1 as an essential tool for calculating integrals. Introducing new variables, that is, choosing appropriate coordinates, is a prominent technique also in higher dimensional integration. Unsurprisingly, the proof of the substitution rule in this case is more difficult. However, we have already laid a foundation in the form of the substitution rule for linear maps, which we derived in Theorem IX.5.25.

Besides proving the general substitution rule for n-dimensional Lebesgue integrals, this section will illustrate its significance by means of some important examples. The same theorem is also the cornerstone of the theory of integration on manifolds, the subject of our last chapter.

In the following, suppose

- X and Y are open subsets of $\mathbb{R}^n$;
  E is a Banach space.

Pulling back the Lebesgue measure

Let $(X, \mathcal{A})$ be a measurable space and $(Y, \mathcal{B}, \nu)$ a measure space. If $f : X \to Y$ is a *bijective* map that satisfies $f(\mathcal{A}) \subset \mathcal{B}$, that is, one whose inverse map is $\mathcal{B}$-$\mathcal{A}$-measurable, one easily verifies that

$$f^*\nu : \mathcal{A} \to [0, \infty] , \quad A \mapsto \nu\big(f(A)\big)$$

defines a measure on $\mathcal{A}$, the **pull back** (or the **inverse image**) of the measure ν by f. In the special case $(X, \mathcal{A}) = \big(\mathbb{R}^n, \mathcal{L}(n)\big)$ and $(Y, \mathcal{B}, \nu) = \big(\mathbb{R}^n, \mathcal{L}(n), \lambda_n\big)$, the particular case of the substitution rule covered in Theorem IX.5.25 describes the pull back of λ_n by automorphisms of $\mathbb{R}^n$:

$$\Phi^* \lambda_n = |\det \Phi| \, \lambda_n \quad \text{for } \Phi \in \mathcal{L}\mathrm{aut}(\mathbb{R}^n) \, .$$

Using this result, we will now determine the pull back of the Lebesgue measure by arbitrary C^1-diffeomorphisms. A technical result is essential to that end:

8.1 Lemma *Suppose $\Phi \in \mathrm{Diff}^1(X, Y)$. Then*

$$\lambda_n\big(\Phi(J)\big) \leq \int_J |\det \partial\Phi| \, dx$$

for every interval $J \subset\subset X$ of the form $[a, b)$, where $a, b \in \mathbb{Q}^n$.

Proof (i) First consider a cube $J = \big[x_0 - (r/2)\mathbf{1}, x_0 + (r/2)\mathbf{1}\big)$ with center $x_0 \in X$ and edge length $r > 0$. Next set $\mathbb{R}^n_\infty := (\mathbb{R}^n, |\cdot|_\infty)$ and

$$K := \max_{x \in \bar{J}} \|\partial\Phi(x)\|_{\mathcal{L}(\mathbb{R}^n_\infty)} \, .$$

It follows from the mean value theorem that

$$|\Phi(x) - \Phi(x_0)|_\infty \leq K\,|x - x_0|_\infty \quad \text{for } x \in J \;.$$

Therefore $\Phi(J)$ is contained in $\bar{\mathbb{B}}_\infty^n(\Phi(x_0), Kr/2)$, and we find

$$\lambda_n(\Phi(J)) \leq (Kr)^n = K^n \lambda_n(J) \;. \tag{8.1}$$

(ii) Suppose $J \subset\subset X$ is of the form $[a, b)$, with $a, b \in \mathbb{Q}^n$. Take $\varepsilon > 0$ and let $M := \max_{x \in \bar{J}} \|[\partial\Phi(x)]^{-1}\|_{\mathcal{L}(\mathbb{R}_\infty^n)}$. Since $\partial\Phi$ is uniformly continuous on $\bar{J}$, there exists $\delta > 0$ such that

$$\|\partial\Phi(x) - \partial\Phi(y)\|_{\mathcal{L}(\mathbb{R}_\infty^n)} \leq \varepsilon/M \tag{8.2}$$

for all $x, y \in \bar{J}$ such that $|x - y| < \delta$. Because $a, b \in \mathbb{Q}^n$, we can decompose J (by edge subdivision) into N disjoint cubes J_k of the form $[\alpha, \beta)^n$ with $0 < \beta - \alpha < \delta$. Now choose $x_k \in \bar{J}_k$ such that

$$|\det \partial\Phi(x_k)| = \min_{y \in \bar{J}_k} |\det \partial\Phi(y)|$$

and set $T_k := \partial\Phi(x_k)$ and $\Phi_k := T_k^{-1} \circ \Phi$. Because

$$\partial\Phi_k(y) = T_k^{-1}\partial\Phi(y) = 1_n + [\partial\Phi(x_k)]^{-1}\,[\partial\Phi(y) - \partial\Phi(x_k)]$$

it follows from (8.2) and the definition of M that

$$\max_{y \in \bar{J}_k} \|\partial\Phi_k(y)\|_{\mathcal{L}(\mathbb{R}_\infty^n)} \leq 1 + \varepsilon \quad \text{for } k \in \{1, \dots, N\} \;. \tag{8.3}$$

By the special case of the substitution rule treated in Theorem IX.5.25, we have

$$\lambda_n(\Phi(J_k)) = \lambda_n(T_k T_k^{-1}\Phi(J_k)) = |\det T_k|\,\lambda_n(\Phi_k(J_k)) \;.$$

Thus (8.1) and (8.3) imply

$$\lambda_n(\Phi(J_k)) \leq (1 + \varepsilon)^n\,|\det T_k|\,\lambda_n(J_k) \quad \text{for } k \in \{1, \dots, N\} \;.$$

Taking into account the bijectivity of Φ and the choice of x_k, we find

$$\lambda_n(\Phi(J)) = \lambda_n\Big(\bigcup_{k=1}^N \Phi(J_k)\Big) = \sum_{k=1}^N \lambda_n(\Phi(J_k))$$

$$\leq (1 + \varepsilon)^n \sum_{k=1}^N |\det T_k|\,\lambda_n(J_k) \leq (1 + \varepsilon)^n \sum_{k=1}^N \int_{J_k} |\det \partial\Phi|\,dx$$

$$= (1 + \varepsilon)^n \int_J |\det \partial\Phi|\,dx \;.$$

The claim follows upon taking the limit $\varepsilon \to 0$. $\blacksquare$

8.2 Proposition *Suppose $\Phi \in \mathrm{Diff}^1(X, Y)$. Then*

$$\Phi^* \lambda_n(A) = \lambda_n(\Phi(A)) = \int_A |\det \partial \Phi| \, dx \quad \text{for } A \in \mathcal{L}(n) \,|\, X \ .$$

Proof (i) From the monotone convergence theorem, it follows easily that

$$\mu_\Phi : \mathcal{L}(n) \,|\, X \to [0, \infty] \ , \quad A \mapsto \int_A |\det \partial \Phi| \, dx$$

is a complete measure (see Exercise 2.11).

(ii) Suppose U is open and compactly contained in X. By Proposition IX.5.6, there is a sequence (J_k) of disjoint intervals of the form $[a, b)$, with $a, b \in \mathbb{Q}^n$, such that $U = \bigcup_k J_k$. From (i) and Lemma 8.1, it follows that

$$\lambda_n(\Phi(U)) = \lambda_n\left(\bigcup_k \Phi(J_k)\right) = \sum_k \lambda_n(\Phi(J_k)) \le \sum_k \int_{J_k} |\det \partial \Phi| \, dx$$

$$= \sum_k \mu_\Phi(J_k) = \mu_\Phi\left(\bigcup_k J_k\right) = \mu_\Phi(U) = \int_U |\det \partial \Phi| \, dx \ .$$

(iii) Let U be open in X. By Remarks 1.16(d) and (e), there is a sequence (U_k) of open subsets of X such that $U_k \subset\subset U_{k+1}$ and $U = \bigcup_k U_k$. From (ii) and the continuity from below of the measures λ_n and μ_Φ, it follows that

$$\lambda_n(\Phi(U)) = \lim_k \lambda_n(\Phi(U_k)) \le \lim_k \mu_\Phi(U_k) = \mu_\Phi(U) = \int_U |\det \partial \Phi| \, dx \ .$$

(iv) Let $A \in \mathcal{L}(n) \,|\, X$ be bounded. Using Corollary IX.5.5, we find a sequence (U_k) of bounded open subsets of X such that $G := \bigcap_k U_k \supset A$ and $\lambda_n(G) = \lambda_n(A)$. From (iii) and the continuity from above of the measures λ_n and μ_Φ, we have

$$\lambda_n(\Phi(G)) = \lim_k \lambda_n\left(\Phi\left(\bigcap_{j=0}^k U_j\right)\right) \le \lim_k \mu_\Phi\left(\bigcap_{j=0}^k U_j\right)$$

$$= \mu_\Phi(G) = \int_G |\det \partial \Phi| \, dx \ .$$

Noting that $A \subset G$ and $\lambda_n(A) = \lambda_n(G)$, we obtain

$$\lambda_n(\Phi(A)) \le \lambda_n(\Phi(G)) \le \int_G |\det \partial \Phi| \, dx = \int_A |\det \partial \Phi| \, dx \ .$$

(v) Take any $A \in \mathcal{L}(n) \,|\, X$, and set $A_k := A \cap k\mathbb{B}^n$ for $k \in \mathbb{N}$. From (iv) and the continuity of the measures from below, we obtain

$$\lambda_n(\Phi(A)) = \lim_k \lambda_n(\Phi(A_k)) \le \lim_k \mu_\Phi(A_k) = \mu_\Phi(A) = \int_A |\det \partial \Phi| \, dx \ .$$

(vi) Let $f \in \mathcal{S}(Y, \mathbb{R}^+)$ have normal form $f = \sum_{j=0}^{k} \alpha_j \chi_{A_j}$. By (v),

$$\int_Y f \, dy = \sum_{j=0}^{k} \alpha_j \lambda_n(A_j) = \sum_{j=0}^{k} \alpha_j \lambda_n\big(\Phi(\Phi^{-1}(A_j))\big)$$

$$\leq \sum_{j=0}^{k} \alpha_j \int_{\Phi^{-1}(A_j)} |\det \partial\Phi| \, dx = \int_X (f \circ \Phi) \, |\det \partial\Phi| \, dx \ .$$

(vii) Suppose X is bounded. Given $f \in \mathcal{L}_0(Y, \mathbb{R}^+)$, let (f_k) be a sequence in $\mathcal{S}(Y, \mathbb{R}^+)$ such that $f_k \uparrow f$ (see Theorem 1.12). Then $f_k \circ \Phi$ belongs to $\mathcal{S}(X, \mathbb{R}^+)$. Because the sequence $(f_k \circ \Phi)_k$ converges increasingly to $f \circ \Phi$, we know that $(f \circ \Phi) \, |\det \partial\Phi|$ belongs to $\mathcal{L}_0(X, \mathbb{R}^+)$. Now (vi) and the monotone convergence theorem imply

$$\int_Y f \, dy = \lim_k \int_Y f_k \, dy \leq \lim_k \int_X (f_k \circ \Phi) \, |\det \partial\Phi| \, dx = \int_X (f \circ \Phi) \, |\det \partial\Phi| \, dx \ .$$

(viii) Let X be arbitrary and take $f \in \mathcal{L}_0(Y, \mathbb{R}^+)$. In view of Remarks 1.16(d) and (e), we can find an ascending sequence of relatively compact open subsets X_k of X such that $X = \bigcup_{k=0}^{\infty} X_k$. According to (vii), $g_k := \chi_{X_k} f \, |\det \Phi|$ belongs to $\mathcal{L}_0(X, \mathbb{R}^+)$, and we have $g_k \uparrow g := f \, |\det \Phi|$. Therefore $g \in \mathcal{L}_0(X, \mathbb{R}^+)$. Setting $Y_k := \Phi(X_k)$, we obtain from (vii) that

$$\int_{Y_k} f \, dy \leq \int_{X_k} (f \circ \Phi) \, |\det \partial\Phi| \, dx \ .$$

Now $Y = \bigcup_{k=0}^{\infty} Y_k$ and the monotone convergence theorem yield

$$\int_Y f \, dy \leq \int_X (f \circ \Phi) \, |\det \partial\Phi| \, dx \ . \tag{8.4}$$

(ix) Suppose $A \in \mathcal{L}(n) \,|\, X$. We swap the roles of X and Y in (viii) and apply (8.4) to the C^1-diffeomorphism $\Phi^{-1} : Y \to X$ and the function $(\chi_{\Phi(A)} \circ \Phi) \, |\det \partial\Phi|$, which belongs to $\mathcal{L}_0(X, \mathbb{R}^+)$. Then

$$\int_X (\chi_{\Phi(A)} \circ \Phi) \, |\det \partial\Phi| \, dx \leq \int_Y \Big[\big((\chi_{\Phi(A)} \circ \Phi) \, |\det \partial\Phi|\big) \circ \Phi^{-1}\Big] \, |\det \partial\Phi^{-1}| \, dy$$

$$= \int_Y \chi_{\Phi(A)} \, \big|\det\big[(\partial\Phi \circ \Phi^{-1})\partial\Phi^{-1}\big]\big| \, dy \ .$$

Further noting that

$$1_n = \partial(\mathrm{id}_Y) = \partial(\Phi \circ \Phi^{-1}) = (\partial\Phi \circ \Phi^{-1})\partial\Phi^{-1} \tag{8.5}$$

and $\chi_{\Phi(A)} \circ \Phi = \chi_A$, we obtain

$$\int_A |\det \partial\Phi| \, dx \leq \int_Y \chi_{\Phi(A)} \, dy = \lambda_n(\Phi(A)) \ .$$

Because of (v), the claim follows. $\blacksquare$

8.3 Example Define $X := \big\{ (r,\varphi) \in \mathbb{R} \times (0,2\pi) \ ; \ 0 < r < \varphi/2\pi \big\}$ and

$$\Phi : X \to \mathbb{R}^2 , \quad (r,\varphi) \mapsto (r\cos\varphi, r\sin\varphi) .$$

Then $Y := \Phi(X)$ is open in $\mathbb{R}^2$, and $\Phi \in \mathrm{Diff}^\infty(X,Y)$ satisfies

$$[\partial\Phi(r,\varphi)] = \begin{bmatrix} \cos\varphi & -r\sin\varphi \\ \sin\varphi & r\cos\varphi \end{bmatrix} .$$

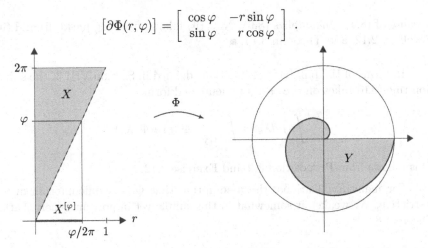

Therefore $\det\partial\Phi(r,\varphi) = r$. Also $\mathrm{pr}_2(X) = (0,2\pi)$, and $X^{[\varphi]} = (0,\varphi/2\pi)$ for $\varphi \in (0,2\pi)$. By Proposition 8.2 and Tonelli's theorem, then,

$$\lambda_2(Y) = \int_X r \, d(r,\varphi) = \int_0^{2\pi} \int_0^{\varphi/2\pi} r \, dr \, d\varphi = \pi/3 .$$

The substitution rule: general case

After these preliminaries, it is no longer difficult to prove the substitution rule for diffeomorphisms. First we consider the scalar case, whose proof is accessible even to readers who skipped over the proof of Fubini's theorem for vector-valued functions. We treat the general case at the end of the section.

8.4 Theorem (substitution rule) *Suppose* $\Phi \in \mathrm{Diff}^1(X,Y)$.

(i) *For* $f \in \mathcal{L}_0(Y,\mathbb{R}^+)$,

$$\int_Y f \, dy = \int_X (f \circ \Phi) \, |\det\partial\Phi| \, dx . \tag{8.6}$$

(ii) *A function* $f : Y \to \mathbb{K}$ *is integrable if and only if* $(f \circ \Phi) \, |\det\partial\Phi|$ *belongs to* $\mathcal{L}_1(X)$. *In this case, (8.6) holds.*

Proof (i) Theorem IX.5.12 implies that $\Phi(\mathcal{L}_X) \subset \mathcal{L}_Y$. Hence $f \circ \Phi$ is measurable, by Corollary 1.5. Since $|\det\partial\Phi|$ is continuous, hence measurable, Remark 1.2(d) implies that $g := (f \circ \Phi) |\det \partial \Phi|$ is measurable also. From (8.5) we obtain $f = (g \circ \Phi^{-1})| \det \partial \Phi^{-1}|$. Now (8.4), with (Y, Φ^{-1}, g) in the role of (X, Φ, f), gives

$$\int_X (f \circ \Phi) \, |\det \partial \Phi| \, dx \le \int_Y f \, dy \ .$$

Because of (8.4), this implies (8.6). Now (ii) follows from (i), parts (ii) and (iii) of Corollary 2.12, and Theorem 3.14. ∎

In terms of the pull back of functions defined in Section VIII.3, the substitution rule (8.6) takes on the easily remembered form

$$\int_Y f \, d\lambda_n = \int_{\Phi^{-1}(Y)} (\Phi^* f) \, d(\Phi^* \lambda_n) \ .$$

This follows from Proposition 8.2 and Exercise 2.12.

For many applications, the assumption that Φ is a diffeomorphism is too restrictive. We weaken it somewhat in this simple yet important generalization of Theorem 8.4:[1]

8.5 Corollary *Let M be a measurable subset of X such that $M \backslash \mathring{M}$ has Lebesgue measure zero. Suppose $\Phi \in C^1(X, \mathbb{R}^n)$ is such that $\Phi \,|\, \mathring{M}$ is a diffeomorphism from $\mathring{M}$ onto $\Phi(\mathring{M})$.*

(i) *For every $f \in \mathcal{L}_0(M, \mathbb{R}^+)$,*

$$\int_{\Phi(M)} f \, dy = \int_M (f \circ \Phi) \, |\det \partial \Phi| \, dx \ . \tag{8.7}$$

(ii) *A function $f : \Phi(M) \to \mathbb{K}$ belongs to $\mathcal{L}_1(\Phi(M))$ if and only if $(f \circ \Phi)\,|\det \partial \Phi|$ belongs to $\mathcal{L}_1(M)$. In this case, (8.7) holds.*

Proof Because $\lambda_n(M \backslash \mathring{M}) = 0$, the set $\Phi(M) \backslash \Phi(\mathring{M}) \subset \Phi(M \backslash \mathring{M})$ also has measure zero, by Corollary IX.5.10. The claims then follow from Lemma 2.15 and Theorem 8.4. ∎

It is clear that this corollary gives a (partial) generalization of the substitution rule of Theorem VI.5.1, though limited to diffeomorphisms. There is one obvious difference from the one-dimensional case considered before: now the derivative term (that is, the functional determinant) appears as an absolute value. The reason is that the prior result used the *oriented* integral.

---

[1] See Exercise 7 for a further generalization.

Plane polar coordinates

A special case of special importance in applications consists of diffeomorphisms induced by polar coordinates, which we now introduce. We begin with the two-dimensional case.

Let
$$f_2 : \mathbb{R}^2 \to \mathbb{R}^2 \ , \quad (r, \varphi) \mapsto (x, y) := (r \cos \varphi, r \sin \varphi)$$
be the (**plane**) **polar coordinate map**[2], and let $V_2 := (0, \infty) \times (0, 2\pi)$.

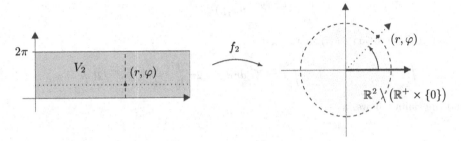

Then f_2 is smooth, and $\det \partial f_2(r, \varphi) = r$, as was shown in Example 8.3. Clearly $\overline{V}_2 \setminus V_2$ has measure zero; moreover

$$f_2(\overline{V}_2) = \mathbb{R}^2 \ , \quad f_2(V_2) = \mathbb{R}^2 \setminus (\mathbb{R}^+ \times \{0\}) \tag{8.8}$$

and

$$f_2 \,|\, V_2 \in \mathrm{Diff}^\infty (V_2, f_2(V_2)) \ . \tag{8.9}$$

Therefore Corollary 8.5 applies with $M := \overline{V}_2$:

8.6 Proposition (integration in polar coordinates)
(i) *For $g \in \mathcal{L}_0(\mathbb{R}^2, \mathbb{R}^+)$, we have*

$$\int_{\mathbb{R}^2} g(x, y)\, d(x, y) = \int_0^{2\pi} \int_0^\infty g(r \cos \varphi, r \sin \varphi) r\, dr\, d\varphi$$
$$= \int_0^\infty r \int_0^{2\pi} g(r \cos \varphi, r \sin \varphi)\, d\varphi\, dr \ . \tag{8.10}$$

(ii) *The function $g : \mathbb{R}^2 \to \mathbb{K}$ is integrable if and only if the map*

$$(0, \infty) \times (0, 2\pi) \to \mathbb{K} \ , \quad (r, \varphi) \mapsto g(r \cos \varphi, r \sin \varphi) r$$

is integrable. Then (8.10) holds.

Proof This follows from Corollary 8.5 together with (8.8), (8.9), and the Fubini–Tonelli theorem. ∎

---

[2] See Conclusion III.6.21(d).

These integrals simplify when f depends only on $|x|$, that is, on r. To illustrate, we present an elegant calculation of the Gaussian error integral, for which knowledge of the Γ-function is not required (compare Application VI.9.7).

8.7 Example $\int_{-\infty}^{\infty} e^{-x^2}\, dx = \sqrt{\pi}.$

Proof Tonelli's theorem implies

$$\left(\int_{-\infty}^{\infty} e^{-x^2}\, dx \right)^2 = \int_{-\infty}^{\infty} e^{-x^2}\, dx \int_{-\infty}^{\infty} e^{-y^2}\, dy = \int_{\mathbb{R}} \left(\int_{\mathbb{R}} e^{-(x^2+y^2)}\, dx \right) dy$$

$$= \int_{\mathbb{R}^2} e^{-(x^2+y^2)}\, d(x,y) \ .$$

Therefore Proposition 8.6(i) shows that

$$\left(\int_{-\infty}^{\infty} e^{-x^2}\, dx \right)^2 = \int_0^{2\pi} \int_0^{\infty} r e^{-r^2}\, dr\, d\varphi = 2\pi \int_0^{\infty} \frac{d}{dr}\left[-e^{-r^2}/2 \right] dr = \pi \ ,$$

and the claim follows. ∎

Polar coordinates in higher dimensions

For $n \geq 1$, we define $h_n : \mathbb{R}^n \to \mathbb{R}^{n+1}$ recursively through

$$h_1(z) := (\cos z, \sin z) \quad \text{for } z \in \mathbb{R} \tag{8.11}$$

and

$$h_{n+1}(z) := \big(h_n(z') \sin z_{n+1}, \cos z_{n+1} \big) \quad \text{for } z = (z', z_{n+1}) \in \mathbb{R}^n \times \mathbb{R} \ . \tag{8.12}$$

Obviously h_n is smooth, and by induction, we verify that

$$|h_n(z)| = 1 \quad \text{for } z \in \mathbb{R}^n \ . \tag{8.13}$$

Now we define $f_n : \mathbb{R}^n \to \mathbb{R}^n$ for $n \geq 2$ by

$$f_n(y) := y_1 h_{n-1}(z) \quad \text{for } y = (y_1, z) \in \mathbb{R} \times \mathbb{R}^{n-1} \ . \tag{8.14}$$

Then f_n is also smooth, and we have

$$h_{n-1}(z) = f_n(1, z) \ , \quad |f_n(y)| = |y_1| \ . \tag{8.15}$$

We will usually follow convention by renaming the y-coordinates as

$$(r, \varphi, \vartheta_1, \ldots, \vartheta_{n-2}) := (y_1, y_2, y_3, \ldots, y_n) \ .$$

By induction, one checks easily that

$$f_n : \mathbb{R}^n \to \mathbb{R}^n \ , \quad (r, \varphi, \vartheta_1, \ldots, \vartheta_{n-2}) \mapsto (x_1, x_2, x_3, \ldots, x_n) \tag{8.16}$$

is given by

$$\left.\begin{array}{rl}
x_1 &= r\cos\varphi\sin\vartheta_1\sin\vartheta_2\cdots\sin\vartheta_{n-2}\ , \\
x_2 &= r\sin\varphi\sin\vartheta_1\sin\vartheta_2\cdots\sin\vartheta_{n-2}\ , \\
x_3 &= r\cos\vartheta_1\sin\vartheta_2\cdots\sin\vartheta_{n-2}\ , \\
&\vdots \\
x_{n-1} &= r\cos\vartheta_{n-3}\sin\vartheta_{n-2}\ , \\
x_n &= r\cos\vartheta_{n-2}\ .
\end{array}\right\} \tag{8.17}$$

Thus f_2 coincides with the plane polar coordinate map, and f_3 is the **spherical co-ordinate map** of Example VII.9.11(a). In the general case, f_n is the **n-dimensional polar coordinate map**. From (8.12) and (8.14), the recursive relation

$$f_n(y) = \big(f_{n-1}(y')\sin y_n, y_1\cos y_n\big) \quad \text{for } y = (y', y_n) \in \mathbb{R}^{n-1}\times\mathbb{R} \tag{8.18}$$

follows for $n \geq 3$. For $n \geq 2$, we set

$$W_{n-1} := (0, 2\pi) \times (0, \pi)^{n-2}\ , \quad V_n := (0, \infty) \times W_{n-1}\ , \tag{8.19}$$

and

$$V_n(r) := (0, r) \times W_{n-1} \quad \text{for } r > 0\ . \tag{8.20}$$

If we denote the closed $(n-1)$-dimensional half-space by

$$H_{n-1} := \mathbb{R}^+ \times \{0\} \times \mathbb{R}^{n-2} \subset \mathbb{R}^n\ , \tag{8.21}$$

we find

$$h_{n-1}(W_{n-1}) = S^{n-1}\backslash H_{n-1}\ , \quad f_n(V_n(r)) = r\mathbb{B}^n\backslash H_{n-1} \tag{8.22}$$

and

$$h_{n-1}\big(\overline{W}_{n-1}\big) = S^{n-1}\ , \quad f_n\big(\overline{V_n(r)}\big) = r\overline{\mathbb{B}}^n\ . \tag{8.23}$$

Also

$$f_n(V_n) = \mathbb{R}^n\backslash H_{n-1}\ , \quad f_n\big(\overline{V}_n\big) = \mathbb{R}^n\ . \tag{8.24}$$

In addition, the maps $h_{n-1}\,|\,W_{n-1}$ and $f_n\,|\,V_n$ are bijective onto their images.

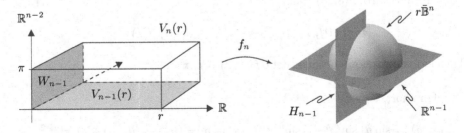

These statements follow easily by induction.

8.8 Lemma For $n \geq 3$ and $r > 0$, the map f_n is a C^∞ diffeomorphism from $V_n(r)$ onto $r\mathbb{B}^n \setminus H_{n-1}$ and from V_n onto $\mathbb{R}^n \setminus H_{n-1}$. Moreover

$$\det \partial f_n(r, \varphi, \vartheta_1, \ldots, \vartheta_{n-2}) = (-1)^n r^{n-1} \sin \vartheta_1 \sin^2 \vartheta_2 \cdots \sin^{n-2} \vartheta_{n-2}$$

for $(r, \varphi, \vartheta_1, \ldots, \vartheta_{n-2}) \in \overline{V}_n$.

Proof In view of the foregoing, we need only calculate the value of the functional determinant $\det \partial f_n(y)$. We do this recursively. From (8.12) and (8.14), we have

$$[\partial f_{n+1}(y)] = \begin{bmatrix} h_{n-1}(z') \sin z_n & \vdots & [r\partial_z(h_{n-1}(z') \sin z_n)] \\ \cdots\cdots\cdots\cdots\cdots & \vdots & \cdots\cdots\cdots\cdots\cdots\cdots\cdots\cdots \\ \cos z_n & \vdots & 0 \quad \cdots \quad 0 \quad -r \sin z_n \end{bmatrix}$$

$$= \begin{bmatrix} [\partial f_n(y') \sin z_n] & \vdots & \begin{matrix} * \\ \vdots \\ * \end{matrix} \\ \cdots\cdots\cdots\cdots\cdots & \vdots & \cdots\cdots\cdots \\ * & \cdots & * \quad \vdots \quad -r \sin z_n \end{bmatrix} ,$$

where $y = (r, z) = (y', z_n)$ and $z = (z', z_n) \in \mathbb{R}^n$. Expanding in the last row, we find

$$\det \partial f_{n+1}(y) = (-1)^n \cos z_n \det S - r \sin^{n+1} z_n \det \partial f_n(y') , \qquad (8.25)$$

where $S := [r\partial_z(h_{n-1}(z') \sin z_n)]$. We can assume that $\sin z_n \neq 0$; otherwise the claim is trivial. In the last column of S we have $r h_{n-1}(z') \cos z_n$. This vector differs only by the factor $r \cot z_n$ from the first column vector, $h_{n-1}(z') \sin z_n$, of the matrix $T := [\partial f_n(y') \sin z_n]$. The first $n-1$ columns of S also agree with the last $n-1$ columns of T, in the same order. Therefore

$$\det S = (-1)^{n-1} r \cot z_n \det T = (-1)^{n-1} r \cos z_n \sin^{n-1} z_n \det \partial_n f(y') .$$

Thus it follows from (8.25) that

$$\det \partial f_{n+1}(y) = -r \sin^{n-1} z_n \det \partial f_n(y') .$$

The claim now follows because $\det \partial f_2(r, \varphi) = r$. ∎

For short, let's set

$$w_n(\vartheta) := \sin \vartheta_1 \sin^2 \vartheta_2 \cdots \sin^{n-2} \vartheta_{n-2} , \quad \vartheta := (\vartheta_1, \ldots, \vartheta_{n-2}) \in [0, \pi]^{n-2} .$$

8.9 Proposition (integration in polar coordinates) *Suppose $n \geq 3$.*

(i) *For $g \in \mathcal{L}_0(\mathbb{R}^n, \mathbb{R}^+)$, we have*

$$\int_{\mathbb{R}^n} g\, dx = \int_{V_n} (g \circ f_n)(r, \varphi, \vartheta) r^{n-1} w_n(\vartheta)\, d(r, \varphi, \vartheta) . \qquad (8.26)$$

(ii) *The map $g : \mathbb{R}^n \to \mathbb{K}$ is integrable if and only if*

$$V_n \to \mathbb{K} , \quad (r, \varphi, \vartheta) \mapsto (g \circ f_n)(r, \varphi, \vartheta) r^{n-1} w_n(\vartheta)$$

is integrable. Then (8.26) holds.

Proof Because $\lambda_n (\overline{V}_n \backslash V_n) = 0$, the claim follows from (8.24), Corollary 8.5, and Lemma 8.8. ∎

8.10 Examples (a) For $g \in \mathcal{L}_0(\mathbb{R}^3, \mathbb{R}^+)$, we have

$$\int_{\mathbb{R}^3} g(x, y, z)\, d(x, y, z)$$
$$= \int_0^\infty \int_0^{2\pi} \int_0^\pi g(r \cos \varphi \sin \vartheta, r \sin \varphi \sin \vartheta, r \cos \vartheta) r^2 \sin \vartheta\, d\vartheta\, d\varphi\, dr . \qquad (8.27)$$

The integrals on the right side can be performed in any order.

Proof This follows from Proposition 8.9(i) and Tonelli's theorem. ∎

(b) A map $g : \mathbb{R}^3 \to \mathbb{K}$ is integrable if and only if

$$V_3 \to \mathbb{K} , \quad (r, \varphi, \vartheta) \mapsto g(r \cos \varphi \sin \vartheta, r \sin \varphi \sin \vartheta, r \cos \vartheta) r^2 \sin \vartheta$$

is integrable. Such a map satisfies (8.27), and the integrals there can be performed in any order.

Proof This is a consequence of Proposition 8.9(ii) and the Fubini–Tonelli theorem. ∎

(c) For $n \geq 3$, we have

$$2\pi \int_{[0,\pi]^{n-2}} w_n(\vartheta)\, d\vartheta = n\omega_n ,$$

where $\omega_n = \pi^{n/2}/\Gamma(1 + n/2)$ is the volume of $\mathbb{B}^n$.

Proof From (8.22), (8.23), Proposition 8.9, and Tonelli's theorem, it follows that

$$\omega_n = \int_{\mathbb{B}^n} dx = \int_{\mathbb{B}^n} 1\, dx = \int_{V_n(1)} (1 \circ f_n)(r, \varphi, \vartheta) r^{n-1} w_n(\vartheta)\, d(r, \varphi, \vartheta)$$
$$= \int_0^1 r^{n-1}\, dr \int_0^{2\pi} d\varphi \int_{[0,\pi]^{n-2}} w_n(\vartheta)\, d\vartheta = \frac{2\pi}{n} \int_{[0,\pi]^{n-2}} w_n(\vartheta)\, d\vartheta ,$$

and we are done. ∎

Integration of rotationally symmetric functions

Suppose $0 \leq r_0 < r_1 \leq \infty$ and set $R(r_0, r_1) := \{ x \in \mathbb{R}^n \ ; \ r_0 < |x| < r_1 \}$. We say that a function $g: R(r_0, r_1) \to E$ is **rotationally symmetric** if there is a map $\overset{\bullet}{g}: (r_0, r_1) \to E$ such that

$$g(x) = \overset{\bullet}{g}(|x|) \quad \text{for } x \in R(r_0, r_1) .$$

This is the case if and only if g is constant on every sphere rS^{n-1} with $r_0 < r < r_1$. For such a function, $\overset{\bullet}{g}$ is uniquely determined by g (and vice versa).

As we saw in Example 8.7, integration problems simplify considerably for rotationally symmetric functions.

8.11 Theorem *Suppose $0 \leq r_0 < r_1 \leq \infty$.*

(i) *If $g \in \mathcal{L}_0\big(R(r_0, r_1), \mathbb{R}^+\big)$ is rotationally symmetric, then*

$$\int_{R(r_0, r_1)} g \, dx = n\omega_n \int_{r_0}^{r_1} \overset{\bullet}{g}(r) \, r^{n-1} \, dr , \tag{8.28}$$

where $\omega_n := \lambda_n(\mathbb{B}^n) = \pi^{n/2}/\Gamma(1 + n/2)$.

(ii) *A rotationally symmetric function $g: R(r_0, r_1) \to \mathbb{K}$ is integrable if and only if*

$$(r_0, r_1) \to \mathbb{K} , \quad r \mapsto \overset{\bullet}{g}(r) \, r^{n-1}$$

is integrable. In this case (8.28) holds.

Proof The case $n = 1$ is clear (see Exercise 5.12). For $n \geq 2$, it follows from (8.15) and the rotational symmetry of g that

$$g \circ f_n(r, \varphi, \vartheta) = \overset{\bullet}{g}(r) \quad \text{for } r_0 < r < r_1 \text{ and } (\varphi, \vartheta) \in W_{n-1} .$$

Now the claim arises from Propositions 8.6 and 8.9 (applied to the trivial extension of g) and Example 8.10(c). $\blacksquare$

8.12 Examples (a) Suppose $f: \mathbb{R}^n \to \mathbb{K}$ is measurable and there are $c \geq 0$, $\rho > 0$, and $\varepsilon > 0$ such that

$$|f(x)| \leq \begin{cases} c \, |x|^{-n+\varepsilon} & \text{if } 0 < |x| \leq \rho , \\ c \, |x|^{-n-\varepsilon} & \text{if } \quad |x| \geq \rho . \end{cases}$$

Then f is integrable.

Proof Set

$$g(x) := c\big(|x|^{-n+\varepsilon} \chi_{\rho\overline{\mathbb{B}}^n}(x) + |x|^{-n-\varepsilon} \chi_{(\rho\mathbb{B}^n)^c}(x)\big) \quad \text{for } x \in \mathbb{R}^n \setminus \{0\} = R(0, \infty) .$$

Then g is rotationally symmetric, and $|f(x)| \leq g(x)$ for $x \in R(0, \infty)$. By Examples VI.8.4(a) and (b), $r \mapsto \overset{\bullet}{g}(r) r^{n-1}$ belongs to $\mathcal{L}_1(\mathbb{R}^+)$. Hence Theorem 8.11 implies that g also belongs to $\mathcal{L}_1\big(R(0, \infty)\big) = \mathcal{L}_1(\mathbb{R}^n)$. Now the claim follows from Theorem 3.14. $\blacksquare$

(b) Let $\mu \in \mathcal{L}_\infty(\mathbb{R}^n)$ have compact support. Also define

$$\frac{1}{r} : \mathbb{R}^n \setminus \{0\} \to \mathbb{R}^+ , \quad x \mapsto \frac{1}{|x|} .$$

Then $(1/r)^\alpha * \mu$ exists for $\alpha < n$, and

$$\left(\frac{1}{r}\right)^\alpha * \mu(x) = \int_{\mathbb{R}^n} \frac{\mu(y)}{|x - y|^\alpha} \, dy \quad \text{for } x \in \mathbb{R}^n .$$

Proof Take $x \in \mathbb{R}^n$ and define $K := \operatorname{supp}(\mu)$ and $g_x(y) := \|\mu\|_\infty |y|^{-\alpha} \chi_{x-K}(y)$ for $y \neq 0$. Then $\tilde{g}_x$ belongs to $\mathcal{L}_0(\mathbb{R}^n)$, and

$$|\mu(x - y)| \, |y|^{-\alpha} \leq g_x(y) \quad \text{for } y \neq 0 .$$

Because $\alpha < n$, part (a) shows that $\tilde{g}_x$ is integrable. The claim now follows from Theorem 7.8(ii). ∎

For $n \geq 3$, the function $u_n := (1/r)^{n-2} * \mu$, in the notation of (b), is called the **Newtonian** or **Coulomb potential** associated with the density μ. From Exercise 3.6 we know that u_n is smooth and harmonic in K^c, and (b) shows that u_n is defined on all of $\mathbb{R}^n$.

The substitution rule for vector-valued functions

We now prove the substitution formula of Theorem 8.4 for vector-valued functions.

8.13 Lemma Let $f \in \mathcal{S}_c(Y, E)$ and $\Phi \in \operatorname{Diff}^1(X, Y)$. Then $(f \circ \Phi) |\det \partial \Phi|$ belongs to $\mathcal{L}_1(X, E)$, and

$$\int_Y f \, dy = \int_X (f \circ \Phi) |\det \partial \Phi| \, dx .$$

Proof Because $\operatorname{supp}(f \circ \Phi) = \Phi^{-1}(\operatorname{supp}(f))$, the support of $f \circ \Phi$ is compact. In particular, $f \circ \Phi$ belongs to $\mathcal{S}_c(X, E)$. It easily follows that $(f \circ \Phi) |\det \partial \Phi|$ is integrable. Also Theorem 2.11(iii) shows that, for $e \in E$ and $g \in \mathcal{L}_1(X, \mathbb{K})$, the function eg belongs to $\mathcal{L}_1(X, E)$ and $e \int_X g \, dx = \int_X eg \, dx$. Letting $\sum_{j=0}^m e_j \chi_{A_j}$ be the normal form of f, we see from Proposition 8.2 that

$$\int_Y f \, dy = \sum_{j=0}^m e_j \lambda_n(A_j) = \sum_{j=0}^m e_j \int_{\Phi^{-1}(A_j)} |\det \partial \Phi| \, dx$$

$$= \sum_{j=0}^m \int_{\Phi^{-1}(A_j)} e_j |\det \partial \Phi| \, dx = \int_X (f \circ \Phi) |\det \partial \Phi| \, dx . \quad ∎$$

8.14 Theorem (substitution rule)   Let $\Phi \in \mathrm{Diff}^1(X, Y)$ and $f \in E^Y$. Then f belongs to $\mathcal{L}_1(Y, E)$ if and only if $(f \circ \Phi) |\det \partial\Phi|$ belongs to $\mathcal{L}_1(X, E)$. In this case, we have

$$\int_Y f \, dy = \int_X (f \circ \Phi) |\det \partial\Phi| \, dx \ .$$

Proof   (i) Let $f \in \mathcal{L}_1(Y, E)$, and take a sequence (f_j) in $\mathcal{S}_c(Y, E)$ converging a.e. in $\mathcal{L}_1(Y, E)$ to f and satisfying $\lim \int_Y f_j = \int_Y f$ (see Lemma 6.18, Remarks 6.19(a) and (c), and Theorem 2.18). Set $g_j := (f_j \circ \Phi) |\det \partial\Phi|$ for $j \in \mathbb{N}$. Thanks to Lemma 8.13, we know that (g_j) is a Cauchy sequence in $\mathcal{L}_1(X, E)$ and that $\int_Y f_j \, dy = \int_X g_j \, dx$. Because $\mathcal{L}_1(X, E)$ is complete, there exists $g \in \mathcal{L}_1(X, E)$ such that $g_j \to g$ in $\mathcal{L}_1(X, E)$. Also, it follows from Theorem 2.18 that $\int_X g_j \, dx$ converges to $\int_X g \, dx$ and that some subsequence $(g_{j_k})_{k \in \mathbb{N}}$ of (g_j) converges a.e. in X to g. Hence g and $(f \circ \Phi) |\det \partial\Phi|$ coincide a.e. in X. By Lemma 2.15, $(f \circ \Phi) |\det \partial\Phi|$ belongs to $\mathcal{L}_1(X, E)$, and $\int_X g = \int_X (f \circ \Phi) |\det \partial\Phi|$. It follows that

$$\int_Y f \, dy = \lim_j \int_Y f_j \, dy = \lim_j \int_X g_j \, dx = \int_X g \, dx = \int_X (f \circ \Phi) |\det \partial\Phi| \, dx \ .$$

(ii) For the converse, suppose $(f \circ \Phi) |\det \partial\Phi|$ belongs to $\mathcal{L}_1(X, E)$. From (8.5) we have

$$f = \big((f \circ \Phi) |\det \partial\Phi| \big) \circ \Phi^{-1} |\det \partial(\Phi^{-1})| \ ,$$

so part (i) shows that f belongs to $\mathcal{L}_1(Y, E)$. ∎

It is clear that Corollary 8.5 is also true for E-valued maps. From this it follows that Propositions 8.6(ii) and 8.9(ii) and Theorem 8.11(ii) also hold for E-valued functions.

Exercises

1 Let $G \in \mathbb{R}^{n \times n}$ be symmetric and positive definite. Prove that

$$\int_{\mathbb{R}^n} e^{-(Gx \,|\, x)} \, dx = \pi^{n/2} / \sqrt{\det G} \ .$$

(Hint: principal axis transformation.)

2 Show that for $p \in \mathbb{C}$ with $\mathrm{Re}\, p > n/2$, we have

$$\int_{\mathbb{R}^n} (1 + |x|^2)^{-p} \, dx = \pi^{n/2} \Gamma(p - n/2) / \Gamma(p) \ .$$

(Hint: Look at Example 6.13(b).)

3 Suppose $D := \{ (x, y) \in \mathbb{R}^2 \ ; \ x, y \geq 0, \ x + y \leq 1 \}$ and $p, q \in (0, \infty)$. Show that for $f : (0, 1) \to \mathbb{R}$, the function

$$D \to \mathbb{R} \ , \quad (x, y) \mapsto x^{p-1} y^{q-1} f(x + y)$$

is integrable if and only if $s \mapsto s^{p+q-1} f(s)$ belongs to $\mathcal{L}_1((0,1))$. In this case, we have

$$\int_D x^{p-1} y^{q-1} f(x+y)\, d(x,y) = \mathsf{B}(p,q) \int_0^1 s^{p+q-1} f(s)\, ds \ .$$

(Hint: Consider $(s,t) \mapsto \big(s(1-t), st\big)$.)

4 Let $0 \le \alpha < \beta \le 2\pi$, and suppose $f : [\alpha, \beta] \to (0, \infty)$ is measurable. Show that

$$S(\alpha, \beta, f) := \big\{ z \in \mathbb{C} \ ; \ \arg_N(z) \in [\alpha, \beta], \ |z| \le f\big(\arg_N(z)\big) \big\}$$

is Lebesgue measurable and that

$$\lambda_2\big(S(\alpha, \beta, f)\big) = \frac{1}{2} \int_\alpha^\beta \big[f(\varphi) \big]^2 \, d\varphi \ .$$

5 Suppose $g \in \mathcal{L}^2_{\mathrm{sym}}(\mathbb{R}^n)$ is positive definite. Calculate the volume of the solid ellipsoid $g^{-1}([0,1])$ enclosed by the surface $g^{-1}(1)$ (see Remark VII.10.18).

6 (Sard's lemma) Suppose $\Phi \in C^1(X, \mathbb{R}^n)$, and let $C := \big\{ x \in X \ ; \ \partial\Phi(x) \notin \mathcal{L}\mathrm{aut}(\mathbb{R}^n) \big\}$ be the set of critical points of Φ. Show that $\Phi(C)$ has measure zero. (Hint: Because C is σ-compact, it suffices to check that $\Phi(C \cap J)$ is has measure zero for every compact n-dimensional cube J. Take $x_0 \in C$ and $r > 0$ such that $J_0 := [x_0 - (r/2)\mathbf{1}, x_0 + (r/2)\mathbf{1}]$ is compactly contained in X, and set

$$\rho(r) := \max_{x \in J_0} \int_0^1 \big\| \partial\Phi\big(x_0 + t(x - x_0)\big) \big\| \, dt \ .$$

Show that there is a $c_n > 0$ such that $\lambda_n(\Phi(J_0)) \le c_n r^n \rho(r)$. Because $\lim_{r \to 0} \rho(r) = 0$, the claim follows by subdividing the edges of J_0.)

7 Suppose $\Phi \in C^1(X, \mathbb{R}^n)$ and $C := \big\{ x \in X \ ; \ \partial\Phi(x) \notin \mathcal{L}\mathrm{aut}(\mathbb{R}^n) \big\}$. Also suppose $\Phi \,|\, (X \backslash C)$ is injective. Prove:

(i) For $f \in \mathcal{L}_0(X, \mathbb{R}^+)$,

$$\int_{\Phi(X)} f \, dy = \int_X (f \circ \Phi) \, |\det \partial\Phi| \, dx \ . \tag{8.29}$$

(ii) The function $f : \Phi(X) \to E$ belongs to $\mathcal{L}_1\big(\Phi(X), E\big)$ if and only if $(f \circ \Phi) \, |\det \partial\Phi|$ lies in $\mathcal{L}_1(X, E)$. In this case, (8.29) holds.

9 The Fourier transform

To conclude this chapter, we introduce the most important integral transformation, called the Fourier transform.[1] The study of its fundamental properties is as it were a recapitulation of Lebesgue integration theory: we will encounter at every turn such cornerstones as the completeness of Lebesgue spaces, the dominated convergence theorem, and the Fubini–Tonelli theorem.

Particularly appealing is the interaction of the Fourier transform with the convolution and with the Hilbert space structure of L_2. We illustrate the former through Fourier multiplication operators and the second via Plancherel's theorem and applications of the position and momentum operators of quantum mechanics.

In this section, we exclusively consider spaces of complex-valued functions defined on all of $\mathbb{R}^n$. For this reason, as in Section 7, we omit $(\mathbb{R}^n, \mathbb{C})$ from our notation and write, for example, $\mathcal{L}_1$ for $\mathcal{L}_1(\mathbb{R}^n, \mathbb{C})$. In addition, $\int f \, dx$ always means $\int_{\mathbb{R}^n} f \, dx$, and we canonically identify $\mathbb{R}^n$ with its dual space, so that $\langle \cdot, \cdot \rangle$ coincides formally with the Euclidean inner product.

Definition and elementary properties

Let $f \in \mathcal{L}_1$. The map $\mathbb{R}^n \to \mathbb{C}$, $x \mapsto e^{-i \langle x, \xi \rangle} f(x)$ belongs to $\mathcal{L}_1$ for every $\xi \in \mathbb{R}^n$. The map $\widehat{f} : \mathbb{R}^n \to \mathbb{C}$ defined by

$$\widehat{f}(\xi) := (2\pi)^{-n/2} \int_{\mathbb{R}^n} e^{-i \langle x, \xi \rangle} f(x) \, dx \quad \text{for } \xi \in \mathbb{R}^n \tag{9.1}$$

is called the **Fourier transform of f**. The map $\mathcal{F} := (f \mapsto \widehat{f})$ is also called the Fourier transform (or, if necessary to avoid confusion, the **Fourier transformation**).

Different conventions intervene in the definition just given; instead of (9.1), one often sees the Fourier transform being defined as[2]

$$\xi \mapsto \int e^{-i \langle x, \xi \rangle} f(x) \, dx \quad \text{or} \quad \xi \mapsto \int e^{-2\pi i \langle x, \xi \rangle} f(x) \, dx \ .$$

Obviously, these differences in normalization are immaterial to the underlying theory; however, they do cause powers of 2π to appear in some of the following expressions. One should be mindful of this when reading the literature. The normalization chosen here has the advantage that such factors appear only in a few places and that Plancherel's theorem takes on a particularly simple form.

9.1 Remarks (a) For $f \in L_1$, set $\mathcal{F}f := \widehat{f} := \mathcal{F}\overset{*}{f}$, where $\overset{*}{f}$ is an arbitrary representative of f. Then $\mathcal{F}f$ is well defined, and $\mathcal{F} \in \mathcal{L}(L_1, BC)$.

---

[1] The contents of this section will not be used in the rest of this book.
[2] See Section VIII.6.

Proof The first statement is obvious. Because

$$|\widehat{f}(\xi)| \leq (2\pi)^{-n/2} \|f\|_1 \quad \text{for } \xi \in \mathbb{R}^n ,$$

the second follows easily from Theorem 3.17 (on the continuity of parameterized integrals) and Theorem VI.2.5. ∎

(b) For $f \in L_1$, we have $\overset{\smile}{\widehat{f}} = \overset{\frown}{\widetilde{f}}$. The function defined by

$$\mathbb{R}^n \to \mathbb{C} , \quad \xi \mapsto \overset{\smile}{\widetilde{f}}(\xi) = (2\pi)^{-n/2} \int e^{i\langle x,\xi\rangle} f(x) \, dx$$

is called the **inverse Fourier transform** of f, for reasons soon to become clear; and the map

$$\mathcal{F} := (f \mapsto \overset{\smile}{\widetilde{f}})$$

is the **inverse Fourier transform(ation)**. Because inversion $(f \mapsto \overset{\smile}{f})$ is a continuous automorphism on $\mathcal{L}_1$, L_1, and BC, the inverse Fourier transform has the same continuity properties as the Fourier transform.

Proof This follows immediately from the substitution rule. ∎

(c) For $\lambda > 0$, we denote by $\sigma_\lambda : \mathbb{R}^n \to \mathbb{R}^n$, $x \mapsto \lambda x$ the **dilation** by the factor λ. We define an action of the group $((0,\infty), \cdot)$ on Funct $:= \mathrm{Funct}(\mathbb{R}^n, \mathbb{C})$,

$$((0,\infty), \cdot) \times \mathrm{Funct} \to \mathrm{Funct} , \quad (\lambda, f) \mapsto \sigma_\lambda f , \tag{9.2}$$

by setting

$$\sigma_\lambda f := f \circ \sigma_{1/\lambda} = (\sigma_{1/\lambda})^* f .$$

If V is a vector subspace of Funct that is invariant under this action (meaning that $\sigma_\lambda(V) \subset V$ for $\lambda > 0$), the map

$$\sigma_\lambda : V \to V , \quad v \mapsto \sigma_\lambda v$$

is linear and satisfies $\sigma_\lambda \sigma_\mu = \sigma_{\lambda\mu}$ and $\sigma_1 = \mathrm{id}_V$ for $\lambda, \mu > 0$. Therefore σ_λ is a vector space automorphism of V, with $(\sigma_\lambda)^{-1} = \sigma_{1/\lambda}$ for $\lambda > 0$. This shows that

$$((0,\infty), \cdot) \to \mathrm{Aut}(V) , \quad \lambda \mapsto \sigma_\lambda$$

is a linear representation of the multiplicative group $((0,\infty), \cdot)$ on V. In particular, $\{\sigma_\lambda \; ; \; \lambda > 0\}$ is a subgroup of $\mathrm{Aut}(V)$, the **group of dilations** on V. Accordingly $\sigma_\lambda v$ is the **dilation of** v by the factor λ. As with the translation group, we say that $((0,\infty), \cdot)$ is **linearly representable** in V if V is invariant under (9.2).

Suppose $1 \leq p \leq \infty$. Then $((0,\infty), \cdot)$ is linearly representable on L_p, and

$$\|\sigma_\lambda f\|_p = \lambda^{n/p} \|f\|_p .$$

Proof This follows from the substitution rule. ∎

(d) $\mathcal{F}\sigma_\lambda = \lambda^n \sigma_{1/\lambda}\mathcal{F}$ for $\lambda > 0$.

Proof Suppose $f \in \mathcal{L}_1$ and $\lambda > 0$. Then

$$\mathcal{F}\sigma_\lambda f(\xi) = (2\pi)^{-n/2} \int e^{-i\langle x,\xi\rangle} f(x/\lambda)\,dx = \lambda^n (2\pi)^{-n/2} \int e^{-\langle x/\lambda,\lambda\xi\rangle} f(x/\lambda)\lambda^{-n}\,dx$$

for $\xi \in \mathbb{R}^n$. But the substitution rule shows that the last expression is equal to $\lambda^n \widehat{f}(\lambda\xi)$. ∎

(e) Suppose $a \in \mathbb{R}^n$. Then $\left(e^{i\langle a,\cdot\rangle}f\right)^{\widehat{}} = \tau_a \widehat{f}$ for $f \in \mathcal{L}_1$. ∎

The space of rapidly decreasing functions

We now introduce a vector subspace of $\mathcal{L}_1$ where the Fourier transform is especially manageable. Using density arguments, we will then be able to broaden the results to larger function spaces.

We say $f \in C^\infty$ is **rapidly decreasing** if for every $(k,m) \in \mathbb{N}^2$, there is a $c_{k,m} > 0$ such that

$$(1 + |x|^2)^k |\partial^\alpha f(x)| \leq c_{k,m} \quad \text{for } x \in \mathbb{R}^n, \quad \alpha \in \mathbb{N}^n, \quad \text{and} \quad |\alpha| \leq m .$$

In other words, $f \in C^\infty$ is rapidly decreasing if, as $|x| \to \infty$, every derivative $\partial^\alpha f$ goes to zero faster than any power of $1/|x|$.

We now set

$$q_{k,m}(f) := \max_{|\alpha|\leq m} \sup_{x\in\mathbb{R}^n} (1 + |x|^2)^{k/2} |\partial^\alpha f(x)| \quad \text{for } f \in C^\infty \text{ and } k,m \in \mathbb{N} .$$

The space

$$\mathcal{S} := \left\{ f \in C^\infty \ ; \ q_{k,m}(f) < \infty \text{ for } k,m \in \mathbb{N} \right\}$$

is called **Schwartz space** or the **space of rapidly decreasing functions**.

9.2 Remarks **(a)** $\mathcal{S}$ is a vector subspace of BUC^∞. Every $q_{k,m}$ is a norm on $\mathcal{S}$.

Proof Let $m \in \mathbb{N}$. Then $\mathcal{S}$ is a vector subspace of BC^m, since $q_{0,m}$ is the norm on BC^m. Let $\alpha \in \mathbb{N}^n$ with $|\alpha| \leq m$. Then it follows easily from the mean values theorem that $\partial^\alpha f$ is uniformly continuous. The proves the first statement. The second is clear. ∎

(b) For $(f,g) \in \mathcal{S} \times \mathcal{S}$, let

$$d(f,g) := \sum_{k,m=0}^{\infty} 2^{-(k+m)} \frac{q_{k,m}(f-g)}{1 + q_{k,m}(f-g)} .$$

Then $(\mathcal{S},d)$ is a metric space.

Proof (i) Clearly the double series $\sum 2^{-(k+m)} q_{k,m}(f)/(1+q_{k,m}(f))$ converges for every $f \in \mathcal{S}$. Thus $d \colon \mathcal{S}\times\mathcal{S} \to \mathbb{R}^+$ is well defined. Also d is symmetric and vanishes identically on the diagonal of $\mathcal{S} \times \mathcal{S}$.

(ii) Because $t \mapsto t/(1+t)$ is increasing on $\mathbb{R}^+$, we have, for $r, s, t \in \mathbb{R}^+$ with $r \leq s+t$,

$$\frac{r}{1+r} \leq \frac{s+t}{1+s+t} = \frac{s}{1+s+t} + \frac{t}{1+s+t} \leq \frac{s}{1+s} + \frac{t}{1+t} \;.$$

Now it follows easily that d satisfies the triangle inequality. $\blacksquare$

(c) For $f \in \mathcal{S}$ and a sequence (f_j) in $\mathcal{S}$, there is equivalence between:
   (i) $\lim f_j = f$ in $(\mathcal{S}, d)$;
   (ii) $\lim(f - f_j) = 0$ in $(\mathcal{S}, d)$;
   (iii) $\lim_j q_{k,m}(f - f_j) = 0$ for $k, m \in \mathbb{N}$.

Thus a sequence (f_j) converges in $\mathcal{S}$ to f if and only if $(f_j - f)$ converges to zero with respect to every seminorm $q_{k,m}$.

Proof "(i)$\Rightarrow$(ii)" This implication is clear.

"(ii)$\Rightarrow$(iii)" Take $\varepsilon \in (0, 1]$ and $k, m \in \mathbb{N}$. There exists an $N \in \mathbb{N}$ such that the inequality $d(f, f_j) < \varepsilon/2^{k+m+1}$ is satisfied for $j \geq N$. Thus

$$\frac{2^{-(k+m)} q_{k,m}(f - f_j)}{1 + q_{k,m}(f - f_j)} < \frac{\varepsilon}{2^{k+m+1}} \;,$$

so $q_{k,m}(f - f_j) < \varepsilon$ for $j \geq N$.

"(iii)$\Rightarrow$(i)" Take $\varepsilon > 0$. There is an $N \in \mathbb{N}$ such that

$$\sum_{k+m=N+1}^{\infty} \frac{2^{-(k+m)} q_{k,m}(f - f_j)}{1 + q_{k,m}(f - f_j)} \leq \sum_{\ell=N+1}^{\infty} 2^{-\ell} < \frac{\varepsilon}{2} \;.$$

By assumption, there is $M \in \mathbb{N}$ such that $q_{k,m}(f - f_j) \leq \varepsilon/4$ for $j \geq M$ and $k+m \leq N$. Therefore

$$d(f, f_j) \leq \sum_{k,m=0}^{N} \frac{2^{-(k+m)} q_{k,m}(f - f_j)}{1 + q_{k,m}(f - f_j)} + \frac{\varepsilon}{2} \leq \varepsilon \quad \text{for } j \geq M \;. \quad \blacksquare$$

(d) $\mathcal{D}$ is a dense vector subspace of $\mathcal{S}$. The function $\mathbb{R}^n \to \mathbb{R}, \; x \mapsto e^{-|x|^2}$ belongs to $\mathcal{S}$ but not to $\mathcal{D}$.

Proof It is clear that $\mathcal{D}$ is a vector subspace of $\mathcal{S}$. Suppose $f \in \mathcal{S}$. We choose $\varphi \in \mathcal{D}$ such that $\varphi \,|\, \overline{\mathbb{B}}^n = 1$ and set

$$f_j(x) := f(x) \varphi(x/j) \quad \text{for } x \in \mathbb{R}^n \;, \quad j \in \mathbb{N}^\times \;.$$

Then f_j belongs to $\mathcal{D}$, and

$$f(x) - f_j(x) = f(x)\big(1 - \varphi(x/j)\big) \quad \text{for } x \in \mathbb{R}^n \;.$$

Therefore $\partial^\alpha (f - f_j)(x) = 0$ for $x \in j\mathbb{B}^n$ and $\alpha \in \mathbb{N}^n$. Now Leibniz's rule shows that there is a $c = c(\varphi, m) > 0$ such that

$$\left| \partial^\alpha (f - f_j)(x) \right| = \left| \sum_{\beta \leq \alpha} \binom{\alpha}{\beta} \partial^\beta f(x) \partial^{\alpha-\beta}(1 - \varphi)(x/j) j^{-|\alpha-\beta|} \right| \leq c \max_{\beta \leq \alpha} |\partial^\beta f(x)|$$

$$\leq c\, q_{k+1,m}(f)(1 + |x|^2)^{-(k+1)/2}$$

for $x \in \mathbb{R}^n$, $j \in \mathbb{N}^\times$, $k \in \mathbb{N}$, and $|\alpha| \leq m$. Setting $C := c\, q_{k+1,m}(f)$, we find

$$q_{k,m}(f - f_j) = \max_{|\alpha| \leq m} \sup_{|x| \geq j} (1 + |x|^2)^{k/2} |\partial^\alpha (f - f_j)(x)|$$

$$\leq c\, q_{k+1,m}(f) \sup_{|x| \geq j} (1 + |x|^2)^{-1/2} \leq C/j \ ,$$

and, as $j \to \infty$, the first claim follows from (c). The second one is clear. ∎

(e) For $m \in \mathbb{N}$, we have $\mathcal{S} \hookrightarrow BUC^m$.

Proof This follows from (a) and (c). ∎

(f) $\mathcal{S}$ is a dense vector subspace of C_0.

Proof Suppose $f \in \mathcal{S}$. Then it follows from (a) and because $|f(x)| \leq q_{1,0}(f)(1 + |x|^2)^{-1/2}$ for $x \in \mathbb{R}^n$ that f belongs to C_0. Therefore $\mathcal{S}$ is a vector subspace of C_0. Theorem 7.13 shows $\mathcal{D}$ is a dense vector subspace of C_0, and therefore the claim follows from the inclusions $\mathcal{D} \subset \mathcal{S} \subset C_0$. ∎

(g) For $k, m \in \mathbb{N}$, there are positive constants c and C such that

$$c \max_{\substack{|\alpha| \leq m \\ |\beta| \leq k}} \sup_{x \in \mathbb{R}^n} \left| \partial^\alpha (x^\beta f(x)) \right| \leq q_{k,m}(f) \leq C \max_{\substack{|\alpha| \leq m \\ |\beta| \leq k}} \sup_{x \in \mathbb{R}^n} |x^\beta \partial^\alpha f(x)| \quad \text{for } f \in \mathcal{S} \ .$$

Proof This follows easily from the Leibniz rule. ∎

(h) Let $f \in \mathcal{S}$ and $\alpha, \beta \in \mathbb{N}^n$. Then $x \mapsto x^\alpha \partial^\beta f(x)$ belongs to $\mathcal{S}$.

Proof This is a consequence of (g). ∎

(i) The inversion $f \mapsto \check{f}$ is a continuous automorphism of $\mathcal{S}$.

Proof This is obvious. ∎

9.3 Theorem Let $p \in [1, \infty)$. Then $\mathcal{S}$ is a dense vector subspace of L_p, and there is a $c = c(n, p) > 0$ such that

$$\|f\|_p \leq c\, q_{n+1,0}(f) \quad \text{for } f \in \mathcal{S} \ . \tag{9.3}$$

Proof For $f \in \mathcal{S}$, we have

$$\int |f|^p \, dx = \int |f(x)|^p (1 + |x|^2)^{(n+1)p/2} (1 + |x|^2)^{-(n+1)p/2} \, dx$$

$$\leq \left(q_{n+1,0}(f) \right)^p \int (1 + |x|^2)^{-(n+1)p/2} \, dx \ . \tag{9.4}$$

Further, by Theorem 8.11(i) and because $(n+1)p > n$, we have

$$\int_{[|x| \geq 1]} |x|^{-(n+1)p} \, dx = n\omega_n \int_1^\infty r^{-((n+1)p - n + 1)} \, dr < \infty \ .$$

Therefore $\int (1 + |x|^2)^{-(n+1)p/2} \, dx$ is also finite, and (9.3) follows from (9.4). In particular, f belongs to L_p, and we see that $\mathcal{S}$ is a vector subspace of L_p. By

Theorem 7.13, $\mathcal{D}$ is a dense vector subspace of L_p, and by Remark 9.2(d), it is contained in $\mathcal{S}$. The claim follows. ∎

The convolution algebra $\mathcal{S}$

By Remark 9.2(a) and Theorem 9.3, $\mathcal{S} \times \mathcal{S}$ is contained in $BUC^\infty \times L_1$. Therefore the convolution is defined on $\mathcal{S} \times \mathcal{S}$, and by Corollary 7.9, we have

$$* : \mathcal{S} \times \mathcal{S} \to BUC^\infty . \tag{9.5}$$

The next result shows that $f * g$ is actually rapidly decreasing for $(f, g) \in \mathcal{S} \times \mathcal{S}$.

9.4 Proposition *The convolution $\mathcal{S} \times \mathcal{S}$ is a continuous and bilinear map into $\mathcal{S}$.*

Proof (i) We verify next that the convolution $\mathcal{S} \times \mathcal{S}$ maps into $\mathcal{S}$. So suppose $(f, g) \in \mathcal{S} \times \mathcal{S}$ and $k, m \in \mathbb{N}$. By (9.5), it suffices to check that $q_{k,m}(f * g)$ is finite. Because

$$|x|^k \leq (|x - y| + |y|)^k = \sum_{j=0}^{k} \binom{k}{j} |x - y|^j |y|^{k-j} \quad \text{for } x, y \in \mathbb{R}^n ,$$

there is a $c_k > 0$ such that

$$|x|^k |f * g(x)| \leq \int \sum_{j=0}^{k} \binom{k}{j} |x - y|^j |f(x - y)| |y|^{k-j} |g(y)| \, dy$$

$$\leq c_k q_{k,0}(f) \int (1 + |y|^2)^{k/2} |g(y)| \, dy .$$

Noting that $\tilde{c}_n := \int (1 + |y|^2)^{-(n+1)/2} \, dy$ is finite, we find

$$|x|^k |f * g(x)| \leq c_k \tilde{c}_n \, q_{k,0}(f) q_{k+n+1,0}(g) .$$

Thus by Remark 9.2(g), there is a $c = c(k, n) \geq 1$ such that

$$q_{k,0}(f * g) \leq c \, q_{k,0}(f) q_{k+n+1,0}(g) . \tag{9.6}$$

Finally by Theorem 7.8(iv), we have

$$q_{k,m}(f * g) = \max_{|\alpha| \leq m} q_{k,0}(\partial^\alpha (f * g)) = \max_{|\alpha| \leq m} q_{k,0}((\partial^\alpha f) * g) ,$$

and (9.6) implies

$$q_{k,m}(f * g) \leq c \max_{|\alpha| \leq m} q_{k,0}(\partial^\alpha f) q_{k+n+1,0}(g) = c \, q_{k,m}(f) q_{k+n+1,0}(g) . \tag{9.7}$$

(ii) It is clear that the convolution is bilinear. Suppose $(f, g) \in \mathcal{S} \times \mathcal{S}$ and $((f_j, g_j))_{j \in \mathbb{N}}$ is a sequence in $\mathcal{S} \times \mathcal{S}$ such that $(f_j, g_j) \to (f, g)$ in $\mathcal{S} \times \mathcal{S}$ as $j \to \infty$.

Also let
$$\alpha := c\big(q_{k,m}(f) + q_{k+n+1,0}(g) + 1\big)$$
where c is the constant from (9.7). Let $\varepsilon \in (0,1]$. By Remark 9.2(c), there is an $N \in \mathbb{N}$ such that
$$q_{k,m}(f - f_j) < \varepsilon/\alpha , \quad q_{k+n+1,0}(g - g_j) < \varepsilon/\alpha \quad \text{for } j \geq N .$$
Because
$$f * g - f_j * g_j = (f - f_j) * g + (f_j - f) * (g - g_j) + f * (g - g_j)$$
it follows from (9.7) that
$$q_{k,m}(f * g - f_j * g_j) \leq c\big(q_{k,m}(f - f_j)q_{k+n+1,0}(g) + q_{k,m}(f - f_j)q_{k+n+1,0}(g - g_j)$$
$$+ q_{k,m}(f)q_{k+n+1,0}(g - g_j)\big) < \varepsilon$$
for $j \geq N$. Thus we are done. ∎

9.5 Corollary $(\mathcal{S}, +, *)$ *is a subalgebra of the commutative algebra* $(L_1, +, *)$.

Proof   This follows from Proposition 9.4 and Theorem 9.3. ∎

Calculations with the Fourier transform

We now derive some rules for the Fourier transformation of derivatives and the derivatives of Fourier transforms. It will simplify the presentation of these formulas to set $\Lambda(x) := (1 + |x|^2)^{1/2}$ for $x \in \mathbb{R}^n$ and
$$D_j := -i\partial_j , \quad j \in \{1, \ldots, n\} \quad \text{for } D^\alpha := D_1^{\alpha_1} \cdots D_n^{\alpha_n} , \quad \alpha \in \mathbb{N}^n ,$$
where i is the imaginary unit. As usual, the polynomial function induced by the polynomial $p \in \mathbb{C}[X_1, \ldots, X_n]$ will also be denoted by p.

9.6 Proposition   *Suppose* $f \in \mathcal{L}_1$.

 (i) *For* $\alpha \in \mathbb{N}^n$, *suppose* $D^\alpha f$ *exists and belongs to* $\mathcal{L}_1$. *Then* $X^\alpha \widehat{f} = \widehat{D^\alpha f}$.

 (ii) *For* $m \in \mathbb{N}$, *suppose* $\Lambda^m f$ *belongs to* $\mathcal{L}_1$. *Then* $\widehat{f}$ *belongs to* BC^m, *and*
$$D^\alpha \widehat{f} = (-1)^{|\alpha|} \widehat{X^\alpha f} \quad \text{for } \alpha \in \mathbb{N}^n , \quad |\alpha| \leq m .$$

Proof   (i) Suppose $\{\varphi_\varepsilon \ ; \ \varepsilon > 0\}$ is a smoothing kernel. By integration by parts (see Exercise 7.10), it follows that
$$\int \xi^\alpha e^{-i\langle x, \xi\rangle}(f * \varphi_\varepsilon)(x)\, dx = (-1)^{|\alpha|} \int D_x^\alpha (e^{-i\langle x,\xi\rangle})(f * \varphi_\varepsilon)(x)\, dx$$
$$= \int e^{-i\langle x,\xi\rangle}\big((D^\alpha f) * \varphi_\varepsilon\big)(x)\, dx . \tag{9.8}$$

Theorem 7.11 and Theorem 2.18(ii) imply that

$$\lim_{\varepsilon \to 0} (2\pi)^{-n/2} \int \xi^\alpha e^{-i\langle x,\xi\rangle} (f * \varphi_\varepsilon)(x)\, dx = \xi^\alpha \widehat{f}(\xi)$$

and

$$\lim_{\varepsilon \to 0} (2\pi)^{-n/2} \int e^{-i\langle x,\xi\rangle} ((D^\alpha f) * \varphi_\varepsilon)(x)\, dx = \widehat{D^\alpha f}(\xi)$$

for $\xi \in \mathbb{R}^n$. Using (9.8), this proves the claim.

(ii) We set $h(x,\xi) := e^{-i\langle x,\xi\rangle} f(x)$ for $(x,\xi) \in \mathbb{R}^n \times \mathbb{R}^n$. Then $h(\,\cdot\,,\xi)$ belongs to $\mathcal{L}_1$ for every $\xi \in \mathbb{R}^n$, and $h(x,\cdot)$ belongs to C^∞ for every $x \in \mathbb{R}^n$. Further, we have

$$D_\xi^\alpha h(x,\xi) = (-1)^{|\alpha|} x^\alpha h(x,\xi) \quad \text{for } (x,\xi) \in \mathbb{R}^{2n} , \quad \alpha \in \mathbb{N}^n ,$$

and thus

$$|D_\xi^\alpha h(x,\xi)| \le (1+|x|^2)^{|\alpha|/2} \, |h(x,\xi)| = \Lambda^{|\alpha|}(x)\, |f(x)| . \tag{9.9}$$

It then follows from the theorem on the differentiation of parametrized integrals that $\widehat{f}$ belongs to C^m and that

$$D^\alpha \widehat{f}(\xi) = (2\pi)^{-n/2} \int D_\xi^\alpha h(x,\xi)\, dx = (2\pi)^{-n/2} (-1)^{|\alpha|} \int x^\alpha h(x,\xi)\, dx$$

$$= (-1)^{|\alpha|} \widehat{X^\alpha f}(\xi)$$

for $\xi \in \mathbb{R}^n$ and $\alpha \in \mathbb{N}^n$ with $|\alpha| \le m$. Finally (9.9) shows

$$\left| D^\alpha \widehat{f}(\xi) \right| \le (2\pi)^{-n/2} \int |D_\xi^\alpha h(x,\xi)|\, dx \le (2\pi)^{-n/2} \|\Lambda^{|m|} f\|_1 < \infty \quad \text{for } \xi \in \mathbb{R}^n .$$

Thus $\widehat{f}$ belongs to BC^m. ∎

9.7 Proposition The Fourier transformation maps $\mathcal{S}$ continuously and linearly into itself.

Proof (i) Suppose $f \in \mathcal{S}$ and $m \in \mathbb{N}$. Then

$$\int \Lambda^m(x)\, |f(x)|\, dx = \int (1+|x|^2)^{(m+n+1)/2}\, |f(x)|\, (1+|x|^2)^{-(n+1)/2}\, dx$$

$$\le q_{m+n+1,0}(f) \int (1+|x|^2)^{-(n+1)/2}\, dx < \infty .$$

We find using Proposition 9.6(ii) that $\widehat{f}$ belongs to BC^m and thus to BC^∞.

(ii) Suppose $k, m \in \mathbb{N}$ and $\alpha, \beta \in \mathbb{N}^n$ with $|\alpha| \le m$ and $|\beta| \le k$. Also suppose $f \in \mathcal{S}$. Then it follows from Remark 9.2(h) and Theorem 9.3 that $\Lambda^m f$

and $D^\beta(X^\alpha f)$ belong to $\mathcal{L}_1$. Therefore Proposition 9.6 implies

$$\xi^\beta D^\alpha \widehat{f}(\xi) = (-1)^{|\alpha|} \xi^\beta \widehat{X^\alpha f}(\xi) = (-1)^{|\alpha|} \big(D^\beta(X^\alpha f)\big)^{\wedge}(\xi) \quad \text{for } \xi \in \mathbb{R}^n . \quad (9.10)$$

Remark 9.2(g) shows there is a $c > 0$ such that

$$\big|\xi^\beta D^\alpha \widehat{f}(\xi)\big| \le (2\pi)^{-n/2} \int |D^\beta(X^\alpha f)(x)|\,(1+|x|^2)^{(n+1)/2}(1+|x|^2)^{-(n+1)/2}\,dx$$

$$\le c\, q_{m+n+1,k}(f)$$

for $|\alpha| \le m$ and $|\beta| \le k$. Hence there is a $C > 0$ such that

$$q_{k,m}\big(\widehat{f}\big) \le C q_{m+n+1,k}(f) . \quad (9.11)$$

Therefore $\widehat{f}$ belongs to $\mathcal{S}$. The continuity of the Fourier transformation now follows easily from (9.11) and Remark 9.2(c). Thus we are done. ∎

9.8 Corollary   For $f \in \mathcal{S}$ and $\alpha \in \mathbb{N}^n$, we have

$$\widehat{D^\alpha f} = X^\alpha \widehat{f} \quad \text{and} \quad \widehat{X^\alpha f} = (-1)^{|\alpha|} D^\alpha \widehat{f} .$$

Proof   These are special cases of (9.10). ∎

Proposition 9.6 and Corollary 9.8 show that the Fourier transformation maps differentiation into multiplication by functions, and conversely. This fact underlies much of its great utility.

It is now easy to improve the statement of Remark 9.1(a) to one saying that the image of L_1 under $\mathcal{F}$ already lies in C_0.

9.9 Proposition[3]   (Riemann–Lebesgue) $\mathcal{F} \in \mathcal{L}(L_1, C_0)$.

Proof   Proposition 9.7 and $\mathcal{S} \subset C_0$ imply $\mathcal{F}(\mathcal{S}) \subset C_0$. From Theorem 9.3, we know that $\mathcal{S}$ is a dense vector subspace of L_1, and Remark 9.1(a) guarantees that $\mathcal{F}$ maps the space L_1 continuously into BC. The claim now follows because C_0 is a closed vector subspace of BC. ∎

9.10 Examples   (a) For $g := g_n : \mathbb{R}^n \to \mathbb{R}$, $x \mapsto e^{-|x|^2/2}$, we have $\widehat{g} = g$.

Proof   (i) A property of the exponential implies

$$g_n(x) = g_1(x_1) \cdot \dots \cdot g_1(x_n) \quad \text{for } x = (x_1, \dots, x_n) \in \mathbb{R}^n .$$

---

[3] Also known as the Riemann–Lebesgue lemma.

For clarity, we denote by $\mathcal{F}_n$ the Fourier transformation on $\mathbb{R}^n$. Then it follows from the Fubini–Tonelli theorem that

$$\mathcal{F}_n(g_n)(\xi) = (2\pi)^{-n/2} \int_{\mathbb{R}^n} e^{-i\langle x,\xi\rangle} e^{-|x|^2/2} \, dx = (2\pi)^{-n/2} \int_{\mathbb{R}^n} \prod_{j=1}^{n} e^{-i x_j \xi_j} e^{-x_j^2/2} \, dx$$

$$= \prod_{j=1}^{n} (2\pi)^{-1/2} \int_{\mathbb{R}} e^{-i x_j \xi_j} e^{-x_j^2/2} \, dx_j = \prod_{j=1}^{n} \mathcal{F}_1(g_1)(\xi_j) \ .$$

This shows that it suffices to treat the one-dimensional case.

(ii) Suppose therefore $n = 1$. For $f := \widehat{g}$, we have from Example 8.7 that

$$f(0) = \widehat{g}(0) = \frac{1}{\sqrt{2\pi}} \int_{-\infty}^{\infty} e^{-x^2/2} \, dx = 1 \ .$$

Because $xe^{-x^2/2} = -\partial(e^{-x^2/2})$, that is, because $Xg = -\partial g = -i\, Dg$, Corollary 9.8 gives

$$\partial f = \partial \widehat{g} = i\, D\widehat{g} = -i\, \widehat{Xg} = -\widehat{Dg} = -X\widehat{g} = -Xf \ .$$

Therefore f solves the linear initial value problem $y'(t) = -t\, y(t)$ with $y(0) = 1$ on $\mathbb{R}$; its unique solution is g. ∎

(b) With the notation of (a) and (7.11), we have

$$\widehat{g(\varepsilon\,\cdot)}(\xi) = g_\varepsilon(\xi) \quad \text{for } \xi \in \mathbb{R}^n \ , \quad \varepsilon > 0 \ .$$

Proof Because $g(\varepsilon\,\cdot) = \sigma_{1/\varepsilon} g$, this follows from (a) and Remark 9.1(d). ∎

(c) Suppose

$$\varphi(x) := (2\pi)^{-n/2} e^{-|x|^2} \quad \text{for } x \in \mathbb{R}^n \ ,$$

and let $\varepsilon > 0$. Then $\widehat{\varphi(\varepsilon\,\cdot)} = k_\varepsilon$, where $k_1 = k$, is the **Gaussian kernel**.

Proof From $\varphi = (2\pi)^{-n/2} \sigma_{1/\sqrt{2}} g$ and Remark 9.1(c), it follows that

$$\varphi(\varepsilon\,\cdot) = \sigma_{1/\varepsilon} \varphi = (2\pi)^{-n/2} \sigma_{1/\sqrt{2}\,\varepsilon} \, g = (2\pi)^{-n/2} g(\sqrt{2}\,\varepsilon\,\cdot) \ .$$

Thus we get from (b) that

$$\widehat{\varphi(\varepsilon\,\cdot)}(x) = (2\pi)^{-n/2} g_{\sqrt{2}\,\varepsilon}(x) = \varepsilon^{-n} (4\pi)^{-n/2} e^{-|x|^2/4\varepsilon^2} = k_\varepsilon(x)$$

for $x \in \mathbb{R}^n$. ∎

The Fourier integral theorem

To prepare for more in-depth study of the Fourier transformation on L_1, we provide the following results.

9.11 Proposition *Suppose* $f, g \in L_1$. *Then* $\widehat{f}g$ *and* $f\widehat{g}$ *belong to* L_1, *and*

$$\int \widehat{f}g\,dx = \int f\widehat{g}\,dx .$$

Proof From Proposition 9.9, it follows easily that $\widehat{f}g$ and $f\widehat{g}$ belong to L_1. Let $\overset{*}{f}$ and $\overset{*}{g}$ be representatives of f and g, respectively. Then Lemma 7.2 shows that

$$h: \mathbb{R}^{2n} \to \mathbb{C} , \quad (x, y) \mapsto e^{-i\langle x, y\rangle} \overset{*}{f}(x)\overset{*}{g}(y) \tag{9.12}$$

is measurable. Because

$$\int\int |h(x, y)|\,dx\,dy = \|f\|_1\,\|g\|_1 , \tag{9.13}$$

we can apply the Fubini–Tonelli theorem to h, and we find

$$\int \widehat{\overset{*}{f}}(y)\overset{*}{g}(y)\,dy = \int (2\pi)^{-n/2} \int e^{-i\langle x, y\rangle}\overset{*}{f}(x)\,dx\,\overset{*}{g}(y)\,dy$$

$$= \int (2\pi)^{-n/2}\int e^{-i\langle x, y\rangle}\overset{*}{g}(y)\,dy\,\overset{*}{f}(x)\,dx = \int \widehat{\overset{*}{g}}(x)\overset{*}{f}(x)\,dx .$$

Then claim now follows after noting $\widehat{f} = \widehat{\overset{*}{f}}$ and $\widehat{g} = \widehat{\overset{*}{g}}$. ∎

We now prove theorems about the inverse of the Fourier transformation for various assumptions on the function and its transform.

9.12 Theorem *For* $f \in L_1$, *these statements are true:*

(i) $$\lim_{\varepsilon \to 0}(2\pi)^{-n/2}\int e^{i\langle\,\cdot\,,\,\xi\rangle}\widehat{f}(\xi)e^{-\varepsilon^2\,|\xi|^2}\,d\xi = f \qquad \text{in } L_1 .$$

(ii) (*Fourier integral theorem for* L_1) *If* $\widehat{f}$ *belongs to* L_1, *then* $f = \overline{\mathcal{F}}(\widehat{f})$, *where* $\overline{\mathcal{F}}$ *is the Fourier cotransformation.*

Proof (i) We use the notation of Example 9.10 and set

$$\varphi^{\varepsilon}(\xi, y) := e^{i\langle\xi, y\rangle}\varphi(\varepsilon\xi) = (2\pi)^{-n/2}e^{i\langle y, \xi\rangle}e^{-\varepsilon^2\,|\xi|^2}$$

for $\xi, y \in \mathbb{R}^n$ and $\varepsilon > 0$. We let $\widehat{\varphi^{\varepsilon}}(\,\cdot\,, y)$ be the Fourier transform of $\xi \mapsto \varphi^{\varepsilon}(\xi, y)$ for $y \in \mathbb{R}^n$. From Example 9.10(c) and Remark 9.1(e), it follows that

$$\widehat{\varphi^{\varepsilon}}(x, y) = k_{\varepsilon}(y - x) \quad \text{for } x, y \in \mathbb{R}^n .$$

Therefore Proposition 9.11 implies

$$(2\pi)^{-n/2}\int \widehat{f}(\xi)e^{i\langle y, \xi\rangle}e^{-\varepsilon^2\,|\xi|^2}\,d\xi = \int \widehat{f}(\xi)\varphi^{\varepsilon}(\xi, y)\,d\xi$$

$$= \int f(x)\widehat{\varphi^{\varepsilon}}(x, y)\,dx = k_{\varepsilon} * f(y)$$

for $y \in \mathbb{R}^n$. The claim now follows from Theorem 7.11 and Example 7.12(a).

(ii) If $\widehat{f}$ belongs to L_1, the dominated convergence theorem shows that

$$\lim_{\varepsilon \to 0} \int e^{i\langle y, \xi \rangle} \widehat{f}(\xi) e^{-\varepsilon^2 |\xi|^2} \, d\xi = \int e^{i\langle y, \xi \rangle} \widehat{f}(\xi) \, d\xi = (2\pi)^{n/2} \mathcal{F}(\widehat{f})^{\vee}(y)$$

for $y \in \mathbb{R}^n$. Thus (i), Remark 9.1(b), and Theorem 2.18(i) finish the proof. ∎

9.13 Corollary

(i) (Fourier integral theorem for $\mathcal{S}$) *The Fourier transformation is a continuous automorphism of $\mathcal{S}$. Its inverse is the Fourier cotransformation.*

(ii) *The Fourier transformation maps L_1 continuously and injectively into C_0 and has a dense image.*

(iii) *For $f \in L_1 \cap BUC$, the equality*[4]

$$f(x) = \lim_{\varepsilon \to 0} (2\pi)^{-n/2} \int e^{i\langle x, \xi \rangle} \widehat{f}(\xi) e^{-\varepsilon^2 |\xi|^2} \, d\xi$$

holds uniformly with respect to $x \in \mathbb{R}^n$.

(iv) *For $f \in L_1 \cap BUC$, suppose $\widehat{f}$ belongs to L_1. Then*

$$f(x) = (2\pi)^{-n/2} \int e^{i\langle x, \xi \rangle} \widehat{f}(\xi) \, d\xi \quad \text{for } x \in \mathbb{R}^n .$$

Proof (i) As in the case of normed vector spaces, we denote by $\mathcal{L}(\mathcal{S})$ the vector space of all continuous endomorphisms of $\mathcal{S}$; similarly, we let $\mathcal{L}aut(\mathcal{S})$ be the automorphisms of $\mathcal{S}$. Then it follows from Remark 9.2(i) and Proposition 9.7 that $\mathcal{F}$ and $\overline{\mathcal{F}}$ belong to $\mathcal{L}(\mathcal{S})$. Because $\mathcal{S} \subset L_1$, Theorem 9.12(ii) therefore shows that $\overline{\mathcal{F}}$ is a left inverse of $\mathcal{F}$ in $\mathcal{L}(\mathcal{S})$. It then follows from $\widetilde{\widehat{u}} = \widehat{\widetilde{u}}$ that $\mathcal{F}\overline{\mathcal{F}}f = \mathcal{F}(\mathcal{F}f)^{\vee} = \widecheck{\mathcal{F}}\mathcal{F}f = \overline{\mathcal{F}}\mathcal{F}f = f$ for $f \in \mathcal{S}$. Therefore $\overline{\mathcal{F}}$ is also a right inverse of $\mathcal{F}$ in $\mathcal{L}(\mathcal{S})$, which proves $\mathcal{F} \in \mathcal{L}aut(\mathcal{S})$.

(ii) If $\widehat{f} = 0$ for $f \in L_1$, then $f = 0$ follows from Theorem 9.12(ii). Therefore $\mathcal{F}$ is injective on L_1, and from the Riemann–Lebesgue lemma, we know that $\mathcal{F}$ belongs to $\mathcal{L}(L_1, C_0)$. Because (i) and $\mathcal{S} \subset L_1$, we have $\mathcal{S} = \mathcal{F}(\mathcal{S}) \subset \mathcal{F}(L_1)$. It then follows from Remark 9.2(f) that $\mathcal{F}(L_1)$ is dense in C_0.

(iii) follows from the proof of Theorem 9.12(i) and Theorem 7.11.

(iv) is now clear. ∎

9.14 Remarks (a) For $f \in \mathcal{S}$, we have $\widehat{\widehat{f}} = \widecheck{f}$.

(b) One can show that L_1 does not have a closed image in C_0 under the Fourier transformation (see [Rud83]). Hence $\mathcal{F} \in \mathcal{L}(L_1, C_0)$ is not surjective. ∎

----
[4]One can show that (iii) and (iv) remain true for $f \in L_1 \cap C$.

Convolutions and the Fourier transform

We now study what happens to convolutions under the Fourier transformations. So we first introduce another space of smooth functions; these will turn out to be particularly significant in the next subsection.

Suppose $\varphi \in C^\infty$. If to every $\alpha \in \mathbb{N}^n$, there are constants $c_\alpha > 0$ and $k_\alpha \in \mathbb{N}$ such that

$$|\partial^\alpha \varphi(x)| \leq c_\alpha (1 + |x|^2)^{k_\alpha} \quad \text{for } x \in \mathbb{R}^n ,$$

then we say φ is **slowly increasing**. We denote by $\mathcal{O}_M$ the set of all functions with this property, the **space of slowly increasing functions**.

9.15 Remarks (a) In the sense of vector subspaces, we have the inclusions $\mathcal{S} \subset \mathcal{O}_M \subset C^\infty$ and $\mathbb{C}[X_1, \ldots, X_n] \subset \mathcal{O}_M$.

(b) $(\mathcal{O}_M, +, \cdot)$ is a commutative algebra with unity.

(c) Suppose $(\varphi, f) \in \mathcal{O}_M \times \mathcal{S}$. Then φf belongs to $\mathcal{S}$, and to every $m \in \mathbb{N}$, there are $c = c(\varphi, m) > 0$ and $k' = k'(\varphi, m) \in \mathbb{N}$ such that $q_{k,m}(\varphi f) \leq c\, q_{k+k',m}(f)$ for $k \in \mathbb{N}$.

Proof Suppose $m \in \mathbb{N}$. Then there are $c = c(\varphi, m) > 0$ and $k' = k'(\varphi, m) \in \mathbb{N}$ such that

$$|\partial^\alpha \varphi(x)| \leq c(1 + |x|^2)^{k'/2} \quad \text{for } x \in \mathbb{R}^n , \quad \alpha \in \mathbb{N}^n , \quad |\alpha| \leq m .$$

Now it follows from the Leibniz rule that

$$q_{k,m}(\varphi f) = \max_{|\alpha| \leq m} \sup_{x \in \mathbb{R}^n} (1 + |x|^2)^{k/2} \left| \sum_{\beta \leq \alpha} \binom{\alpha}{\beta} \partial^\beta \varphi(x) \partial^{\alpha-\beta} f(x) \right|$$

$$\leq c \max_{|\alpha| \leq m} \sup_{x \in \mathbb{R}^n} (1 + |x|^2)^{(k+k')/2} |\partial^\alpha f(x)| = c\, q_{k+k',m}(f)$$

for $f \in \mathcal{S}$ and $k \in \mathbb{N}$. ∎

(d) Suppose $\varphi \in \mathcal{O}_M$. Then $f \mapsto \varphi f$ is a linear and continuous map of $\mathcal{S}$ into itself.

Proof This follows from (c) and Remark 9.2(c). ∎

(e) For every $s \in \mathbb{R}$, Λ^s belongs to $\mathcal{O}_M$. ∎

We can now prove another important property of the Fourier transformation.

9.16 Theorem (convolution theorem)

(i) $(f * g)\widehat{} = (2\pi)^{n/2} \widehat{f}\widehat{g}$ for $(f, g) \in \mathcal{L}_1 \times \mathcal{L}_1$.

(ii) $\widehat{\varphi} * \widehat{f} = (2\pi)^{n/2} \widehat{\varphi f}$ for $(\varphi, f) \in \mathcal{S} \times \mathcal{L}_1$.

Proof (i) By (9.12) and (9.13), we see that the Fubini–Tonelli theorem can be applied. It then follows from Corollary 7.9 that

$$(f * g)\widehat{}(\xi) = (2\pi)^{-n/2} \int e^{-i\langle x,\xi\rangle} \int f(x-y)g(y)\, dy\, dx$$
$$= (2\pi)^{-n/2} \int g(y) \int e^{-i\langle x,\xi\rangle} f(x-y)\, dx\, dy \ .$$

Because

$$\int e^{-i\langle x,\xi\rangle} f(x-y)\, dx = e^{-i\langle y,\xi\rangle} \int e^{-i\langle z,\xi\rangle} f(z)\, dz = e^{-i\langle y,\xi\rangle}(2\pi)^{n/2}\widehat{f}(\xi) \ ,$$

we then get

$$(f * g)\widehat{}(\xi) = (2\pi)^{-n/2} \int (2\pi)^{n/2}\widehat{f}(\xi)e^{-i\langle y,\xi\rangle}g(y)\, dy = (2\pi)^{n/2}\widehat{f}(\xi)\widehat{g}(\xi) \ .$$

(ii) Suppose $(\varphi, f) \in \mathcal{S} \times \mathcal{L}_1$. By Theorem 9.3, we find a sequence (f_j) in $\mathcal{S}$ such that $f_j \to f$ in $\mathcal{L}_1$. Propositions 9.4 and 9.7 imply that $\widehat{\varphi} * \widehat{f_j}$ belongs to $\mathcal{S}$. By Remark 9.15(c), φf_j also belongs to $\mathcal{S}$, so it follows from (i) and Remark 9.14(a) that

$$(\widehat{\varphi} * \widehat{f_j})\widehat{} = (2\pi)^{n/2}\widehat{\widehat{\varphi}}\widehat{\widehat{f_j}} = (2\pi)^{n/2}(\varphi f_j)\widecheck{} \quad \text{for } j \in \mathbb{N} \ .$$

By Theorem 9.12(ii), we then get

$$\widehat{\varphi} * \widehat{f_j} = (2\pi)^{n/2}\widehat{\varphi f_j} \quad \text{for } j \in \mathbb{N} \ . \tag{9.14}$$

Because $f_j \to f$ in L_1, it follows from Remark 9.1(a) that $\widehat{f_j} \to \widehat{f}$ in BC. Therefore Corollary 7.9 implies, because $\widehat{\varphi} \in \mathcal{S} \subset L_1$, that the sequence $(\widehat{\varphi} * \widehat{f_j})$ converges in BC to $\widehat{\varphi} * \widehat{f}$. Because $\varphi f_j \to \varphi f$ clearly holds in L_1, we deduce from Proposition 9.9 that the sequence $(\widehat{\varphi f_j})$ converges in BC to $\widehat{\varphi f}$. Then the claim follows from Remark 9.1(a). ∎

As an application of the convolution theorem, we prove a lemma, which forms the basis for the L_2-theory of the Fourier transformation.

9.17 Lemma For $f \in \mathcal{L}_1 \cap \mathcal{L}_2$, $\widehat{f}$ belongs to $C_0 \cap \mathcal{L}_2$, and $\|f\|_2 = \|\widehat{f}\|_2$.

Proof Suppose $f \in \mathcal{L}_1 \cap \mathcal{L}_2$. Because $\widehat{f}$ belongs to C_0 by the Riemann–Lebesgue lemma, it suffices to verify $\|f\|_2 = \|\widehat{f}\|_2$. So we set $g := f * \overline{\widetilde{f}}$. By Theorem 7.3(ii) and Exercise 7.2, we know g belongs to $\mathcal{L}_1 \cap C_0$, and

$$g(0) = \int f(y)\overline{\widetilde{f}}(0-y)\, dy = \int f\overline{f} = \|f\|_2^2 \ .$$

From Corollary 9.13(iii), it follows that

$$\|f\|_2^2 = g(0) = \lim_{\varepsilon \to 0} (2\pi)^{-n/2} \int \widehat{g}(\xi) e^{-\varepsilon^2 |\xi|^2} \, d\xi \ . \tag{9.15}$$

Now we note

$$\widehat{\widehat{\overline{f}}} = (2\pi)^{-n/2} \int e^{-i\langle x, \xi \rangle} \overline{\widehat{f}}(-x) \, dx = (2\pi)^{-n/2} \overline{\int e^{-i\langle -x, \xi \rangle} f(-x) \, dx} = \overline{\overline{\widehat{f}}} \ ,$$

which follows from the Euclidean invariance of integrals. Then Theorem 9.16(i) shows

$$\widehat{g} = \left(f * \overline{\widetilde{f}} \right)^{\widehat{}} = (2\pi)^{n/2} \widehat{f} \, \overline{\widehat{f}} = (2\pi)^{n/2} \left| \widehat{f} \right|^2 \ .$$

In particular $\widehat{g}$, is not negative. Therefore (9.15) and monotone convergence theorem imply $\|f\|_2 = \|\widehat{f}\|_2$. ∎

Fourier multiplication operators

To illustrate the significance of the mapping properties of the Fourier transformation, we now consider linear differential operators with constant coefficients and show that they are represented "in the Fourier domain" by multiplication operators.

For $m \in \mathbb{N}$, we denote by $\mathbb{C}_m[X_1, \ldots, X_n]$ the vector subspace $\mathbb{C}[X_1, \ldots, X_n]$ consisting of all polynomials of degree $\leq m$. For

$$p = \sum_{|\alpha| \leq m} a_\alpha X^\alpha \in \mathbb{C}_m[X_1, \ldots, X_n] \ ,$$

we let

$$p(D) := \sum_{|\alpha| \leq m} a_\alpha D^\alpha \ ,$$

which is a linear differential operator of order $\leq m$ with *constant coefficients*. Here p is called the **symbol** of $p(D)$. In the following, we set

$$\mathcal{D}\text{iffop}^0 := \left\{ p(D) \ ; \ p \in \mathbb{C}[X_1, \ldots, X_n] \right\} \ ,$$

and $\mathcal{D}\text{iffop}_m^0$ is the subset of all constant-coefficient, linear differential operators of order not higher than m.

9.18 Remarks (a) $p(D) \in \mathcal{D}\text{iffop}^0$ is a linear and continuous map of $\mathcal{S}$ into itself, that is, $p(D) \in \mathcal{L}(\mathcal{S})$.

Proof This follows from Remarks 9.2(c) and (h). ∎

(b) The map

$$\mathbb{C}[X_1, \ldots, X_n] \to \mathcal{L}(\mathcal{S}) \ , \quad p \mapsto p(D) \tag{9.16}$$

is linear and injective.

Proof The linearity is obvious. Suppose $p = \sum_{|\alpha| \le m} a_\alpha X^\alpha \in \mathbb{C}[X_1, \ldots, X_n]$ and that $p(D)f = 0$ for all $f \in \mathcal{S}$. We choose a $\varphi \in \mathcal{D}$ such that $\varphi \,|\, \mathbb{B}^n = 1$. For $\beta \in \mathbb{N}^n$, it follows from the Leibniz rule that

$$D^\alpha(\varphi X^\beta) = \varphi D^\alpha X^\beta + \sum_{\gamma < \alpha} \binom{\alpha}{\gamma} D^{\alpha - \gamma} \varphi D^\gamma X^\beta .$$

Because $\varphi(x) = 1$ for $|x| < 1$, we then derive

$$D^\alpha(\varphi X^\beta)(0) = D^\alpha X^\beta(0) = \begin{cases} \beta! & \text{if } \alpha = \beta , \\ 0 & \text{otherwise} \end{cases}$$

for $\alpha \in \mathbb{N}^n$. Because $\varphi X^\beta \in \mathcal{D} \subset \mathcal{S}$, we thus find $0 = p(D)(\varphi X^\beta) = \beta! \, a_\beta$ for $\beta \in \mathbb{N}^n$ with $|\beta| \le m$; therefore $p = 0$. This proves the claimed injectivity. ∎

(c) $p(D)$ is formally self-adjoint if and only if p has real coefficients.

Proof Letting

$$\mathcal{A}(\partial) := p(D) = \sum_{|\alpha| \le m} a_\alpha (-i)^{|\alpha|} \partial^\alpha$$

it follows from Proposition 7.24 that

$$\mathcal{A}^\sharp(\partial) = \sum_{|\alpha| \le m} (-1)^{|\alpha|} \overline{a_\alpha (-i)^{|\alpha|}} \, \partial^\alpha = \sum_{|\alpha| \le m} (-i)^{|\alpha|} \bar{a}_\alpha \, \partial^\alpha ,$$

which finishes the proof. ∎

By Remark 9.18(b), we can identify $\mathcal{D}\mathrm{iffop}^0$ [or $\mathcal{D}\mathrm{iffop}^0_m$] with the image of $\mathbb{C}[X_1, \ldots, X_n]$ [or $\mathbb{C}_m[X_1, \ldots, X_n]$] under the map (9.16). In other words, in the sense of vector subspaces, we have

$$\mathcal{D}\mathrm{iffop}^0_m \subset \mathcal{D}\mathrm{iffop}^0 \subset \mathcal{L}(\mathcal{S}) \quad \text{for } m \in \mathbb{N} .$$

For $a \in \mathcal{O}_M$ and $f \in \mathcal{S}$, it follows from Corollary 9.13(i) and Remark 9.15(d) that $\left(f \mapsto a\hat{f} \right) \in \mathcal{L}(\mathcal{S})$. Then it follows again from Corollary 9.13(i) that

$$a(D) := \mathcal{F}^{-1} a \mathcal{F} : \mathcal{S} \to \mathcal{S} , \quad f \mapsto \mathcal{F}^{-1}\left(a\hat{f} \right)$$

is a well defined element of $\mathcal{L}(\mathcal{S})$, a **Fourier multiplication operator** with **symbol** a. We set

$$\mathrm{Op} := \left\{ a(D) \in \mathcal{L}(\mathcal{S}) \; ; \; a \in \mathcal{O}_M \right\} .$$

9.19 Proposition   Op *is a commutative algebra of* $\mathcal{L}(\mathcal{S})$ *with unity, and the map*

$$ev \colon (\mathcal{O}_M, +, \cdot) \to \mathrm{Op}, \quad a \mapsto a(D)$$

is an algebra isomorphism.

Proof   It is clear that $\mathcal{O}_M := (\mathcal{O}_M, +, \cdot)$ is a commutative subalgebra with unity of the algebra $\mathbb{C}^{(\mathbb{R}^n)}$. It is also easy to verify that ev maps the vector space $\mathcal{O}_M$ linearly in $\mathcal{L}(\mathcal{S})$.

For $a, b \in \mathcal{O}_M$ and $f \in \mathcal{S}$, we have

$$(ab)(D)f = \mathcal{F}^{-1}\big(ab\widehat{f}\big) = \mathcal{F}^{-1}\big(a\mathcal{F}\mathcal{F}^{-1}(b\widehat{f})\big) = \mathcal{F}^{-1}\big(a\widehat{b(D)f}\big) = a(D) \circ b(D)f \ .$$

Therefore ev is a surjective algebra homomorphism.

Finally let $a, b \in \mathcal{O}_M$ with $a(D) = b(D)$. Let $\xi \in \mathbb{R}^n$, and denote by $\varphi \in \mathcal{D}$ a cutoff function for $\overline{\mathbb{B}}^n(\xi, 1)$. Then $f := \mathcal{F}^{-1}\varphi$ belongs to $\mathcal{S}$ with $\widehat{f}(\xi) = 1$, and it follows from Corollary 9.13(i) that

$$a(\xi) = \big(a\widehat{f}\big)(\xi) = \mathcal{F}\big(a(D)f\big)(\xi) = \mathcal{F}\big(b(D)f\big)(\xi) = \big(b\widehat{f}\big)(\xi) = b(\xi) \ .$$

Because this is true for every $\xi \in \mathbb{R}^n$, we have $a = b$. Therefore ev is injective. ∎

9.20 Corollary

(i)   *For* $a, b \in \mathcal{O}_M$, *we have* $ab(D) = a(D)b(D) = b(D)a(D)$.

(ii) $\mathbf{1}(D) = 1_{\mathcal{L}(\mathcal{S})}$.

(iii) Diffop^0 *is the image of* $\mathbb{C}[X_1, \ldots, X_n]$ *under* ev. *In particular,* Diffop^0 *is a commutative subalgebra of* Op *with unity.*

Proof   (i) and (ii) are special cases of Proposition 9.19.

(iii) For $p \in \mathbb{C}[X_1, \ldots, X_n] \subset \mathcal{O}_M$ with $p = \sum_{|\alpha| \le m} a_\alpha X^\alpha$, we get from Proposition 9.6(i) that

$$ev(p)f = \mathcal{F}^{-1}p\mathcal{F}f = \mathcal{F}^{-1}(p\widehat{f}) = \sum_{|\alpha| \le m} a_\alpha \mathcal{F}^{-1}\big(X^\alpha \widehat{f}\big)$$

$$= \sum_{|\alpha| \le m} a_\alpha \mathcal{F}^{-1}\big(\widehat{D^\alpha f}\big) = \sum_{|\alpha| \le m} a_\alpha D^\alpha f$$

for $f \in \mathcal{S}$. Therefore $ev(p) = \sum_{|\alpha| \le m} a_\alpha D^\alpha$, from which the claim follows. ∎

This corollary implies that the Fourier transformation can be used to solve linear differential equations with constant coefficients by reducing them to simple algebraic equations. This fact is part of the fundamental significance of the Fourier transformation. The following examples give a first glimpse into these methods.

9.21 Examples (a) Suppose the polynomial $p \in \mathbb{C}[X_1, \ldots, X_n]$ has no real zeros. Then $p(D) \in \mathcal{L}(\mathcal{S})$ is an automorphism of $\mathcal{S}$, and $[p(D)]^{-1} = (1/p)(D)$.

Proof We see easily that $1/p$ belongs to $\mathcal{O}_M$. Now we deduce from Corollary 9.20 that

$$1_{\mathcal{L}(\mathcal{S})} = \mathbf{1}(D) = (p \cdot 1/p)(D) = p(D)(1/p)(D) = (1/p)(D)p(D) .$$

Because $a(D) \in \mathcal{L}(\mathcal{S})$ for $a \in \mathcal{O}_M$, this proves the claim. ∎

(b) $1 - \Delta \in \mathcal{L}\mathrm{aut}(\mathcal{S})$, and $(1 - \Delta)^{-1} = \Lambda^{-2}(D)$.

Proof Because $1 - \Delta = \Lambda^2(D)$, this follows from (a). ∎

Example 9.21(b) says that the partial differential equation

$$-\Delta u + u = f \tag{9.17}$$

has a unique solution $u \in \mathcal{S}$ for every $f \in \mathcal{S}$ and that u depends continuously on f in the topology of $\mathcal{S}$. Also we can obtain the solution $u \in \mathcal{S}$ of (9.17) by first "Fourier transforming" this equation. This, according to Proposition 9.6, gives the equation $(|\xi|^2 + 1)\widehat{u}(\xi) = \Lambda^2(\xi)\widehat{u}(\xi) = \widehat{f}(\xi)$ for $\xi \in \mathbb{R}^n$. This equation can then be solved for $\widehat{u}$, giving $\widehat{u} = \Lambda^{-2}\widehat{f}$, and then "reverse Fourier transformed", giving $u = \mathcal{F}^{-1}(\Lambda^{-2}\widehat{f}) = \Lambda^{-2}(D)f$. This "method of Fourier transformation" plays a prominent role in the theory of partial differential equations. Note that $\Lambda^{-2}(D)$ or, more generally, $(1/p)(D)$, is not a differential operator.

Plancherel's theorem

To conclude this chapter, we show that the Fourier transformation can also be defined on L_2, and we explain a few consequences of this fact.

Suppose H is a Hilbert space. We say $T \colon H \to H$ is **unitary** if T is an isometric isomorphism.

9.22 Remarks Suppose H is a (real or complex) Hilbert space and $T \colon H \to H$ is linear.

(a) If T is unitary, then T belongs to $\mathcal{L}\mathrm{aut}(H)$, and

$$(Tx \mid Ty) = (x \mid y) \quad \text{for } x, y \in H .$$

Proof The first statement is clear. Because T is an isometry, we have

$$4\operatorname{Re}(Tx \mid Ty) = \|T(x + y)\|^2 - \|T(x - y)\|^2 = \|x + y\|^2 - \|x - y\|^2 = 4\operatorname{Re}(x \mid y) ,$$

and therefore $\operatorname{Re}(Tx \mid Ty) = \operatorname{Re}(x \mid y)$ for $x, y \in H$. Replacing y in this identity by iy, we get

$$\operatorname{Im}(Tx \mid Ty) = \operatorname{Re}(Tx \mid Tiy) = \operatorname{Re}(x \mid iy) = \operatorname{Im}(x \mid y) ,$$

and thus $(Tx \mid Ty) = (x \mid y)$ for $x, y \in H$. ∎

(b) If H is finite-dimensional, then the following statements are equivalent:

(i) T is unitary.

(ii) $(Tx \mid Ty) = (x \mid y)$ for $x, y \in H$.

(iii) $T^*T = \mathrm{id}_H$.

Proof "(i)⟹(ii)" is a consequence of (a).

"(ii)⟹(iii)" Let $\{b_1, \ldots, b_m\}$ be an orthonormal basis of H. Then every $y \in H$ can be expanded as $y = \sum_{j=1}^m (y \mid b_j) b_j$ (see Exercise II.3.12 and Theorem VI.7.14). From Exercise VII.1.5 and (ii), it follows that

$$ T^*Tx = \sum_{j=1}^m (T^*Tx \mid b_j) b_j = \sum_{j=1}^m (Tx \mid Tb_j) b_j = \sum_{j=1}^m (x \mid b_j) b_j = x $$

for every $x \in H$.

"(iii)⟹(i)" Because $T^*T = \mathrm{id}_H$, we know T is injective and is therefore also surjective by the rank formula of linear algebra. For $x \in H$, we also have

$$ \|Tx\|^2 = (Tx \mid Tx) = (T^*Tx \mid x) = (x \mid x) = \|x\|^2 \ . $$

Therefore T is an isometry. ∎

9.23 Theorem (Plancherel) *The Fourier transformation has a unique extension from $L_1 \cap L_2$ to a unitary operator on L_2.*

Proof Denote by X_2 the vector subspace $L_1 \cap L_2$ of the Hilbert space L_2. Then it follows from Lemma 9.17 that $\mathcal{F}$ belongs to $\mathcal{L}(X_2, L_2)$ and is an isometry. Because X_2 contains the space $\mathcal{S}$, Theorem 9.3 and VI.2.6 imply the existence of a unique isometric extension $\mathfrak{F} \in \mathcal{L}(L_2)$. As an isometry, $\mathfrak{F}$ has a closed image, which by Corollary 9.13(i) contains the space $\mathcal{S}$. Therefore Proposition V.4.4 implies that $\mathfrak{F}$ is surjective and therefore unitary. ∎

As usual, we reuse the symbol $\mathcal{F}$ for the unique continuous extension of $\mathfrak{F}$ and likewise call it the **Fourier transformation**.[5]

The next proposition describes the Fourier transform $\mathcal{F}f$ for an arbitrary $f \in L_2$.

9.24 Proposition *For $f \in L_2$, we have*

$$ \mathcal{F}f = \lim_{R \to \infty} \mathcal{F}(\chi_{R\overline{\mathbb{B}}^n} f) = \lim_{R \to \infty} (2\pi)^{-n/2} \int_{[|x| \le R]} e^{-i\langle x, \cdot \rangle} f(x) \, dx \qquad \text{in } L_2 \ . $$

---

[5]On L_2, the Fourier transformation will sometimes also be called the Fourier–Plancherel, or Plancherel, transformation.

Proof For $R > 0$, the element $f_R := \chi_{R\bar{\mathbb{B}}^n} f$ belongs to $L_1 \cap L_2$, and the dominated convergence theorem implies

$$\int |f - f_R|^2 \, dx = \int |f|^2 \, (1 - \chi_{R\bar{\mathbb{B}}^n})^2 \, dx \to 0 \quad (R \to \infty) \ .$$

Therefore $\lim_{R \to \infty} f_R = f$ in L_2. Then by Plancherel's theorem, $\mathcal{F} f_R$ converges in L_2 to $\mathcal{F} f$. Because

$$\mathcal{F}(f_R)(\xi) = (2\pi)^{-n/2} \int_{[|x| \le R]} e^{-i\langle x, \xi \rangle} f(x) \, dx \quad \text{for } \xi \in \mathbb{R}^n \ ,$$

the claim follows. ∎

9.25 Example Suppose $n = 1$ and $a > 0$. Also let $f := \chi_{[-a,a]} \in \mathcal{L}_1(\mathbb{R})$. Then

$$\widehat{f}(\xi) = \frac{1}{\sqrt{2\pi}} \int_{-a}^{a} e^{-ix\xi} \, dx = \frac{-1}{\sqrt{2\pi}\, i\xi} \big(e^{-i\xi a} - e^{i\xi a}\big) = \sqrt{\frac{2}{\pi}}\, a\, \frac{\sin(a\xi)}{a\xi}$$

for $\xi \in \mathbb{R}$. Because $\int |f|^2 \, dx = 2a$, Plancherel's theorem gives

$$\int_{-\infty}^{\infty} \Big[\frac{\sin(ax)}{ax}\Big]^2 \, dx = \frac{\pi}{a} \quad \text{for } a > 0 \ .$$

Note that $x \mapsto \sin(x)/x$ does not belong to $\mathcal{L}_1(\mathbb{R})$. ∎

Symmetric operators

Suppose E is a Banach space over $\mathbb{K}$. By a **linear operator A in E**, we mean a map $A \colon \operatorname{dom}(A) \subset E \to E$ such that $\operatorname{dom}(A)$ is a vector subspace of E and such that A is linear. For linear operators $A_j \colon \operatorname{dom}(A_j) \subset E \to E$ and $\lambda \in \mathbb{K}^\times$, we define $A_0 + \lambda A_1$ by

$$\operatorname{dom}(A_0 + \lambda A_1) := \operatorname{dom}(A_0) \cap \operatorname{dom}(A_1) \ , \quad (A_0 + \lambda A_1)x := A_0 x + \lambda A_1 x \ .$$

The **product** $A_0 A_1$ is defined by

$$\operatorname{dom}(A_0 A_1) := \big\{ x \in \operatorname{dom}(A_1) \ ; \ A_1 x \in \operatorname{dom}(A_0) \big\} \ , \quad (A_0 A_1)x := A_0(A_1 x) \ .$$

Finally, the operator defined by

$$\operatorname{dom}([A_0, A_1]) := \operatorname{dom}(A_0 A_1 - A_1 A_0) \ , \quad [A_0, A_1]x := (A_0 A_1 - A_1 A_0)x$$

is called the **commutator** of A_0 and A_1. Obviously $A_0 + \lambda A_1$, $A_0 A_1$, and $[A_0, A_1]$ are linear operators in E, for which

$$A_0 + \lambda A_1 = \lambda A_1 + A_0 \ , \quad \lambda A_0 = A_0(\lambda \operatorname{id}_E) \ , \quad [A_0, A_1] = -[A_1, A_0] \ .$$

Suppose now H is a Hilbert space, and $A: \operatorname{dom}(A) \subset H \to H$ is a linear operator on H. If

$$(Au \,|\, v) = (u \,|\, Av) \quad \text{for } u, v \in \operatorname{dom}(A) ,$$

we say A is **symmetric**.

9.26 Remarks (a) Suppose H is a complex Hilbert space and A is a linear operator on H. Then these statements are equivalent:

(i) A is symmetric.

(ii) $(Au \,|\, u) \in \mathbb{R}$ for $u \in \operatorname{dom}(A)$.

Proof "(i)$\Rightarrow$(ii)" Because A is symmetric, it follows that

$$(Au \,|\, u) = (u \,|\, Au) = \overline{(Au \,|\, u)} \quad \text{for } u \in \operatorname{dom}(A) ,$$

and therefore $\operatorname{Im}(Au \,|\, u) = 0$.

"(ii)$\Rightarrow$(i)" For $u, v \in \operatorname{dom}(A)$, we have

$$\bigl(A(u + v) \,|\, u + v\bigr) = (Au \,|\, u) + (Av \,|\, u) + (Au \,|\, v) + (Av \,|\, v) . \tag{9.18}$$

Because of (ii), it follows that $\operatorname{Im}(Au \,|\, v) = -\operatorname{Im}(Av \,|\, u)$, and therefore

$$\operatorname{Im}(Au \,|\, v) = -\operatorname{Im}(Av \,|\, u) = -\operatorname{Im} \overline{(u \,|\, Av)} = \operatorname{Im}(u \,|\, Av) .$$

Replacing u in (9.18) by iu, we get

$$\operatorname{Re}(Au \,|\, v) = \operatorname{Im}\bigl(A(iu) \,|\, v\bigr) = \operatorname{Im}(iu \,|\, Av) = \operatorname{Re}(u \,|\, Av) .$$

Therefore $(Au \,|\, v) = (v \,|\, Au)$. ∎

(b) Suppose $p \in \mathbb{C}[X_1, \ldots, X_n]$ and P is the linear operator on L_2 such that $\operatorname{dom}(P) = \mathcal{S}$ and $Pu := p(D)u$ for $u \in \mathcal{S}$. Then these statements are equivalent:

(i) P is symmetric.

(ii) $p(D)$ is formally self-adjoint.

(iii) p has real coefficients.

Proof "(i)$\Rightarrow$(ii)" That P is symmetric implies

$$\bigl(p(D)u \,|\, v\bigr) = (Pu \,|\, v) = (u \,|\, Pv) = \bigl(u \,|\, p(D)v\bigr) \quad \text{for } u, v \in \mathcal{D} ,$$

which in turn implies (ii) by the uniqueness of formally adjoint operators.

"(ii)$\Rightarrow$(iii)" Remark 9.18(c).

"(iii)$\Rightarrow$(i)" Suppose $p = \sum_{|\alpha| \le m} a_\alpha X^\alpha$. Then Corollary 9.20(iii) and Plancherel's theorem imply

$$(Pu \,|\, u) = \bigl(p(D)u \,|\, u\bigr) = (p\widehat{u} \,|\, \widehat{u}) = \sum_{|\alpha| \le m} a_\alpha \int \xi^\alpha \,|\widehat{u}(\xi)|^2 \, d\xi \quad \text{for } u \in \mathcal{S} .$$

Therefore $(Pu \,|\, u)$ is real, and the claim follows from (a). ∎

(c) With $\mathcal{S}$ as their domain, the Laplace, wave, and Schrödinger operators are symmetric in L_2.

Proof This follows from (b) and Examples 7.25(a) and (e). ∎

The Heisenberg uncertainty relation

As another application of Plancherel's theorem, we close this section by discussing several important properties of the position and momentum operators of quantum mechanics. So we fix $j \in \{1, \ldots, n\}$ and set

$$\mathrm{dom}(A_j) := \{\, u \in L_2 \; ; \; X_j \widehat{u} \in L_2 \,\} \,, \quad \mathrm{dom}(B_j) := \{\, u \in L_2 \; ; \; X_j u \in L_2 \,\} \,.$$

Then we define linear operators in L_2, the **momentum operator** A_j and the **position operator** B_j (for the j-th coordinate), by

$$A_j u := \mathcal{F}^{-1}(X_j \widehat{u}) \quad \text{and} \quad B_j v := X_j v \quad \text{for } u \in \mathrm{dom}(A_j) \,, \quad v \in \mathrm{dom}(B_j) \,.$$

9.27 Remarks **(a)** We have $\mathcal{S} \subset \mathrm{dom}(A_j)$, and

$$A_j u = X_j(D) u = D_j u = -i \partial_j u \quad \text{for } u \in \mathcal{S} \,.$$

Proof This follows from Proposition 9.7 and Corollary 9.8. ∎

(b) We have $\mathcal{F}(\mathrm{dom}(A_j)) = \mathrm{dom}(B_j)$ and a commutative diagram:

$$
\begin{array}{ccc}
\mathrm{dom}(A_j) & \xrightarrow{\;\;A_j\;\;} & L_2 \\[4pt]
{\scriptstyle \mathcal{F}}\big\downarrow & & \big\downarrow{\scriptstyle \mathcal{F}} \\[4pt]
\mathrm{dom}(B_j) & \xrightarrow{\;\;B_j\;\;} & L_2
\end{array}
$$

In particular,

$$A_j u = \mathcal{F}^{-1} B_j \mathcal{F} u \,, \quad u \in \mathrm{dom}(A_j) \quad \text{and} \quad B_j u = \mathcal{F} A_j \mathcal{F}^{-1} u \,, \quad u \in \mathrm{dom}(B_j) \,.$$

Proof These are consequences of Plancherel's theorem. ∎

(c) The position and momentum operators of quantum mechanics are symmetric.

Proof Let $u \in \mathrm{dom}(A_j)$. Then (b) and Plancherel's theorem imply

$$(A_j u \,|\, u) = (\mathcal{F}^{-1} B_j \mathcal{F} u \,|\, u) = (B_j \widehat{u} \,|\, \widehat{u}) = \int \xi_j \, |\widehat{u}(\xi)|^2 \, d\xi \,,$$

Now the claim follows from Remark 9.26(a). ∎

(d) For $u \in \mathrm{dom}([A_j, B_j])$, we have $([A_j, B_j] u \,|\, u) = 2i \, \mathrm{Im}(A_j B_j u \,|\, u)$.

Proof By (b), (c), and Plancherel's theorem, we get for $u \in \mathrm{dom}([A_j, B_j])$ that

$$
\begin{aligned}
([A_j, B_j] u \,|\, u) &= (A_j B_j u - B_j A_j u \,|\, u) \\
&= (\mathcal{F}^{-1} B_j \mathcal{F} B_j u - B_j \mathcal{F}^{-1} B_j \mathcal{F} u \,|\, u) \\
&= (\mathcal{F} B_j u \,|\, B_j \mathcal{F} u) - (B_j \mathcal{F} u \,|\, \mathcal{F} B_j u) \\
&= 2i \, \mathrm{Im}(\mathcal{F} B_j u \,|\, B_j \mathcal{F} u) = 2i \, \mathrm{Im}(\mathcal{F}^{-1} B_j \mathcal{F} B_j u \,|\, u) \\
&= 2i \, \mathrm{Im}(A_j B_j u \,|\, u) \,. \quad \blacksquare
\end{aligned}
$$

(e) The operator $i[A_j, B_j]$ is symmetric in L_2.

Proof This follows from (d). ∎

(f) We have $\mathcal{S} \subset \mathrm{dom}([A_j, B_j])$, and $[A_j, B_j]u = -iu$ on $u \in \mathcal{S}$.

Proof The first statement follows easily from Proposition 9.7 and Remark 9.2(h). Also (a) shows that
$$[A_j, B_j]u = D_j(X_j u) - X_j D_j u = (D_j X_j)u = -iu$$
for $u \in \mathcal{S}$. ∎

(g) (Heisenberg uncertainty relation for $\mathcal{S}$) For $j \in \{1, \ldots, n\}$, we have
$$\|u\|_2^2 \leq 2 \|\partial_j u\|_2 \|X_j u\|_2 \quad \text{for } u \in \mathcal{S} .$$

Proof Let $u \in \mathcal{S}$. By (d) and (f), we have
$$-i \|u\|_2^2 = -i(u \,|\, u) = ([A_j, B_j]u \,|\, u) = 2i \, \mathrm{Im}(A_j B_j u \,|\, u) .$$
The Cauchy–Schwarz inequality therefore gives
$$\|u\|_2^2 = 2 \left| \mathrm{Im}(A_j B_j u \,|\, u) \right| \leq 2 \left| (A_j B_j u \,|\, u) \right| = 2 \left| (B_j u \,|\, A_j u) \right| \leq 2 \|A_j u\|_2 \|B_j u\|_2 ,$$
and thus the claim follows because of (a). ∎

We conclude this section by extending the validity of the Heisenberg uncertainty relation on $\mathcal{S}$ to $\mathrm{dom}(A_j) \cap \mathrm{dom}(B_j)$. We first need a lemma.

9.28 Lemma *For every $u \in \mathrm{dom}(A_j) \cap \mathrm{dom}(B_j)$, there is a sequence (u_m) in $\mathcal{S}$ such that*
$$\lim_{m \to \infty} (u_m, A_j u_m, B_j u_m) = (u, A_j u, B_j u) \quad \text{in } L_2^3 .$$

Proof (i) Suppose $u \in \mathrm{dom}(A_j) \cap \mathrm{dom}(B_j)$, and let $\{ k_\varepsilon ; \ \varepsilon > 0 \}$ be the Gaussian approximation kernel. We set $u^\varepsilon := k_\varepsilon * u$. By Exercise 8(iv), u^ε belongs to $\mathcal{S}$, and Theorem 7.11 shows $\lim_{\varepsilon \to 0} u^\varepsilon = u$ in L_2.

(ii) Because $\check{k} = k$, it follows from Example 9.10(c) that
$$\widehat{k_\varepsilon}(\xi) := \check{k}_\varepsilon(\xi) = \mathcal{F}^{-1} k_\varepsilon(\xi) = \varphi(\varepsilon \xi) = (2\pi)^{-n/2} e^{-\varepsilon^2 |\xi|^2} \quad \text{for } \xi \in \mathbb{R}^n .$$

According to Theorem 9.3, we can find a sequence (v_m) in $\mathcal{S}$ such that $\lim_m v_m = u$ in L_2. The convolution theorem therefore shows
$$(k_\varepsilon * v_m)\widehat{\ }(\xi) = (2\pi)^{n/2} \widehat{k}_\varepsilon(\xi) \widehat{v}_m(\xi) = e^{-\varepsilon^2 |\xi|^2} \widehat{v}_m(\xi) \quad \text{for } \xi \in \mathbb{R}^n .$$

The limit $m \to \infty$ then gives $\widehat{u^\varepsilon} = e^{-\varepsilon^2 |\cdot|^2} \widehat{u}$ (see Corollary 7.9 and Theorem 9.23). Because
$$\|A_j u - A_j u^\varepsilon\|_2^2 = \|X_j \widehat{u} - X_j \widehat{u^\varepsilon}\|_2^2 = \int |\xi_j \widehat{u}(\xi)|^2 \left(1 - e^{-\varepsilon^2 |\xi|^2}\right)^2 d\xi ,$$
it follows from the dominated convergence theorem that $\lim_{\varepsilon \to 0} A_j u^\varepsilon = A_j u$ in L_2.

(iii) Let $\overset{*}{u}$ be a representative of u. We set

$$d_\varepsilon(x,z) := x_j\big(\overset{*}{u}(x) - \overset{*}{u}(x - \varepsilon z)\big)\,,\quad g_\varepsilon(x,z) := d_\varepsilon(x,z)k(z)$$

for $\varepsilon > 0$ and $(x,z) \in \mathbb{R}^n \times \mathbb{R}^n$. Then it follows, as in (7.7) (or from the Minkowski inequality for integrals), that

$$\|X_j u - X_j u^\varepsilon\|_2 \le \left(\int\Big[\int |g_\varepsilon(x,z)|\,dz\Big]^2 dx\right)^{1/2} \le \int \|d_\varepsilon(\,\cdot\,,z)\|_2\, k(z)\,dz\,,\quad (9.19)$$

where for the last inequality we used $g_\varepsilon = \big(d_\varepsilon\sqrt{k}\big)\sqrt{k}$ and $\int k\,dx = 1$ together with the Cauchy–Schwarz inequality. Further noting

$$d_\varepsilon(\,\cdot\,,z) = X_j\overset{*}{u} - \tau_{\varepsilon z}(X_j\overset{*}{u}) - \varepsilon z_j\tau_{\varepsilon z}\overset{*}{u}\,,$$

it follows from the strong continuity of the translation group on L_2 and the translation invariance of integrals that

$$\lim_{\varepsilon\to 0} \|d_\varepsilon(\,\cdot\,,z)\|_2\, k(z) = 0\quad\text{for } z \in \mathbb{R}^n\,,$$

and

$$\|d_\varepsilon(\,\cdot\,,z)\|_2\, k(z) \le 2\max\{\|X_j u\|_2, \|u\|_2\}(1 + |z_j|)k(z)\quad\text{for } \varepsilon \in (0,2]\,,\quad z \in \mathbb{R}^n\,.$$

Because $z \mapsto (1 + |z_j|)k(z)$ belongs to $\mathcal{L}_1$, the claim is implied by (9.19) and the dominated convergence theorem. ∎

9.29 Corollary (Heisenberg uncertainty relation) *For $1 \le j \le n$, we have*

$$\|u\|_2^2 \le 2\,\|A_j u\|_2\, \|B_j u\|_2\quad\text{for } u \in \operatorname{dom}(A_j) \cap \operatorname{dom}(B_j)\,.$$

Proof This follows from Remarks 9.27(a) and (g) and Lemma 9.28. ∎

From Remark 9.27(a) and Lemma 9.28 it easily follows, as in the proof of Theorem 7.27, that the distributional derivative $\partial_j u$ belongs to L_2 for $u \in \operatorname{dom}(A_j)$ and is therefore a weak L_2-derivative. Also $A_j u = -i\,\partial_j u$. Consequently, we can also write the Heisenberg uncertainty relation for $u \in \operatorname{dom}(A_j) \cap \operatorname{dom}(B_j)$ in the form

$$\left(\frac{1}{2}\int |u|^2\,dx\right)^2 \le \int |\partial_j u|^2\,dx \int |X_j u|^2\,dx$$

if we interpret $\partial_j u$ in the weak sense. The significance of this broadened interpretation of the operators A_j and B_j is clarified in the theory of unbounded self-adjoint operators on Hilbert spaces, as developed in functional analysis. Self-adjoint operators built from the position and momentum operators, in particular the Schrödinger operators, are used in the mathematical construction of quantum mechanics (for example [RS72]). For an interpretation of the Heisenberg uncertainty relation, we refer you to the physics literature.

Exercises

1   Let $a > 0$. Determine the Fourier transform of

$$\text{(i)}\ \sin(ax)/x\,, \qquad \text{(ii)}\ 1/(a^2 + x^2)\,, \qquad \text{(iii)}\ e^{-a\,|x|}\,,$$

$$\text{(iv)}\ (1 - |x|/a)\chi_{[-a,a]}(x)\,, \qquad \text{(v)}\ \big(\sin(ax)/x\big)^2\,.$$

(Hint: See Section VIII.6.)

2   Let $f(x) := (\sin(x)/x)^2$ and $g(x) := e^{2ix}f(x)$ for $x \in \mathbb{R}^{\times}$. Then show $f * g = 0$.
(Hint: Apply Exercise 1 and Theorem 9.16.)

3   Show that if $f \in \mathcal{L}_1$ satisfies either $f * f = f$ or $f * f = 0$, then $f = 0$.

4   Let $\{\,\varphi_\varepsilon\ ;\ \varepsilon > 0\,\}$ be an approximation to the identity, and let (ε_j) be a null sequence. Show that $(\mathcal{F}(\varphi_{\varepsilon_j}))$ converges in $\mathcal{D}'(\mathbb{R}^n)$ to $(2\pi)^{-n/2}\mathbf{1}$.

5   For $a, f \in \mathcal{S}$, show $a(D)f = \widehat{a} * f$.

6   For $s \geq 0$, define $H^s := \{\,u \in L_2\ ;\ \Lambda^s \widehat{u} \in L_2\,\}$ and $(u\,|\,v)_{H^s} := (\Lambda^s \widehat{u}\,|\,\widehat{v})_{L_2}$ for $u, v \in H^s$.
Show

(i) $H^s := \big(H^s\ ;\ (\cdot\,|\,\cdot)_{H^s}\big)$ is a Hilbert space with $H^0 = L_2$, and

$$\mathcal{S} \overset{d}{\hookrightarrow} H^s \overset{d}{\hookrightarrow} H^t \overset{d}{\hookrightarrow} L_2 \quad \text{for } s > t > 0\,;$$

(ii) $H^m = W_2^m$ for $m \in \mathbb{N}$.

7   For $s > n/2$, show

(i) $\mathcal{F}(H^s) \subset L_1$ and

(ii) $H^2 \overset{d}{\hookrightarrow} C_0$ (**Sobolev embedding theorem**).

(Hints: (i) Apply the Cauchy–Schwarz inequality to $\Lambda^s\,|\widehat{u}|\,\Lambda^{-s}$.
(ii) The Riemann–Lebesgue theorem.)

8   Suppose $\sigma \geq 0$, and let $\{\,k_\varepsilon\ ;\ \varepsilon > 0\,\}$ be the Gaussian approximating kernel. Prove:

(i) $T(t) := [f \mapsto k_{\sqrt{t}} * f]$ belongs to $\mathcal{L}(H^\sigma)$ for every $t > 0$.

(ii) $T(t + s) = T(t)T(s),\ \ s, t > 0$.

(iii) $\lim_{t \to 0} T(t)f = f$ for $f \in H^\sigma$.

(iv) $T(t)(L_2) \subset \mathcal{S},\ \ t > 0$.

(v) For $f \in L_2 \cap C$, let $u(t, x) := T(t)f(x)$ for $(t, x) \in [0, \infty) \times \mathbb{R}^n$. Show that u solves the initial value problem of the heat equation in $\mathbb{R}^n$, that is,

$$\partial_t u - \Delta u = 0 \text{ in } (0, \infty) \times \mathbb{R}^n \quad \text{and} \quad u(0, \cdot) = f \text{ on } \mathbb{R}^n\,, \qquad (9.20)$$

in the sense that $u \in C^\infty\big((0, \infty) \times \mathbb{R}^n\big) \cap C(\mathbb{R}^+ \times \mathbb{R}^n)$ and that u satisfies (9.20) pointwise.

Remark Let $T(0) := \mathrm{id}_{H^\sigma}$. Then $\{T(t) \; ; \; t \geq 0\}$ is called the **Gauss–Weierstrass semigroup** (of H^σ).
(Hint: (v) To get an initial value problem for an ordinary differential equation, apply to (9.20) the Fourier transformation with respect to $x \in \mathbb{R}^n$.)

9 Let $n = 1$ and $p_y(x) := \sqrt{2/\pi}\, y/(x^2 + y^2)$ for $(x, y) \in \mathbb{H}^2$. Also let $\sigma \geq 0$. Prove these statements:

(i) $P(y) := [f \mapsto p_y * f]$ belongs to $\mathcal{L}(H^\sigma)$ for every $t > 0$.

(ii) $P(y + z) = P(y)P(z)$ for $y, z > 0$.

(iii) $\lim_{y \to 0} P(y)f = f$ for $f \in H^\sigma$.

(iv) $P(y)(L_2) \subset \mathcal{S}$.

(v) For $f \in L_2 \cap C$, let

$$u(x, y) := \big(P(y)f\big)(x) \quad \text{for } (x, y) \in \mathbb{H}^2 \; .$$

Then u belongs to $C^2(\mathbb{H}^2) \cap C(\overline{\mathbb{H}}^2)$ and solves the Dirichlet boundary value problem for the half plane given by

$$\Delta u = 0 \text{ in } \mathbb{H}^2 \quad \text{and} \quad u(\cdot, 0) = f \text{ on } \mathbb{R} \; .$$

Remark With $P(0) := \mathrm{id}_{H^\sigma}$, we call $\{P(y) \; ; \; y \geq 0\}$ the **Poisson semigroup** (of H^σ).
(Hints: (ii) Exercise 1. (v) Example 9.21(b).)

10 Suppose X is open in $\mathbb{R}^n$ and (X_k) is an ascending sequence of relatively compact open subsets of X with $X = \bigcup_k X_k$ (see Remarks 1.16(d) and (e)). Also let

$$q_k(f) := \max_{|\alpha| \leq k} \|\partial^\alpha f\|_{\infty, \overline{X}_k} \quad \text{for } f \in C^\infty(X) \text{ and } k \in \mathbb{N} \; ,$$

and

$$d(f, g) := \sum_{k=0}^\infty 2^{-k} \frac{q_k(f - g)}{1 + q_k(f - g)} \quad \text{for } f, g \in C^\infty(X) \; .$$

Show that $(C^\infty(X), d)$ is a complete metric space. (Hint: To prove the completeness, apply the diagonal sequence principle (Remark III.3.11(a)).)

11 Show that $\mathcal{D} \overset{d}{\hookrightarrow} C^\infty$ and $\mathcal{S} \overset{d}{\hookrightarrow} C^\infty$.
(Hint: Consider $\varphi(\varepsilon \cdot)$ with a cutoff function φ for $\overline{\mathbb{B}}^n$.)

12 For $f \in \mathcal{D}$, let

$$F(z) := \int e^{-i(z \mid x)_{\mathbb{C}^n}} f(x)\, dx \quad \text{for } z \in \mathbb{C} \; .$$

Show then that F belongs to $C^\omega(\mathbb{C}, \mathbb{C})$.
(Hint: With Remark V.3.4(c) in mind, apply Corollary 3.19.)

13 Show that $\hat{f}$ does not belong to $\mathcal{D}$ for $f \in \mathcal{D} \setminus \{0\}$. (Hint: Recall Exercise 12 and the identity theorem for analytic functions (Theorem V.3.13).)

Chapter XI

Manifolds and differential forms

In Chapter VIII, we learned about Pfaff forms and saw that differential forms of first degree are closely connected with the theory of line integrals. In this chapter, we will treat the higher-dimensional analogue of line integrals, in which differential forms of higher degree are integrated over certain submanifolds of $\mathbb{R}^n$. So this chapter will deal with the theory of differential forms.

In Section 1, we generalize what we know about manifolds. In particular, we explore the concept of a submanifold of a given manifold, and we introduce manifolds with boundary.

In Section 2, we compile the needed results from multilinear algebra. They form the algebraic foundation for the theory of differential forms: In Section 3, we treat differential forms on open subsets of $\mathbb{R}^n$. In Section 4, we make this theory global and then discuss the orientability of manifolds.

Because we always consider submanifolds of Euclidean spaces, we can naturally endow them with a Riemannian metric. In Section 5, we look more closely at this additional structure and explain several basic facts of Riemannian geometry. To accommodate the needs of physics, we also treat semi-Riemannian metrics; in the examples, we will always confine ourselves to Minkowski space.

Section 6, which concludes this chapter, makes the connection between the theory of differential forms and classical vector analysis. In particular, we study the operators gradient, divergence, and curl, and we derive their basic properties. We give their local coordinate representations and calculate these explicitly in several important examples.

In Section 2, which otherwise concerns linear algebra, we also introduce the Hodge star operator, which we will need in later sections to define the codifferential. Then we will be able unify the various operators of vector analysis into the language of the Hodge calculus. This material can be skipped on first reading: For this reason, we wait for the end of each section to discuss any material that uses Hodge theory.

In the entire book, we restrict to submanifolds of $\mathbb{R}^n$. However, apart from the definition of the tangent space, we structure all proofs so that they remain true or can be easily modified for abstract manifolds. Thus Chapters XI and XII give a first introduction to differential topology and differential geometry; though they sometimes lack the full elegance of the general theory, the many examples we consider do form a solid foundation for further study of the subject.

1 Submanifolds

In this section,

- M is an m-dimensional manifold and N is an n-dimensional manifold.

More precisely, this means M is an m-dimensional C^∞ submanifold of $\mathbb{R}^{\overline{m}}$ for some $\overline{m} \geq m$; a like statement holds for N.

For simplicity and to emphasize the essential, we restrict to the study of smooth maps. In particular, we always understand a **diffeomorphism** to be a C^∞ diffeomorphism, and we set

$$\mathrm{Diff}(M, N) := \mathrm{Diff}^\infty(M, N) .$$

However, whenever anything is proved in the following, it will also hold for C^k manifolds and C^k maps, where, if necessary, $k \in \mathbb{N}^\times$ must be restricted appropriately. We will usually put these adjustments in remarks[1] and leave their verification to you.

Definitions and elementary properties

Let $0 \leq \ell \leq m$. A subset L of M is called an (**ℓ-dimensional**) **submanifold of M** if for every $p \in L$ there is a chart (φ, U) of M around p such that[2]

$$\varphi(U \cap L) = \varphi(U) \cap \left(\mathbb{R}^\ell \times \{0\}\right) .$$

Every such chart is a **submanifold chart of M for L**. The number $m - \ell$ is called the **codimension of L in M**.

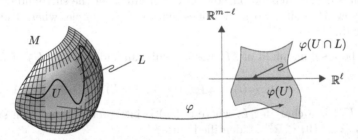

Clearly this definition directly generalizes of the idea of a submanifold of $\mathbb{R}^m$.

In the context of submanifolds, immersions play an important role. They will be introduced in analogy to the definition given Section VII.9.

Let $k \in \mathbb{N}^\times \cup \{\infty\}$. Then $f \in C^k(M, N)$ is a **C^k immersion** if $T_p f : T_p M \to T_{f(p)} N$ is injective for every $p \in M$. We call a C^k immersion f a **C^k embedding**

---

[1] In small print sections entitled "regularity".

[2] To avoid bothersome special cases, we interpret the empty set as a submanifold of dimension ℓ for every $\ell \in \{0, \ldots, m\}$ (see Section VII.9).

of M in N if f is a homeomorphism from M to $f(M)$ (where $f(M)$ is naturally provide with the relative topology of N). Instead of C^∞ immersion [or C^∞
embedding], we say for short **immersion** [or **embedding**].

1.1 Remarks (a) If L is an ℓ-dimensional submanifold of M and M is submanifold
of N, then L is an ℓ-dimensional submanifold of N.

Proof Let $p \in L$, and let (φ, U) be a submanifold chart of M for L around p. Also let
(ψ, V) be a submanifold chart of N for M around p. We can also assume $U = V \cap M$.
Letting $X := \varphi(U) \subset \mathbb{R}^m$ and $Y := \mathrm{pr} \circ \psi(V) \subset \mathbb{R}^m$, where $\mathrm{pr} : \mathbb{R}^m \times \mathbb{R}^{n-m} \to \mathbb{R}^m$
denotes the canonical projection, we have

$$\chi := \mathrm{pr} \circ \psi \circ \varphi^{-1} \in \mathrm{Diff}(X, Y) \ .$$

Now we define $\Phi \in \mathrm{Diff}(Y \times \mathbb{R}^{n-m}, X \times \mathbb{R}^{n-m})$ by

$$\Phi(y, z) := \left(\chi^{-1}(y), z \right) \quad \text{for } (y, z) \in Y \times \mathbb{R}^{n-m} \ ,$$

and set $\Psi := \Phi \circ \psi$. Then $\Psi(V)$ is open in $\mathbb{R}^n$, and $\Psi \in \mathrm{Diff}(V, \Psi(V))$ with

$$\Psi(V \cap L) = \left(\varphi(U \cap L) \times \{0\} \right) \cap \left(\mathbb{R}^\ell \times \{0\} \right) = \Psi(V) \cap \left(\mathbb{R}^\ell \times \{0\} \right) \subset \mathbb{R}^n \ ,$$

as one can easily check. Therefore (Ψ, V) is a submanifold chart of N for L around p. ∎

(b) Because the $\mathbb{R}^{\overline{m}} = \mathbb{R}^{\overline{m}} \times \{0\} \subset \mathbb{R}^n$ is a submanifold of $\mathbb{R}^n$ for $n \geq \overline{m}$, it
follows from (a) that M is an m-dimensional submanifold of $\mathbb{R}^n$ for every $n \geq \overline{m}$.
This shows that the "surrounding space" $\mathbb{R}^{\overline{m}}$ of M does not play an important
role so long as we are only interested in the "inside properties" of M, that is, in
properties that are described only with the help of charts and tangent spaces of
M and which do not depend on how M is "situated" in the surrounding space.[3]
However, how M is situated in $\mathbb{R}^{\overline{m}}$ does matter, for example, when defining the
normal bundle $T^\perp M$.

(c) Let L be a submanifold of M. For the submanifold chart (φ, U) of M for L,
we set

$$(\varphi_L, U_L) := (\varphi \,|\, U \cap L, U \cap L) \ .$$

Then (φ_L, U_L) is a chart for L, where $\varphi(U_L)$ is interpreted as an open subset of
$\mathbb{R}^\ell$, that is, $\mathbb{R}^\ell \times \{0\} \subset \mathbb{R}^m$ is identified with $\mathbb{R}^\ell$.

If $\mathcal{A} := \left\{ (\varphi_\lambda, U_\lambda) \;;\; \lambda \in \Lambda \right\}$ is a set of submanifold charts of M for L such
that L is covered by the coordinate patches (charted territories) $\{ U_\lambda \;;\; \lambda \in \Lambda \}$,
then $\left\{ (\varphi_{\lambda, L}, U_{\lambda, L}) \;;\; \lambda \in \Lambda \right\}$ is an atlas for L, the **atlas induced by** $\mathcal{A}$.

Proof We leave the simple verifications to you. ∎

(d) Suppose L and K are respectively ℓ- and k-dimensional submanifolds of M
and N. Then $L \times K$ is an $(\ell + k)$-dimensional submanifold of the manifold $M \times N$,
which is $(m+n)$-dimensional.

---

[3]In Section 4, it will be clear that tangent spaces also have an "inside" characterization.

Proof This follows simply from the definitions. We again leave the proof to you.[4] ■

(e) Let L be a submanifold of M. Then

$$i : L \to M \ , \quad p \mapsto p$$

is an embedding, the **natural embedding** of L in M; we write it as $i : L \hookrightarrow M$. We identify T_pL for $p \in L$ with its image in T_pM under the injection T_pi, that is, we regard T_pL as a vector subspace of T_pM: $T_pL \subset T_pM$.

Proof Let (φ, U) be a submanifold chart of M for L. Then i has the local representation

$$\varphi \circ i \circ \varphi_L^{-1} : \varphi_L(U_L) \to \varphi(U) \ , \quad x \mapsto (x, 0) \ .$$

Now the claim is clear. ■

(f) If $f : M \to N$ is an immersion, then $m \leq n$.

(g) Let L be a submanifold of M of dimension ℓ, and suppose f belongs to $\mathrm{Diff}(M, N)$. Then $f(L)$ is an ℓ-dimensional submanifold of N.

Proof We leave the simple check to you. ■

(h) Every open subset of M is an m-dimensional submanifold of M.

(i) If (φ, U) is a chart of M, then $\varphi : U \to \mathbb{R}^m$ is an embedding, and φ is a diffeomorphism from U to $\varphi(U)$.

(j) Suppose L and K are respectively submanifolds of M and N, and $i_L : L \hookrightarrow M$ and $i_K : K \hookrightarrow N$ are their respective natural embeddings. Let $k \in \mathbb{N} \cup \{\infty\}$ and $f \in C^k(M, N)$ with $f(L) \subset K$. Then the restriction of f to L satisfies

$$f \,|\, L := f \circ i_L \in C^k(L, K) \ ,$$

and the diagrams

$$
\begin{array}{ccc}
L & \overset{i_L}{\lhook\joinrel\longrightarrow} & M \\
{\scriptstyle f|L}\big\downarrow & & \big\downarrow{\scriptstyle f} \\
K & \underset{i_K}{\lhook\joinrel\longrightarrow} & N
\end{array}
\qquad
\begin{array}{ccc}
T_pL & \overset{T_pi_L}{\longrightarrow} & T_pM \\
{\scriptstyle T_p(f|L)}\big\downarrow & & \big\downarrow{\scriptstyle T_pf} \\
T_{f(p)}K & \underset{T_{f(p)}i_K}{\longrightarrow} & T_{f(p)}N
\end{array}
$$

commute. Identifying T_pL with its image in T_pM under T_pi_L, that is, regarding T_pL in the canonical way as a vector subspace of T_pM, we have in particular $T_p(f\,|\,L) = (T_pf)\,|\,T_pL$.

Proof This follows from obvious changes to the proof of Example VII.10.10(b), which is generalized by this statement. ■

---

[4]See Exercise VII.9.4.

(k) (regularity) Analogous definitions and statements hold when M is a C^k manifold for $k \in \mathbb{N}^\times$. In this case L is also a C^k manifold, and the natural inclusion $i : L \hookrightarrow M$ belongs to the class C^k. ∎

The next theorem, a generalization of Proposition VII.9.10, shows that we can generate submanifolds using embeddings.

1.2 Theorem

(i) *Suppose $f : M \to N$ is an immersion. Then f is locally an embedding, that is, for every p in M, there is a neighborhood U such that $f\,|\,U$ is an embedding.*

(ii) *If $f : M \to N$ is an embedding, then $f(M)$ is an m-dimensional submanifold of N, and f is a diffeomorphism from M to $f(M)$.*

Proof (i) Let $p \in M$, and suppose (φ, U_0) and (ψ, V) are respectively charts of M around p and of N around $f(p)$ such that $f(U_0) \subset V$. Then

$$f_{\varphi,\psi} := \psi \circ f \circ \varphi^{-1} : \varphi(U_0) \to \psi(V)$$

is an immersion by Remark 1.1(i). By the immersion theorem (Theorem VII.9.7), there is an open neighborhood X of $\varphi(p)$ in $\varphi(U_0)$ such that $f_{\varphi,\psi}(X)$ is an m-dimensional submanifold of $\mathbb{R}^n$. Then $\psi \in \mathrm{Diff}(V, \psi(V))$ and Remark 1.1(g) imply that $f(U)$, with $U := \varphi^{-1}(X)$, is an m-dimensional submanifold of N.

By appropriately shrinking X, Remark VII.9.9(d) shows that $f_{\varphi,\psi}$ is a diffeomorphism from $X = \varphi(U)$ to $f_{\varphi,\psi}(X) = \psi \circ f(U)$. Therefore f is a diffeomorphism from U to $f(U)$, where $f(U)$ is provided with the topology induced by N. Therefore $f\,|\,U$ is an embedding.

(ii) Suppose f is an embedding. For $q \in f(M)$, suppose (ψ, V) is a chart of N around q and (φ, U) is a chart of M around $p := f^{-1}(q)$ with $f(U) \subset V$. Because f is topological from M to $f(M)$, we know $f(U)$ is open in $f(M)$. Therefore we can assume that $f(U) = f(M) \cap V$. Now it follows from the proof of (i) that $f(M) \cap V$ is an m-dimensional submanifold of N. Because this is true for every $q \in f(M)$, we conclude $f(M)$ is an m-dimensional submanifold of N.

By (i), f is a local diffeomorphism from M to $f(M)$. Because f is topological, it follows that $f \in \mathrm{Diff}(M, f(M))$. ∎

From Remark VII.9.9(c), we know that the image of an injective immersion is generally not a submanifold. The following theorem gives a simple sufficient condition which tells whether an injective immersion is an embedding.

1.3 Theorem *Suppose M is compact and $f : M \to N$ is an injective immersion. Then f is an embedding, $f(M)$ is an m-dimensional submanifold of N, and $f \in \mathrm{Diff}(M, f(M))$.*

Proof Because M compact and $f(M)$ is a metric space, the bijective continuous map $f : M \to f(M)$ is topological (see Exercise III.3.3). Now the claim follows from Theorem 1.2. ∎

1.4 Remark (regularity) Let $k \in \mathbb{N}^\times$. Then corresponding versions of Theorems 1.2 and 1.3 remain true when M and N are C^k manifolds and f belongs to the class C^k. ∎

1.5 Examples (a) Suppose $1 \le \ell < m$, and let (x, y) denote a general point of $\mathbb{R}^{\ell+1} \times \mathbb{R}^{m-\ell} = \mathbb{R}^{m+1}$. Then

$$L_y := \sqrt{1 - |y|^2}\, S^\ell \times \{y\}$$

is an ℓ-dimensional submanifold of the m-sphere S^m for every $y \in \mathbb{B}^{m-\ell}$. It is diffeomorphic to S^ℓ. The tangent space at the point $p \in L_y$ satisfies

$$T_p L_y = T_p S^m \cap \left(p, \mathbb{R}^{\ell+1} \times \{0\}\right) \subset T_p \mathbb{R}^{m+1} . \tag{1.1}$$

Proof For $y \in \mathbb{B}^{m-\ell}$, the map

$$F_y : \mathbb{R}^{\ell+1} \to \mathbb{R}^{m+1} , \quad x \mapsto \left(\sqrt{1 - |y|^2}\, x, y\right) \tag{1.2}$$

is a smooth immersion. Because S^ℓ and S^m are respectively submanifolds of $\mathbb{R}^{\ell+1}$ and $\mathbb{R}^{m+1}$ and because $F_y(S^\ell) \subset S^m$, Remark 1.1(j) with $i_\ell : S^\ell \hookrightarrow \mathbb{R}^{\ell+1}$ gives

$$f_y := F_y \,|\, S^\ell = F_y \circ i_\ell \in C^\infty(S^\ell, S^m) . \tag{1.3}$$

Clearly f_y is injective, and the chain rule of Remark VII.10.9(b) implies

$$T_p f_y = T_p F_y \circ T_p i_\ell \quad \text{for } p \in S^\ell .$$

Therefore $T_p f_y$ is injective (see Exercise I.3.3), that is, f_y is an immersion. Because S^ℓ is compact, Theorem 1.3 shows that $L_y = f_y(S^\ell)$ is an ℓ-dimensional submanifold of S^m and is diffeomorphic to S^ℓ. Then (1.1) is a simple consequence of (1.2) and (1.3). ∎

(b) (torus-like hypersurfaces of rotation) Let

$$\gamma : S^1 \to (0, \infty) \times \mathbb{R} , \quad t \mapsto \left(\rho(t), \sigma(t)\right)$$

be an injective immersion and therefore by Theorem 1.3 an embedding. Also let $i : S^m \hookrightarrow \mathbb{R}^{m+1}$, and define

$$f : S^m \times S^1 \to \mathbb{R}^{m+1} \times \mathbb{R} , \quad (q, t) \mapsto \left(\rho(t) i(q), \sigma(t)\right) .$$

Then f is an embedding, and

$$T^{m+1} := f(S^m \times S^1)$$

is a hypersurface in $\mathbb{R}^{m+2}$, which is diffeomorphic to $S^m \times S^1$.

In the case $m = 0$, the set T^1 consists of two copies of the closed, smooth curve $\gamma(S^1)$, which has no points of self-intersection[5] and reflects symmetrically about the y-axis.

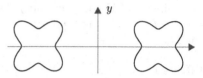

For $m = 1$, T^2 is the surface of rotation in $\mathbb{R}^3$ generated by rotating the meridional curve

$$\Gamma := \big\{ \big(\rho(t), 0, \sigma(t)\big) \; ; \; t \in S^1 \big\}$$

around the z-axis (see Example VII.9.11(e)). T^2 "is a 2-torus", that is, it is diffeomorphic to $\mathsf{T}^2 := S^1 \times S^1$. In particular, $\mathsf{T}^2_{a,r}$, the 2-torus from Example VII.9.11(f), is diffeomorphic to T^2.

In the general case, we call T^{m+1} a **torus-like hypersurface of rotation**.

Proof By Example VII.9.5(b), S^m and S^1 are m- and 1-dimensional manifolds, respectively. Therefore $S^m \times S^1$ is an $(m+1)$-dimensional manifold.

Suppose $(\varphi \times \psi, U \times V)$ is a product chart[6] of $S^m \times S^1$. Because γ is an immersion, its local representation with respect to ψ (and the trivial chart $\mathrm{id}_{\mathbb{R}^2}$ of $\mathbb{R}^2$), that is, $\gamma_\psi = (r, s)$ with $r := \rho \circ \psi^{-1}$ and $s := \sigma \circ \psi^{-1}$, satisfies

$$\big(\dot{r}(y), \dot{s}(y)\big) \neq (0,0) \quad \text{for } y \in \psi(V) . \tag{1.4}$$

Further, the local representation of f with respect to $\varphi \times \psi$ has the form

$$f_{\varphi \times \psi}(x, y) = \big(r(y)g(x), s(y)\big) \quad \text{for } (x, y) \in \varphi(U) \times \psi(V) ,$$

where $g := i \circ \varphi^{-1}$ is the parametrization of S^m belonging to φ. From this is follows that

$$\big[\partial f_{\varphi \times \psi}(x, y)\big] = \left[\begin{array}{c:c} r(y)\partial g(x) & \dot{r}(y)g(x) \\ \hdashline 0 & \dot{s}(y) \end{array} \right] \in \mathbb{R}^{(m+2) \times (m+1)} .$$

Because $r(y) > 0$ and because $\partial g(x)$ is injective, the first m columns of this matrix are linearly independent. If $\dot{s}(y) \neq 0$, then the matrix has rank $m + 1$. If $\dot{s}(y) = 0$, then we have $\dot{r}(y) \neq 0$ by (1.4). From $|g(x)|^2 = (g(x) \,|\, g(x)) = 1$ for $x \in \varphi(U)$, it follows that $(g(x) \,|\, \partial_j g(x)) = 0$ for $1 \leq j \leq m$ and $x \in \varphi(U)$. This shows that the matrix has rank $m + 1$ in this case as well. Therefore f is an immersion.

We now consider the equation $f(q, t) = (y, s)$ for some $(y, s) \in T^{m+1}$. From the relations $\rho(t)i(q) = y$ and $|i(q)| = 1$, it follows that $\rho(t) = |y|$. Because γ is injective, there is exactly one $t \in S^1$ such that $(\rho(t), \sigma(t)) = (|y|, s)$. Likewise, there is exactly one $q \in S^m$ with $i(q) = y/|y|$. Therefore the equation $(\rho(t)i(q), \sigma(t)) = (y, s)$, with (y, s) as above, has a unique solution (since $y = |y| \, (y/|y|)$). Hence f is an injective immersion

---

[5] Here and in the following, "curve" means a one-dimensional manifold (see Remark 1.19(a)).

[6] That is, (φ, U) and (ψ, V) are respectively charts of S^m and S^1, and $\varphi \times \psi(q, t) := (\varphi(q), \psi(t))$.

of $S^m \times S^1$ in $\mathbb{R}^{m+2}$. Now all the claims follow from Theorem 1.3 because $S^m \times S^1$ is compact. ∎

(c) Suppose L and M are submanifolds of N with $L \subset M$. Then L is a submanifold of M.

Proof Because $\mathrm{id}_N \in \mathrm{Diff}(N,N)$, we know $i := \mathrm{id}_N \mid L$ is an immersion of L in N with $i(L) \subset M$. Therefore it follows from Remark 1.1(j) that i is a bijective immersion of L in M. Because L and M carry the topology induced by N and because M induces the same topology on L, we know i, as a restriction of a diffeomorphism, is topological. Therefore i is an embedding, and the claim follows from Theorem 1.2. ∎

(d) Suppose the assumptions of (b) are satisfied with $m = 1$. Then for every $(q_0, t_0) \in S^1 \times S^1$, the images of

$$f(\,\cdot\,, t_0) \colon S^1 \to \mathbb{R}^3$$

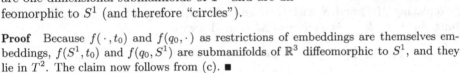

and

$$f(q_0, \cdot) \colon S^1 \to \mathbb{R}^3$$

are one-dimensional submanifolds of T^2 and are diffeomorphic to S^1 (and therefore "circles").

Proof Because $f(\,\cdot\,, t_0)$ and $f(q_0, \cdot)$ as restrictions of embeddings are themselves embeddings, $f(S^1, t_0)$ and $f(q_0, S^1)$ are submanifolds of $\mathbb{R}^3$ diffeomorphic to S^1, and they lie in T^2. The claim now follows from (c). ∎

Submersions

Suppose $f \in C^1(M, N)$. Then we say $p \in M$ is a **regular point** of f if $T_p f$ is surjective. Otherwise p is a **singular point**. A point $q \in N$ is said to be a **regular value** of f if every $p \in f^{-1}(q)$ is a regular point. If every point of M is regular, we say f is a **regular map** or a **submersion**.

These definitions generalize concepts introduced in Section VII.8.

1.6 Remarks **(a)** If p is a regular point of f, then $m \geq n$. Every $q \in N \setminus f(M)$ is a regular value of f.

(b) The point $p \in M$ is a regular point of $f = (f^1, \ldots, f^n) \in C^1(M, \mathbb{R}^n)$ if and only if the cotangent vectors[7]

$$df^j(p) := d_p f^j = \mathrm{pr}_2 \circ T_p f^j \in T_p^* M \quad \text{for } 1 \leq j \leq n$$

are linearly independent.

(c) A singular point of $f \in C^1(M, \mathbb{R})$ is also called a **critical point**. Therefore $p \in M$ is a critical point of f if and only if $df(p) = 0$.[8] ∎

---

[7]See Section VIII.3.
[8]See Remark VII.3.14(a).

The following theorem generalizes the regular value theorem to the case of maps between manifolds.

1.7 Theorem (regular value) *Suppose $q \in N$ is a regular value of the map $f \in C^\infty(M, N)$. Then $L := f^{-1}(q)$ is a submanifold of M of codimension n. For $p \in L$, the kernel of $T_p f$ is $T_p L$.*

Proof Let $p_0 \in f^{-1}(q)$. Let (φ, U) be a chart of M around p_0, and let (ψ, V) be a chart of N around q with $f(U) \subset V$. Then it follows from the chain rule that for every $p \in U \cap f^{-1}(q)$, the point $\varphi(p)$ is a regular point of the local representation

$$f_{\varphi,\psi} := \psi \circ f \circ \varphi^{-1} \in C^\infty(\varphi(U), \mathbb{R}^n) \ .$$

In other words, $y := \psi(q)$ is a regular value of $f_{\varphi,\psi}$. Therefore Theorem VII.9.3 guarantees that $(f_{\varphi,\psi})^{-1}(y)$ is an $(m-n)$-dimensional submanifold of $\mathbb{R}^m$. Hence there are open sets X and Y of $\mathbb{R}^m$ and a $\Phi \in \mathrm{Diff}(X, Y)$ such that

$$\Phi(X \cap (f_{\varphi,\psi})^{-1}(y)) = Y \cap (\mathbb{R}^{m-n} \times \{0\}) \ .$$

By replacing $\varphi(U)$ and X with their intersection, we can assume that $\varphi(U) = X$. But then $\varphi_1 := \Phi \circ \varphi$ is a chart of M around p with

$$\varphi_1\big(f^{-1}(q) \cap U\big) = \Phi \circ \varphi\big(f^{-1} \circ \psi^{-1}(y) \cap U\big)$$
$$= \Phi\big((f_{\varphi,\psi})^{-1}(y) \cap X\big) = Y \cap \big(\mathbb{R}^{m-n} \times \{0\}\big)$$

and is therefore a submanifold chart of M for $f^{-1}(q)$. The second claim now follows from an obvious modification of the proof of Theorem VII.10.7. ∎

1.8 Remarks (a) Theorem 1.7 has a converse that says that every submanifold of M can be represented locally as the fiber of a regular map. More precisely, it says that if L is an ℓ-dimensional submanifold of M, then for every $p \in L$ there are a neighborhood U in M and an $f \in C^\infty(U, \mathbb{R}^{m-\ell})$ such that $f^{-1}(0) = U \cap L$, and 0 is a regular value of f.

Proof Suppose (φ, U) is a submanifold chart of M around p for L. Then the function defined by $f(q) := (\varphi^{\ell+1}(q), \ldots, \varphi^m(q))$ for $q \in U$ belongs to $C^\infty(U, \mathbb{R}^{m-\ell})$ and satisfies $f^{-1}(0) = U \cap L$. Because φ is a diffeomorphism, 0 is a regular value f. ∎

(b) (regularity) If q is a regular value of $f \in C^k(M, N)$ for some $k \in \mathbb{N}^\times$, then $f^{-1}(q)$ is a C^k submanifold of M. In this case it suffices to assume that M is itself a C^k manifold. ∎

1.9 Examples (a) Suppose X is open in $\mathbb{R}^m \times \mathbb{R}^n$ and $q \in \mathbb{R}^n$ is a regular value of $f \in C^\infty(X, \mathbb{R}^n)$ with $M := f^{-1}(q) \neq \emptyset$. Then M is an m-dimensional submanifold of X. For

$$\pi := \mathrm{pr} \,|\, M : M \to \mathbb{R}^m$$

with

$$\mathrm{pr} : \mathbb{R}^m \times \mathbb{R}^n \to \mathbb{R}^m \ , \quad (x, y) \mapsto x \ ,$$

we have $\pi \in C^\infty(M, \mathbb{R}^m)$. Finally let $p \in M$, and suppose $D_1 f(p) \in \mathcal{L}(\mathbb{R}^m, \mathbb{R}^n)$ is surjective.[9] Then p is regular point of π if and only if $D_2 f(p)$ is bijective.

Proof The regular value theorem guarantees that M is an m-dimensional submanifold of X with $T_p M = \ker(T_p f)$ for $p \in M$. Because π is the restriction of a linear and therefore smooth map, it follows from Remark 1.1(j) that $\pi \in C^\infty(M, \mathbb{R}^m)$ and $T_p \pi = T_p \operatorname{pr} | T_p M$.

It follows from $T_p \operatorname{pr} = (p, \partial \operatorname{pr}(p))$ and $\partial \operatorname{pr}(p)(h, k) = h$ for $(h, k) \in \mathbb{R}^m \times \mathbb{R}^n$ that $T_p \pi$ is surjective if and only if for every $y \in \mathbb{R}^m$ there is an $(h, k) \in \mathbb{R}^m \times \mathbb{R}^n$ such that

$$\partial f(p)(h, k) = D_1 f(p) h + D_2 f(p) k = 0$$

and $h = y$. This is because $D_1 f(p)$ is surjective if and only if for every $z \in \mathbb{R}^n$ there is a $k \in \mathbb{R}^n$ such that $D_2 f(p) k = z$ or, equivalently, if and only if $D_2 f(p)$ itself is surjective. Because $D_2 f(p) \in \mathcal{L}(\mathbb{R}^n)$, this finishes the proof. ∎

(b) ("cusp catastrophe") For

$$f : \mathbb{R}^2 \times \mathbb{R} \to \mathbb{R} , \quad ((u, v), x) \mapsto u + vx + x^3 ,$$

we have

$$[D_1 f(w, x)] = [1, x] \in \mathbb{R}^{1 \times 2} , \quad \text{where} \quad w := (u, v) .$$

Therefore 0 is a regular value of f, and $M := f^{-1}(0)$ is a surface in $\mathbb{R}^3$. Because $D_2 f(w, x) = v + 3x^2$, we know by (a) that

$$K := \{ ((u, v), x) \in M ; v + 3x^2 = 0 \}$$

is the set of singular points of the projection $\pi : M \to \mathbb{R}^2$. It satisfies

$$K = \gamma(\mathbb{R}) \quad \text{with}$$
$$\gamma : \mathbb{R} \to \mathbb{R}^3 , \quad t \mapsto (2t^3, -3t^2, t) . \tag{1.5}$$

In particular, K is a 1-dimensional submanifold of M, a smoothly embedded curve. Its projection $B := \pi(K)$ is the image of

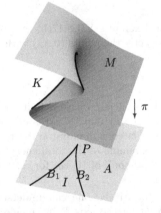

$$\sigma : \mathbb{R} \to \mathbb{R}^2 , \quad t \mapsto (2t^3, -3t^2) ,$$

a Neil parabola.[10] It is the union of the 0-dimensional manifold $P := \{(0, 0)\} \in \mathbb{R}^2$, the "cusp", and the two one-dimensional manifolds $B_1 := \sigma((-\infty, 0))$ and $B_2 := \sigma((0, \infty))$.

Proof The point $(u, v, x) \in \mathbb{R}^3$ belongs to K if and only if it satisfies the equations

$$u + vx + x^3 = 0 \quad \text{and} \quad v + 3x^2 = 0 . \tag{1.6}$$

---

[9]We apply the notations of Section VII.8.
[10]See Remark VII.9.9(a).

By eliminating v from the first equation, we see that (1.6) is equivalent to

$$2x^3 = u \quad \text{and} \quad 3x^2 = -v \ .$$

This proves (1.5). For the derivative of the map

$$g : \mathbb{R}^3 \to \mathbb{R}^2 \ , \quad (u, v, x) \mapsto (u - 2x^3, v + 3x^2) \ ,$$

we find

$$[\partial g(u, v, x)] = \begin{bmatrix} 1 & 0 & -6x^2 \\ 0 & 1 & 6x \end{bmatrix} \in \mathbb{R}^{2 \times 3} \ .$$

This matrix has rank 2, which shows that 0 is a regular value of g. Therefore by the regular value theorem, $K = g^{-1}(0)$ is a 1-dimensional submanifold of $\mathbb{R}^3$. Because $K \subset M$ it follows from Remark 1.5(c) that K is a submanifold of M. The rest is obvious. ∎

1.10 Remark (catastrophe theory) We consider now a point particle of mass 1 moving along the real axis with potential energy U and total energy

$$E(\dot{x}, x) = \frac{\dot{x}^2}{2} + U(x) \quad \text{for } x \in \mathbb{R} \ .$$

According to Example VII.6.14(a), Newton's equation of motion is

$$\ddot{x} = -U'(x) \ .$$

From Examples VII.8.17(b) and (c), we know that the critical points of the energy E are exactly the points $(0, x_0)$ with $U'(x_0) = 0$. Because the Hessian matrix of E has the form

$$\begin{bmatrix} 1 & 0 \\ 0 & U''(x_0) \end{bmatrix}$$

at $(0, x_0)$, it is positive definite if and only if $U''(x_0) > 0$. Hence it follows from Theorem VII.5.14 that $(0, x_0)$ is an isolated minimum of the total energy if and only if x_0 is an isolated minimum of the potential energy.[11] It is graphically clear that an isolated minimum of the total energy is "stable" in the sense that $(\dot{x}(t), x(t))$ stays in "the neighborhood" of $(0, x_0)$ for all $t \in \mathbb{R}^+$ if this is true as its motion begins, that is, at $t = 0$.

Intuitively, one can understand how x will move along the axis $\mathbb{R}$ by imagining a small ball rolling without friction along the graph of U while experiencing the force of gravity. If it lies on the "bottom of a potential well", that is, at a local minimum, then it will not move because $\dot{x}(t) = U'(x_0) = 0$. If the ball is released near a local minimum then ball will roll downhill past the minimum and up the other "slope of the valley" until

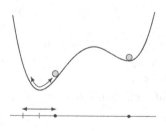

----

[11] We consider only the "generic" case in which $U''(x_0) \neq 0$ is satisfied if $U'(x_0) = 0$.

it "runs out" of kinetic energy at its original height. Then it will reverse course and roll until it again comes to instantaneous rest where it was initially released. Thus the ball will execute a periodic oscillation about x_0.[12]

Now we assume that U depends continuously on additional "control parameters" $u, v, \ldots$. By varying these parameters, we can vary the graph of U continuously. In this way, it can happen that a local minimum merges first into a saddle point and then ceases to be a critical point. A ball that had previously been confined to the neighborhood of the local minimum would then leave this neighborhood and oscillate about another resting point.

Now consider an observer who can see the ball move but is unaware of the mechanism underlying the process. She would see that the ball, which had before rested peacefully at a certain place, would suddenly, "for no apparent reason", begin to roll and oscillate periodically about another (fictitious) center. It would seem to be a sudden and drastic change of the situation, a "catastrophe".

In order to understand such catastrophes (and avoid them if necessary), one must understand the mechanism by which they occur. In the situation described above, this boils down to understanding how the critical points of the potential (and in particular the relative minima) depend on the control parameters.

To illustrate, we consider the potential

$$U_{(u,v)} : \mathbb{R} \to \mathbb{R} , \quad x \mapsto ux + vx^2/2 + x^4/4$$

for $(u, v) \in \mathbb{R}^2$. The critical points of $U_{(u,v)}$ are just the zeros of the function f from Example 1.9(b). Therefore the manifold M, the *catastrophe manifold*, describes all critical points of the two parameter set $\{ U_{(u,v)} ; (u, v) \in \mathbb{R}^2 \}$ of potentials. Of particular interest is that subset of M, the *catastrophe set* K, consisting of all singular points of the projection π from M to the parameter space. In our example, K is a curve smoothly embedded in M, the *fold curve*, because the catastrophe manifold is "folded" along K. The image of K under π, that is, the projection of the fold curve onto the parameter plane, is the *bifurcation set* B. Every point of $\mathbb{R}^2 \setminus B$ is a regular point of π. The fiber $\pi^{-1}(u, v)$ consists of exactly one point for $(u, v) \in A \cup P$, exactly two points for $(u, v) \in B_1 \cup B_2$, and exactly three points for $(u, v) \in I$, where A and I are depicted in the illustration to Example 1.9(b). The following pictures show the qualitative form of the potential $U_{(u,v)}$ when (u, v) belongs to these sets.

---

[12]This plausible scenario can be proved using the theory of ordinary differential equations; see for example [Ama95].

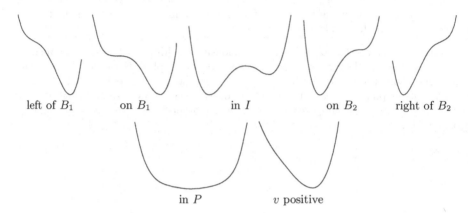

| left of B_1 | on B_1 | in I | on B_2 | right of B_2 |

| in P | v positive |

Now consider a continuous curve C in the parameter space that begins in A and ends in I (or the reverse), while staying in $B_1 \cup B_2$. While moving continuously along this curve, the number of points in the inverse image of π will change suddenly from 1 to 3 (or from 3 to 1). As illustrated at right, one such curve C is obtained by projecting a curve Γ on the catastrophe manifold M that "jumps" when crossing the fold curve. In short, the value of x experiences a "catastrophe".

These facts have led to many inter-pretations of "catastrophe theory" which —not least because of its name— have been leveraged to great popularity and, especially in the popularized science lit-erature, have kindled exaggerated hopes that the subject will somehow explain or help prevent real-world catastrophes. We refer to [Arn84] for a critical, nontechnical introduction to catastrophe theory, and we recommend [PS78] for a detailed pre-sentation and several applications of the mathematical theory of singularities, of which catastrophe theory is a part. ∎

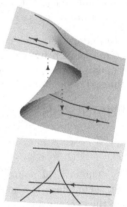

Submanifolds with boundary

We know that the open unit ball $\mathbb{B}^m$ and its boundary, the $(m-1)$-sphere S^{m-1}, are respectively m- and $(m-1)$-dimensional submanifolds of $\mathbb{R}^m$. However, the closed ball $\overline{\mathbb{B}}^m = \mathbb{B}^m \cup S^{m-1}$ is not a manifold, because a point $p \in \partial\mathbb{B}^m = S^{m-1}$ has no neighborhood U in $\overline{\mathbb{B}}^m$ that is mapped topologically onto an open set V of $\mathbb{R}^m$; such a neighborhood U, as the homeomorphic image of an open set V, would likewise need to be open in $\mathbb{R}^m$, which is not true. In the neighborhood of p, that is, "by viewing it with a very strong microscope", $\overline{\mathbb{B}}^m$ does not look like $\mathbb{R}^m$, but rather like a half-space. To capture such situations also, we must generalize the idea of a manifold by allowing subsets of half-spaces to be parameter sets.

In the following, $m \in \mathbb{N}^{\times}$, and

$$\mathbb{H}^m := \mathbb{R}^{m-1} \times (0, \infty)$$

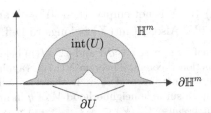

is the **open upper half-space** of $\mathbb{R}^m$. We identify its boundary $\partial \mathbb{H}^m = \mathbb{R}^{m-1} \times \{0\}$ with $\mathbb{R}^{m-1}$ if there is no fear of misunderstanding. If U is an open subset of $\overline{\mathbb{H}}^m := \overline{\mathbb{H}^m} = \mathbb{R}^{m-1} \times \mathbb{R}^+$, we call $\mathrm{int}(U) := U \cap \mathbb{H}^m$ the **interior** and $\partial U := U \cap \partial \mathbb{H}^m$ the **boundary** of U. Note that the boundary ∂U is not the topological boundary[13] of U either in $\overline{\mathbb{H}}^m$ or in $\mathbb{R}^m$ (unless $U = \mathbb{H}^m$ in the latter case).

Suppose X is open in $\overline{\mathbb{H}}^m$ and E is a Banach space. Then $f : X \to E$ is said to be **differentiable** at the boundary point $x_0 \in \partial X$ if there is a neighborhood U of x_0 in $\mathbb{R}^m$ and a differentiable function $f_U : U \to E$ that agrees with f in $U \cap X$. Then it follows from Proposition VII.2.5 that

$$\partial_j f_U(x_0) = \lim_{t \to 0+} \big(f_U(x_0 + te_j) - f_U(x_0)\big)/t$$
$$= \lim_{t \to 0+} \big(f(x_0 + te_j) - f(x_0)\big)/t$$

for $1 \le j \le m$, where $(e_1, \dots, e_m)$ is the standard basis of $\mathbb{R}^m$. This and Proposition VII.2.8 show that $\partial f_U(x_0)$ is already determined by f. Therefore the **derivative**

$$\partial f(x_0) := \partial f_U(x_0) \in \mathcal{L}(\mathbb{R}^m, E)$$

of f is well defined at x_0, that is, independent of the choice of the local continuation f_U of f.

A map $f : X \to E$ is said to be **continuously differentiable** if f is differentiable at every point of X and if the map

$$\partial f : X \to \mathcal{L}(\mathbb{R}^m, E), \quad x \mapsto \partial f(x)$$

is continuous.[14]

The higher derivatives of f are defined analogously, and these are also independent of the particular local continuation. For $k \in \mathbb{N}^{\times} \cup \{\infty\}$, the C^k maps of X to E form a vector space, which, as in the case of open subsets of $\mathbb{R}^m$, we denote by $C^k(X, E)$.

Suppose Y is open in $\overline{\mathbb{H}}^m$. Then $f : X \to Y$ is also called a C^k **diffeomorphism**, and we write $f \in \mathrm{Diff}^k(X, Y)$, if f is bijective and if f and f^{-1} belong to the class C^k. In particular, $\mathrm{Diff}(X, Y) := \mathrm{Diff}^{\infty}(X, Y)$ is the set of all smooth, that is, C^{∞}, diffeomorphisms from X to Y.

---

[13]From this point on, we use the symbol ∂M exclusively for boundaries, and, for clarity, we write $\mathrm{Rd}(M)$ for the topological boundary of a subset M of a topological space, that is, we put $\mathrm{Rd}(M) := \overline{M} \setminus \mathring{M}$.

[14]Naturally, we say f is differentiable at $x_0 \in \mathrm{int}(X)$ if $f \,|\, \mathrm{int}(X)$ is differentiable at x_0.

1.11 Remarks   Suppose X and Y are open in $\overline{\mathbb{H}}^m$ and $f : X \to Y$ is a C^k diffeomorphism for some $k \in \mathbb{N}^\times \cup \{\infty\}$.

(a) If ∂X is not empty, then $\partial Y \neq \emptyset$, and $f \mid \partial X$ is a C^k diffeomorphism from ∂X to ∂Y. Also, $f \mid \mathrm{int}(X)$ belongs to $\mathrm{Diff}^k\big(\mathrm{int}(X), \mathrm{int}(Y)\big)$.

Proof   Suppose $p \in \partial X$ and $q := f(p)$ belongs to $\mathrm{int}(Y)$. Then it follows from the inverse function theorem, Theorem VII.7.3, (applied to a local extension of f) that $\partial f^{-1}(q)$ is an automorphism of $\mathbb{R}^m$. It therefore follows, again from Theorem VII.7.3, that f^{-1} maps a suitable neighborhood V of q in $\mathrm{int}(Y)$ to an open neighborhood U of p in $\mathbb{R}^m$. But because $f^{-1}(V) \subset X \subset \overline{\mathbb{H}}^m$ and $p = f^{-1}(q) \in \partial X$, this is not possible. Therefore $f(\partial X) \subset \partial Y$. Analogously we find $f^{-1}(\partial Y) \subset \partial X$. This shows $f(\partial X) = \partial Y$.

Because X and Y in $\overline{\mathbb{H}}^m$ are open, both ∂X and ∂Y are open in $\partial \mathbb{H}^m = \mathbb{R}^{m-1}$, and $f \mid \partial X$ is a bijection from ∂X to ∂Y. Because $f \mid \partial X$ and $f^{-1} \mid \partial Y$ obviously belongs to the class C^k, we know $f \mid \partial X$ is a C^k diffeomorphism from ∂X to ∂Y. The last statement is now clear. ∎

(b) For $p \in \partial X$, we have $\partial f(p)(\partial \mathbb{H}^m) \subset \partial \mathbb{H}^m$ and $\partial f(p)(\pm \mathbb{H}^m) \subset \pm \mathbb{H}^m$.

Proof   From $f(\partial X) = \partial Y$, it follows that $f^m \mid \partial X = 0$ for the m-th coordinate function f^m of f. From this we get $\partial_j f^m(p) = 0$ for $1 \leq j \leq m-1$. Therefore the Jacobi matrix of f has at p the form

$$
[\partial f(p)] = \left[
\begin{array}{ccc:c}
 & & & \partial_m f^1(p) \\
 & \partial(f \mid \partial \mathbb{H}^m)(p) & & \vdots \\
 & & & \partial_m f^{m-1}(p) \\
\hdashline
0 & \cdots & 0 & \partial_m f^m(p)
\end{array}
\right] .
\tag{1.7}
$$

Because $f(X) \subset Y \subset \overline{\mathbb{H}}^m$, the inequality $f^m(q) \geq 0$ holds for $q \in X$. Hence we find

$$
\partial_m f^m(p) = \lim_{t \to 0+} t^{-1}\big(f^m(p + te_m) - f^m(p)\big) = \lim_{t \to 0+} t^{-1} f^m(p + te_m) \geq 0 .
$$

Since $\partial f(p) \in \mathcal{L}\mathrm{aut}(\mathbb{R}^m)$ (see Remark VII.7.4(d)) and since $\partial_m f^m(p) \geq 0$, we have $\partial_m f^m(p) > 0$. From (1.7), we read off

$$
\big(\partial f(p)x\big)^m = \partial_m f^m(p)t \quad \text{for } x := (y, t) \in \mathbb{R}^{m-1} \times \mathbb{R} .
$$

Therefore the sign of the m-th coordinate of $\partial f(p)x$ agrees with $\mathrm{sign}(t)$, and we are done. ∎

We can now define the concept of submanifold with boundary. A subset B of the n-dimensional manifold N is said to be a **b-dimensional submanifold of N with boundary** if for every $p \in B$ there is a chart (ψ, V) of N around p, a **submanifold chart** of N around p for B, such that

$$
\psi(V \cap B) = \psi(V) \cap \big(\overline{\mathbb{H}}^b \times \{0\}\big) \subset \mathbb{R}^n .
\tag{1.8}
$$

Here we say p is a **boundary point** of B if $\psi(p)$ lies in $\partial\mathbb{H}^b := \partial\mathbb{H}^b \times \{0\}$.

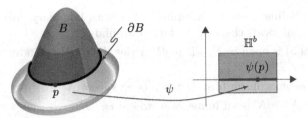

The set of all boundary points forms the **boundary**[15] ∂B of B. The set $\text{int}(B) := B \backslash \partial B$ is called the **interior** of the submanifold B with boundary. Finally B is a **hypersurface in N with boundary** if $b = n - 1$.

1.12 Remarks

(a) Every submanifold M of N, in the sense given in the beginning of this section, is a submanifold with boundary, but with an empty boundary. We call such objects **(sub)manifolds without boundary**.

(b) The boundary ∂B and the interior $\text{int}(B)$ are well defined, that is, independent of charts.

Proof Suppose (χ, W) is another submanifold chart of N around p for B. Also let f be the restriction of the transition function $\chi \circ \psi^{-1}$ to $\psi(V \cap W) \cap (\overline{\mathbb{H}}^b \times \{0\})$, understood as an open subset of $\overline{\mathbb{H}}^b$. Then it follows from Remark 1.11(a) that $\chi(p)$ belongs to $\partial\mathbb{H}^b$ if and only if $\psi(p)$ does. ∎

(c) Suppose $p \in \text{int}(B)$. Then (1.8) implies

$$\psi\big(V \cap \text{int}(B)\big) = \psi(V) \cap \big(\mathbb{H}^b \times \{0\}\big) .$$

Because $\mathbb{H}^b$ is diffeomorphic to $\mathbb{R}^b$, this shows that $\text{int}(B)$ is a b-dimensional submanifold of N without boundary.

(d) In the case $p \in \partial B$, it follows from (1.8) that

$$\psi(V \cap \partial B) = \psi(V) \cap \big(\mathbb{R}^{b-1} \times \{0\}\big) .$$

Therefore ∂B is a $(b-1)$-dimensional submanifold of N without boundary.

(e) Every b-dimensional submanifold of N with boundary is a b-dimensional submanifold of $\mathbb{R}^{\bar{n}}$ with boundary.

Proof This follows in analogy to the proof of Remark 1.1(a). ∎

(f) (regularity) It is clear how C^k submanifolds with boundary are defined for $k \in \mathbb{N}^\times$, and that the analogues of (a)–(c) remain true. ∎

---

[15]Note that the boundary ∂B and the interior $\text{int}(B)$ are generally different from the topological boundary $\text{Rd}(B)$ and the topological interior $\mathring{B}$ of B. In the context of statements about manifolds, we will understand "boundary" and "interior" in the sense of the definitions above.

Local charts

Suppose B is a b-dimensional submanifold of N with boundary. We call the map φ a (b-dimensional local) **chart** of (or for) B around p if

- $U := \mathrm{dom}(\varphi)$ is open in B, where B carries the topology induced by N (and therefore by $\mathbb{R}^{\overline{n}}$).

- φ is a homeomorphism from U to an open subset X of $\overline{\mathbb{H}}^b$.

- $i_B \circ \varphi^{-1} : X \to N$ is an immersion, where $i_B : B \to N$, $p \mapsto p$ denotes the injection.

Note that except for the fact that $\varphi(U)$ is open *in* $\overline{\mathbb{H}}^b$ and $\mathbb{R}^n$ is replaced by N, this definition agrees literally with the definition of a C^∞ chart of a submanifold of $\mathbb{R}^n$ (see Section VII.9).

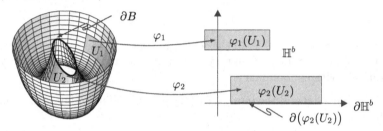

1.13 Remarks   (a) If (ψ, V) is a submanifold chart of N for B, the intersected chart $(\varphi, U) := (\psi \,|\, V \cap B, V \cap B)$ is a b-dimensional chart for B.

(b) If (φ_1, U_1) and (φ_2, U_2) are charts of B around $p \in B$, then $\varphi_j(U_1 \cap U_2)$ is open in $\overline{\mathbb{H}}^b$ for $j = 1, 2$, and **transition function** $\varphi_2 \circ \varphi_1^{-1}$ satisfies

$$\varphi_2 \circ \varphi_1^{-1} \in \mathrm{Diff}\big(\varphi_1(U_1 \cap U_2), \varphi_2(U_1 \cap U_2)\big) \ .$$

(c) Suppose (φ, U) is a chart for B around $p \in \partial B$. Then

$$(\varphi_{\partial B}, U_{\partial B}) := (\varphi \,|\, U \cap \partial B, U \cap \partial B)$$

is a chart for ∂B, a $(b-1)$-dimensional submanifold of N without boundary.

(d) All concepts and definitions, for example, differentiability of maps and local representations, that can be described using charts of manifolds, carry over straightforwardly to submanifolds with boundary. In particular, $i_B : B \hookrightarrow N$, that is, the natural embedding $p \mapsto p$ of B in N, is a smooth map.

(e) If C is a submanifold of M with boundary and $f \in \mathrm{Diff}(B, C)$, then $f(\partial B) = \partial C$, and $f \,|\, \partial B$ is a diffeomorphism from ∂B to ∂C.

Proof   This follows from Remark 1.11(a). ∎

(f) Suppose B is a b-dimensional submanifold of N with boundary, and $f \in C^\infty(B, M)$ is an **embedding**, that is, f is a bijective immersion and a homeomorphism from B to $f(B)$. Then $f(B)$ is a b-dimensional submanifold M with boundary satisfying $\partial f(B) = f(\partial B)$, and f is a diffeomorphism from B to $f(B)$.

Proof The proof of Theorem 1.2(ii) also applies here. ■

(g) (regularity) All previous statements transfer literally to C^k submanifolds with boundary. ■

Naturally, we again say a family $\{(\varphi_\alpha, U_\alpha) \; ; \; \alpha \in A\}$ of charts of B with $B = \bigcup_\alpha U_\alpha$ is an **atlas** of B.

Tangents and normals

Suppose B is a submanifold of N with boundary, and let $p \in \partial B$. Also suppose (φ, U) is a chart of B around p. Then we define the **tangent space** $T_p B$ of B at the point p by

$$T_p B := T_{\varphi(p)}(i_B \circ \varphi^{-1})(T_{\varphi(p)} \mathbb{R}^b) \;,$$

where $b := \dim(B)$. Therefore $T_p B$ is a ("full") b-dimensional vector subspace of the tangent space $T_p N$ of N at p (and not, say, a half-space). An obvious modification of the proof of Remark VII.10.3(a) shows that $T_p B$ is well defined, that is, independent of which chart is used. In this case, we define the **tangent bundle** TB of B by $TB := \bigcup_{p \in B} T_p B$.

1.14 Remarks **(a)** For $p \in \partial B$, $T_p \partial B$ is a $(b-1)$-dimensional vector subspace of $T_p B$.

Proof This is a simple consequence of Remarks 1.12(d) and 1.13(c). ■

(b) Suppose $p \in \partial B$ and (φ, U) is chart of B around p. Letting

$$T_p^\pm B := T_{\varphi(p)}(i_B \circ \varphi^{-1})(\varphi(p), \pm \overline{\mathbb{H}}^b) \;,$$

we have $T_p B = T_p^+ B \cup T_p^- B$ and $T_p^+ B \cap T_p^- B = T_p(\partial B)$. The vector v is an **inward pointing** [or an **outward pointing**] tangent vector if and only if v belongs to the set $T_p^+ B \backslash T_p(\partial B)$ [or $T_p^- B \backslash T_p(\partial B)$]. This is the case if and only if the b-th component of $(T_p \varphi) v$ is positive [or negative].

Proof From Remarks 1.11(b) and 1.13(b), it follows easily that $T_p^\pm B$ is defined in a coordinate-independent way. ■

(c) Let C be a submanifold of M with or without boundary. For $f \in C^1(C, N)$, the **tangential** $T_p f$ of f at $p \in C$ is defined as in the case of manifolds without boundary. Then the analogues of Remarks VII.10.9 remain true. ■

Suppose $p \in \partial B$. Then $T_p(\partial B)$ is a $(b-1)$-dimensional vector subspace of the b-dimensional vector space $T_p B$. As a vector subspace of $T_p N$ (and therefore of $T_p \mathbb{R}^{\bar{n}}$), $T_p B$ is an inner product space with the inner product $(\,\cdot\,|\,\cdot\,)_p$ induced by the Euclidean scalar product on $\mathbb{R}^{\bar{n}}$. Hence there is exactly one unit vector $\nu(p)$ in $T_p^- B$ that is orthogonal to $T_p(\partial B)$, and we call it the **outward (unit) normal vector** of ∂B at p. Clearly $-\nu(p) \in T_p^+ B$ is the unique inward pointing vector of $T_p B$ that is orthogonal to $T_p(\partial B)$, and we call it the **inward (unit) normal** vector ∂B at p.

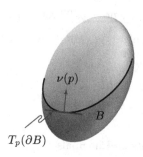

The regular value theorem

We have already seen that submanifolds without boundary can be represented in many cases (actually always, locally) as fibers of regular maps. We will now extend this important and simple criterion to the case of submanifolds with boundary.

1.15 Theorem (regular value) *Suppose c is a regular value of $f \in C^\infty(N, \mathbb{R})$. Then*

$$B := f^{-1}\big((-\infty, c]\big) = \big\{\, p \in N \;;\; f(p) \le c \,\big\}$$

is an n-dimensional submanifold of N with boundary with $\partial B = f^{-1}(c)$ and $\mathrm{int}(B) = f^{-1}\big((-\infty, c)\big)$. For $p \in \partial B$, we have $T_p(\partial B) = \ker(d_p f)$, and the outward unit normal $\nu(p)$ on ∂B is given by $\nabla_p f(p) / |\nabla_p f|_p$.

Proof Because $f^{-1}((-\infty, c))$ is open in N and is therefore an n-dimensional submanifold of N, it suffices to consider $p \in f^{-1}(c)$.

Therefore let $p \in f^{-1}(c)$, and let (ψ, V) be a chart of N around p such that $\psi(p) = 0$. Then $g := c - f \circ \psi^{-1}$ belongs to $C^\infty(\psi(V), \mathbb{R})$ and satisfies $g(0) = 0$ and $g(x) \ge 0$ if and only if x lies in $\psi(V \cap B)$. Also 0 is a regular point of g. By renaming the coordinates (that is, by composing ψ with a permutation), we can assume that $\partial_n g(0) \ne 0$ and therefore $\partial_n g(0) > 0$.

Consider the map $\varphi \in C^\infty(\psi(V), \mathbb{R}^n)$ defined by $\varphi(x) := (x^1, \ldots, x^{n-1}, g(x))$. It satisfies $\varphi(0) = 0$ and

$$\partial \varphi = \left[\begin{array}{ccc:ccc} & & & & & 0 \\ & 1_{n-1} & & & & \vdots \\ & & & & & 0 \\ \hdashline \partial_1 g & \cdots & \partial_{n-1} g & & : & \partial_n g \end{array}\right].$$

Therefore $\partial\varphi(0)$ is an automorphism of $\mathbb{R}^n$, and Theorem VII.7.3 (the inverse function theorem) guarantees the existence of open neighborhoods U and W of 0 in $\psi(V)$ such that $\varphi\,|\,U$ is a diffeomorphism from U to W.

Letting $V_0 := \psi^{-1}(U)$ and $\chi := \varphi \circ \psi\,|\,V_0$, we see that (χ, V_0) is a chart of N around p with $\chi(p) = 0$ and $\chi(B \cap V_0) = \chi(V_0) \cap \overline{\mathbb{H}}^n$. This shows that B is a submanifold of N with boundary with $\partial B = f^{-1}(c)$ and $\text{int}(B) = f^{-1}((-\infty, c))$. Thus we get from Theorem 1.7 that

$$T_p(\partial B) = \ker(T_p f) = \ker(d_p f) \quad \text{for } p \in \partial B . \tag{1.9}$$

Because $\langle d_p f, v \rangle_p = (\nabla_p f \,|\, v)_p$ for $v \in T_p N$, it follows from (1.9) that $\nabla_p f$ is orthogonal to $T_p(\partial B)$.

Finally, let $\lambda \colon (-\varepsilon, \varepsilon) \to N$ be a C^1 path in N with $\lambda(0) = p$ and $\dot{\lambda}(0) = \nabla_p f$ (see Theorem VII.10.6). Then

$$(f \circ \lambda)^{\cdot}(0) = \langle d_p f, \nabla_p f \rangle = |\nabla_p f|_p^2 > 0 .$$

Therefore we derive from the Taylor formula of Corollary IV.3.3 that

$$f(\lambda(t)) = c + t\,|\nabla_p f|_p^2 + o(t) \quad (t \to 0) .$$

Therefore $f(\lambda(t)) > c$, that is, $f(\lambda(t)) \notin B$ for sufficiently small positive t. This implies that $\nabla_p f$ is an outward pointing tangent vector of B at p. Now the last claim is also clear. $\blacksquare$

1.16 Remarks (a) Because we can locally represent submanifolds as fibers of regular maps (see Remark 1.8(a)), we can also locally represent submanifolds with boundary as inverse images of half open intervals. More precisely, suppose B is an n-dimensional submanifold of N with boundary. Then there is for every point $p \in B$ a neighborhood U in N and a function $f \in C^\infty(U, \mathbb{R})$ such that $B \cap U = f^{-1}((-\infty, 1))$ if $p \in \text{int}(B)$ but $f(p) = 0$ and $B \cap U = f^{-1}((-\infty, 0])$ if $p \in \partial B$, and for which 0 is a regular value.

Proof Suppose (φ, U) is a submanifold chart of N around p for B with $\varphi(p) = 0$. We can assume that $\varphi(U)$ is contained in $\mathbb{B}_\infty^n$. If p is an interior point of B, we set $f(q) := \varphi^n(q)$ for $q \in U$. Then f belongs to $C^\infty(U, \mathbb{R})$, and $f^{-1}((-\infty, 1)) = U$. If p belongs to ∂B, we set $f(q) := -\varphi^n(q)$ for $q \in U$. Then $f(p) = 0$, and $f^{-1}((-\infty, 0]) = U \cap B$. Because $\varphi \in \text{Diff}(U, \varphi(U))$, we know f is a submersion. Therefore 0 is a regular value of f. $\blacksquare$

(b) (regularity) Suppose c is a regular value of $f \in C^k(N, \mathbb{R})$ for some $k \in \mathbb{N}^\times$. Then $f^{-1}((-\infty, c])$ is an n-dimensional submanifold in N with boundary. In this case, one need only assume that N is a C^k manifold. $\blacksquare$

1.17 Examples (a) For every $r > 0$, $\overline{\mathbb{B}}_r^n := r\overline{\mathbb{B}}^n = \{ x \in \mathbb{R}^n \,;\, |x| \leq r \}$ is an n-dimensional submanifold of $\mathbb{R}^n$ with boundary. Its boundary coincides with the topological boundary and therefore with the $(n-1)$-sphere of radius r, that is, $\partial\overline{\mathbb{B}}_r^n = rS^{n-1}$. The outward normal $\nu(p)$ at $p \in \partial\overline{\mathbb{B}}_r^n$ is given by $(p, p/|p|)$.

In the case $n = 1$, the ball $\bar{\mathbb{B}}_r^1$ is the closed interval $[-r, r]$ in $\mathbb{R}$, and the 0-sphere with radius r is given by $S_r^0 = \{-r\} \cup \{r\}$. The outward normal has $\nu(-r) = (-r, -1)$ and $\nu(r) = (r, 1)$.

$$\nu(-r) \qquad\qquad \bar{\mathbb{B}}_r^1 \qquad\qquad \nu(r)$$

$$\underset{-r}{\longleftarrow} \qquad\qquad \underset{0}{} \qquad\qquad \underset{r}{\longrightarrow}$$

Proof This follows from Theorem 1.15 with $N := \mathbb{R}^n$ and $f(x) := |x|^2$ for $x \in \mathbb{R}^n$. ∎

(b) Suppose $A \in \mathbb{R}_{\mathrm{sym}}^{(n+1)\times(n+1)}$ and $c \in \mathbb{R}^\times$. Also suppose

$$V_c := \left\{ x \in \mathbb{R}^{n+1} \ ; \ (Ax \,|\, x) \leq c \right\}$$

is not empty. If A is positive definite and $c > 0$, then V_c is an $(n+1)$-dimensional solid whose boundary is the n-dimensional ellipsoid

$$K_c := \left\{ x \in \mathbb{R}^{n+1} \ ; \ (Ax \,|\, x) = c \right\} .$$

If A is negative definite and $c < 0$, then V_c is the complement of the interior of V_{-c}, and the boundary of V_{-c} is the n-dimensional ellipsoid K_{-c}. If A is indefinite but invertible, then V_c is the "interior" or "exterior" of an appropriate n-dimensional hyperboloid K_c that bounds V_c. In every case, $Ax/|Ax|$ is the outward normal of V_c at K_c. (Compare this with Remark VII.10.18, and interpret the pictures there accordingly.)

(c) Suppose $A \in \mathbb{R}^{(n+1)\times(n+1)}$ is symmetric and $c \in \mathbb{R}^\times$ with $K_c \neq \emptyset$. Also suppose $v \in \mathbb{R}^{n+1}\setminus\{0\}$ and $\alpha, \beta \in \mathbb{R}$ with $\alpha < \beta$. Then

$$B := \left\{ x \in K_c \ ; \ \alpha \leq (v \,|\, x) \leq \beta \right\}$$

is the part of K_c that lies between the two parallel hyperplanes

$$H_\gamma := \left\{ x \in \mathbb{R}^{n+1} \ ; \ (v \,|\, x) = \gamma \right\} \quad \text{for } \gamma \in \{\alpha, \beta\} .$$

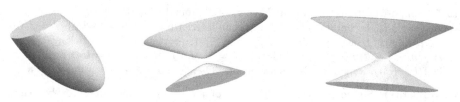

If H_α and H_β are not tangent hyperplanes of K_c, then B is an n-dimensional submanifold of K_c with boundary with

$$\partial B = \left\{ x \in K_c \ ; \ (v \,|\, x) \in \{\alpha, \beta\} \right\} .$$

Proof Because the map $g := (v \,|\, \cdot) \,|\, K_c : K_c \to \mathbb{R}$ is smooth by Remark 1.1(j), we know $g^{-1}((\alpha, \beta))$ is open in K_c. Therefore $g^{-1}((\alpha, \beta))$ is an n-dimensional submanifold of K_c.

Hence it suffices to show that every $p \in g^{-1}(\{\alpha, \beta\})$ is a boundary point of B. So let V be an open neighborhood in K_c of $p \in g^{-1}(\beta)$ such that $g^{-1}(\alpha) \cap V = \emptyset$. The assumption at H_β is not a tangent hyperplane implies that β is a regular value of $f := g \,|\, V$ (prove this!). The claim now follows from Theorem 1.15 applied to the manifold V and the function f. A similar argument shows that every $p \in g^{-1}(\alpha)$ is a boundary point of B. ∎

(d) (cylinder-like rotational hypersurfaces) Suppose

$$\gamma \colon [0,1] \to (0,\infty) \times \mathbb{R} \ , \quad t \mapsto \big(\rho(t), \sigma(t)\big)$$

is a smooth embedding. Also let $i \colon S^m \hookrightarrow \mathbb{R}^{m+1}$ and

$$f \colon S^m \times [0,1] \to \mathbb{R}^{m+1} \times \mathbb{R} \ , \quad (q,t) \mapsto \big(\rho(t)i(q), \sigma(t)\big) \ .$$

Then f is a smooth embedding, and

$$Z^{m+1} := f\big(S^m \times [0,1]\big)$$

is a hypersurface in $\mathbb{R}^{m+2}$ with boundary which is diffeomorphic to the "spherical cylinder" $S^m \times [0,1]$.

In the case $m = 0$, Z^1 consists of two copies of smooth, non-self-intersecting, compact curves[16] $\gamma([0,1])$ that are symmetric about the y-axis.

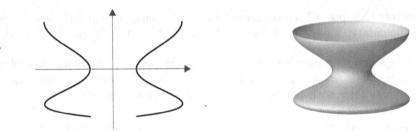

For $m = 1$, Z^2 is the surface of rotation in $\mathbb{R}^3$ obtained by rotating the **meridian curve**

$$\Gamma := \{\, (\rho(t), 0, \sigma(t)) \ ; \ t \in [0,1] \,\}$$

around the z-axis.

In the general case, we call Z^{m+1} a **cylinder-like surface of rotation with boundary**. Its boundary satisfies

$$\partial Z^{m+1} = f\big(S^m \times \{0\}\big) \cup f\big(S^m \times \{1\}\big) \ ,$$

while its interior has

$$\mathrm{int}(Z^{m+1}) = f\big(S^m \times (0,1)\big) \ .$$

---

[16]That is, one-dimensional manifolds with boundary.

In particular, $\mathrm{int}(Z^{m+1})$ is a cylinder-type hypersurface of rotation without boundary. In the case $m = 1$, it is generated by rotating the meridian curve

$$\mathrm{int}(\Gamma) := \big\{ \, (\rho(t), 0, \sigma(t)) \; ; \; 0 < t < 1 \big\}$$

around the z-axis.

Proof It is easy to see[17] that $S^m \times [0,1]$ is a submanifold of $\mathbb{R}^{m+2}$ with boundary and that its boundary is $(S^m \times \{0\}) \cup (S^m \times \{1\})$. An obvious modification of the proof of Example 1.5(b) shows that f is an embedding. Now the claims follow from Remark 1.13(f). ∎

One-dimensional manifolds

Obviously every perfect interval J in $\mathbb{R}$ is a one-dimensional submanifold of $\mathbb{R}^n$ with or without boundary, depending on whether J is open or not. Also, we already know that the 1-sphere S^1 is a one-dimensional submanifold of $\mathbb{R}^n$, provided $n \geq 2$. It is easy to see[18] that a nonempty perfect interval is diffeomorphic to $(0,1)$ if it is open, to $[0,1)$ if it is closed on one side, and to $[0,1]$ if it is compact. The following important classification theorem shows that these intervals and S^1 are, up to diffeomorphism, the only one-dimensional connected manifolds.

1.18 Theorem *Suppose C is a connected one-dimensional submanifold N with [or without] boundary. Then C is diffeomorphic to $[0,1]$ or $[0,1)$ [or to $(0,1)$] or S^1.*

Proof For a proof, we refer to Section 3.4 of [BG88], which treats manifolds without boundary. An obvious modifications of the arguments there also covers the case of manifolds with boundary (see the appendix in [Mil65]). ∎

1.19 Remarks (a) We understand a (smooth) **curve C embedded** in N to be the image of a perfect interval of S^1 under a (smooth) embedding. In the last case, we also call C the **1-sphere embedded in N**. Then Theorem 1.18 says that every connected one-dimensional submanifold of N with or without boundary is an embedded curve, and conversely.

(b) (regularity) Theorem 1.18 remains true for C^1 manifolds. ∎

Partitions of unity

We conclude this section by proving a technical result which will be particularly helpful in the transition from local to global (and conversely).

---

[17] See Exercise 4.
[18] See Exercise 7.

Suppose X is an n-dimensional submanifold of $\mathbb{R}^{\bar{n}}$ with or without boundary for some $\bar{n} \in \mathbb{N}^{\times}$. Also let $\{ U_\alpha \; ; \; \alpha \in \mathsf{A} \}$ be an open cover of X. Then we say that the family $\{ \pi_\alpha \; ; \; \alpha \in \mathsf{A} \}$ is a **smooth partition of unity of unity** subordinate to this cover if it satisfies the properties

  (i) $\pi_\alpha \in C^\infty(X, [0,1])$ with $\mathrm{supp}(\pi_\alpha) \subset\subset U_\alpha$ for $\alpha \in \mathsf{A}$;

  (ii) the family $\{ \pi_\alpha \; ; \; \alpha \in \mathsf{A} \}$ is **locally finite**, that is, for every $p \in X$ there is an open neighborhood V such that $\mathrm{supp}(\pi_\alpha) \cap V = \emptyset$ for all but finitely many $\alpha \in \mathsf{A}$;

 (iii) $\sum_{\alpha \in \mathsf{A}} \pi_\alpha(p) = 1$ for every $p \in X$.

1.20 Proposition *Every open cover X has a smooth partition of unity subordinate to it.*

Proof (i) Let (φ, U) be a chart around $p \in X$. Then $\varphi(U)$ is open in $\overline{\mathbb{H}}^n$. Hence there is a compact neighborhood K' of $\varphi(p)$ in $\overline{\mathbb{H}}^n$ such that $K' \subset \varphi(U)$. Because φ is topological, $K := \varphi^{-1}(K')$ is a compact neighborhood of p in X with $K \subset U$, and $(\varphi \,|\, \mathring{K}, \mathring{K})$ is a chart around p. In particular, X is locally compact.

Proposition X.7.14 implies the existence of a $\chi' \in C^\infty(\varphi(U), [0,1])$ with $\chi' \,|\, K' = 1$ and $\mathrm{supp}(\chi') \subset\subset \varphi(U)$. We set $\chi(q) := \varphi^*\chi'(q)$ if $q \in U$ and $\chi(q) := 0$ if q belongs to $X \backslash U$. Then χ lies in $C^\infty(X, [0,1])$ and has compact support, which is contained in U.

(ii) By Corollary IX.1.9(ii) and Remark X.1.16(e), there exists a countable cover $\{ V_j \; ; \; j \in \mathbb{N} \}$ of X consisting of relatively compact open sets. We set $K_0 := \overline{V}_0$. Then there are $i_0, \ldots, i_m \in \mathbb{N}$ such that K_0 is covered by $\{ V_{i_0}, \ldots, V_{i_m} \}$. In addition, we set $j_1 := \max\{ i_0, \ldots, i_m \} + 1$ and $K_1 := \bigcup_{i=0}^{j_1} \overline{V}_i$. The set K_1 is compact, and $K_0 \subset\subset K_1$. We then inductively obtain a sequence (K_j) of compact sets with $K_j \subset\subset K_{j+1}$, and $\bigcup_{j=0}^{\infty} K_j = \bigcup_{j=0}^{\infty} V_j = X$.

(iii) We first assume that $K_j \neq K_{j+1}$ for $j \in \mathbb{N}$, and we set $W_j := K_j \backslash \mathring{K}_{j-1}$ for $j \in \mathbb{N}$ with $K_{-1} := \emptyset$. Then W_j is compact, and $W_j \cap W_k = \emptyset$ for $|j - k| \geq 2$. We also have $\bigcup_{j=0}^{\infty} W_j = X$.

Let $\mathcal{U} := \{ U_\alpha \; ; \; \alpha \in \mathsf{A} \}$ be an open cover of X. From (i) and the compactness of W_j, it follows that for every $j \in \mathbb{N}$, there is a finite cover $\{ \widetilde{U}_{j,i} \in \mathcal{U} \; ; \; 0 \leq i \leq m(j) \}$ of W_j. We set

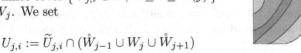

$$U_{j,i} := \widetilde{U}_{j,i} \cap (\mathring{W}_{j-1} \cup W_j \cup \mathring{W}_{j+1})$$

and choose functions $\chi_{j,i} \in C^\infty(U_{j,i}, [0,1])$ so that

$$\mathrm{supp}(\chi_{j,i}) \subset\subset U_{j,i} \subset \mathring{W}_{j-1} \cup W_j \cup \mathring{W}_{j+1} \quad \text{for } 0 \leq i \leq m(j) ,$$

with $W_{-1} := \emptyset$ and

$$\bigcup_{i=0}^{m(j)} [\chi_{j,i} > 0] \supset W_j$$

for $j \in \mathbb{N}$. Then $\{ \chi_{j,i} ; 0 \leq i \leq m(j), j \in \mathbb{N} \}$ is a locally finite family. Therefore

$$\chi := \sum_{j=0}^{\infty} \sum_{i=0}^{m(j)} \chi_{j,i}$$

is defined, belongs to $C^\infty(X, [0,1])$, and satisfies $\chi(p) > 0$ for $p \in X$. Now we set

$$\pi_\alpha := \sum_\alpha \chi_{j,i}/\chi \quad \text{for } \alpha \in \mathsf{A} ,$$

where $\sum_\alpha$ means the sum over all index pairs (j,i) for which $U_{j,i}$ is contained in U_α. Then $\{ \pi_\alpha ; \alpha \in \mathsf{A} \}$ is a smooth partition of unity subordinate to the cover $\{ U_\alpha ; \alpha \in \mathsf{A} \}$.

(iv) If there is a $j \in \mathbb{N}$ such that $K_j = K_{j+1}$, then $X = K_j$. Therefore X is compact. In this case, the claim follows by a simple modification of (iii) (as only a single compact set, namely X, must be considered). ∎

Remark (a), below, shows that Proposition 1.20 is a wide-reaching generalization of Theorem X.7.16.

1.21 Remarks   **(a)** Suppose K is a compact subset of the manifold X, and suppose $\{ U_j ; 1 \leq j \leq m \}$ is an open cover of K. Then there are functions $\pi_j \in C^\infty(X, [0,1])$ such that $\operatorname{supp}(\pi_j) \subset\subset U_j$ for $1 \leq j \leq m$, and $\sum_{j=1}^m \pi_j(p) = 1$ for $p \in K$.

Proof   Let $U_0 := X \backslash K$. Then $\{ U_j ; 0 \leq j \leq m \}$ is an open cover of X. Now the claim follows easily from Proposition 1.20. ∎

(b) The proof of Proposition 1.20 shows that every submanifold of $\mathbb{R}^{\overline{m}}$ with or without boundary is locally compact, has a countable basis, and is σ-compact.

(c) (regularity) Suppose $k \in \mathbb{N}^\times$. Replacing $\pi_\alpha \in C^\infty(X, [0,1])$ by $\pi_\alpha \in C^k(X, [0,1])$ in part (i) of the definition above, we obtain a C^k partition of unity subordinate to the cover $\{ U_\alpha ; \alpha \in \mathsf{A} \}$. Then Proposition 1.20 remains true if one replaces "smooth partition" by "C^k partition". In this case, it suffices to assume that X belongs to the class C^k. ∎

Convention In the rest of this book, we understand every manifold to be a smooth submanifold with boundary in a suitable "surrounding space" $\mathbb{R}^{\overline{m}}$.

Exercises

1 Suppose $f : M \to N$ is a submersion. Show that f "locally looks like a projection", that is, for every $p \in M$, there are charts (φ, U) of M around p and (ψ, V) of N around

$f(p)$ with $f(U) \subset V$ that satisfy

$$f_{\varphi,\psi} : \mathbb{R}^n \times \mathbb{R}^{m-n} \to \mathbb{R}^n , \quad (x,y) \mapsto x .$$

2 Suppose $f : M \to N$ is an immersion. Prove that f locally looks like the canonical injection $\mathbb{R}^m \to \mathbb{R}^m \times \mathbb{R}^{n-m}$, $x \mapsto (x,0)$.

3 Show that every diffeomorphism from M to N locally looks like the identity in $\mathbb{R}^m$.

4 Suppose B is a submanifold of N with boundary. Show that $M \times B$ is a submanifold of $M \times N$ with boundary with $\partial(M \times B) = M \times \partial B$.

5 Show that both the cylinder $[0,1] \times M$ with "cross section' M and the "filled" torus $S^1 \times \overline{\mathbb{B}}^2$ are manifolds with boundary. Determine the dimension and boundary of each.

6 Show that the closed r-ball $r\overline{\mathbb{B}}^n$ in $\mathbb{R}^n$ is diffeomorphic to the closed unit ball $\overline{\mathbb{B}}^n$.

7 Show that a perfect interval in $\mathbb{R}$ is diffeomorphic to $(0,1)$, $[0,1)$, or $[0,1]$.

8 Suppose B is a nonempty k-dimensional submanifold of M (with or without boundary). Show that the Hausdorff dimension of B equals k.
(Hints: Exercises 4–6 of IX.3 and Remark 1.21(b).)

9 Suppose B is a submanifold of M with boundary and $f \in C^\infty(B,N)$. Show that graph(f) is a submanifold of $M \times N$ with boundary and determine its boundary.

10 Suppose X is an n-dimensional submanifold of $\mathbb{R}^{\overline{n}}$ with or without boundary, and let $U := \{ U_\alpha \; ; \; \alpha \in \mathsf{A} \}$ and $V := \{ V_\beta \; ; \; \beta \in \mathsf{B} \}$ denote open covers of X. We call V a **refinement** of U if there is a $j : \mathsf{B} \to \mathsf{A}$ such that $V_\beta \subset U_{j(\beta)}$ for $\beta \in \mathsf{B}$. Show that every smooth partition of unity subordinate to V induces a smooth partition of unity subordinate to U.

2 Multilinear algebra

To construct and understand the calculus of differential forms of higher degree, we need several results from linear (more precisely, multilinear) algebra, which we provide in this section.

2.1 Remarks Suppose V is a finite-dimensional vector space.

(a) V can be provided with an inner product $(\cdot\,|\,\cdot)_V$, so that $(V,(\cdot\,|\,\cdot)_V)$ is a Hilbert space. All norms on V are equivalent.

Proof By Remark I.12.5, there is a vector space isomorphism $T : \mathbb{K}^m \to V$ such that $m := \dim(V)$. Then

$$(v\,|\,w)_V := (T^{-1}v\,|\,T^{-1}w) \quad \text{for } v, w \in V$$

defines a scalar product on V, where $(\cdot\,|\,\cdot)$ denotes the Euclidean inner product in $\mathbb{K}^m$. Thus $(V,(\cdot\,|\,\cdot)_V)$ is a finite-dimensional inner product space and therefore a Hilbert space, as we know from Remark VII.1.7(b). The second claim follows from Corollary VII.1.5. ∎

(b) As usual (in functional analysis), we denote by V^* the space of all (continuous) conjugate linear maps from V to $\mathbb{C}$, while V' is the space dual to V, the space of all (continuous) linear forms on V. Then it follows from (a) and the Riesz representation theorem (Theorem VII.2.14) that the map

$$V \to V^* , \quad v \mapsto (v\,|\,\cdot)_V \tag{2.1}$$

is an isometric isomorphism, whereas

$$V \to V' , \quad v \mapsto (\cdot\,|\,v)_V \tag{2.2}$$

is conjugate linear. If $\mathbb{K} = \mathbb{R}$, then $V^* = V'$, and the maps (2.1) and (2.2) are identical because every real scalar product is symmetric. In the following, we will exclusively treat the real case, and so, for this and some historical reasons, we will write V^* instead of V'. ∎

In this section, let

- V and W be finite-dimensional real vector spaces.

Exterior products

For $r \in \mathbb{N}$, we denote by $\mathcal{L}^r(V,\mathbb{R})$ the vector space of all r-linear maps $V^r \to \mathbb{R}$. By Remark 2.1(b) and Theorem VII.4.2(iii), this notation is consistent with that introduced in Section VII.4. In particular, we have

$$\mathcal{L}^0(V,\mathbb{R}) = \mathbb{R} \quad \text{and} \quad \mathcal{L}^1(V,\mathbb{R}) = V^* .$$

An r-linear map $\alpha \colon V^r \to W$ is said to be **alternating** if $r \geq 2$ and

$$\alpha(v_{\sigma(1)}, \ldots, v_{\sigma(r)}) = \operatorname{sign}(\sigma)\, \alpha(v_1, \ldots, v_r) \quad \text{for } v_1, \ldots, v_r \in V$$

and for every permutation $\sigma \in \mathsf{S}_r$ (see Exercise I.9.6). We set

$$\textstyle\bigwedge^0 V^* := \mathcal{L}^0(V, \mathbb{R}) = \mathbb{R} \quad \text{and} \quad \bigwedge^1 V^* := \mathcal{L}^1(V, \mathbb{R}) = V^*$$

and

$$\textstyle\bigwedge^r V^* := \left\{ \alpha \in \mathcal{L}^r(V, \mathbb{R}) \; ; \; \alpha \text{ is alternating} \right\} \quad \text{for } r \geq 2 \; .$$

Here $\bigwedge^r V^*$ is called the **r-fold exterior product of V^*** for $r \in \mathbb{N}$, and $\alpha \in \bigwedge^r V^*$ is an **alternating r-form on V** (or, for short, simply an **r-form**).

2.2 Remarks   (a) $\bigwedge^r V^*$ is a vector subspace of $\mathcal{L}^r(V, \mathbb{R})$, the **vector space of alternating r-forms** on V.

(b) Let $r \geq 2$ and $\alpha \in \mathcal{L}^r(V, \mathbb{R})$. These four statements are equivalent:

(i) $\alpha \in \bigwedge^r V^*$.

(ii) $\alpha(v_1, \ldots, v_r) = 0$ if $v_j = v_k$ for any a pair (j, k) with $j \neq k$.

(iii) $\alpha(\ldots, v_j, \ldots, v_k, \ldots) = -\alpha(\ldots, v_k, \ldots, v_j, \ldots)$ for $j \neq k$, that is, if two entries in $\alpha(v_1, \ldots, v_r)$ are exchanged, its sign reverses.

(iv) If $v_1, \ldots, v_r \in V$ are linearly independent, then $\alpha(v_1, \ldots, v_r) = 0$.

Proof   The implication "(i)$\Rightarrow$(iii)$\Rightarrow$(ii)" is obvious.

"(ii)$\Rightarrow$(iv)" Suppose $v_1, \ldots, v_r \in V$ are linearly independent. This means there are $k \in \{1, \ldots, r\}$ and $\lambda_1, \ldots, \lambda_r \in \mathbb{R}$ such that $\lambda_k = 0$ and $v_k = \sum_{j=1}^r \lambda_j v_j$. Now it follows from the linearity of α in its k-th variable and from (ii) that

$$\alpha(v_1, \ldots, v_r) = \sum_{j=1}^r \lambda_j \alpha(v_1, \ldots, \underset{(k)}{v_j}, \ldots, v_r) = 0 \; .$$

"(iv)$\Rightarrow$(iii)" From (iv) and the multilinearity, we get

$$\begin{aligned}
0 &= \alpha(\ldots, v_j + v_k, \ldots, v_j + v_k, \ldots) \\
&= \alpha(\ldots, v_j, \ldots, v_j, \ldots) + \alpha(\ldots, v_j, \ldots, v_k, \ldots) \\
&\quad + \alpha(\ldots, v_k, \ldots, v_j, \ldots) + \alpha(\ldots, v_k, \ldots, v_k, \ldots) \\
&= \alpha(\ldots, v_j, \ldots, v_k, \ldots) + \alpha(\ldots, v_k, \ldots, v_j, \ldots) \; ,
\end{aligned}$$

and which proves the claim.

"(iii)$\Rightarrow$(i)" This follows from the fact that every permutation can be written as a product of transpositions (see Exercise I.9.6). $\blacksquare$

(c) $\bigwedge^r V^* = \{0\}$ for $r > \dim(V)$.

Proof   This follows from (iv) of (b). $\blacksquare$

For $r \in \mathbb{N}^\times$ and $\varphi^1, \ldots, \varphi^r \in V^*$, the **exterior product**[1]

$$\varphi^1 \wedge \cdots \wedge \varphi^r$$

is defined by

$$\varphi^1 \wedge \cdots \wedge \varphi^r(v_1, \ldots, v_r) := \det\big[\langle \varphi^j, v_k\rangle\big] = \det \begin{bmatrix} \langle \varphi^1, v_1\rangle & \cdots & \langle \varphi^1, v_r\rangle \\ \vdots & & \vdots \\ \langle \varphi^r, v_1\rangle & \cdots & \langle \varphi^r, v_r\rangle \end{bmatrix} \quad (2.3)$$

for $v_1, \ldots, v_r \in V$. It is known from linear algebra that the determinant of an $(r \times r)$-matrix is an alternating r-form in its column vectors. From this and the linearity of $\varphi^1, \ldots, \varphi^r$, it follows immediately that $\varphi^1 \wedge \cdots \wedge \varphi^r$ belongs to $\bigwedge^r V^*$: The exterior product $\varphi^1 \wedge \cdots \wedge \varphi^r$ is an alternating r-form on V.

2.3 Proposition

(i) Let $m := \dim(V) > 0$. If $(e_1, \ldots, e_m)$ is a basis[2] of V and $(\varepsilon^1, \ldots, \varepsilon^m)$ is the associated dual basis of V^*, then

$$\big\{ \varepsilon^{j_1} \wedge \cdots \wedge \varepsilon^{j_r} \; ; \; 1 \le j_1 < j_2 < \cdots < j_r \le m \big\}$$

is a basis of $\bigwedge^r V^*$ for $1 \le r \le m$.

(ii) $\dim(\bigwedge^r V^*) = \binom{m}{r}$ for $r \in \mathbb{N}$.

Proof For short, we set

$$\mathbb{J}_r := \mathbb{J}_r^m := \big\{ (j) := (j_1, \ldots, j_r) \in \mathbb{N}^r \; ; \; 1 \le j_1 < j_2 < \cdots < j_r \le m \big\} \; .$$

Also, for an ordered multiindex $(j) \in \mathbb{J}_r$, let

$$\varepsilon^{(j)} := \varepsilon^{j_1} \wedge \cdots \wedge \varepsilon^{j_r} \; .$$

(i) Let α be an alternating r-form. Because every vector $v \in V$ has the basis representation $v = \sum_{k=1}^m \langle \varepsilon^k, v\rangle e_k$, it follows from Remark 2.2(b) that

$$\alpha(v_1, \ldots, v_r) = \sum_{k_1=1}^m \cdots \sum_{k_r=1}^m \langle \varepsilon^{k_1}, v_1\rangle \cdots \langle \varepsilon^{k_r}, v_r\rangle \alpha(e_{k_1}, \ldots, e_{k_r})$$

$$= \sum_{(j) \in \mathbb{J}_r} a_{(j)} \sum_{\sigma \in S_r} \text{sign}(\sigma)\langle \varepsilon^{\sigma(j_1)}, v_1\rangle \cdots \langle \varepsilon^{\sigma(j_r)}, v_r\rangle \; ,$$

where

$$a_{(j)} := \alpha(e_{j_1}, \ldots, e_{j_r}) \; . \quad (2.4)$$

---

[1] The exterior product is also called the **wedge** product.

[2] If $\{e_1, \ldots, e_m\}$ is an ordered basis, that is, the order of its elements is fixed, we write $(e_1, \ldots, e_m)$.

By Remark VII.1.19(a) and because the determinant of a square matrix does not change when it is transposed, we can rewrite the inner sum of the last expression as

$$\det\left(\left[\langle \varepsilon^{j_\mu}, v_\nu\rangle\right]_{1\le\mu,\nu\le r}\right) = \varepsilon^{(j)}(v_1,\ldots,v_r) \ .$$

Therefore

$$\alpha(v_1,\ldots,v_r) = \sum_{(j)\in\mathbb{J}_r} a_{(j)}\varepsilon^{(j)}(v_1,\ldots,v_r) \quad \text{for } v_1,\ldots,v_r\in V \ ,$$

and hence

$$\alpha = \sum_{(j)\in\mathbb{J}_r} a_{(j)}\varepsilon^{(j)} \ . \tag{2.5}$$

This shows that the set $\{\,\varepsilon^{(j)} \ ; \ (j)\in\mathbb{J}_r\,\}$ spans the vector space $\bigwedge^r V^*$.

Now suppose

$$\alpha = \sum_{(j)\in\mathbb{J}_r} b_{(j)}\varepsilon^{(j)}$$

with $b_{(j)}\in\mathbb{R}$ is another representation of α. Then we have in particular that

$$\alpha(e_{k_1},\ldots,e_{k_r}) = \sum_{(j)\in\mathbb{J}_r} b_{(j)}\varepsilon^{(j)}(e_{k_1},\ldots,e_{k_r}) \quad \text{for } (k)\in\mathbb{J}_r \ .$$

Because

$$\varepsilon^{(j)}(e_{k_1},\ldots,e_{k_r}) = \det\left(\left[\delta^{j_\mu}_{k_\nu}\right]_{1\le\mu,\nu\le r}\right) = \begin{cases} 1 & \text{if } (j)=(k) \ , \\ 0 & \text{otherwise} \ , \end{cases}$$

it follows that $b_{(j)}=a_{(j)}$ for $(j)\in\mathbb{J}_r$. Therefore the representation (2.5) is unique.

(ii) This statement is now clear because an m element set contains exactly $\binom{m}{r}$ subsets with r elements (see Exercise I.6.3). ∎

In the following, let

$$\alpha^1\wedge\cdots\wedge\widehat{\alpha^j}\wedge\cdots\wedge\alpha^r$$

for $1\le j\le r$ denote the $(r-1)$-form one gets by omitting the linear form α^j from $\alpha^1\wedge\cdots\wedge\alpha^r$. We use like notation, for example

$$\alpha^1\wedge\cdots\wedge\widehat{\alpha^j}\wedge\cdots\wedge\widehat{\alpha^k}\wedge\cdots\wedge\alpha^r \ ,$$

when more linear forms are omitted.

2.4 Examples (a) The one-dimensional vector spaces $\bigwedge^0 V^* = \mathbb{R}$ and $\bigwedge^m V^*$ have bases 1 and $\varepsilon^1 \wedge \cdots \wedge \varepsilon^m$, respectively.

(b) $\{\varepsilon^1, \ldots, \varepsilon^m\}$ is a basis of $\bigwedge^1 V^* = V^*$.

(c) $\{\varepsilon^1 \wedge \cdots \wedge \widehat{\varepsilon^j} \wedge \cdots \wedge \varepsilon^m \;;\; 1 \leq j \leq m\}$ is a basis of $\bigwedge^{m-1} V^*$.

(d) For the basis representation

$$\alpha^j = \sum_{k=1}^m a_k^j \varepsilon^k \in V^* \quad \text{for } 1 \leq j \leq r \,,$$

and

$$\alpha^1 \wedge \cdots \wedge \alpha^r = \sum_{(j) \in \mathbb{J}_r} a_{(j)} \varepsilon^{(j)} \in \bigwedge^r V^* \,,$$

we have $a_k^i = \langle \alpha^i, e_k \rangle$ for $1 \leq i \leq r$ and $1 \leq k \leq m$. Also

$$a_{(j)} = \det\big([a_{j_k}^i]_{1 \leq i,k \leq r}\big) \quad \text{for } (j) = (j_1, \ldots, j_r) \in \mathbb{J}_r \,.$$

Proof This follows from (2.3) and (2.4). ∎

(e) For $r \geq 1$, we have

$$\bigwedge^r V^* = \mathrm{span}\{\varphi^1 \wedge \cdots \wedge \varphi^r \;;\; \varphi^j \in V^*, \; 1 \leq j \leq r\} \,.$$

(f) For $r \geq 2$, $\varphi^1 \wedge \cdots \wedge \varphi^r = 0$ if and only if $\varphi^1, \ldots, \varphi^r$ are linearly independent.

Proof This follows from (2.3). ∎

As the next proposition shows, we can define a bilinear map from $\bigwedge^r V^* \times \bigwedge^s V^*$ to $\bigwedge^{r+s} V^*$ using the basis representation.

2.5 Proposition Let $r, s \in \mathbb{N}^\times$.

(i) *There is exactly one map*

$$\wedge : \bigwedge^r V^* \times \bigwedge^s V^* \to \bigwedge^{r+s} V^* \,, \quad (\alpha, \beta) \mapsto \alpha \wedge \beta \,, \tag{2.6}$$

the **exterior product**, *with the properties that*
  (α) $\wedge$ *is bilinear;*
  (β) *for* $\varphi^1, \ldots, \varphi^r, \psi^1, \ldots, \psi^s \in V^*$,

$$(\varphi^1 \wedge \cdots \wedge \varphi^r) \wedge (\psi^1 \wedge \cdots \wedge \psi^s) = \varphi^1 \wedge \cdots \wedge \varphi^r \wedge \psi^1 \wedge \cdots \wedge \psi^s \,. \tag{2.7}$$

(ii) *Given the basis representations*

$$\alpha = \sum_{(j) \in \mathbb{J}_r} a_{(j)} \varepsilon^{(j)} \quad \text{and} \quad \beta = \sum_{(k) \in \mathbb{J}_s} b_{(k)} \varepsilon^{(k)} \,, \tag{2.8}$$

we have

$$\alpha \wedge \beta = \sum_{\substack{(j) \in \mathbb{J}_r \\ (k) \in \mathbb{J}_s}} a_{(j)} b_{(k)} \varepsilon^{(j)} \wedge \varepsilon^{(k)} . \qquad (2.9)$$

(iii) *The exterior product is associative and* **graded anticommutative**, *that is,*

$$\alpha \wedge \beta = (-1)^{rs} \beta \wedge \alpha \quad \text{for } \alpha \in \bigwedge^r V^* \text{ and } \beta \in \bigwedge^s V^* .$$

Proof If $\wedge$ is some linear map from $\bigwedge^r V^* \times \bigwedge^s V^*$ to $\bigwedge^{r+s} V^*$ satisfying (2.7), it follows immediately from (2.8) that (2.9) is true. Hence we can use (2.9) and a given basis to uniquely define (that is, by the bilinear continuation of the basis elements to the entire space) the bilinear map (2.6) with the properties (α) and (β). By (2.3), (2.7), and Example 2.4(e), $\wedge$ is independent of chosen basis. (iii) is now an immediate consequence of the properties of the determinant. ∎

2.6 Remarks Suppose E_k for $k \in \mathbb{N}$ are vector spaces on the same field K.

(a) The **direct sum**

$$E := \bigoplus_{k=0}^{\infty} E_k =: \bigoplus_{k \geq 0} E_k$$

is defined as follows:

E is the set of all sequences (x_k) in $\bigcup_{k=0}^{\infty} E_k$ with $x_k \in E_k$ for $k \in \mathbb{N}$ that satisfy $x_k = 0$ for almost all $k \in \mathbb{N}$. On E, addition $+$ and multiplication by scalars are defined by

$$(x_k) + \lambda(y_k) := (x_k + \lambda y_k) \quad \text{for } (x_k), (y_k) \in E \text{ and } \lambda \in \mathbb{K} .$$

Then E is a K-vector space.[3] In addition, E_k will be identified with a vector subspace by means of the linear map

$$E_k \to E , \quad x_k \mapsto (0, \ldots, 0, x_k, 0, \ldots) ,$$

where x_k occupies the k-th entry in the sequence at right. Obviously

$$E = \text{span}\{ E_k ; k \in \mathbb{N} \} \quad \text{and} \quad E_k \cap E_j = \{0\} \quad \text{for } k \neq j ,$$

which justifies the name "direct sum" (see Example I.12.3(l)).

(b) Letting $E := \bigoplus_{k \geq 0} E_k$, we define a multiplication

$$E \times E \to E , \quad (v, w) \mapsto v \odot w$$

---

[3] E is the vector space of all maps $f : \mathbb{N} \to \bigcup_{k=0}^{\infty} E_k$ with compact support, with $f(k) \in E_k$ and $k \in \mathbb{N}$, endowed with the pointwise product of Example I.12.3(e).

so that $(E, +, \odot)$ is an algebra (see Section I.12). We call E the **graded algebra** (over K), and we say the multiplication is **graded** if

$$E_k \odot E_\ell \subset E_{k+\ell} \quad \text{for } k, \ell \in \mathbb{N} \,.$$

If the relations

$$v_k \odot v_\ell = (-1)^{k\ell} v_\ell \odot v_k \quad \text{for } k, \ell \in \mathbb{N}$$

are also satisfied, then both the multiplication and the algebra are said to be **graded anticommutative**. ∎

We set

$$\textstyle\bigwedge V^* := \bigoplus_{r \geq 0} \bigwedge^r V^*$$

and extend the definition of the exterior product by defining

$$\alpha \wedge \beta := \beta \wedge \alpha := \alpha\beta \quad \text{for } \alpha \in \textstyle\bigwedge^0 V^* = \mathbb{R} \,, \quad \beta \in \textstyle\bigwedge V^* \,. \qquad (2.10)$$

We also let $\mathbb{J}_0 := \{0\}$.

2.7 Theorem

(i) *There is exactly one bilinear, associative, and graded anticommutative map*

$$\textstyle\bigwedge V^* \times \bigwedge V^* \to \bigwedge V^*$$

*that extends the exterior product (2.6) and/or (2.10) to all of $\bigwedge V^* \times \bigwedge V^*$. It also will be denoted by $\wedge$ and called the **exterior product** on $\bigwedge V^*$.*

(ii) $\dim(\bigwedge V^*) = 2^{\dim(V)}$.

Proof   (i) This follows immediately from Proposition 2.5 and definition (2.10) by the natural bilinear extension.

(ii) Because $\bigwedge^r V^* = \{0\}$ for $r > \dim(V)$ and because $\bigwedge V^*$ is a direct sum of vector subspaces $\bigwedge^r V^*$, it follows from Proposition 2.3(ii) and the binomial theorem that

$$\dim(\textstyle\bigwedge V^*) = \sum_{r=0}^{m} \binom{m}{r} = 2^m$$

with $m := \dim(V)$. ∎

This theorem shows that $\bigwedge V^*$, when provided with the natural vector space structure and the exterior product, is an associative, graded anticommutative, real algebra of dimension $2^{\dim(V)}$. It is called the **Grassmann algebra** (or the **exterior algebra**) of V^*.

2.8 Remark Because V is finite-dimensional, V can be identified with V^{**} by means of the **canonical isomorphism**

$$\kappa: V \to V^{**} := (V^*)^*$$

defined by

$$\langle \kappa(v), v^* \rangle_{V^*} := \langle v^*, v \rangle \quad \text{for } v \in V, \quad v^* \in V^*.$$

Therefore the Grassmann algebra

$$\bigwedge V := \bigoplus_{r \geq 0} \bigwedge^r V$$

is also well defined on V.

Proof Clearly $\kappa: V \to V^{**}$ is linear. Suppose $\{e_1, \ldots, e_m\}$ is a basis of V, and suppose $\{\varepsilon_1, \ldots, \varepsilon_m\}$ is the associated dual basis of V^*. For $v \in \ker(\kappa)$, we have

$$\langle \varepsilon_j, v \rangle = \langle \kappa(v), \varepsilon_j \rangle_{V^*} = 0 \quad \text{for } j = 1, \ldots, m.$$

Then $v = \sum_{j=1}^m \langle \varepsilon_j, v \rangle e_j$ implies $v = 0$, so κ is injective. Now $\dim(V^{**}) = m$ (see Theorem VII.2.14) implies κ is an isomorphism. $\blacksquare$

Pull backs

For $A \in \mathcal{L}(V, W)$ and $\alpha \in \bigwedge^r W^*$, we define $A^*\alpha$ by

$$A^*\alpha(v_1, \ldots, v_r) := \alpha(Av_1, \ldots, Av_r) \quad \text{for } v_1, \ldots, v_r \in V$$

if $r \geq 1$ and by

$$A^*\alpha := \alpha \quad \text{for } \alpha \in \bigwedge^0 W^* = \mathbb{R}$$

if $r = 0$. Then we call $A^*\alpha$ the **pull back of α by A on V**.

2.9 Remarks (a) For $\alpha \in \bigwedge^r W^*$, the pull back $A^*\alpha$ belongs to $\bigwedge^r V^*$, and the map A^* is linear:

$$A^* \in \mathcal{L}(\bigwedge W^*, \bigwedge V^*) \quad \text{with} \quad A(\bigwedge^r W^*) \subset \bigwedge^r V^* \text{ and } r \in \mathbb{N}.$$

We call A^* the **pull back transformation** (or usually the **pull back**) by A.

In the case $r = 1$, A^* is the map dual to A (denoted in Section VIII.3 by $A^\top$). Note also that A maps the vector space V to W, while A^* maps $\bigwedge W^*$ to $\bigwedge V^*$ and therefore "in the reverse direction":

$$V \xrightarrow{A} W$$
$$\bigwedge V^* \xleftarrow{A^*} \bigwedge W^*$$

(b) If X is another finite-dimensional real vector space and $B \in \mathcal{L}(W, X)$, then

$$(BA)^* = A^* B^* \quad \text{and} \quad (\mathrm{id}_V)^* = \mathrm{id}_{\bigwedge V^*} .$$

In other words, the map $A \mapsto A^*$ is **contravariant**.

(c) We have

$$A^*(\alpha \wedge \beta) = A^*\alpha \wedge A^*\beta \quad \text{for } \alpha, \beta \in \textstyle\bigwedge W^* .$$

Therefore A^* is an algebra homomorphism from $\bigwedge W^*$ to $\bigwedge V^*$.

Proof These statements follow obviously from the definitions of the pull back and the exterior product. ∎

Let $m := \dim(V)$ and $W := V$, and let $\alpha \in \bigwedge^m V^*$. According to Proposition 2.3(ii), $\bigwedge^m V^*$ is one-dimensional, and hence $A^*\alpha$ must be proportional to α; we determine the multiple next.

2.10 Proposition *For $m := \dim(V)$ and $A \in \mathcal{L}(V)$,*

$$A^*\alpha = \det(A)\alpha \quad \text{for } \alpha \in \textstyle\bigwedge^m V^* .$$

Proof Let $\{e_1, \ldots, e_m\}$ be a basis of V, and let $[a_k^j] \in \mathbb{R}^{m \times m}$ be the matrix of A in this basis (see Section VII.1). Then

$$Ae_k = \sum_{j=1}^m a_k^j e_j \quad \text{for } 1 \le k \le m .$$

From this and the properties of $\alpha \in \bigwedge^m V^*$, it follows that

$$
\begin{aligned}
A^*\alpha(e_1, \ldots, e_m) &= \alpha(Ae_1, \ldots, Ae_m) \\
&= \sum_{j_1=1}^m \cdots \sum_{j_m=1}^m a_1^{j_1} \cdots \cdot a_m^{j_m}\, \alpha(e_{j_1}, \ldots, e_{j_m}) \\
&= \sum_{\sigma \in S_m} \mathrm{sign}(\sigma)\, a_1^{\sigma(1)} \cdots \cdot a_m^{\sigma(m)}\, \alpha(e_1, \ldots, e_m) \\
&= \det(A)\, \alpha(e_1, \ldots, e_m) ,
\end{aligned}
$$

where in the last step we have used the signature formula of Remark VII.1.19 and the fact that $\det(A^\top) = \det(A)$. Now the claim follows from the multilinearity of α. ∎

The volume element

Suppose $\mathcal{O}r$ now is an orientation of V, that is, $V := (V, \mathcal{O}r)$ is an oriented vector space. For short, we a call positively oriented ordered basis of V (which is therefore an element of $\mathcal{O}r$) a **positive basis** (see Remark VIII.2.4). Also let $m := \dim(V)$.

We call every $\alpha \in \bigwedge^m V^* \backslash \{0\}$ a **volume form** on V. Two volume forms α and β are **equivalent** if there is a $\lambda > 0$ such that $\alpha = \lambda\beta$. We check easily that this definition induces an equivalence relation $\sim$ on the set of all volume forms on V. Because $\dim(\bigwedge^m V^*) = 1$, there are exactly two equivalence classes.

2.11 Remarks **(a)** Suppose $(e_1, \ldots, e_m)$ is a positive basis of V and $(\varepsilon^1, \ldots, \varepsilon^m)$ is the associated dual basis. If $(\widetilde{e}_1, \ldots, \widetilde{e}_m)$ is a basis of V and $\alpha \in \bigwedge^m V^* \backslash \{0\}$ with $\alpha \sim \varepsilon^1 \wedge \cdots \wedge \varepsilon^m$, then $(\widetilde{e}_1, \ldots, \widetilde{e}_m)$ is positive if and only if $\alpha(\widetilde{e}_1, \ldots, \widetilde{e}_m) > 0$. This means that the two equivalence classes of $\bigwedge^m V^* \backslash \{0\}$ can be identified with the two orientations of V. In other words, the volume form α **determines the orientation** $\mathcal{O}r$ **of** V through the requirement

$$\alpha(e_1, \ldots, e_m) > 0 \iff (e_1, \ldots, e_m) \in \mathcal{O}r .$$

Proof Suppose $B \in \mathcal{L}(V)$ is the change of basis from $(e_1, \ldots, e_m)$ to $(\widetilde{e}_1, \ldots, \widetilde{e}_m)$, that is, $\widetilde{e}_j = Be_j$ for $1 \leq j \leq m$. Then it follows from Proposition 2.10 that

$$\alpha(\widetilde{e}_1, \ldots, \widetilde{e}_m) = \det(B)\, \alpha(e_1, \ldots, e_m) = \det(B)\lambda$$

where $\alpha = \lambda \varepsilon^1 \wedge \cdots \wedge \varepsilon^m$ and $\lambda > 0$. $\blacksquare$

(b) We say an automorphism A of V is **orientation preserving** [or **reversing**] if $\det(A) > 0$ [or $\det(A) < 0$]. We set

$$\mathcal{L}aut^+(V) := GL^+(V) := \left\{ A \in \mathcal{L}aut(V) \ ; \ \det(A) > 0 \right\} .$$

(i) The following statements are equivalent for $A \in \mathcal{L}aut(V)$:

   (α) $A \in \mathcal{L}aut^+(V)$.

   (β) For every basis $(b_1, \ldots, b_m)$, the bases $(b_1, \ldots, b_m)$ and $(Ab_1, \ldots, Ab_m)$ have the same orientation.

   (γ) For every $\alpha \in \bigwedge^m V^* \backslash \{0\}$, the volume forms α and $A^*\alpha$ determine the same orientation of V.

(ii) $\mathcal{L}aut^+(V)$ is a subgroup of $\mathcal{L}aut(V) =: GL(V)$.

Proof (i) This follows from $A^*\alpha = \det(A)\alpha$ and the definition of orientation.

   (ii) The map

$$\mathcal{L}aut(V) \to (\mathbb{R}^\times, \cdot) , \quad A \mapsto \det(A)$$

is a homomorphism. According to Exercise I.7.5, $\mathcal{L}aut^+(V)$, as the inverse image of the subgroup $((0, \infty), \cdot)$ of $(\mathbb{R}^\times, \cdot)$, is a subgroup of $\mathcal{L}aut(V)$. $\blacksquare$

Suppose now $(V, (\cdot \mid \cdot), \mathcal{O}r)$ is an oriented inner product space. Also let $(e_1, \ldots, e_m)$ be a positive orthonormal basis (ONB), and let $(\varepsilon^1, \ldots, \varepsilon^m)$ be the associated dual basis of V^*. Then

$$\omega := \omega_V := \varepsilon^1 \wedge \cdots \wedge \varepsilon^m$$

is called the **volume element** of V.

2.12 Remarks **(a)** For every positive ONB $(\widetilde{e}_1, \ldots, \widetilde{e}_m)$ of V, we have

$$\omega(\widetilde{e}_1, \ldots, \widetilde{e}_m) = 1 .$$

Proof Let B by the basis change specified by $\widetilde{e}_j = Be_j$ for $1 \leq j \leq m$. Then B belongs to $\mathcal{L}\mathrm{aut}^+(V) \cap O(m)$, and thus $\det(B) = 1$ (see Exercise VII.9.2). Therefore it follows from Proposition 2.10 that

$$\omega(\widetilde{e}_1, \ldots, \widetilde{e}_m) = B^*\omega(e_1, \ldots, e_m) = \det(B)\varepsilon^1 \wedge \cdots \wedge \varepsilon^m(e_1, \ldots, e_m) = 1 ,$$

which proves the claim. ∎

(b) The volume element of V is the unique volume form that assigns the value 1 to any, and thus every, positive ONB.

Proof This follows from (a). ∎

(c) For $v_1, \ldots, v_m \in \mathbb{R}^m$, let

$$P(v_1, \ldots, v_m) := \left\{ \sum_{j=1}^m t^j v_j \; ; \; 0 \leq t^j \leq 1 \right\} ,$$

that is, $P(v_1, \ldots, v_m)$ is the parallelepiped spanned by $v_1, \ldots, v_m$. Then

$$|\omega_{\mathbb{R}^m}(v_1, \ldots, v_m)| = \mathrm{vol}_m(P(v_1, \ldots, v_m)) := \lambda_m(P(v_1, \ldots, v_m)) .$$

In other words, the volume element assigns every m-tuple of vectors the **oriented volume**[4] of the parallelogram they span.

Proof We define $B \in \mathcal{L}(\mathbb{R}^m)$ by $v_j = Be_j$ for $1 \leq j \leq m$. Then

$$P(v_1, \ldots, v_m) = B([0,1]^m) .$$

Then it follows from Proposition 2.10 and (a) that

$$\omega_{\mathbb{R}^m}(v_1, \ldots, v_m) = B^*\omega_{\mathbb{R}^m}(e_1, \ldots, e_m) = \det(B) .$$

From Theorem IX.5.25, we know that $\lambda_m(B([0,1]^m)) = |\det(B)|$, as desired. ∎

In the following proposition, we represent the volume element ω in terms of an arbitrary positive basis of V.

2.13 Proposition *Suppose* $(b_1, \ldots, b_m)$ *is a positive basis of V and* $(\beta^1, \ldots, \beta^m)$ *is its dual basis. Then*

$$\omega = \sqrt{G}\, \beta^1 \wedge \cdots \wedge \beta^m ,$$

where $G := \det\big[(b_j \,|\, b_k)\big]$ *is the* **Gram determinant**. *In particular,*

$$\omega(b_1, \ldots, b_m) = \sqrt{G} .$$

---

[4]An oriented volume is positive if and only if $(v_1, \ldots, v_m)$ is positive; it is negative if and only if $(v_1, \ldots, v_m)$ belongs to $-\mathcal{O}r$.

Proof Let $(e_1, \ldots, e_m)$ be a positive ONB of V, and define $B \in \mathcal{L}(V)$ by $b_j = Be_j$ for $1 \leq j \leq m$. According to (i) of Remark 2.11(b), we have $\det(B) > 0$. From Remark 2.12(a) and (2.3), we get

$$\omega(e_1, \ldots, e_m) = 1 = \beta^1 \wedge \cdots \wedge \beta^m (b_1, \ldots, b_m)$$
$$= B^*(\beta^1 \wedge \cdots \wedge \beta^m)(e_1, \ldots, e_m) .$$

Therefore

$$\omega = \det(B)\beta^1 \wedge \cdots \wedge \beta^m ,$$

because an m-form is determined by its value on a basis of V, and because of Proposition 2.10. Also

$$(b_j \,|\, b_k) = (Be_j \,|\, Be_k) = (B^*Be_j \,|\, e_k) \quad \text{for } 1 \leq j, k \leq m \tag{2.11}$$

(see Exercise VII.1.5). Because $(e_1, \ldots, e_m)$ is an ONB, any $v \in V$ has the representation $v = \sum_{k=1}^m (v \,|\, e_k)e_k$. From this it follows that

$$Te_j = \sum_{k=1}^m (Te_j \,|\, e_k)e_k \quad \text{for } 1 \leq j \leq m \text{ and } T \in \mathcal{L}(V) .$$

Hence (2.11) shows that $\big[(b_j \,|\, b_k)\big] \in \mathbb{R}^{m \times m}$ is the matrix of B^*B in the basis $(e_1, \ldots, e_m)$. Therefore

$$G = \det\big[(b_j \,|\, b_k)\big] = \det(B^*B) = (\det(B))^2$$

because $\det(B^*) = \det(B)$. The claim follows. ∎

The Riesz isomorphism

Suppose $(V, (\cdot \,|\, \cdot))$ is an inner product space and $m := \dim(V)$. We denote the Riesz isomorphism (2.2) by

$$\Theta := \Theta_V : V \to V^* , \quad v \mapsto (\cdot \,|\, v) ,$$

that is,

$$\langle \Theta v, w \rangle = (w \,|\, v) \quad \text{for } v, w \in V . \tag{2.12}$$

Then

$$(\alpha \,|\, \beta)_* := (\Theta^{-1}\alpha \,|\, \Theta^{-1}\beta) \quad \text{for } \alpha, \beta \in V^* \tag{2.13}$$

defines an inner product on V^*, the **scalar product dual to** $(\cdot \,|\, \cdot)$. In the following, we always provide V^* with this inner product, so that $V^* := (V^*, (\cdot \,|\, \cdot)_*)$ is an inner product space.

2.14 Remarks Suppose $\{e_1, \ldots, e_m\}$ is a basis in V and $\{\varepsilon^1, \ldots, \varepsilon^m\}$ is its dual basis in V^*.

(a) We set

$$g_{jk} := (e_j \,|\, e_k) \quad \text{for } 1 \leq j, k \leq m \text{ , where } [g^{jk}] := [g_{jk}]^{-1} \in \mathbb{R}^{m \times m} \ .$$

Then

$$\Theta e_j = \sum_{k=1}^{m} g_{jk}\varepsilon^k \text{ and } \Theta^{-1}\varepsilon^j = \sum_{k=1}^{m} g^{jk} e_k \quad \text{for } 1 \leq j \leq m \ .$$

Proof From the basis expansion $\Theta e_j = \sum_{k=1}^{m} a_{jk}\varepsilon^k$ for $1 \leq j \leq m$ and from (2.12), we get

$$(e_j \,|\, v) = \langle \Theta e_j, v \rangle = \sum_{k=1}^{m} a_{jk}\langle \varepsilon^k, v \rangle \quad \text{for } v \in V \text{ and } 1 \leq j \leq m \ .$$

Replacing v by each of $e_1, \ldots, e_m$, we find $a_{jk} = (e_j \,|\, e_k)$, which proves the first statement. The representation of $\Theta^{-1}\varepsilon^j$ is obvious. ∎

(b) For $v = \sum_{j=1}^{m} \xi^j e_j \in V$ and $w = \sum_{j=1}^{m} \eta^j e_j \in V$, we have

$$(v \,|\, w) = \sum_{j,k=1}^{m} g_{jk}\xi^j \eta^k \ .$$

For $\alpha = \sum_{j=1}^{m} a_j\varepsilon^j \in V^*$ and $\beta = \sum_{j=1}^{m} b_j\varepsilon^j \in V^*$, we have the relation

$$(\alpha \,|\, \beta)_* = \sum_{j,k=1}^{m} g^{jk} a_j b_k \ .$$

Proof The first statement is obvious. From (a) and (2.13), we derive

$$(\varepsilon^i \,|\, \varepsilon^\ell)_* = (\Theta^{-1}\varepsilon^i \,|\, \Theta^{-1}\varepsilon^\ell) = \sum_{j,k=1}^{m} g^{ij}g^{\ell k}(e_j \,|\, e_k) = \sum_{j,k=1}^{m} g^{ij}g^{\ell k}g_{jk} = g^{i\ell}$$

for $1 \leq i, \ell \leq m$. Now the second claim follows from the bilinearity of $(\cdot \,|\, \cdot)_*$. ∎

(c) If $\{e_1, \ldots, e_m\}$ is an ONB, then $\Theta e_j = \varepsilon^j$ for $1 \leq j \leq m$, and $\{\varepsilon^1, \ldots, \varepsilon^m\}$ is likewise an ONB.

(d) You may have noticed that we have used upper indices to label the coefficients of a vector in a basis representation, whereas we used lower indices for the expansion coefficients of a 1-form. That is,

$$v = \sum_{j=1}^{m} \xi^j e_j \in V \text{ and } \alpha = \sum_{j=1}^{m} a_j\varepsilon^j \in V^* \ .$$

From (a) and the symmetry of $[g_{jk}]$, it follows that

$$\Theta v = \sum_{j=1}^{m} b_j \varepsilon^j \in V^* \quad \text{and} \quad \Theta^{-1} \alpha = \sum_{j=1}^{m} \eta^j e_j \in V$$

with

$$b_j := \sum_{k=1}^{m} g_{jk} \xi^k \quad \text{and} \quad \eta^j := \sum_{k=1}^{m} g^{jk} a_k \quad \text{for } 1 \le j \le m .$$

The application of Θ [or Θ^{-1}] formally effects a **lowering** [or **raising**] of indices. On these grounds, we may borrow the musical notations

$$g^\flat := \Theta \quad \text{and} \quad g^\sharp := \Theta^{-1} ,$$

or simply $v^\flat := \Theta v$ for $v \in V$ and $\alpha^\sharp := \Theta^{-1} \alpha$ for $\alpha \in V^*$. ∎

The Hodge star operator[5]

Suppose $(V, (\cdot | \cdot), Or)$ is an oriented inner product space, $m := \dim(V)$, and ω is the volume element of V. Also let $\{e_1, \dots, e_m\}$ be an ONB of V, and let $\{\varepsilon^1, \dots, \varepsilon^m\}$ be its dual basis.

We now define a scalar product $(\cdot | \cdot)_r$ on $\bigwedge^r V^*$ as follows:
For $r = 0$, let

$$(\alpha | \beta)_0 := \alpha \beta \quad \text{for } \alpha, \beta \in \bigwedge^0 V^* = \mathbb{R} . \tag{2.14}$$

For $1 \le r \le m$, let

$$\alpha = \sum_{(j) \in \mathbb{J}_r} a_{(j)} \varepsilon^{(j)} \quad \text{and} \quad \beta = \sum_{(j) \in \mathbb{J}_r} b_{(j)} \varepsilon^{(j)} ,$$

which, according to Proposition 2.3, are valid basis representations of $\alpha, \beta \in \bigwedge^r V^*$. Then we set

$$(\alpha | \beta)_r := \sum_{(j) \in \mathbb{J}_r} a_{(j)} b_{(j)} . \tag{2.15}$$

It is clear that $(\cdot | \cdot)_r$ is a scalar product on $\bigwedge^r V^*$ for $0 \le r \le m$. By Remarks 2.14(b) and (c), we have $(\cdot | \cdot)_1 = (\cdot | \cdot)_*$.

2.15 Remarks (a) The basis $\{ \varepsilon^{(j)} ; (j) \in \mathbb{J}_r \}$ is an ONB of $(\bigwedge^r V^*, (\cdot | \cdot)_r)$ for $1 \le r \le m$.

(b) For $\alpha^1, \dots, \alpha^r, \beta^1, \dots, \beta^r \in V^*$, we have

$$(\alpha^1 \wedge \cdots \wedge \alpha^r | \beta^1 \wedge \cdots \wedge \beta^r)_r = \det [(\alpha^j | \beta^k)_*] .$$

---

[5]This section and the next may be skipped on first reading.

Proof Suppose

$$\alpha^j = \sum_{i=1}^m a_i^j \varepsilon^i \quad \text{and} \quad \beta^k = \sum_{i=1}^m b_i^k \varepsilon^i \quad \text{for } 1 \le j, k \le r .$$

Also let

$$\alpha^1 \wedge \cdots \wedge \alpha^r = \sum_{(j) \in \mathbb{J}_r} a_{(j)} \varepsilon^{(j)} \quad \text{and} \quad \beta^1 \wedge \cdots \wedge \beta^r = \sum_{(k) \in \mathbb{J}_r} b_{(k)} \varepsilon^{(k)}$$

be basis representations. Then according to Example 2.4(d), we have

$$a_{(j)} = \det\big([a_{j_k}^i]_{1 \le i,k \le r}\big) \quad \text{and} \quad b_{(k)} = \det\big([b_{k_\ell}^i]_{1 \le i,\ell \le r}\big)$$

for $(j) = (j_1, \dots, j_r) \in \mathbb{J}_r$ and $(k) = (k_1, \dots, k_r) \in \mathbb{J}_r$.

By the bilinearity and symmetry of $(\cdot \mid \cdot)_*$ and the fact that the determinant is an alternating r-form in its row vectors, we find (see the proof of Proposition 2.3(i))

$$
\begin{aligned}
\det\big[(\alpha^j \mid \beta^k)_*\big] &= \sum_{(j) \in \mathbb{J}_r} \sum_{\sigma \in S_r} \mathrm{sign}(\sigma)\, a_{j_{\sigma(1)}}^1 \cdot \cdots \cdot a_{j_{\sigma(r)}}^r \det\big([(\varepsilon^{j_k} \mid \beta^\ell)_*]_{1 \le k,\ell \le r}\big) \\
&= \sum_{(j) \in \mathbb{J}_r} \det\big([a_{j_k}^i]_{1 \le i,k \le r}\big) \det\big([(\varepsilon^{j_k} \mid \beta^\ell)_*]_{1 \le k,\ell \le r}\big) \\
&= \sum_{(j) \in \mathbb{J}_r} a_{(j)} \det\big([(\varepsilon^{j_k} \mid \beta^\ell)_*]_{1 \le k,\ell \le r}\big) \\
&= \sum_{(j) \in \mathbb{J}_r} \sum_{(k) \in \mathbb{J}_r} a_{(j)} b_{(k)} \det\big([(\varepsilon^{j_i} \mid \varepsilon^{k_\ell})_*]_{1 \le i,\ell \le r}\big) .
\end{aligned}
$$

By (a), we have

$$\det\big[(\varepsilon^{j_i} \mid \varepsilon^{k_\ell})_*\big] = \det\big([\delta^{j_i,k_\ell}]_{1 \le i,\ell \le r}\big) = \begin{cases} 1 & \text{if } (j) = (k) , \\ 0 & \text{otherwise} . \end{cases}$$

Thus we get

$$\det\big[(\alpha^j \mid \beta^k)_*\big] = \sum_{(j) \in \mathbb{J}_r} a_{(j)} b_{(j)} .$$

By (2.15), this finishes the proof. ∎

(c) The scalar product $(\cdot \mid \cdot)_r$ on $\bigwedge^r V^*$ does not depend on the special choice of ONB or its orientation, but rather only on the inner product $(\cdot \mid \cdot)$ on V.

Proof This follows from (b), Example 2.4(e), and the scalar product's bilinearity. ∎

Because

$$\dim(\textstyle\bigwedge^r V^*) = \binom{m}{r} = \binom{m}{m-r} = \dim(\textstyle\bigwedge^{m-r} V^*) , \tag{2.16}$$

$\bigwedge^r V^*$ and $\bigwedge^{m-r} V^*$ are isomorphic vector spaces for $0 \le r \le m$. We now introduce a special (natural) isomorphism from $\bigwedge^r V^*$ to $\bigwedge^{m-r} V^*$, the Hodge star operator.

We first note that for every $\alpha \in \bigwedge^r V^*$, Proposition 2.5 implies

$$(\beta \mapsto \alpha \wedge \beta) \in \mathcal{L}(\textstyle\bigwedge^{m-r} V^*, \textstyle\bigwedge^m V^*) \ . \tag{2.17}$$

Because $\bigwedge^m V^*$ is one-dimensional, there exists exactly one $f_\alpha(\beta) \in \mathbb{R}$ such that

$$\alpha \wedge \beta = f_\alpha(\beta)\omega_V \quad \text{for } \beta \in \textstyle\bigwedge^{m-r} V^* \ .$$

By (2.17), f_α belongs to $\mathcal{L}(\bigwedge^{m-r} V^*, \mathbb{R})$. Then according to the Riesz representation theorem, there is exactly one $*\alpha \in \bigwedge^{m-r} V^*$ with $f_\alpha(\beta) = (*\alpha \mid \beta)_{m-r}$ for $\beta \in \bigwedge^{m-r} V^*$. In other words, every $\alpha \in \bigwedge^r V^*$ has a unique element $*\alpha \in \bigwedge^{m-r} V^*$ such that

$$\alpha \wedge \beta = (*\alpha \mid \beta)_{m-r}\omega_V \quad \text{for } \beta \in \textstyle\bigwedge^{m-r} V^* \ . \tag{2.18}$$

Therefore $*\alpha = \Theta^{-1} f_\alpha$, where Θ denotes the Riesz isomorphism Θ of the space $\bigwedge^{m-r} V^*$. Hence

$$(\alpha \mapsto *\alpha) \in \mathcal{L}(\textstyle\bigwedge^r V^*, \textstyle\bigwedge^{m-r} V^*) \ . \tag{2.19}$$

This map is called the **Hodge star operator** (or simply the Hodge star).

2.16 Remarks   (a) The Hodge star is an isomorphism.

Proof   From $*\alpha = 0$ and (2.18), it follows that $\alpha \wedge \beta = 0$ for every $\beta \in \bigwedge^{m-r} V^*$. For the special choice $\beta := \varepsilon^{r+1} \wedge \cdots \wedge \varepsilon^m$, it follows from

$$\alpha = \sum_{(j) \in \mathbb{J}_r} a_{(j)}\varepsilon^{(j)} \quad \text{with } (j_0) := (1,\dots,r)$$

that $0 = \alpha \wedge \beta = a_{(j_0)}\omega_V$, and therefore $a_{(j_0)} = 0$. Analogously we find that $a_{(j)} = 0$ for $(j) \in \mathbb{J}_r$. Therefore (2.19) is injective. Now the claim is implied by (2.16). ∎

(b) The Hodge star depends on the scalar product and the orientation of V. ∎

2.17 Examples   (a) For $1 \leq j \leq m$, we have $*\varepsilon^j = (-1)^{j-1}\varepsilon^1 \wedge \cdots \wedge \widehat{\varepsilon^j} \wedge \cdots \wedge \varepsilon^m$.

Proof   From the alternating property, the associativity of the exterior product, and Example 2.4(f) it follows that

$$\varepsilon^j \wedge (\varepsilon^1 \wedge \cdots \wedge \widehat{\varepsilon^k} \wedge \cdots \wedge \varepsilon^m) = (-1)^{j-1}\delta^{jk}\varepsilon^1 \wedge \cdots \wedge \varepsilon^m = (-1)^{j-1}\delta^{jk}\omega$$

for $1 \leq k \leq m$. Now the claim is implied by (2.18) and the fact that, according to Remark 2.15(a),

$$\{\varepsilon^1 \wedge \cdots \wedge \widehat{\varepsilon^k} \wedge \cdots \wedge \varepsilon^m \ ; \ 1 \leq k \leq m \}$$

is an ONB of $\bigwedge^{m-1} V^*$. ∎

(b) For $1 \leq j \leq m$, we have

$$*(\varepsilon^1 \wedge \cdots \wedge \widehat{\varepsilon^j} \wedge \cdots \wedge \varepsilon^r) = (-1)^{m-j}\varepsilon^j \ .$$

Proof   Because

$$(\varepsilon^1 \wedge \cdots \wedge \widehat{\varepsilon^j} \wedge \cdots \wedge \varepsilon^m) \wedge \varepsilon^k = (-1)^{m-j}\delta^{jk}\omega \ ,$$

the statement follows as in the previous proof. ∎

(c) $*1 = \omega$ and $*\omega = 1$.

(d) We now consider the general case covering both (a) and (b). Suppose therefore $1 \le r \le m - 1$ and $(j) \in \mathbb{J}_r$. Then there is exactly one $(j^c) \in \mathbb{J}_{m-r}$ such that $(j) \vee (j^c) := (j_1, \ldots, j_r, j_1^c, \ldots, j_{m-r}^c)$ is a permutation of $\{1, \ldots, m\}$. Putting $s(j) := \mathrm{sign}((j) \vee (j^c))$, we then have

$$*\varepsilon^{(j)} = s(j)\varepsilon^{(j^c)} . \tag{2.20}$$

It follows for $\alpha = \sum_{(j) \in \mathbb{J}_r} a_{(j)} \varepsilon^{(j)} \in \bigwedge^r V^*$ that

$$*\alpha = \sum_{(j) \in \mathbb{J}_r} s(j) a_{(j)} \varepsilon^{(j^c)} .$$

Proof   For $(k) \in \mathbb{J}_{m-r}$ with $(k) \ne (j^c)$, we have $\varepsilon^{(j)} \wedge \varepsilon^{(k)} = 0$, because at least one ε^{j_i} occurs twice in this product. For $(k) = (j^c)$, we derive from (2.3) that

$$\varepsilon^{(j)} \wedge \varepsilon^{(j^c)} = s(j)\omega . \tag{2.21}$$

Now (2.20) follows from (2.18) and Remark 2.15(a). ∎

(e) For $\alpha \in \bigwedge^r V^*$ with $0 \le r \le m$, we have $**\alpha := *(*\alpha) = (-1)^{r(m-r)}\alpha$.

Proof   For $(j), (k) \in \mathbb{J}_r$, it follows from (2.18), (d), and Proposition 2.5(iii) that

$$(**\varepsilon^{(j)} \,|\, \varepsilon^{(k)})_r \omega = (*\varepsilon^{(j)}) \wedge \varepsilon^{(k)} = s(j)\varepsilon^{(j^c)} \wedge \varepsilon^{(k)} = (-1)^{r(m-r)} s(j)\varepsilon^{(k)} \wedge \varepsilon^{(j^c)} .$$

Then because $\varepsilon^{(k)} \wedge \varepsilon^{(j^c)} = 0$ for $(k) \ne (j)$ and using (2.21), we find

$$(**\varepsilon^{(j)} \,|\, \beta)_r = (-1)^{r(m-r)}(\varepsilon^{(j)} \,|\, \beta)_r \quad \text{for } \beta \in \bigwedge^r V^* .$$

Hence $**\varepsilon^{(j)} = (-1)^{r(m-r)}\varepsilon^{(j)}$ for $(j) \in \mathbb{J}_r$, which, because of Proposition 2.3(i), proves the claim. ∎

(f) For $\alpha, \beta \in \bigwedge^r V^*$, we have the relationship

$$\alpha \wedge *\beta = \beta \wedge *\alpha = (\alpha \,|\, \beta)_r \omega . \tag{2.22}$$

Proof   Suppose $\alpha := \beta := \varepsilon^{(j)}$. Then by (d) and (2.21), we have

$$\alpha \wedge *\beta = \alpha \wedge *\alpha = \beta \wedge *\alpha = s(j)\varepsilon^{(j)} \wedge \varepsilon^{(j^c)} = \omega = (\varepsilon^{(j)} \,|\, \varepsilon^{(j)})_r \omega = (\alpha \,|\, \beta)_r \omega .$$

Letting $\alpha := \varepsilon^{(j)}$ and $\beta := \varepsilon^{(k)}$ with $(j) \ne (k)$, we obtain from (d) that

$$\alpha \wedge *\beta = s(k)\varepsilon^{(j)} \wedge \varepsilon^{(k^c)} = 0 = s(j)\varepsilon^{(k)} \wedge \varepsilon^{(j^c)} = \beta \wedge *\alpha .$$

In addition, from Remark 2.15(a) we have $(\alpha \,|\, \beta)_r = 0$. Therefore (2.22) also holds in this case. Now this claim also follows from Proposition 2.3(i). ∎

Indefinite inner products

For several applications, particularly in physics, one must drop the assumption that the scalar product is positive definite. So we will now shortly go over the modifications needed to handle this case.

A bilinear form $\mathfrak{b} : V \times V \to \mathbb{R}$ is said to be **nondegenerate** if for every $y \in V \backslash \{0\}$ there is an $x \in V$ such that $\mathfrak{b}(x, y) \neq 0$. It is **symmetric** if

$$\mathfrak{b}(x, y) = \mathfrak{b}(y, x) \quad \text{for } x, y \in V .$$

Suppose $\mathfrak{b} : V \times V \to \mathbb{R}$ is a nondegenerate symmetric bilinear form on $V :=$ $(V, (\cdot \mid \cdot))$. In the following remarks, we list some basic properties of $\mathfrak{b}$.

2.18 Remarks   (a) There is a $\mathfrak{b}$-**orthonormal basis** ($\mathfrak{b}$-ONB) of V, that is, there is a basis $\{b_1, \ldots, b_m\}$ of V such that $\mathfrak{b}(b_j, b_k) = \pm \delta_{jk}$ for $1 \leq j, k \leq m$. If r is the number of plus sign and s is the number of minuses, then $r + s = m$. The number $t := r - s$ is called the **signature** of $\mathfrak{b}$. The signature, as well as r and s, is independent of the choice of $\mathfrak{b}$-ONB. In particular,[6]

$$(-1)^s = \text{sign}(\mathfrak{b}) := \text{sign}\big(\det [\mathfrak{b}(b_j, b_k)]\big) .$$

Proof   Theorem VII.4.2(iii) clearly implies that $\mathfrak{b}$ is continuous. Then $\mathfrak{b}(x, \cdot) : V \to \mathbb{R}$ is a continuous linear form on V. Therefore, the Riesz representation theorem (Theorem VII.2.14) guarantees the existence of a unique $\mathfrak{B}x \in V$ such that

$$\mathfrak{b}(x, y) = (\mathfrak{B}x \mid y) \quad \text{for } y \in V .$$

From the linearity of $\mathfrak{b}(\cdot, y)$, it follows that $x \mapsto \mathfrak{B}x$ is linear. Then by Theorem VII.1.6, $\mathfrak{B}$ belongs to $\mathcal{L}(V)$, and

$$\mathfrak{b}(x, y) = (\mathfrak{B}x \mid y) \quad \text{for } x, y \in V .$$

The map $\mathfrak{B}$ is called the **representation operator** of $\mathfrak{b}$ with respect to $(\cdot \mid \cdot)$. Because $\mathfrak{b}$ is nondegenerate, $\mathfrak{B}$ is an automorphism of V (and conversely), and because $\mathfrak{b}$ is symmetric, so is $\mathfrak{B}$.

Remark 2.1(a) allows us to identify V with $\mathbb{R}^m$. Therefore the principal axis transformation theorem[7] guarantees the existence of an ONB $\{v_1, \ldots, v_m\}$ of V and eigenvalues $\lambda_1 \geq \cdots \geq \lambda_m$ of $\mathfrak{B}$ such that

$$\mathfrak{B}v_j = \lambda_j v_j \quad \text{for } 1 \leq j \leq m . \tag{2.23}$$

Because $\mathfrak{b}$ is nondegenerate, we have $\lambda_j \neq 0$ for $1 \leq j \leq m$. We set $b_j := v_j / \sqrt{|\lambda_j|}$. Then $\{b_1, \ldots, b_m\}$ is a basis of V, and it follows from (2.23) that

$$\mathfrak{b}(b_j, b_k) = (\mathfrak{B}b_j \mid b_k) = (\mathfrak{B}v_j \mid v_k) / \sqrt{|\lambda_j \lambda_k|} = \lambda_j (v_j \mid v_k) / \sqrt{|\lambda_j \lambda_k|} = \text{sign}(\lambda_j) \delta_{jk}$$

for $1 \leq j, k \leq m$.

---

[6] $\text{sign}(\mathfrak{b})$ should not be confused with the signature t. Obviously $2\,\text{sign}(\mathfrak{b}) = m - t$.
[7] See Example VII.10.17(b).

To show that $t = r - s$ is independent of the choice of $\mathfrak{b}$-ONB, it suffices, because $r + s = m$, to prove this is true of r. Suppose therefore $\{c_1, \ldots, c_m\}$ is a $\mathfrak{b}$-ONB of V such that $\mathfrak{b}(c_j, c_j) = 1$ for $1 \leq j \leq \rho$ and $\mathfrak{b}(c_j, c_j) = -1$ for $\rho + 1 \leq j \leq m$. We want to show that the vectors $b_1, \ldots, b_r, c_{\rho+1}, \ldots, c_m$ are linearly independent, since this would imply $r + (m - \rho) \leq m$ and therefore $r \leq \rho$; also, by exchanging the two $\mathfrak{b}$-ONB, we would analogously obtain $\rho \leq r$, which then determines r.

Suppose therefore

$$\beta_1 b_1 + \cdots + \beta_r b_r = \gamma_{\rho+1} c_{\rho+1} + \cdots + \gamma_m c_m$$

with real numbers $\beta_1, \ldots, \beta_r, \gamma_{\rho+1}, \ldots, \gamma_m$. Every linear dependence relation of the set $\{b_1, \ldots, b_r, c_{\rho+1}, \ldots, c_m\}$ can be so written. Then for $v := \beta_1 b_1 + \cdots + \beta_r b_r$, we have

$$\mathfrak{b}(v, v) = \sum_{j=1}^{r} \beta_r^2 = -\sum_{j=\rho+1}^{m} \gamma_j^2 \ ,$$

which implies $\beta_1 = \cdots = \beta_r = \gamma_{\rho+1} = \cdots = \gamma_m = 0$. The last claim is now clear. ∎

(b) (Riesz representation theorem) To every $v^* \in V^*$, there is exactly one $v \in V$ with $\mathfrak{b}(v, w) = \langle v^*, w \rangle$ for $w \in W$. The map

$$\Theta_\mathfrak{b} : V \to V^* \ , \quad v \mapsto \mathfrak{b}(v, \cdot)$$

is a vector space isomorphism, the **Riesz isomorphism** with respect to $\mathfrak{b}$. The statements of Remark 2.14(a) also hold in this case.

Proof With the representation operator $\mathfrak{B}$ of $\mathfrak{b}$ and the Riesz isomorphism Θ of V, Theorem VII.2.14 implies

$$\mathfrak{b}(v, w) = (\mathfrak{B}v \,|\, w) = \langle \Theta \mathfrak{B}v, w \rangle \quad \text{for } v, w \in V \ .$$

The claim then follows after putting $\Theta_\mathfrak{b} := \Theta \mathfrak{B}$. ∎

(c) For every basis $\{v_1, \ldots, v_m\}$ of V, the **Gram determinant** with respect to $\mathfrak{b}$, that is,

$$G_\mathfrak{b} := \det\big([\mathfrak{b}(v_j, v_k)]\big) \ ,$$

is nonzero.

Proof The determinant $G_\mathfrak{b}$ is zero if and only if the system of linear equations

$$\sum_{k=1}^{m} \mathfrak{b}(v_j, v_k)\xi^k = 0 \quad \text{for } 1 \leq j \leq m \ , \tag{2.24}$$

has a nontrivial solution. If $v := \sum_{k=1}^{m} \xi^k v_k$, then (2.24) is equivalent to $\mathfrak{b}(v_j, v) = 0$ for $1 \leq j \leq m$. Because $\{v_1, \ldots, v_m\}$ is a basis of V and $\mathfrak{b}$ is nondegenerate, it follows that $v = 0$, and we are done. ∎

(d) Suppose $(b_1, \ldots, b_m)$ is a positive basis of V and $(\beta^1, \ldots, \beta^m)$ is its dual basis. Also assume $\mathfrak{B}$ is unitary. Then

$$\varepsilon^1 \wedge \cdots \wedge \varepsilon^m = \sqrt{|G_\mathfrak{b}|}\, \beta^1 \wedge \cdots \wedge \beta^m \ .$$

Proof The first part of the proof of Proposition 2.13 shows that

$$\varepsilon^1 \wedge \cdots \wedge \varepsilon^m = \det(B)\beta^1 \wedge \cdots \wedge \beta^m ,$$

where $B \in \mathcal{L}(V)$ is the change of basis from $(e_1, \ldots, e_m)$ to $(b_1, \ldots, b_m)$. With the representation $\mathfrak{B}$ of $\mathfrak{b}$, we find

$$\mathfrak{b}(b_j, b_k) = (\mathfrak{B}b_j \,|\, b_k) = (\mathfrak{B}Be_j \,|\, Be_k) = (B^*\mathfrak{B}Be_j \,|\, e_k) \quad \text{for } 1 \leq j, k \leq m .$$

As in the proof of Proposition 2.13, this implies

$$G_\mathfrak{b} = \det\big[\mathfrak{b}(b_j, b_k)\big] = \det(B^*\mathfrak{B}B) = \det(\mathfrak{B})(\det(B))^2 .$$

Because $|\det(\mathfrak{B})| = 1$, it follows that $(\det(B))^2 = |G_\mathfrak{b}|$, which implies the claim. ∎

Suppose now $\mathcal{O}r$ is an orientation of V and $\mathfrak{b}$ is a nondegenerate symmetric bilinear form on V. Also let $(e_1, \ldots, e_m)$ be a positive $\mathfrak{b}$-ONB of V whose dual basis is $(\varepsilon^1, \ldots, \varepsilon^m)$.

On V^*, we define by

$$\mathfrak{b}_*(v^*, w^*) := \mathfrak{b}(\Theta_\mathfrak{b}^{-1}v^*, \Theta_\mathfrak{b}^{-1}w^*) \quad \text{for } v^*, w^* \in V^* .$$

the nondegenerate symmetric bilinear form $\mathfrak{b}_*$. For $\alpha^1, \ldots, \alpha^r, \beta^1, \ldots, \beta^r \in V^*$, we set

$$\mathfrak{b}_r(\alpha^1 \wedge \cdots \wedge \alpha^r, \beta^1 \wedge \cdots \wedge \beta^r) := \det\big[\mathfrak{b}_*(\alpha^j, \beta^k)\big] ,$$

and therefore $\mathfrak{b}_1 = \mathfrak{b}_*$; we also define

$$\mathfrak{b}_r : \textstyle\bigwedge^r V^* \times \bigwedge^r V^* \to \mathbb{R} \quad \text{for } r \geq 1$$

by bilinear extension using the basis representation of Proposition 2.3(i). As in (2.16)–(2.19), it follows (with $\mathfrak{b}_0 := (\,\cdot\,|\,\cdot\,)_0$) that there is a linear map

$$\textstyle\bigwedge^r V^* \to \bigwedge^{m-r} V^* , \quad \alpha \mapsto *\alpha$$

for $0 \leq r \leq m$, called the **Hodge star operator**, that is characterized by

$$\alpha \wedge \beta = \mathfrak{b}_{m-r}(*\alpha, \beta)\varepsilon^1 \wedge \cdots \wedge \varepsilon^m \quad \text{for } \beta \in \textstyle\bigwedge^{m-r} V^* . \tag{2.25}$$

2.19 Remarks (a) The Hodge star is an isomorphism that depends only on the bilinear form $\mathfrak{b}$ and the orientation, not on the $\mathfrak{b}$-ONB.

(b) For $1 \leq r \leq m$, $\{\varepsilon^{(j)} \; ; \; (j) \in \mathbb{J}_r\}$ is a $\mathfrak{b}_r$-ONB, and for $\omega := \varepsilon^1 \wedge \cdots \wedge \varepsilon^m$, we have $\mathfrak{b}_m(\omega, \omega) = \text{sign}(\mathfrak{b})$.

Proof The first statement follows easily from the definition of $\mathfrak{b}_r$. Because

$$\mathfrak{b}_m(\omega, \omega) = \det\big(\text{diag}\big[\mathfrak{b}_*(\varepsilon^1, \varepsilon^1), \ldots, \mathfrak{b}_*(\varepsilon^m, \varepsilon^m)\big]\big)$$
$$= \det\big(\text{diag}\big[\mathfrak{b}(e_1, e_1), \ldots, \mathfrak{b}(e_m, e_m)\big]\big) ,$$

the second statement is also true. ∎

(c) We have $*1 = \operatorname{sign}(\mathfrak{b})\omega$ and $*\omega = 1$. Also

$$*\varepsilon^{(j)} = s(j)\mathfrak{b}_{m-r}(\varepsilon^{(j^c)}, \varepsilon^{(j^c)})\varepsilon^{(j^c)} \quad \text{for } (j) \in \mathbb{J}_r \ ,$$

for $1 \leq r \leq m - 1$.

Proof First $\omega = 1 \wedge \omega = \mathfrak{b}_m(*1, \omega)\omega$ implies $\mathfrak{b}_m(*1, \omega) = 1$. Next $\dim(\bigwedge^m V^*) = 1$ gives $*1 = a\omega$ with $a \in \mathbb{R}$. From this we obtain with (b) that

$$1 = \mathfrak{b}_m(*1, \omega) = a\mathfrak{b}_m(\omega, \omega) = a\operatorname{sign}(\mathfrak{b}) \ ,$$

and therefore $a = \operatorname{sign}(\mathfrak{b})$. This proves the first claim. Analogously, we find $*\omega = 1$.

Suppose $1 \leq r \leq m - 1$ and $(j) \in \mathbb{J}_r$. Then

$$\omega = s(j)\varepsilon^{(j)} \wedge \varepsilon^{(j^c)} = s(j)\mathfrak{b}_{m-r}(*\varepsilon^{(j)}, \varepsilon^{(j^c)})\omega \ ,$$

and therefore $\mathfrak{b}_{m-r}(*\varepsilon^{(j)}, \varepsilon^{(j^c)}) = s(j)$. Note $\{\varepsilon^{(k)} \ ; \ (k) \in \mathbb{J}_{m-r}\}$ is a $\mathfrak{b}_{m-r}$-ONB of $\bigwedge^{m-r} V^*$. Also $*\varepsilon^{(j)} \in \bigwedge^{m-r} V^*$, and $\mathfrak{b}_{m-r}(*\varepsilon^{(j)}, \varepsilon^{(k)}) = 0$ for $(k) \neq (j^c)$. It follows that $*\varepsilon^{(j)} = a\varepsilon^{(j^c)}$ with $a \in \mathbb{R}$, and therefore

$$a\mathfrak{b}_{m-r}(\varepsilon^{(j^c)}, \varepsilon^{(j^c)}) = \mathfrak{b}_{m-r}(*\varepsilon^{(j)}, \varepsilon^{(j^c)}) = s(j) \ .$$

This implies $a = s(j)\mathfrak{b}_{m-r}(\varepsilon^{(j^c)}, \varepsilon^{(j^c)})$. Now the last claim is clear. ∎

(d) For $\alpha \in \bigwedge^r V^*$ with $0 \leq r \leq m$, we have $**\alpha = \operatorname{sign}(\mathfrak{b})\,(-1)^{r(m-r)}\alpha$.

Proof As in the proof of (c), we obtain from

$$\mathfrak{b}_r(*\varepsilon^{(j^c)}, \varepsilon^{(j)})\omega = \varepsilon^{(j^c)} \wedge \varepsilon^{(j)} = (-1)^{r(m-r)}\varepsilon^{(j)} \wedge \varepsilon^{(j^c)} = s(j)(-1)^{r(m-r)}\omega \ ,$$

that $*\varepsilon^{(j^c)} = s(j)(-1)^{r(m-r)}\mathfrak{b}_r(\varepsilon^{(j)}, \varepsilon^{(j)})\varepsilon^{(j)}$. Therefore we find by (c) that

$$\begin{aligned}
(\varepsilon^{(j)}) &= *\big(s(j)\mathfrak{b}_{m-r}(\varepsilon^{(j^c)}, \varepsilon^{(j^c)})\varepsilon^{(j^c)}\big) \\
&= s(j)^2(-1)^{r(m-r)}\mathfrak{b}_r(\varepsilon^{(j)}, \varepsilon^{(j)})\mathfrak{b}_{m-r}(\varepsilon^{(j^c)}, \varepsilon^{(j^c)})\varepsilon^{(j)} \ ,
\end{aligned}$$

from which the claim follows. ∎

(e) For $\alpha, \beta \in \bigwedge^r V^*$, we have

$$\alpha \wedge *\beta = \beta \wedge *\alpha = \operatorname{sign}(\mathfrak{b})\, \mathfrak{b}_r(\alpha, \beta)\omega \ .$$

Proof This is true by an obvious modification of the proof of Example 2.17(f). ∎

An important use of these ideas is the **Minkowski space** $\mathbb{R}^4_{1,3} := \big(\mathbb{R}^4, (\cdot \mid \cdot)_{1,3}\big)$, that is, the "spacetime" of special relativity with the **Minkowski metric**

$$(x \mid y)_{1,3} := x_0 y_0 - x_1 y_1 - x_2 y_2 - x_3 y_3 \ .$$

(In relativity theory, the "0-th coordinate" is the time.) We will elaborate on this later.

An indefinite nondegenerate symmetric bilinear form $\mathfrak{b}$ is also called an **indefinite inner product**; accordingly, $(V, \mathfrak{b})$ is an **indefinite inner product space**.

Tensors

For the sake of completeness, we now briefly introduce the concept of general tensors, which we will encounter in several later sections. Suppose $r, s \in \mathbb{N}$. An $(r + s)$-linear map

$$\gamma : \underbrace{V^* \times \cdots \times V^*}_{r} \times \underbrace{V \times \cdots \times V}_{s} \to \mathbb{R}$$

is called a **tensor on V of type (r, s)** or an (r, s)**-tensor**. In particular, γ is **contravariant of order r** and **covariant of order s** (or r**-contravariant** and s**-covariant**). We denote by $T_s^r(V)$ the normed[8] vector space of all (r, s)-tensors on V.

For $\gamma_1 \in T_{s_1}^{r_1}(V)$ and $\gamma_2 \in T_{s_2}^{r_2}(V)$, the **tensor product** $\gamma_1 \otimes \gamma_2$ is defined by

$$\gamma_1 \otimes \gamma_2(\alpha^1, \ldots, \alpha^{r_1}, \beta^1, \ldots, \beta^{r_2}, v_1, \ldots, v_{s_1}, w_1, \ldots, w_{s_2})$$
$$:= \gamma_1(\alpha^1, \ldots, \alpha^{r_1}, v_1, \ldots, v_{s_1}) \gamma_2(\beta^1, \ldots, \beta^{r_2}, w_1, \ldots, w_{s_2})$$

with $\alpha^1, \ldots, \alpha^{r_1}, \beta^1, \ldots, \beta^{r_2} \in V^*$ and $v_1, \ldots, v_{s_1}, w_1, \ldots, w_{s_2} \in V$.

In the following and as usual, we identify V^{**} with V using the canonical isomorphism κ of Remark 2.8.

2.20 Remarks (a) $T_0^1(V) = V$, $T_1^0(V) = V^*$, and $T_2^0(V) = \mathcal{L}^2(V, \mathbb{R})$.

(b) For $\gamma \in T_1^1(V)$, there exists exactly one $C \in \mathcal{L}(V)$ with

$$\gamma(v^*, v) = \langle v^*, Cv \rangle \quad \text{for } v \in V , \quad v^* \in V^* . \tag{2.26}$$

The map

$$T_1^1(V) \to \mathcal{L}(V) , \quad \gamma \mapsto C$$

is an isometric isomorphism.

Proof For $v \in V$, the map $\gamma(\,\cdot\,, v)$ belongs to $V^{**} = V$. Because γ is bilinear, we have

$$C := \big(v \mapsto \gamma(\,\cdot\,, v)\big) \in \mathcal{L}(V)$$

with $\langle v^*, Cv \rangle = \gamma(v^*, v)$ for $(v, v^*) \in V \times V^*$. Conversely, every $C \in \mathcal{L}(V)$ defines by virtue of (2.26) a $\gamma \in T_1^1(V)$. The last claim is now clear. ∎

(c) The tensor product is bilinear and associative.

(d) Letting $m := \dim(V)$, we have $\dim\big(T_s^r(V)\big) = m^{r+s}$. If $(e_1, \ldots, e_m)$ is a basis of V and $(\varepsilon^1, \ldots, \varepsilon^m)$ is its dual basis, then

$$\big\{ e_{j_1} \otimes \cdots \otimes e_{j_r} \otimes \varepsilon^{k_1} \otimes \cdots \otimes \varepsilon^{k_s} \ ; \ j_i, k_i \in \{1, \ldots, m\} \big\}$$

is a basis of $T_s^r(V)$.

---

[8]See Theorem VII.4.2.

Proof We leave the simple proof to you. ∎

(e) $\bigwedge^r V^*$ is a vector subspace of $T_r^0(V)$.

(f) The dual pairing $\langle \cdot , \cdot \rangle : V^* \times V \to \mathbb{R}$ is a $(1,1)$-tensor on V. ∎

Exercises

1 For $T \in \mathcal{L}^r(V, \mathbb{R})$, the **alternator**, $\mathrm{Alt}(T)$, is defined by

$$\mathrm{Alt}(T)(v_1, \ldots, v_r) := \frac{1}{r!} \sum_{\sigma \in S_r} \mathrm{sign}(\sigma) T(v_{\sigma(1)}, \ldots, v_{\sigma(r)})$$

for $v_1, \ldots, v_r \in V$. Show that
(a) $\mathrm{Alt} \in \mathcal{L}(\mathcal{L}^r(V, \mathbb{R}), \bigwedge^r V^*)$;
(b) $\mathrm{Alt}^2 = \mathrm{Alt}$.

2 For $S \in \mathcal{L}^s(V, \mathbb{R})$ and $T \in \mathcal{L}^t(V, \mathbb{R})$, define $S \otimes T \in \mathcal{L}^{s+t}(V, \mathbb{R})$ by

$$S \otimes T(v_1, \ldots, v_s, v_{s+1}, \ldots, v_{s+t}) := S(v_1, \ldots, v_s) T(v_{s+1}, \ldots, v_{s+t}) ,$$

where $v_1, \ldots, v_{s+t} \in V$. Show that for $\alpha \in \bigwedge^r V^*$ and $\beta \in \bigwedge^s V^*$,

$$\alpha \wedge \beta = \frac{(r+s)!}{r!\, s!} \mathrm{Alt}(\alpha \otimes \beta) .$$

In Exercises 3–8, let $(V, (\cdot \, , \, | \cdot), \mathcal{O}r)$ be an oriented inner product space, let ω be its volume element, and let Θ be the Riesz isomorphism.

3 Let $\dim(V) = 3$. Then the **vector** or **cross product** $\times$ on V is defined by[9]

$$\times : V \times V \to V , \quad (v, w) \mapsto v \times w := \Theta^{-1} \omega(v, w, \cdot) .$$

Show the following:
(a) $(v \times w \,|\, u) = \omega(v, w, u)$ for $u, v, w \in V$.
(b) The vector product is bilinear and alternating.
(c) The vector $v \times w$ is different from zero if and only if v and w are linearly independent.
(d) If v and w are linear independent, then $(v, w, v \times w)$ is a positive basis of V.
(e) The vector $v \times w$ is orthogonal to v and w.
(f) For $v, w \in V \setminus \{0\}$, we have

$$|v \times w| = \sqrt{|v|^2 \, |w|^2 - (v \,|\, w)^2} = |v| \, |w| \sin \varphi ,$$

where $\varphi \in [0, \pi]$ is the (unoriented) angle between the vectors v and w.

---

[9]See Remarks VIII.2.14.

(g) Let (e_1, e_2, e_3) be a positive ONB of V. Then for $v = \sum_j \xi^j e_j$ and $w = \sum_j \eta^j e_j$, we have

$$v \times w = (\xi^2 \eta^3 - \xi^3 \eta^2)e_1 + (\xi^3 \eta^1 - \xi^1 \eta^3)e_2 + (\xi^1 \eta^2 - \xi^2 \eta^1)e_3 \ .$$

(h) (Grassmann identity) $v_1 \times (v_2 \times v_3) = (v_1 \,|\, v_3)v_2 - (v_1 \,|\, v_2)v_3$.

(i) The vector product is not associative.

(j) $(v_1 \times v_2) \times (v_3 \times v_4) = \omega(v_1, v_2, v_4)v_3 - \omega(v_1, v_2, v_3)v_4$.

(k) (Jacobi identity) $v_1 \times (v_2 \times v_3) + v_2 \times (v_3 \times v_1) + v_3 \times (v_1 \times v_2) = 0$.

(Hints: (f) Recall Proposition 2.13 and (a).   (h) The vector product is determined by its values in the basis (e_1, e_2, e_3).)

4   For $0 \le r \le m$, verify the following formulas:

(a) $(*\alpha \,|\, \beta)_{m-r} = (-1)^{r(m-r)}(\alpha \,|\, *\beta)_r$ for $\alpha \in \bigwedge^r V^*$ and $\beta \in \bigwedge^{m-r} V^*$.

(b) $(*\alpha) \wedge \beta = (*\beta) \wedge \alpha$ for $\alpha, \beta \in \bigwedge^r V^*$.

(c) $*(\Theta v \wedge *\Theta w) = (v \,|\, w)$ for $v, w \in V$.

5   Let $(b_1, \dots, b_m)$ be a positive basis of V with $(\beta^1, \dots, \beta^m)$ its dual basis. Prove these:

(a) $\beta^j \wedge *\beta^k = g^{jk}\sqrt{G}\,\beta^1 \wedge \cdots \wedge \beta^m$ for $1 \le j, k \le m$.

(b) $*\beta^j = \sum_{k=1}^{m}(-1)^{k-1}g^{jk}\sqrt{G}\,\beta^1 \wedge \cdots \wedge \widehat{\beta^k} \wedge \cdots \wedge \beta^m$ for $1 \le j \le m$. If V is three-dimensional, show that

$$*(\beta^j \wedge \beta^k) = \frac{1}{\sqrt{G}}\,\mathrm{sign}(j, k, \ell) \sum_{i=1}^{3} g_{\ell i}\beta^i = \frac{1}{G}\,\mathrm{sign}(j, k, \ell)\Theta b_\ell$$

for $(j, k, \ell) \in S_3$.

6   In the case $\dim(V) = 3$, show $v \times w = \Theta^{-1}\big(*(\Theta v \wedge \Theta w)\big)$ for $v, w \in V$.

7   Let (b_1, b_2, b_3) be a positive basis of V with dual basis $(\beta^1, \beta^2, \beta^3)$. Show that

$$b_j \times b_k = \sqrt{G}\,\mathrm{sign}(j, k, \ell) \sum_{i=1}^{3} g^{\ell i}b_i = \sqrt{G}\,\mathrm{sign}(j, k, \ell)\Theta^{-1}\beta^\ell$$

for $(j, k, \ell) \in S_3$.

8   Let $\big(W, (\cdot \,|\, \cdot)_W, \mathcal{O}r(W)\big)$ be an oriented inner product, and let $A \in \mathcal{L}(V, W)$ be an orientation-preserving isometry (that is, $A^*\omega_W = \omega_V$). Then show that the diagram

$$
\begin{array}{ccc}
\bigwedge^r V^* & \xrightarrow{\ *\ } & \bigwedge^{m-r} V^* \\[4pt]
{\scriptstyle A^*}\big\uparrow & & \big\uparrow{\scriptstyle A^*} \\[4pt]
\bigwedge^r W^* & \xrightarrow{\ *\ } & \bigwedge^{m-r} W^*
\end{array}
$$

commutes for $0 \le r \le m$.

9   Formulate and prove the claims of Exercises 4 and 5 for indefinite inner product spaces.

10 For $k \in \mathbb{N}$, let $\mathbb{K}_k[X]$ be the vector space of all polynomials of degree $\leq k$ over $\mathbb{K}$. Show that $\mathbb{K}[X] = \bigoplus_{k \geq 0} \mathbb{K}_k[X]$ is a graded commutative algebra with respect to the usual multiplication of polynomials, that is, with respect to the convolution of Section I.8.

11 Let $(\varepsilon^0, \varepsilon^1, \varepsilon^2, \varepsilon^3)$ be the basis dual to the standard basis of $\mathbb{R}^4$. For $c, E_j, H_j \in \mathbb{R}$, set

$$\alpha := (E_1\varepsilon^1 + E_2\varepsilon^2 + E_3\varepsilon^3) \wedge c\varepsilon^0 + (H_1\varepsilon^2 \wedge \varepsilon^3 + H_2\varepsilon^3 \wedge \varepsilon^1 + H_3\varepsilon^1 \wedge \varepsilon^2),$$
$$\beta := -(H_1\varepsilon^1 + H_2\varepsilon^2 + H_3\varepsilon^3) \wedge c\varepsilon^0 + (E_1\varepsilon^2 \wedge \varepsilon^3 + E_2\varepsilon^3 \wedge \varepsilon^1 + E_3\varepsilon^1 \wedge \varepsilon^2)$$

and calculate $*\alpha$ and $*\beta$ with respect to $(\,\cdot\,|\,\cdot\,)_{1,3}$.

3 The local theory of differential forms

In Section VIII.3, we learned much about differential forms of degree 1, the Pfaff forms, and we developed a calculus that forms the foundation for the theory of line integral. Now we extend these ideas to more dimensions. In a first step, to which this section is given, we introduce differential forms of arbitrary degree on open subsets of Euclidean space, and we provide the calculus of differential forms in this "local" situation. In the sections thereafter, we consider the general situation, namely, differential forms on manifolds.

A differential form of degree r on an open subset X of $\mathbb{R}^m$ is nothing other than a set consisting of an alternating r-form on the tangent space T_xX for each $x \in X$. For this reason, the first part of this section is really only a reformulation of the results of linear algebra provided in Section 2. Rather than formulating new theorems, we will explain the definitions with remarks and examples. Analysis will come into play when we introduce an operation on differential forms, the exterior derivative. The exterior derivative makes use of concepts from analysis and goes beyond linear algebra.

In this entire section
- X is open in $\mathbb{R}^m$ and $\mathbb{K} = \mathbb{R}$.

Definitions and basis representations

For $x \in X$, the cotangent space $T_x^*X = \{x\} \times (\mathbb{R}^m)^*$ is the space dual to the tangent space $T_xX = \{x\} \times \mathbb{R}^m$. Therefore the exterior product

$$\textstyle\bigwedge^r T_x^*X = \{x\} \times \bigwedge^r(\mathbb{R}^m)^* \quad \text{for } r \in \mathbb{N} \tag{3.1}$$

and the Grassmann algebra

$$\textstyle\bigwedge T_x^*X = \{x\} \times \bigwedge(\mathbb{R}^m)^*$$

are well defined on T_x^*X. We can generalize the tangent and cotangent bundle by defining the **bundle of alternating r-forms** on X by

$$\textstyle\bigwedge^r T^*X := \bigcup_{x \in X} \bigwedge^r T_x^*X = X \times \bigwedge^r(\mathbb{R}^m)^*$$

and by defining the **Grassmann bundle** of X by

$$\textstyle\bigwedge T^*X := \bigcup_{x \in X} \bigwedge T_x^*X = X \times \bigwedge(\mathbb{R}^m)^* \ .$$

A map
$$\boldsymbol{\alpha} \colon X \to \textstyle\bigwedge^r T^*X \quad \text{with} \quad \boldsymbol{\alpha}(x) \in \textstyle\bigwedge^r T_x^*X \text{ and } x \in X \ ,$$

that is, a **section**[1] of the Grassmann bundle, is called a **differential form of degree** r (for short, an **r-form**) **on X**. By (3.1), every r-form on X has a unique representation

$$\boldsymbol{\alpha}(x) = (x, \alpha(x)) \quad \text{for } x \in X$$

whose **r-covector part** (for short, covector part) is

$$\alpha : X \to \bigwedge^r (\mathbb{R}^m)^* \, .$$

Let $k \in \mathbb{N} \cup \{\infty\}$. The r-form $\boldsymbol{\alpha}$ belongs to the **class C^k** (or is k-times **continuously differentiable**,[2] or **smooth** in case $k = \infty$) if this is true for its covector part, that is, if

$$\alpha \in C^k \big(X, \bigwedge^r (\mathbb{R}^m)^* \big) \, . \tag{3.2}$$

This definition is meaningful because, according to Remark 2.2(a), $\bigwedge^r (\mathbb{R}^m)^*$ is a (closed) vector subspace of $\mathcal{L}^r(\mathbb{R}^m, \mathbb{R})$.

For simplicity and in order to concentrate on the essential aspects of the theory, we consider almost exclusively smooth r-forms and smooth vector fields. We treat the C^k case only briefly in remarks, whose verification we leave to you.

We denote the set of all smooth r-forms on X by $\Omega^r(X)$. For short we set

$$\mathcal{E}(X) := C^\infty(X) \quad \text{and} \quad \mathcal{V}(X) := \mathcal{V}^\infty(X) \, .$$

If $v_1, \ldots, v_r$ are vector fields on X with corresponding vector parts $v_1, \ldots, v_r$, that is, if $\boldsymbol{v}_j(x) = (x, v_j(x))$ for $x \in X$ and $1 \leq j \leq r$, then we set

$$\boldsymbol{\alpha}(\boldsymbol{v}_1, \ldots, \boldsymbol{v}_r)(x) := \boldsymbol{\alpha}(x)\big(\boldsymbol{v}_1(x), \ldots, \boldsymbol{v}_r(x) \big) \quad \text{for } x \in X \, . \tag{3.3}$$

Then it follows from (VIII.3.1) that

$$\boldsymbol{\alpha}(x)(\boldsymbol{v}_1(x), \ldots, \boldsymbol{v}_r(x)) = \alpha(x)(v_1(x), \ldots, v_r(x)) \quad \text{for } x \in X \, ,$$

that is,

$$\boldsymbol{\alpha}(\boldsymbol{v}_1, \ldots, \boldsymbol{v}_r) = \alpha(v_1, \ldots, v_r) \, . \tag{3.4}$$

This shows that, without causing misunderstanding, we can identify an r-form $\boldsymbol{\alpha}$ with its covector part α and a vector field $\boldsymbol{v}$ with its vector part v. For this reason, we will from now on write differential forms and vector fields in a normal font (not boldface). In each instance, you will be able to decide without trouble whether a symbol describes a form or its covector part (or whether it means a vector field or its vector part).

---

[1]We apply the language of the theory of "vector bundles". We will not elaborate on these here (but see for example [Con93], [Dar94], or [HR72]), although it would lead to a unification of various ideas.

[2]Naturally, we say an r-form of class C^0 is **continuous**.

Addition

$$\Omega^r(X) \times \Omega^r(X) \to \Omega^r(X) , \quad (\alpha, \beta) \mapsto \alpha + \beta$$

and the **exterior product**

$$\wedge : \Omega^r(X) \times \Omega^s(X) \to \Omega^{r+s}(X) , \quad (\alpha, \beta) \mapsto \alpha \wedge \beta$$

are performed pointwise:

$$(\alpha + \beta)(x) := \alpha(x) + \beta(x) \quad \text{and} \quad (\alpha \wedge \beta)(x) := \alpha(x) \wedge \beta(x) \quad \text{for } x \in X .$$

These maps are obviously well defined.

3.1 Remarks (a) $\Omega^0(X) = \mathcal{E}(X)$.

(b) $\Omega^1(X) = \Omega_{(\infty)}(X)$, that is, the smooth 1-forms of X are C^∞ Pfaff forms on X.

(c) $\Omega^r(X) = \{0\}$ for $r > m$.

(d) $\Omega^r(X)$ for $0 \le r \le m$ is an infinite-dimensional real vector space and a free $\mathcal{E}(X)$-module of dimension $\binom{m}{r}$ (with respect to pointwise multiplication). A module basis for $\Omega^r(X)$ is given by

$$\left\{ dx^{(j)} := dx^{j_1} \wedge \cdots \wedge dx^{j_m} ; \ (j) \in \mathbb{J}_r \right\} . \tag{3.5}$$

Proof Because of (a) and the canonical identification of $\mathbb{R}$ with the subring[3] $\mathbb{R}\mathbf{1}$ of $\mathcal{E}(X)$, we have the relation $\alpha \wedge \beta = \alpha\beta$ for $\alpha \in \mathbb{R}$ and $\beta \in \Omega(X)$. The first statement then follows immediately from Remark 2.2(a) and Example I.12.3(e).

According to Remark VIII.3.3, $(dx^1(x), \ldots, dx^m(x))$ is the basis dual to the canonical basis $((e_1)_x, \ldots, (e_m)_x)$ of $T_x X$. Then the remaining claim follows from Proposition 2.3. ∎

(e) An r-form α on X belongs to the class C^k if and only if every r-tuple $v_1, \ldots, v_r$ in $\mathcal{V}^k(X)$ satisfies

$$\alpha(v_1, \ldots, v_r) \in C^k(X) . \tag{3.6}$$

This is the case if and only if the coefficients $a_{(j)}$ of the canonical basis representation[4]

$$\alpha = \sum_{(j) \in \mathbb{J}_r} a_{(j)} dx^{(j)} \tag{3.7}$$

satisfy the relation

$$a_{(j)} \in C^k(X) \quad \text{for } (j) \in \mathbb{J}_r . \tag{3.8}$$

---

[3]$\mathbf{1}(x) = 1$ for $x \in X$.
[4]It follows from Proposition 2.3, as in the proof of (d), that (3.5) is a basis of the $\mathbb{R}^X$-module of all r-forms on X.

Proof When α belongs to the class C^k, it follows easily from (3.2) and Corollary VII.4.7 that (3.6) is true. Then because (2.4) implies

$$a_{(j)} = \alpha(e_{j_1}, \ldots, e_{j_r}) \,, \tag{3.9}$$

(3.8) follows from (3.6). If (3.8) is satisfied, we conclude from (3.7) and the constancy of the basis forms $dx^{(j)}$ that α belongs to the class C^k. ∎

(f) The exterior product[5] is bilinear, associative, and graded anticommutative. Therefore

$$\Omega(X) := \bigoplus_{r \geq 0} \Omega^r(X)$$

is an (infinite-dimensional) associative, graded anticommutative algebra (with respect to the product $\wedge$). Also $\Omega(X)$ is a free $\mathcal{E}(X)$-module of dimension 2^m (with respect to pointwise multiplication); we call it the **module of differential forms** on X.

Proof These are simple consequences of Theorem 2.7 and (d). ∎

(g) Every $\alpha \in \Omega^r(X)$ is an alternating r-form on $\mathcal{V}(X)$.

Proof This follows immediately from the definition (3.3). ∎

(h) (regularity) For $k \in \mathbb{N}$, let $\Omega^r_{(k)}(X)$ be the set of r-forms of class C^k on X. Then the previous statements hold analogously for $\Omega^r_{(k)}(X)$ when $\mathcal{E}(X)$ is replaced everywhere by $C^k(X)$. ∎

In the following, we will generally not state that the coefficients $a_{(j)}$ of the canonical basis representation (3.7) of $\alpha \in \Omega^r(X)$ belong to $\mathcal{E}(X)$. This will be deemed self-evident.

3.2 Examples **(a)** As we already know, every Pfaff Form $\alpha \in \Omega^1(X)$ has the canonical basis representation

$$\alpha = \sum_{j=1}^m a_j \, dx^j \,.$$

(b) For $\alpha \in \Omega^{m-1}(X)$, the basis representation has the form

$$\alpha = \sum_{j=1}^m (-1)^{j-1} a_j \, dx^1 \wedge \cdots \wedge \widehat{dx^j} \wedge \cdots \wedge dx^m \,.$$

(c) In the case $m = 3$, any $\alpha \in \Omega^2(X)$ has the basis representation[6]

$$\alpha = a_1 \, dx^2 \wedge dx^3 + a_2 \, dx^3 \wedge dx^1 + a_3 \, dx^1 \wedge dx^2 \,.$$

Proof Because $dx^3 \wedge dx^2 = -dx^2 \wedge dx^3$, this follows from (b). ∎

---

[5]Sometimes we say **wedge product** instead of exterior product.
[6]Note the cyclic permutation of the indices.

(d) Every $\alpha \in \Omega^m(X)$ has the form $a\,dx^1 \wedge \cdots \wedge dx^m$ with $a \in \mathcal{E}(X)$.

(e) For $m = 3$, the wedge product of

$$\alpha = a_1\,dx^1 + a_2\,dx^2 + a_3\,dx^3 \text{ and } \beta = b_1\,dx^1 + b_2\,dx^2 + b_3\,dx^3$$

is

$$\alpha \wedge \beta = (a_2 b_3 - a_3 b_2)\,dx^2 \wedge dx^3 + (a_3 b_1 - a_1 b_3)\,dx^3 \wedge dx^1$$
$$+ (a_1 b_2 - a_2 b_1)\,dx^1 \wedge dx^2 \ .$$

Proof This follows from Remark 3.1(f). ∎

Pull backs

Let Y be open in $\mathbb{R}^n$ and $\varphi \in C^\infty(X,Y)$. In a generalization of the pull back of Pfaff forms, we introduce the **pull back** of differential forms by φ. It is a map

$$\varphi^* : \Omega(Y) \to \Omega(X) \tag{3.10}$$

defined by

$$(\varphi^*\beta)(x) := (T_x\varphi)^*\beta(\varphi(x)) \quad \text{for } x \in X \text{ and } \beta \in \Omega(Y) \ . \tag{3.11}$$

If $\beta \in \Omega^r(Y)$, then, because $T_x\varphi \in \mathcal{L}(T_x X, T_{\varphi(x)}Y)$ and by Remark 2.9(a), both $(T_x\varphi)^*\beta(\varphi(x))$ and $\beta(\varphi(x)) \in \bigwedge^r T_{\varphi(x)}^* Y$ lie in $\bigwedge^r T_x^* X$. From $T_x\varphi = (\varphi(x), \partial\varphi(x))$ and

$$\partial\varphi \in C^\infty(X, \mathcal{L}(\mathbb{R}^m, \mathbb{R}^n))$$

and also because (3.4) implies

$$\varphi^*\beta(v_1, \ldots, v_r) = (\beta \circ \varphi)((\partial\varphi)v_1, \ldots, (\partial\varphi)v_r) \quad \text{for } v_1, \ldots, v_r \in \mathcal{V}(X) \ ,$$

we see by Remark 3.1(e) that $\varphi^*\beta$ belongs to $\Omega^r(X)$. Therefore (3.10) is well defined through (3.11).

3.3 Remarks **(a)** The map (3.10) is $\mathbb{R}$-linear and satisfies

$$(\psi \circ \varphi)^* = \varphi^* \circ \psi^* \quad \text{and} \quad (\mathrm{id}_X)^* = \mathrm{id}_{\Omega(X)} \ ,$$

that is, the pull back operates contravariantly. It is also compatible with the exterior product, that is,

$$\varphi^*(\alpha \wedge \beta) = \varphi^*\alpha \wedge \varphi^*\beta \quad \text{for } \alpha, \beta \in \Omega(Y) \ .$$

Therefore φ^* is an algebra homomorphism from $\Omega(Y)$ to $\Omega(X)$.

Proof This follows from Remarks 2.9 and the chain rule given in Remark VII.10.2(b). ∎

(b) (regularity) The pull back can also be naturally defined for $\varphi \in C^{k+1}(X,Y)$. If $1 \le r \le m$, then an r-form of class C^{k+1} generally becomes an r-form only of class C^k, while an r-form of class C^k remains in the same class. In the case $r = 0$, the pull back preserves the regularity. ∎

3.4 Examples   Let $(x^1, \ldots, x^m)$ and $(y^1, \ldots, y^n)$ be the Euclidean coordinates of X and Y, respectively.

(a) $\varphi^* \, dy^j = d\varphi^j = \displaystyle\sum_{k=1}^{m} \partial_k \varphi^j \, dx^k$ and $1 \leq j \leq n$.

Proof   See Example VIII.3.14(a). ∎

(b) For
$$\beta = \sum_{(j) \in \mathbb{J}_r} b_{(j)} \, dy^{(j)} \in \Omega^r(Y) ,$$

we have
$$\varphi^* \beta = \sum_{(j) \in \mathbb{J}_r} (\varphi^* b_{(j)}) \, d\varphi^{(j)} .$$

Proof   This is a consequence of (a) and Remark 3.3(a). ∎

(c) In the case $m = n$, we have
$$\varphi^*(dy^1 \wedge \cdots \wedge dy^m) = d\varphi^1 \wedge \cdots \wedge d\varphi^m = (\det \partial\varphi) \, dx^1 \wedge \cdots \wedge dx^m .$$

Proof   The first equality follows from (b). Because $\det T_x \varphi = \det \partial\varphi(x)$ for $x \in X$, the claim follows from Proposition 2.10 and the constancy of the basis form $dx^1 \wedge \cdots \wedge dx^m$ on X. ∎

(d) Let $m = 2$ and $n = 3$, and let (u, v) and (x, y, z) be respective Euclidean coordinates of X and Y. Then[7]
$$\varphi^*(a \, dy \wedge dz + b \, dz \wedge dx + c \, dx \wedge dy)$$
$$= \left[a \circ \varphi \, \frac{\partial(\varphi^2, \varphi^3)}{\partial(u, v)} + b \circ \varphi \, \frac{\partial(\varphi^3, \varphi^1)}{\partial(u, v)} + c \circ \varphi \, \frac{\partial(\varphi^1, \varphi^2)}{\partial(u, v)} \right] du \wedge dv .$$

Proof   Because $d\varphi^j = \varphi^j_u \, du + \varphi^j_v \, dv$ for $1 \leq j \leq 3$ and because
$$\frac{\partial(\varphi^2, \varphi^3)}{\partial(u, v)} = \det \begin{bmatrix} \varphi^2_u & \varphi^2_v \\ \varphi^3_u & \varphi^3_v \end{bmatrix} = \varphi^2_u \varphi^3_v - \varphi^3_u \varphi^2_v$$

etc., the claim follows from (b) and Example 3.2(e). ∎

(e) (plane polar coordinates)   Let
$$f_2 \colon \mathbb{R}^2 \to \mathbb{R}^2 , \quad (r, \varphi) \mapsto (x, y) := (r \cos \varphi, r \sin \varphi)$$

be the polar coordinate map. Then
$$f_2^*(dx \wedge dy) = r \, dr \wedge d\varphi .$$

Proof   This follows from (c) and Example X.8.7. ∎

---

[7]See Remark VII.7.9.

(f) (spherical coordinates) For the spherical coordinate map

$$f_3 : \mathbb{R}^3 \to \mathbb{R}^3 , \quad (r, \varphi, \vartheta) \mapsto (x, y, z) := (r \cos \varphi \sin \vartheta, r \sin \varphi \sin \vartheta, r \cos \vartheta) ,$$

we have

$$f_3^* (dx \wedge dy \wedge dz) = -r^2 \sin \vartheta \, dr \wedge d\varphi \wedge d\vartheta .$$

Proof Lemma X.8.8 and (c). ■

(g) (m-dimensional polar coordinates) Let

$$f_m : \mathbb{R}^m \to \mathbb{R}^m , \quad (r, \varphi, \vartheta_1, \ldots, \vartheta_{m-2}) \mapsto (x^1, \ldots, x^m)$$

be the m-dimensional polar coordinate map (X.8.17). Then

$$f_m^* \, dx^1 \wedge \cdots \wedge dx^m = (-1)^m r^{m-1} w_m(\vartheta) \, dr \wedge d\varphi \wedge d\vartheta_1 \wedge \cdots \wedge d\vartheta_{m-2} ,$$

where $w_m(\vartheta) := \sin \vartheta_1 \sin^2 \vartheta_2 \cdots \sin^{m-2} \vartheta_{m-2}$.

Proof This follows from Lemma X.8.8. ■

(h) (cylindrical coordinates) Let

$$f : \mathbb{R}^3 \to \mathbb{R}^3 , \quad (r, \varphi, z) \mapsto (x, y, z) := (r \cos \varphi, r \sin \varphi, z)$$

be the cylindrical coordinate map. Then

$$f^* (dx \wedge dy \wedge dz) = r \, dr \wedge d\varphi \wedge dz .$$

Proof Example VII.9.11(c) and (c). ■

(i) If φ is a constant map, then $\varphi^* \alpha = 0$ for $\alpha \in \Omega^r(Y)$ with $r \geq 1$.

Proof Because $d\varphi^j = 0$ for $1 \leq j \leq n$, the claim is a consequence of (b). ■

(j) Let $m \leq n$, and let $i : \mathbb{R}^m \hookrightarrow \mathbb{R}^n$ be the natural embedding that identifies $\mathbb{R}^m$ with $\mathbb{R}^m \times \{0\} \subset \mathbb{R}^n$. Also let Y be open in $\mathbb{R}^n$ with

$$Y \cap (\mathbb{R}^m \times \{0\}) \supset i(X) .$$

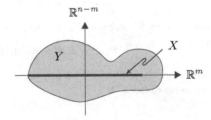

Note that X is an m-dimensional submanifold of Y.

For $\alpha \in \Omega^r(Y)$, define $\alpha \,|\, X$, the **restriction** of α to X, by

$$(\alpha \,|\, X)(x) := \alpha(x, 0) \,|\, (T_x X)^r \quad \text{for } x \in X .$$

In other words, when α has the basis representation

$$\alpha = \sum_{(j) \in \mathbb{J}_r^n} a_{(j)} \, dx^{(j)} ,$$

it follows that

$$(\alpha \mid X)(x) = \sum_{\substack{(j) \in \mathbb{J}_r^n \\ j_r \leq m}} a_{(j)}(x, 0) \, dx^{(j)} \quad \text{for } x \in X .$$

Then $i^* \alpha = \alpha \mid X$.

Proof Because of the linearity of i^* and that of the restriction map

$$\Omega^r(Y) \to \Omega^r(X) , \quad \alpha \mapsto \alpha \mid X ,$$

it suffices to consider the case $\alpha = a \, dx^{(j)}$ for $(j) \in \mathbb{J}_r^n$. Then it follows from (b) that

$$i^* \alpha = (i^* a) \, di^{(j)} .$$

By $(i^* a)(x) = a(x, 0)$ and $i^k = \mathrm{pr}_k \, i = 0$ for $m + 1 \leq k \leq n$ (where $\mathrm{pr}_k : \mathbb{R}^n \to \mathbb{R}$ is the canonical projection), we have $di^{(j)} = 0$ for $j_r > m$. For $j_r \leq m$, we find $di^{(j)} = dx^{(j)}$. Now the claim is obvious. ∎

(k) Let $(q, p) \in \mathbb{R}^m \times \mathbb{R}^m = \mathbb{R}^{2m}$ be any point of $\mathbb{R}^{2m}$. We define the (standard) **symplectic form** on $\mathbb{R}^{2m}$ by

$$\sigma := \sum_{j=1}^{m} dp^j \wedge dq^j .$$

We denote by $\mathrm{Sp}(2m)$ the set of all $S \in \mathcal{L}(\mathbb{R}^{2m})$ with $S^* \sigma = \sigma$. Then $\mathrm{Sp}(2m)$ is a subgroup of $\mathcal{L}\mathrm{aut}(\mathbb{R}^{2m})$, the symplectic group. Any $S \in \mathrm{Sp}(2m)$ satisfies $\det(S) = 1$.

Proof We define $\alpha \in \Omega^{2m}(\mathbb{R}^{2m})$ by $\alpha := \sigma \wedge \cdots \wedge \sigma$ (with m factors). Then there is an $a \in \mathbb{R}^\times$ such that $\alpha = a\omega$, where ω denotes the volume element of $\mathbb{R}^{2m}$. Suppose now $S \in \mathrm{Sp}(2m)$. Then it follows from $S^* \sigma = \sigma$ and Remark 3.3(a) that

$$S^* \alpha = S^* \sigma \wedge \cdots \wedge S^* \sigma = \sigma \wedge \cdots \wedge \sigma = \alpha .$$

Because $S^* \alpha = S^*(a\omega) = a S^* \omega$ and from (c), we find

$$\alpha = S^* \alpha = a \det(S)\omega = \det(S)\alpha ,$$

and therefore $\det(S) = 1$. We leave the proof that $\mathrm{Sp}(2m)$ is a subgroup of $\mathcal{L}\mathrm{aut}(\mathbb{R}^{2m})$ to you as an exercise. ∎

The exterior derivative

In Section VIII.3, we saw that the differential df of a function $f \in \mathcal{E}(X) = \Omega^0(X)$ is a smooth Pfaff form and therefore an element of $\Omega^1(X)$. Obviously $d : \Omega^0(X) \to \Omega^1(X)$ is linear. In addition, we know from Proposition VIII.3.12 that d commutes with pull backs. The following theorem shows that d can be extended to an $\mathbb{R}$-linear map from the module $\Omega(X)$ of differential forms to itself; this map likewise commutes with pull backs.

3.5 Theorem *There is exactly one map*

$$d : \Omega(X) \to \Omega(X) \,,$$

the **exterior derivative**,[8] *with the properties* (i)–(iv):

(i) *d is $\mathbb{R}$-linear and maps $\Omega^r(X)$ to $\Omega^{r+1}(X)$.*

(ii) *d satisfies the* **product rule**

$$d(\alpha \wedge \beta) = d\alpha \wedge \beta + (-1)^r \alpha \wedge d\beta \quad \text{for } \alpha \in \Omega^r(X) \text{ and } \beta \in \Omega(X) \,.$$

(iii) *$d^2 := d \circ d = 0$.*

(iv) *The exterior derivative df for $f \in \mathcal{E}(X)$ equals the differential of f.*

If Y is open in $\mathbb{R}^n$ and $\varphi \in C^\infty(X,Y)$, then

$$d \circ \varphi^* = \varphi^* \circ d \,, \tag{3.12}$$

that is, the exterior derivative commutes with the pull back.

Proof (a) (uniqueness) For

$$\alpha = \sum_{(j) \in \mathbb{J}_r} a_{(j)} \, dx^{(j)} \in \Omega^r(X) \,, \tag{3.13}$$

it follows easily from (i)–(iv) that

$$d\alpha = \sum_{(j) \in \mathbb{J}_r} da_{(j)} \wedge dx^{(j)} \in \Omega^{r+1}(X) \,. \tag{3.14}$$

This implies that at most one map can satisfy the properties (i)–(iv).

(b) (existence) For $\alpha \in \Omega^r(X)$ expanded as in (3.13), we defined $d\alpha$ by (3.14). Then d obviously satisfies the demands (i) and (iv).

To show (ii), realize that (i) means we need only consider the case $\alpha = a \, dx^{(j)}$ and $\beta = b \, dx^{(k)}$ with $(j) \in \mathbb{J}_r$ and $(k) \in \mathbb{J}_s$. Then it follows from (3.14), the properties of the exterior product, and the ordinary product rule of Corollary VII.3.8 that

$$\begin{aligned}
d(\alpha \wedge \beta) = d(ab \, dx^{(j)} \wedge dx^{(k)}) &= d(ab) \wedge dx^{(j)} \wedge dx^{(k)} \\
&= da \wedge dx^{(j)} \wedge b \, dx^{(k)} + (-1)^r a \, dx^{(j)} \wedge db \wedge dx^{(k)} \\
&= d(a \, dx^{(j)}) \wedge b \, dx^{(k)} + (-1)^r a \, dx^{(j)} \wedge d(b \, dx^{(k)}) \\
&= da \wedge \beta + (-1)^r \alpha \wedge d\beta \,,
\end{aligned}$$

as desired.

---

[8]Sometimes the exterior derivative is called the **Cartan derivative**.

For the proof of (iii), we can use the linearity of d to again restrict to the case $\alpha = a\,dx^{(j)}$ with $(j) \in \mathbb{J}_r$. Then it follows from (3.14) and (ii) that

$$d(d\alpha) = d(da \wedge dx^{(j)}) = d^2 a \wedge dx^{(j)} - da \wedge d(dx^{(j)}) \; .$$

By successive application of the product rule (ii) to $d(dx^{(j)})$, we see that the claim will follow if we can show $d^2 a = 0$ for $a \in \Omega^0(X) = \mathcal{E}(X)$.

Suppose therefore $a \in \mathcal{E}(X)$. Then we may use (i), (ii), and (iv) to derive the relation

$$d(da) = d\left(\sum_{k=1}^{m} \partial_k a\,dx^k\right) = \sum_{k=1}^{m} d(\partial_k a) \wedge dx^k$$

$$= \sum_{j,k=1}^{m} \partial_j \partial_k a\,dx^j \wedge dx^k = \sum_{1 \le j < k \le m} (\partial_j \partial_k a - \partial_k \partial_j a)\,dx^j \wedge dx^k = 0 \; ,$$

where the last equality follows from Schwarz's theorem (Corollary VII.5.5). Therefore (iii) is satisfied.

(c) Suppose $\varphi \in C^\infty(X, Y)$ and $(j) \in \mathbb{J}_r^n$. Let $\beta = b\,dy^{(j)} \in \Omega^r(Y)$. Then according to Example 3.4(b), we have

$$\varphi^* \beta = \varphi^* b\,d\varphi^{(j)} \in \Omega^r(X) \; . \tag{3.15}$$

From (3.14) and the property of the pull back explained in Remark 3.3(a), we get

$$\varphi^* d\beta = \varphi^*(db \wedge dy^{(j)}) = \varphi^* db \wedge \varphi^* dy^{(j)} = \varphi^* db \wedge d\varphi^{(j)} \; .$$

Proposition VIII.3.12 implies $\varphi^* db = d(\varphi^* b)$. Therefore we find using (i), (iii), and (3.15) that

$$\varphi^* d\beta = d(\varphi^* b) \wedge d\varphi^{(j)} = d(\varphi^* b) \wedge d\varphi^{(j)} + (-1)^1 \varphi^* b \wedge d(d\varphi^{(j)})$$
$$= d(\varphi^* b \wedge d\varphi^{(j)}) = d(\varphi^* \beta) \; .$$

Now (3.12) follows from the linearity of φ^* and d and from Remark 3.1(e). ∎

3.6 Remarks (a) For $\alpha = \sum_{(j) \in \mathbb{J}_r} a_{(j)}\,dx^{(j)} \in \Omega^r(X)$, we have

$$d\alpha = \sum_{(j) \in \mathbb{J}_r} da_{(j)} \wedge dx^{(j)} \; .$$

Proof   This is the statement (3.13), (3.14). ∎

(b) For $\varphi \in C^\infty(X, Y)$ and $r \in \mathbb{N}$, the diagram

$$
\begin{array}{ccc}
\Omega^r(Y) & \xrightarrow{\quad d \quad} & \Omega^{r+1}(Y) \\
\varphi^* \downarrow & & \downarrow \varphi^* \\
\Omega^r(X) & \xrightarrow{\quad d \quad} & \Omega^{r+1}(X)
\end{array}
$$

commutes.

Proof   This is (3.12). ∎

(c) (regularity) If α is an r-form of class C^{k+1}, then $d\alpha$ is obviously an $(r+1)$-form of class C^k. However, for $\alpha = a\,dx^{(j)}$ with $(j) \in \mathbb{J}_r^m$, we have

$$
d\alpha = da \wedge dx^{(j)} = \sum_i \partial_i a \, dx^i \wedge dx^{(j)} \ ,
$$

where we only sum over the indices $i \in \{1, \dots, m\}$ with $i \neq j_k$ for $1 \leq k \leq r$, because $dx^i \wedge dx^{(j)} = 0$ is true of the remaining indices. Hence there is an r-form α of class C^k for which $d\alpha$ also belongs to the class C^k. ∎

3.7 Examples   **(a)** For $\alpha = \sum_{j=1}^m a_j \, dx^j \in \Omega^1(X)$, we have

$$
d\alpha = \sum_{1 \leq j < k \leq m} (\partial_j a_k - \partial_k a_j) \, dx^j \wedge dx^k \ .
$$

(b) For $\alpha = \sum_{j=1}^m (-1)^{j-1} a_j \, dx^1 \wedge \cdots \wedge \widehat{dx^j} \wedge \cdots \wedge dx^m \in \Omega^{m-1}(X)$, we get

$$
d\alpha = \left(\sum_{j=1}^m \partial_j a_j \right) dx^1 \wedge \cdots \wedge dx^m \ .
$$

(c) $d\alpha = 0$ for $\alpha \in \Omega^m(X)$. ∎

The Poincaré lemma

A differential form $\alpha \in \Omega(X)$ is said to be **closed** if $d\alpha = 0$. We say it is **exact** if there an **antiderivative** $\beta \in \Omega(X)$ such that[9] $d\beta = \alpha$.

3.8 Remarks and examples   **(a)** Example 3.7(a) says $\alpha = \sum_{j=1}^m a_j \, dx^j$ is closed if and only if $\partial_j a_k = \partial_k a_j$ for $1 \leq j, k \leq m$. Therefore this extended notion of closedness reduces to the definition of Section VIII.3 in the case of Pfaff forms.

---

[9]Saying that a form is exact implies that it has degree at least 1.

(b) Every exact form is closed.

Proof This follows from $d^2 = 0$. ∎

(c) Every m-form on X is closed.

Proof Example 3.7(c). ∎

(d) (regularity) The definition of closed is clearly meaningful for forms of class C^1; the notion of exact makes sense for continuous differential forms. ∎

In Theorem VIII.3.8, we have seen that every closed Pfaff form is exact if X is star shaped. In the following, we will show that this "lemma" of Poincaré is also true in the general case.

Let $I := [0, 1]$, and let t be a generic point in I. For $\ell \in \{0, 1\}$, the injection

$$i_\ell : X \to I \times X , \quad x \mapsto (\ell, x)$$

is smooth. Obviously i_0 and i_1 identify the X with the "bottom" $\{0\} \times X$ and the "top" $\{1\} \times X$, respectively, of the cylinder $I \times X$ over X. Therefore[10]

$$i_\ell^* : \Omega^r(I \times X) \to \Omega^r(X)$$

is defined. For $\alpha \in \Omega(I \times X)$ the form $i_0^* \alpha$ [or $i_1^* \alpha$] is a restriction of α to X. It is obtained by replacing (t, x) by $(0, x)$ [or $(1, x)$] in the coefficients of the canonical basis representation of α, and by removing all terms in which dt occurs (see Example 3.4(j)).

We define a linear map

$$K : \Omega^{r+1}(I \times X) \to \Omega^r(X)$$

by

$$K\alpha := \sum_{(j) \in \mathbb{J}_r} \int_0^1 a_{(j)}(t, \cdot) \, dt \, dx^{(j)} \tag{3.16}$$

for

$$\alpha = \sum_{(j) \in \mathbb{J}_r} a_{(j)} \, dt \wedge dx^{(j)} + \sum_{(k) \in \mathbb{J}_{r+1}} b_{(k)} \, dx^{(k)} . \tag{3.17}$$

3.9 Lemma K is well defined and satisfies

$$K \circ d + d \circ K = i_1^* - i_0^* . \tag{3.18}$$

Proof The theorem about the differentiability of parameter-dependent integrals (Theorem X.3.18) implies easily that $K\alpha$, defined for the α of (3.17) by (3.16), belongs to $\Omega^r(X)$. Clearly the map K is also linear.

---

[10]Because the partial derivative ∂_t is defined on I, it is clear how differential forms are defined on $I \times X$. Note that $I \times X$ is a manifold with boundary, and see Section 4.

To show (3.18), it suffices to consider the cases $\alpha = a\, dt \wedge dx^{(j)}$ and $\alpha = b\, dx^{(k)}$ with $(j) \in \mathbb{J}_r$ and $(k) \in \mathbb{J}_{r+1}$.

(i) Let $\alpha = a\, dt \wedge dx^{(j)}$. Then $i_0^* \alpha = i_1^* \alpha = 0$. We also get

$$K\, d\alpha = K\big(da \wedge dt \wedge dx^{(j)}\big) = K\Big(\sum_{\ell=1}^{m} \partial_{x^\ell} a\, dx^\ell \wedge dt \wedge dx^{(j)}\Big)$$
$$= -\sum_{\ell=1}^{m} \int_0^1 \partial_{x^\ell} a(t, \cdot)\, dt\, dx^\ell \wedge dx^{(j)} ,$$

where we have used $dt \wedge dt \wedge dx^{(j)} = 0$. On the other hand, Theorem X.3.18 gives

$$d(K\alpha) = d\Big(\int_0^1 a(t, \cdot)\, dt\, dx^{(j)} \Big) = \sum_{\ell=1}^{m} \int_0^1 \partial_{x^\ell} a(t, \cdot)\, dt\, dx^\ell \wedge dx^{(j)} .$$

This proves the claim in this case.

(ii) Let $\alpha = b\, dx^{(k)}$ with $(k) \in \mathbb{J}_{r+1}$. Then $K\alpha = 0$, and therefore $dK\alpha = 0$. Also, we find

$$d\alpha = \partial_t b\, dt \wedge dx^{(k)} + \sum_{\ell=1}^{m} \partial_{x^\ell} b\, dx^\ell \wedge dx^{(k)}$$

and

$$K\, d\alpha = \int_0^1 \partial_t b(\tau, \cdot)\, d\tau\, dx^{(k)} = \big(b(1, \cdot) - b(0, \cdot)\big)\, dx^{(k)} = i_1^* \alpha - i_0^* \alpha ,$$

so the claim holds in this case also. ∎

Let M and N be manifolds. Two maps $f_0, f_1 \in C^\infty(M, N)$ are said to be **homotopic in N** if there is a map[11] $h \in C^\infty(I \times M, N)$, a **homotopy**, such that $h(j, \cdot) = f_j$ for $j = 0, 1$. A map $f \in C^\infty(M, N)$ is **null-homotopic in N** if it is homotopic in N to a constant map. Finally, we say M is **contractible** if the identity map from M to M is null-homotopic.

3.10 Remarks (a) The statement "f_1 is homotopic in N to f_2" defines an equivalence relation in $C^\infty(M, N)$ (or, more generally, in $C^k(M, N)$).

(b) The concept of a (continuous) homotopy obviously generalizes the idea of a loop homotopy (see Section VIII.4).

(c) Every star shaped open set is contractible.

Proof Let X be star shaped with respect to $x_0 \in X$. Then

$$h : I \times X \to X , \quad (t, x) \mapsto x_0 + t(x - x_0)$$

is obvious a homotopy with $h(0, \cdot) = x_0$ and $h(1, \cdot) = \mathrm{id}_X$. ∎

---

[11] Note that $I \times M$ is a manifold with boundary.

(d) (regularity) For $k \in \mathbb{N}^\times$, the definitions above are meaningful for C^k manifolds M and N if all the functions that appear belong to the class C^k. They are then also meaningful if M and N are topological spaces and all functions considered are continuous. ∎

We can now easily prove the generalized Poincaré lemma.

3.11 Theorem (Poincaré lemma) *If X is contractible, then every closed differential form on X is exact.*

Proof Suppose $\alpha \in \Omega^{r+1}(X)$ is closed. Because X is contractible, there exists an $h \in C^\infty(I \times X, X)$ such that $h(1, \cdot) = \mathrm{id}_X$ and $h(0, \cdot) = p$ for some $p \in X$. Because α is closed, $h^*\alpha \in \Omega^{r+1}(I \times X)$ is also closed because $d \circ h^* = h^* \circ d$. Therefore it follows from Lemma 3.9 that

$$d(Kh^*\alpha) = i_1^* h^*\alpha - i_0^* h^*\alpha = (h \circ i_1)^*\alpha = \alpha \ .$$

This is because $h \circ i_1 = \mathrm{id}_X$ and because $i_0^* h^*\alpha = (h \circ i_0)^*\alpha$ is a null form according to Example 3.4(i). ∎

We should point out that the proof of the Poincaré lemma gives an explicit procedure for constructing an antiderivative of a given closed differential form. The situation is particularly simple when X is star shaped, where we can assume without loss of generality (by applying a suitable translation) that X is star shaped with respect to 0.

3.12 Corollary *Suppose X is star shaped with respect to 0. Suppose with $r \in \mathbb{N}^\times$ that*

$$\alpha = \sum_{(j) \in \mathbb{J}_r} a_{(j)} \, dx^{(j)} \in \Omega^r(X)$$

is closed. Also let

$$\beta := \sum_{(j) \in \mathbb{J}_r} \sum_{k=1}^r (-1)^{k-1} \int_0^1 t^{r-1} a_{(j)}(tx) \, dt \, x^{j_k} \, dx^{j_1} \wedge \cdots \wedge \widehat{dx^{j_k}} \wedge \cdots \wedge dx^{j_r} \ . \qquad (3.19)$$

Then β belongs to $\Omega^{r-1}(X)$, and $d\beta = \alpha$.

Proof In this case, $h(t, x) := tx$ for $(t, x) \in I \times X$ defines a "contraction of X to 0". From $dh^j = x^j \, dt + t \, dx^j$ and Example 3.4(b), it follows that

$$h^*\alpha(t, x) = \sum_{(j) \in \mathbb{J}_r} a_{(j)}(tx) t^r \, dx^{(j)}$$

$$+ \sum_{(j) \in \mathbb{J}_r} \sum_{k=1}^r (-1)^{k-1} a_{(j)}(tx) t^{r-1} x^{j_k} \, dt \wedge dx^{j_1} \wedge \cdots \wedge \widehat{dx^{j_k}} \wedge \cdots \wedge dx^{j_r} \ ,$$

because those terms in which dt occurs at least twice vanish. From (3.16) and (3.17), it follows that $\beta = Kh^*\alpha$, and the claim now follows from the proof of the Poincaré lemma. ∎

3.13 Remarks (a) In the case $r = 1$, that is, when α is a Pfaff form, the formula for β is the case as the one in (VIII.3.4).

(b) Let $m = 3$ and

$$\alpha = a_1\, dx^2 \wedge dx^3 + a_2\, dx^3 \wedge dx^1 + a_3\, dx^1 \wedge dx^2 \in \Omega^2(X) \ .$$

Then the problem of finding a $\beta = \sum_{j=1}^{3} b_j\, dx^j$ with $d\beta = \alpha$ is equivalent to the problem of finding three functions $b_1, b_2, b_3 \in \mathcal{E}(X)$ that satisfy the system

$$\begin{aligned}
\partial_1 b_2 - \partial_2 b_1 &= a_3 \ , \\
\partial_2 b_3 - \partial_3 b_2 &= a_1 \ , \\
\partial_3 b_1 - \partial_1 b_3 &= a_2
\end{aligned} \tag{3.20}$$

of partial differential equations in X. Then, for given $a_j \in \mathcal{E}(X)$,

$$\partial_1 a_1 + \partial_2 a_2 + \partial_3 a_3 = 0 \tag{3.21}$$

is required for (3.20) to have a solution. If X is contractible (for example $X = \mathbb{R}^3$), then (3.21) is also sufficient.

Proof By Example 3.7(a), (3.20) is equivalent to $d\beta = \alpha$. Example 3.7(b) shows that (3.21) is equivalent to $d\alpha = 0$. Now the claim follows from $d^2 = 0$ and the Poincaré lemma. ∎

From Corollary 3.12, it follows in particular that in the case of star shaped domains, (3.20) can be solved by quadrature using the formula (3.19). In the general case, the equation $d\beta = \alpha$ can clearly also be reformulated as an equivalent system of partial differential equations.

Of course, (3.20) does not have a unique solution, because a closed form can be added to β, that is, one can add any solution (b_1, b_2, b_3) of the homogeneous system obtained by zeroing the right side of (3.20).

Tensors

Let $r, s \in \mathbb{N}$. For $x \in X$, we set

$$T_s^r(T_x X) := \{x\} \times T_s^r(\mathbb{R}^m) \tag{3.22}$$

and call $\gamma \in T_s^r(T_xX)$ an r-contravariant and s-covariant **tensor**, or **tensor of type (r,s)** on T_xX. The **bundle of (r,s)-tensors** on X is defined by .

$$T_s^r(X) := \bigcup_{x \in X} T_s^r(T_xX) = X \times T_s^r(\mathbb{R}^m) \ .$$

A map

$$\gamma : X \to T_s^r(X) \qquad \text{with} \quad \gamma(x) \in T_s^r(T_xX) \ ,$$

that is, a section of the tensor bundle $T_s^r(X)$, is called an **(r,s)-tensor (field)** or tensor of **type (r,s)** on X. By (3.22), every (r,s)-tensor γ on X has the unique representation

$$\boldsymbol{\gamma}(x) = \big(x, \gamma(x)\big) \quad \text{for } x \in X \ ,$$

with the **principal part**[12]

$$\gamma : X \to T_s^r(\mathbb{R}^m) \ .$$

Let $k \in \mathbb{N} \cup \{\infty\}$. An (r,s)-tensor $\boldsymbol{\gamma}$ belongs to the **class C^k** (or is k-times **continuously differentiable** or **smooth** if $k = \infty$) if this is true of its principal part, that is, if[13]

$$\gamma \in C^k\big(X, \mathcal{L}^{r+s}(\mathbb{R}^m, \mathbb{R})\big) \ .$$

We denote the set of all smooth (r,s)-tensors on X by

$$\mathcal{T}_s^r(X) \ .$$

If $\boldsymbol{\alpha}_1, \ldots, \boldsymbol{\alpha}_r$ are Pfaff forms and $\boldsymbol{v}_1, \ldots, \boldsymbol{v}_s$ are vector fields on X with corresponding principal parts $\alpha_1, \ldots, \alpha_r$ and $v_1, \ldots, v_s$, then we will set

$$\boldsymbol{\gamma}(\boldsymbol{\alpha}_1, \ldots, \boldsymbol{\alpha}_r, \boldsymbol{v}_1, \ldots, \boldsymbol{v}_s)(x) := \gamma(x)\big(\alpha_1(x), \ldots, \alpha_r(x), v_1(x), \ldots, v_s(x)\big)$$

for $x \in X$. (This is clearly consistent with (3.4).) For these reasons, we can use the same notational conventions as before with vector fields and differential forms, that is, we identify tensors with their principal parts, and from now on use the ordinary font instead of boldface.

Addition

$$\mathcal{T}_s^r(X) \times \mathcal{T}_s^r(X) \to \mathcal{T}_s^r(X) \ , \quad (\gamma, \delta) \mapsto \gamma + \delta \ ,$$

multiplication by functions

$$\mathcal{E}(X) \times \mathcal{T}_s^r(X) \to \mathcal{T}_s^r(X), \ , \quad (f, \gamma) \mapsto f\gamma$$

and the **tensor product**

$$\mathcal{T}_{s_1}^{r_1}(X) \times \mathcal{T}_{s_2}^{r_2}(X) \to \mathcal{T}_{s_1+s_2}^{r_1+r_2}(X) \ , \quad (\gamma, \delta) \mapsto \gamma \otimes \delta \tag{3.23}$$

---

[12] In the case $s = 0$, we called this the vector part, and for $r = 0$, we called it the covector part. A tensor combines vectors and covectors, so such terminology is no longer possible.

[13] As usual, we identify $T_x\mathbb{R}^m$ and $T_x^*\mathbb{R}^m$ with $\mathbb{R}^m$.

will also be defined pointwise:

$$(\gamma + \delta)(x) := \gamma(x) + \delta(x) , \quad (f\gamma)(x) := f(x)\gamma(x) , \quad (\gamma \otimes \delta)(x) := \gamma(x) \otimes \delta(x) .$$

The following remarks are simple consequences of Remarks 2.20 and the chain rule. We leave the detailed proofs to you as exercises.

3.14 Remarks (a) $T_0^1(X) = \mathcal{V}(X)$ and $T_1^0(X) = \Omega^1(X)$. Also

$$T_2^0(X) = C^\infty(X, \mathcal{L}^2(\mathbb{R}^m)) ,$$

where we have used the canonical identification of a tensor with its principal part.

(b) The tensor product is $\mathcal{E}(X)$-bilinear and associative.

(c) $T_s^r(X)$ is an infinite-dimensional $\mathbb{R}$-vector space and an m^{r+s}-dimensional $\mathcal{E}(X)$-module. With the canonical basis $(\partial/\partial x^1, \ldots, \partial/\partial x^m)$ of $\mathbb{R}^m$,

$$\left\{ \frac{\partial}{\partial x^{j_1}} \otimes \cdots \otimes \frac{\partial}{\partial x^{j_r}} \otimes dx^{k_1} \otimes \cdots \otimes dx^{k_s} \; ; \; j_i, k_i \in \{1, \ldots, m\} \right\} \tag{3.24}$$

is a module basis of $T_s^r(X)$.

(d) An (r, s)-tensor γ on X belongs to $T_s^r(X)$ if and only if every r-tuple $\alpha_1, \ldots, \alpha_r$ in $\Omega^1(X)$ and every s-tuple $v_1, \ldots, v_s$ in $\mathcal{V}(X)$ satisfy

$$\gamma(\alpha_1, \ldots, \alpha_r, v_1, \ldots, v_s) \in \mathcal{E}(X) .$$

This is the case if and only if the coefficients of γ in basis (3.24) belong to $\mathcal{E}(X)$.

(e) (regularity) The definitions and claims above have obvious analogues which remain true for tensors of class C^k. ∎

Exercises

1 Let $\alpha, \beta \in \Omega(\mathbb{R}^4)$ be given by

$$\alpha := dx^1 + x^2\, dx^2 \quad \text{and} \quad \beta := \sin(x^2)\, dx^1 \wedge dx^3 + \cos(x^3)\, dx^2 \wedge dx^4 ,$$

and define $h \in C^\infty(\mathbb{R}^4, \mathbb{R}^4)$ by $h(x) := (x^1, x^2, x^3 x^4, x^4)$.
Calculate:

(i) $\gamma := \alpha \wedge \beta$;

(ii) $h^*\gamma$;

(iii) $h^*\gamma(0)(e_1, e_2, e_3 + e_4)$, where (e_1, e_2, e_3, e_4) is the standard basis in $\mathbb{R}^4$;

(iv) $d\alpha, d\beta, d\gamma, d(h^*\gamma)$.

2 Let $f_3 : \mathbb{R}^3 \to \mathbb{R}^3$, $(r, \varphi, \vartheta) \mapsto (x, y, z)$ be the spherical coordinate map.
Calculate

(a) $f_3^*\, dx, f_3^*\, dy, f_3^*\, dz$;

(b) $f_3^*(dy \wedge dz)$;

(c) $f_3^* \, dx \wedge f_3^* (dy \wedge dz)$.

3   A simple thermodynamic system (for example, an ideal gas) is characterized by its volume V and its temperature T (here $V, T \in \mathbb{R}$). The state of such a system is then described by the pressure $p := p(V, T)$ and the internal energy $E := E(V, T)$. By the second law of thermodynamics, the system has another state function $S := S(V, T)$, the entropy, whose differential is given by

$$dS := \frac{dE + p \, dV}{T} \quad \text{for } T > 0 .$$

Show the following facts:

(a) E and p satisfy the relation

$$\frac{\partial E}{\partial V} = T \frac{\partial p}{\partial T} - p .$$

(b) The internal energy of an ideal gas, which satisfies the equation of state $pV = RT$ with $R \in \mathbb{R}$ the (universal gas) constant, is independent of the volume, that is, $E = E(T)$.

(c) For van der Waals gas, which has the equation of state

$$\left(p + \frac{a}{V^2} \right)(V - b) = cT \quad \text{for } a, b, c \in \mathbb{R}^\times , \tag{3.25}$$

the internal energy does depend on volume.

(Hints: (a) $d^2 = 0$.   (c) $(3.25) \Rightarrow T \, \partial p / \partial T = p + a/V^2$.)

Remark   In the physics literature, $d\alpha$ is often written $\delta\alpha$ when the 1-form α is not exact.

4   An r-form $\alpha \in \Omega^r(X)$ is said to be **decomposable** if there are $\alpha_1, \ldots, \alpha_r \in \Omega^1(X)$ such that

$$\alpha = \alpha_1 \wedge \alpha_2 \wedge \cdots \wedge \alpha_r .$$

Let $\alpha, \beta \in \Omega^r(X)$ be decomposable. Calculate $(\alpha + \beta) \wedge (\alpha + \beta)$.

5   Suppose $\alpha = \sum_{j \le k} a_{jk} \, dx^j \wedge dx^k \in \Omega^2(X)$. Show that α is decomposable if and only if

$$a_{ij} a_{k\ell} + a_{jk} a_{i\ell} + a_{ki} a_{j\ell} = 0 \quad \text{for } 1 \le i, j, k, \ell \le n ,$$

where $a_{jk} := -a_{kj}$ for $j \ge k$.

6   Let $\alpha = \sum_{i \le j} a_{ij} \, dx^i \wedge dx^j \in \Omega^2(X)$. Show

$$d\alpha = \sum_{i < j < k} \left(\frac{\partial a_{ij}}{\partial x^k} + \frac{\partial a_{jk}}{\partial x^i} + \frac{\partial a_{ki}}{\partial x^j} \right) dx^i \wedge dx^j \wedge dx^k .$$

7   Calculate the exterior derivatives of

(a) $d\alpha \wedge \beta - \alpha \wedge d\beta$ and

(b) $d\alpha \wedge \beta \wedge \gamma + \alpha \wedge d\beta \wedge \gamma + \alpha \wedge \beta \wedge d\gamma$, where in (b) α and β are of even degree.

8   Find $d\alpha$ if $\alpha := \sum_{j=1}^m (-1)^{j-1} x^j / |x|^m \, dx^1 \wedge \cdots \wedge \widehat{dx^j} \wedge \cdots \wedge dx^m \in \Omega^{m-1}(\mathbb{R}^m \setminus \{0\})$.

9 Let $\alpha := 2xz \, dy \wedge dz + dz \wedge dx - (z^2 + e^x) \, dx \wedge dy \in \Omega^2(\mathbb{R}^3)$. Show that α is exact and determine an antiderivative.

10 Suppose $\omega \in \Omega^2(X)$ is nondegenerate. Show that

$$\Theta_\omega : \mathcal{V}(X) \to \Omega^1(X) \,, \quad v \mapsto \omega(v, \cdot)$$

is an $\mathcal{E}(X)$-module isomorphism.

11 Prove these three statements:

(a) The symplectic form $\sigma \in \Omega^2(\mathbb{R}^{2m})$ is nondegenerate and closed.

(b) The m-fold product $\sigma^m := \sigma \wedge \cdots \wedge \sigma \in \Omega^{2m}(\mathbb{R}^{2m})$ satisfies $\sigma^m \neq 0$.

(c) According to Exercise 10 and (b) the **symplectic gradient** $\operatorname{sgrad} f := \Theta_\sigma^{-1} \, df \in \mathcal{V}(\mathbb{R}^{2m})$ is defined for every $f \in \mathcal{E}(\mathbb{R}^{2m})$. Calculate $\operatorname{sgrad} f$ in the coordinates $(q, p) \in \mathbb{R}^m \times \mathbb{R}^m$.

12 If σ is the symplectic form on $\mathbb{R}^{2m}$, then

$$\{\cdot, \cdot\} : \mathcal{E}(\mathbb{R}^{2m}) \times \mathcal{E}(\mathbb{R}^{2m}) \to \mathcal{E}(\mathbb{R}^{2m}) \,, \quad (f, g) \mapsto \sigma(\operatorname{sgrad} f, \operatorname{sgrad} g)$$

is called the **Poisson bracket**.

For $f, g, h \in \mathcal{E}(\mathbb{R}^{2m})$ and $c \in \mathbb{R}$ prove

(i) in local coordinates $(q_1, \ldots, q_m, p_1, \ldots, p_m)$, the Poisson bracket reads

$$\{f, g\} = \sum_{j=1}^m \left(\frac{\partial f}{\partial p_j} \frac{\partial g}{\partial q_j} - \frac{\partial f}{\partial q_j} \frac{\partial g}{\partial p_j} \right) ;$$

(ii) $\{f, cg + h\} = c\{f, g\} + \{f, h\}$;

(iii) $\{f, g\} = -\{g, f\}$;

(iv) $\{f, \{g, h\}\} + \{g, \{h, f\}\} + \{h, \{f, g\}\} = 0$ (Jacobi identity);

(v) $\{f, gh\} = g\{f, h\} + h\{f, g\}$;

(vi) $\operatorname{sgrad}\{f, g\} = (\operatorname{sgrad} f \mid \operatorname{sgrad} g)_{\mathbb{R}^{2m}}$.

13 Show that the Poisson bracket is related to the symplectic form σ on $\mathbb{R}^{2m}$ by the relation

$$df \wedge dg \wedge \sigma^{m-1} = \frac{1}{m} \{f, g\} \sigma^m \,.$$

4 Vector fields and differential forms

This section is devoted to the global theory of differential forms, that is, to differential forms on manifolds. The first part, which is essentially a simple transfer of the local theory, requires us to focus on the problem of regularity. With help from a theorem about partitions of unity, we can then extend the important concept of the exterior derivative to the case of manifolds and show that the rules we developed for the local theory still apply.

The global theory brings up an important new idea, the orientability of a manifold. We present various ways to characterize this central concept and consider numerous examples. To prepare for the theory of integration on manifolds, we give explicit representations of the volume elements of many important manifolds.

In this entire section,

- M is an m-dimensional, and N is an n-dimensional manifold;
- $r \in \mathbb{N}$.

Vector fields

By a **vector field** v on M, we mean a map

$$v : M \to TM \qquad \text{with} \quad v(p) \in T_p M \text{ for } p \in M ,$$

that is, a section of the tangent bundle. If v is a vector field on M, then we can "transplant" it using a diffeomorphism from M to N. So we define for $\varphi \in \mathrm{Diff}^1(M, N)$ the **push forward** $\varphi_* v$ of v by φ by letting

$$\varphi_* v(q) := (T_{\varphi^{-1}(q)} \varphi) v(\varphi^{-1}(q)) , \qquad q \in N .$$

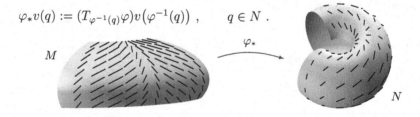

Therefore $\varphi_* v$ is a vector field on N. For functions on M, the **push forward** by a bijection $\psi : M \to N$ is the assignment

$$\psi_* : \mathbb{R}^M \to \mathbb{R}^N , \qquad a \mapsto \psi_* a := a \circ \psi^{-1} .$$

4.1 Remarks (a) For functions, the push forward ψ_* is obviously the same as the pull back ψ^{-1}: $\psi_* = (\psi^{-1})^*$. Note however that, in contrast to the pull back, the push forward is only defined for bijections. In particular, it must be true that $\dim(M) = \dim(N)$.[1]

(b) Let $\varphi \in \mathrm{Diff}^1(M, N)$. Then

$$\varphi_*(a + b) = \varphi_* a + \varphi_* b , \quad \varphi_*(v + w) = \varphi_* v + \varphi_* w ,$$

and

$$\varphi_*(av) = \varphi_* a \, \varphi_* v$$

for $a, b \in \mathbb{R}^M$ and vector fields v and w on M.

(c) Let $\varphi \in \mathrm{Diff}(M, N)$ and $\psi \in \mathrm{Diff}(N, L)$, where L is another manifold. Then

$$(\psi \circ \varphi)_* = \psi_* \circ \varphi_* \quad \text{and} \quad (\mathrm{id}_M)_* = \mathrm{id}_{\mathcal{F}(M)} \tag{4.1}$$

for $\mathcal{F}(M) := \mathcal{E}(M)$ or $\mathcal{F}(M) := \mathcal{V}(M)$. The rule (4.1) means that the push forward operates **covariantly**.

Proof The statement is obvious for push forwards of functions. For vector fields, (4.1) follows from the chain rule of Remark VII.10.9(b) and from Remark 1.14(c). ∎

Let $k \in \mathbb{N} \cup \{\infty\}$. The vector field v on M belongs to the class C^k (that is, it is k-times **continuously differentiable**, or **smooth** in case $k = \infty$) if every point p of M has a chart (φ, U) around p such that[2] $\varphi_* v \in \mathcal{V}^k(\varphi(U))$. We denote the set of all vector fields on M of class C^k by $\mathcal{V}^k(M)$. For simplicity of notation, we set

$$\mathcal{V}(M) := \mathcal{V}^\infty(M) \quad \text{and} \quad \mathcal{E}(M) := C^\infty(M) .$$

4.2 Remarks **(a)** The definition of C^k vector fields is coordinate-independent. If v is a C^k vector field and (ψ, V) is an arbitrary chart of M, then $\psi_* v$ belongs to the class C^k.

Proof Suppose therefore (ψ, V) is a chart of M. Then we need to show that $\psi_* v$ belongs to the class C^k. Every $q \in V$ has a chart (φ, U) of M around it such that $\varphi_* v \in \mathcal{V}^k(\varphi(U))$. Then $\psi_* v = (\psi \circ \varphi^{-1})_* \varphi_* v$ follows from (4.1). Because

$$\psi \circ \varphi^{-1} \in \mathrm{Diff}\big(\varphi(U \cap V), \psi(U \cap V)\big)$$

and $\varphi_* v \in \mathcal{V}^k(\varphi(U))$, we find $\psi_* v \in \mathcal{V}^k(\psi(U \cap V))$. Because this holds for every $q \in V$ and because differentiability is a local property, we get $\psi_* v \in \mathcal{V}^k(\psi(V))$. ∎

(b) The pointwise-defined operations

$$\mathcal{V}(M) \times \mathcal{V}(M) \to \mathcal{V}(M) , \quad (v, w) \mapsto v + w$$

---

[1] See Exercise VII.10.9.

[2] See Section VIII.3. What was said there holds without change for open subsets of $\overline{\mathbb{H}}^m$ as well.

and
$$\mathcal{E}(M) \times \mathcal{V}(M) \to \mathcal{V}(M) , \quad (a, v) \mapsto av$$

make $\mathcal{V}(M)$ into an $\mathcal{E}(M)$-module. In particular, $\mathcal{E}(M)$ and $\mathcal{V}(M)$ are (infinite-dimensional) $\mathbb{R}$-vector spaces.

If $\varphi \in \mathrm{Diff}(M, N)$, then φ_* is a module isomorphism from $\mathcal{E}(M)$ to $\mathcal{E}(N)$ and from $\mathcal{V}(M)$ to $\mathcal{V}(N)$.

Proof It follows from (4.1) that

$$\mathrm{id}_M = (\varphi^{-1} \circ \varphi)_* = (\varphi^{-1})_* \varphi_* \quad \text{and} \quad \mathrm{id}_N = (\varphi \circ \varphi^{-1})_* = \varphi_* (\varphi^{-1})_* .$$

Therefore φ_* is bijective, and $\varphi_*^{-1} = (\varphi^{-1})_*$. The remaining claims are simple consequences of Remark 4.1(b) and the properties of vector fields on open subsets of $\overline{\mathbb{H}}^m$ (see Section VIII.3). ∎

(c) Let X_0 and X_1 be open in $\mathbb{R}^m$, and suppose $\varphi \in \mathrm{Diff}(X_0, X_1)$. Also denote by $\Theta_j : \mathcal{V}(X_j) \to \Omega^1(X_j)$ for $j = 0, 1$ the canonical module isomorphism that was defined in Remark VIII.3.3(g). Then

$$(\varphi^{-1})^* \circ \Theta_0 = \Theta_1 \circ \varphi_* ,$$

that is, the diagram

$$
\begin{array}{ccc}
\mathcal{V}(X_0) & \xrightarrow{\varphi_*} & \mathcal{V}(X_1) \\
\Theta_0 \downarrow & & \downarrow \Theta_1 \\
\Omega^1(X_0) & \xrightarrow{(\varphi^{-1})^*} & \Omega^1(X_1)
\end{array}
$$

commutes.

(d) (regularity) Let $k \in \mathbb{N}$. For $\varphi \in \mathrm{Diff}^{k+1}(M, N)$ and $0 \le \ell \le k$, the push forward φ_* maps $C^\ell(M)$ to $C^\ell(N)$ and $\mathcal{V}^\ell(M)$ to $\mathcal{V}^\ell(N)$, but this statement (without the inequality) does not hold for $\ell = k + 1$.

If M is a C^{k+1} manifold, then the $C^\ell(M)$-modules $C^\ell(M)$ and $\mathcal{V}^\ell(M)$ are defined for $0 \le \ell \le k$; however[3] the modules $C^{k+1}(M)$ and $\mathcal{V}^{k+1}(M)$ are not.

Proof This is because the tangential "loses" one derivative. ∎

Local basis representation

Let (φ, U) be a chart of M around p. Then we denote by

$$\partial_j|_p = \frac{\partial}{\partial x^j}\Big|_p \in T_p M \quad \text{for } 1 \le j \le m$$

---

[3]except for trivial cases

the basis vectors of T_pM corresponding to the local coordinates $\varphi = (x^1, \dots, x^m)$. In other words, $\partial_j|_p$ is the tangent vector on the coordinate path $t \mapsto \varphi^{-1}(\varphi(p) + te_j)$ at the point[4] p, that is,

$$\partial_j|_p := (T_p\varphi)^{-1}(\varphi(p), e_j) \quad \text{for } 1 \le j \le m, \tag{4.2}$$

where $(e_1, \dots, e_m)$ is the canonical basis of $\mathbb{R}^m$.

4.3 Remarks (a) Let $i_M : M \hookrightarrow \mathbb{R}^{\overline{m}}$, and let $g_\varphi := i_M \circ \varphi^{-1} : \varphi(U) \to \mathbb{R}^{\overline{m}}$ be the parametrization belonging to φ. Then

$$(T_p i_M)\partial_j|_p = \big(p, \partial_j g_\varphi(\varphi(p))\big) \in T_p\mathbb{R}^{\overline{m}} \quad \text{for } 1 \le j \le m.$$

This means that, if we identify $\partial_j|_p \in T_pM$ with its image in $T_p\mathbb{R}^{\overline{m}}$ under the canonical injection

$$T_p i_M : T_pM \to T_p\mathbb{R}^{\overline{m}},$$

then we find $\partial_j|_p = \big(p, \partial_j g_\varphi(\varphi(p))\big)$.

Proof From Example VII.10.9(b) and Remark 1.14(c), we get

$$T_{\varphi(p)}g_\varphi = T_{\varphi(p)}(i_M \circ \varphi^{-1}) = T_p i_M \circ T_{\varphi(p)}(\varphi^{-1}) = T_p i_M \circ (T_p\varphi)^{-1}.$$

Then it follows from (4.2) that

$$(T_p i_M)\partial_j|_p = (T_{\varphi(p)}g_\varphi)(\varphi(p), e_j) = (p, \partial g_\varphi(\varphi(p))e_j) = \big(p, \partial_j g_\varphi(\varphi(p))\big). \quad \blacksquare$$

(b) The maps

$$\partial_j = \frac{\partial}{\partial x^j} : U \to TU, \quad p \mapsto \partial_j|_p \quad \text{for } 1 \le j \le m$$

are smooth vector fields on U.

Proof This is clear because

$$(\varphi_* \partial_j)(\varphi(p)) = (T_p\varphi)(T_p\varphi)^{-1}(\varphi(p), e_j) = (\varphi(p), e_j) \quad \text{for } 1 \le j \le m$$

for $p \in U$. $\blacksquare$

---

[4]If p is in the interior of M. If p is a boundary point, we must make φ a submanifold chart of $\mathbb{R}^{\overline{m}}$ around p for M.

(c) For $p \in U$, we have a basis $(\partial_1|_p, \ldots, \partial_m|_p)$ of T_pM and a module basis $(\partial_1, \ldots, \partial_m)$ of $\mathcal{V}(U)$. A vector field v on U belongs to $\mathcal{V}(U)$ if and only if the coefficients v_j of the basis representation

$$v = \sum_{j=1}^{m} v^j \partial_j$$

all belong to $\mathcal{E}(U)$.

Proof The first statement follows from Remark VII.10.5 and the definition of the tangent space at a boundary point. The second claim is a consequence of

$$\varphi_* v = \varphi_* \left(\sum_{j=1}^{m} v^j \partial_j \right) = \sum_{j=1}^{m} (\varphi_* v^j) \varphi_* \partial_j \ ,$$

of (b), and of Remark VIII.3.3(c). ∎

(d) (regularity) Let $k \in \mathbb{N}$, and let M be a C^{k+1} manifold. In this case, $(\partial_1, \ldots, \partial_m)$ is a $C^k(U)$-module basis of $\mathcal{V}^k(U)$. A vector field v on U belongs to $\mathcal{V}^k(U)$ if and only if its coefficients with respect to this basis representation lie in $C^k(U)$. ∎

Differential forms

To generalize the cotangent space T_p^*X and the cotangent bundle T^*X of an open subset X of $\mathbb{R}^m$, we now define the **cotangent space** of M at the point p by

$$T_p^*M := (T_pM)^* = \mathcal{L}(T_pM, \mathbb{R}) \ .$$

We define the **cotangent bundle** of M by

$$T^*M := \bigcup_{p \in M} T_p^*M \ .$$

We denote by

$$\langle \cdot, \cdot \rangle_p : T_p^*M \times T_pM \to \mathbb{R} \quad \text{for } p \in M$$

the dual pairing[5] and call

$$\langle \cdot, \cdot \rangle : T^*M \times TM \to \mathcal{E}(M) \ , \quad (\alpha, v) \mapsto \left[p \mapsto \langle \alpha(p), v(p) \rangle_p \right]$$

the **dual pairing** as well.

Because T_pM is an m-dimensional vector space, so is T_p^*M. Hence for $r \in \mathbb{N}$ and $p \in M$, the r-fold exterior product $\bigwedge^r T_p^*M$ of T_p^*M and the Grassmann

---

[5]See Section VIII.3.

algebra

$$\wedge T_p^* M = \bigoplus_{r \geq 0} \wedge^r T_p^* M$$

of $T_p^* M$ are defined. To extend the concepts introduced in the previous section, we define the **bundle of alternating r-forms** on M by

$$\wedge^r T^* M := \bigcup_{p \in M} \wedge^r T_p^* M \ .$$

We define the **Grassmann bundle** of M by

$$\wedge T^* M := \bigcup_{p \in M} \wedge T_p^* M \ .$$

A **differential form** on M is then a map

$$\alpha : M \to \wedge T^* M \qquad \text{with} \quad \alpha(p) \in \wedge T_p^* M \text{ for } p \in M \ ,$$

that is, a section of the Grassmann bundle. It has **degree r** (or is called an **r-form**) if $\alpha(M) \subset \wedge^r T^* M$. Sometimes we call a 1-form a **Pfaff form**.

If α and β are differential forms on M, then the **sum** $\alpha + \beta$ and the **exterior product**[6] $\alpha \wedge \beta$ are defined pointwise:

$$(\alpha + \beta)(p) := \alpha(p) + \beta(p) \text{ and } \alpha \wedge \beta(p) := \alpha(p) \wedge \beta(p) \quad \text{for } p \in M \ .$$

If α is an r-form on M, then its effect on vector fields is also defined pointwise:

$$\alpha(v_1, \ldots, v_r)(p) := \alpha(p)\big(v_1(p), \ldots, v_r(p)\big) \quad \text{for } p \in M \text{ and } v_1, \ldots, v_r \in \mathcal{V}(M) \ .$$

Finally let $\varphi \in C^1(M, N)$, and let β be a differential form on N. Then the **pull back** of β by φ is again defined pointwise:

$$\varphi^* \beta(p) := (T_p \varphi)^* \beta\big(\varphi(p)\big) \quad \text{for } p \in M \ .$$

Obviously $\varphi^* \beta$ is a differential form on M, the pull back of β by φ. If φ is a C^1 diffeomorphism from M to N, then

$$\varphi_* \alpha := (\varphi^{-1})^* \alpha$$

is the **push forward** of the differential form α on M.

Let $k \in \mathbb{N} \cup \{\infty\}$. The differential form α on M belongs to the **class C^k** (or is k-times **continuously differentiable**,[7] or **smooth** in the case $k = \infty$) if there is a

---

[6] $\wedge$ is also called the **wedge product**.

[7] Of course, we say a differential form of class C^0 is continuous.

chart (φ, U) around every point of M such that $\varphi_* \alpha$ is a differential form of class C^k on $\varphi(U)$. We denote the set of all r-forms of class C^k on M by

$$\Omega_{(k)}^r(M) \ ,$$

and

$$\Omega^r(M) := \Omega_{(\infty)}^r(M)$$

is the set of all smooth r-forms on M. Finally

$$\Omega(M) := \Omega_{(\infty)}(M)$$

is the set of all smooth differential forms on M.

Following our treatment of vector fields, we will generally restrict our attention to the study of smooth differential forms. We leave it to you to prove that all the statements we prove about smooth forms also hold analogously for forms of class C^k provided k has been restricted as the case may require.

4.4 Remarks   (a) The above notion of differentiability of differential forms is coordinate-independent.

If α is an r-form of class C^k on M and (ψ, V) is a chart on M, then $\psi_* \alpha$ is an r-form of class C^k on $\psi(V)$.

Proof   Every $p \in M$ has a chart (φ, U) around it with $\varphi_* \alpha \in \Omega_{(k)}^r(\varphi(U))$. From Remark 3.3(a) and the pointwise definition of the push forward, it follows that

$$\psi_* \alpha = (\psi \circ \varphi^{-1})_* \varphi_* \alpha \ .$$

After this, the claim follows in analogy to the proof of Remark 4.2(a). ∎

(b) $\Omega(M)$ and $\Omega^r(M)$ are $\mathcal{E}(M)$-modules and therefore in particular $\mathbb{R}$-vector spaces. Also

$$\Omega(M) = \bigoplus_{r \geq 0} \Omega^r(M) \ .$$

The exterior product is $\mathbb{R}$-bilinear, associative, and graded anticommutative, that is, it satisfies these rules:

(i) The map

$$\Omega^r(M) \times \Omega^s(M) \to \Omega^{r+s}(M) \ , \quad (\alpha, \beta) \mapsto \alpha \wedge \beta$$

is well defined and $\mathbb{R}$-bilinear.

(ii) $\alpha \wedge (\beta \wedge \gamma) = (\alpha \wedge \beta) \wedge \gamma$ for $\alpha, \beta, \gamma \in \Omega(M)$.

(iii) $\alpha \wedge \beta = (-1)^{rs} \beta \wedge \alpha$ for $\alpha \in \Omega^r(M)$ and $\beta \in \Omega^s(M)$.

Proof   This follows from the definition of smoothness, from the pointwise definition of $\wedge$, and from Theorem 2.7. ∎

(c) Every $\alpha \in \Omega^r(M)$ is an alternating r-form on $\mathcal{V}(M)$.

(d) $\Omega^0(M) = \mathcal{E}(M)$, and $\Omega^r(M) = \{0\}$ for $r > m$.

(e) For $h \in C^\infty(M, N)$, the pull back $h^* : \Omega(N) \to \Omega(M)$ is an algebra homomorphism, that is,

$$h^*(\alpha + \beta) = h^*\alpha + h^*\beta \ , \quad h^*(\alpha \wedge \beta) = h^*\alpha \wedge h^*\beta$$

for $\alpha, \beta \in \Omega(N)$. If $\alpha \in \Omega^r(N)$, then $h^*\alpha$ belongs to $\Omega^r(M)$. Also

$$(k \circ h)^* = h^* \circ k^* \text{ and } (\mathrm{id}_M)^* = \mathrm{id}_{\Omega(M)} \ .$$

If h is a diffeomorphism, then h^* is bijective, and $(h^*)^{-1} = (h^{-1})^* = h_*$.

Proof We leave the simple checks to you. ∎

(f) Suppose M is a submanifold of N and $i : M \hookrightarrow N$ is the natural embedding.[8]
Then for $\alpha \in \Omega^r(N)$,

$$\alpha \,|\, M := i^*\alpha \in \Omega^r(M)$$

is the **restriction**[9] of α to M. Let $p \in M$. Because the tangent space T_pM can be regarded as a vector subspace of T_pN, we have $(\alpha \,|\, M)(p) = \alpha(p) \,|\, (T_pM)^r$. ∎

Local representations

Suppose $f \in C^1(M) := C^1(M, \mathbb{R})$. As in Section VII.10, we define the **differential** df of f by

$$df(p) := \mathrm{pr} \circ T_p f \quad \text{for } p \in M \ ,$$

where

$$\mathrm{pr} := \mathrm{pr}_2 : T_{f(p)}\mathbb{R} = \{f(p)\} \times \mathbb{R} \to \mathbb{R}$$

is the canonical projection.

Let (φ, U) be a chart around $p \in M$. Then it follows from the definitions of $df(p)$ and $\partial_j|_p$ as well as the chain rule of Remarks VII.10.9(b) and 1.14(c) that

$$
\begin{aligned}
\langle df(p), \partial_j|_p \rangle_p &= \langle df(p), (T_{\varphi(p)}\varphi^{-1})(\varphi(p), e_j) \rangle_p \\
&= \mathrm{pr} \circ T_p f \circ T_{\varphi(p)}\varphi^{-1}(\varphi(p), e_j) \\
&= \mathrm{pr} \circ T_{\varphi(p)}(f \circ \varphi^{-1})(\varphi(p), e_j) \\
&= \partial(f \circ \varphi^{-1})(\varphi(p))e_j \\
&= \partial_j(f \circ \varphi^{-1})(\varphi(p)) \\
&= \partial_j(\varphi_* f)(\varphi(p))
\end{aligned}
$$

for $1 \le j \le m$. With the abbreviation

$$\partial_j f(p) := \frac{\partial f}{\partial x^j}(p) := \partial_j(f \circ \varphi^{-1})(\varphi(p)) = \partial_j(\varphi_* f)(\varphi(p)) \tag{4.3}$$

---

[8]In this situation, we always assume that N is without boundary.
[9]See Example 3.4(j).

for $1 \leq j \leq m$ and $p \in U$, we thus have

$$\langle df(p), \partial_j|_p \rangle_p = \partial_j f(p) \quad \text{for } 1 \leq j \leq m , \quad p \in U . \tag{4.4}$$

Therefore

$$\langle df, \partial_j \rangle = \partial_j f \quad \text{for } 1 \leq j \leq m . \tag{4.5}$$

Note that the usual partial derivative $\partial_j f$ on M (in the sense of Remark VII.2.7(a)) is not defined when M is not "flat", that is, not an open subset of $\mathbb{R}^m$. Because derivatives of functions on manifolds can only be defined in terms of local representations, $\partial_j f$ in (4.5) is meaningless unless it is interpreted as the partial derivative of the function "pushed down" by φ to the parameter domain $\varphi(U)$, that is, the partial derivative of the $\varphi_* f$ appearing in (4.3). This rules out any misinterpretation in practice. The notation $\partial f/\partial x^j$ has the advantage that it gives the "name of the coordinates" $(x^1, \ldots, x^m) = \varphi$ in which f is locally written.

In Section VII.2, for the case of open subsets of $\mathbb{R}^m$, we defined the partial derivative $\partial_j f(p)$ as the image of the j-th coordinate unit vector e_j under the (total) derivative $\partial f(p)$ (that is, the linearization of f at p). Since $df(p)$ is just the tangent part of the tangential $T_p f$ and therefore the "linearization of f at the point p", and since $\partial_j|_p$ is the j-th coordinate basis vector of $T_p M$, (4.4) shows that $\partial_j f(p)$ is the tangent part of the image of these coordinate vectors under the tangential of f. Therefore (4.3) is indeed the correct generalization of the concept of partial derivative to functions defined on manifolds.

Finally, it is clear that (4.3) agrees with the classical partial derivative when M is open in $\mathbb{R}^m$ and φ denotes the trivial chart id_M.

4.5 Remarks Let (φ, U) be a chart of M.

(a) For $f \in \mathcal{E}(M) = \Omega^0(M)$, the differential df belongs to $\Omega^1(M)$. The map

$$d: \Omega^0(M) \to \Omega^1(M) , \quad f \mapsto df$$

is $\mathbb{R}$-linear.

(b) Let $(x^1, \ldots, x^m) = \varphi$ be the local coordinates on U induced by φ, so that

$$x^j := \mathrm{pr}_j \circ \varphi \in \mathcal{E}(U) \quad \text{for } 1 \leq j \leq m ,$$

where $\mathrm{pr}_j : \mathbb{R}^m \to \mathbb{R}$ are the canonical projections. Then $\Omega^1(U)$ is a free $\mathcal{E}(U)$-module of dimension m, and $(dx^1, \ldots, dx^m)$ is a module basis with

$$\left\langle dx^j, \frac{\partial}{\partial x^k} \right\rangle = \delta_k^j \quad \text{for } 1 \leq j, k \leq m \tag{4.6}$$

and is the **dual basis** to the basis $(\partial/\partial x^1, \ldots, \partial/\partial x^m)$ of $\mathcal{V}(U)$. The basis representations

$$v = \sum_{j=1}^m v^j \frac{\partial}{\partial x^j} \in \mathcal{V}(U) \quad \text{and} \quad \alpha = \sum_{j=1}^m a_j \, dx^j \in \Omega^1(U) \tag{4.7}$$

require the relations

$$v^j = \langle dx^j, v \rangle \in \mathcal{E}(U) \quad \text{and} \quad a_j = \Big\langle \alpha, \frac{\partial}{\partial x^j} \Big\rangle \in \mathcal{E}(U) \tag{4.8}$$

for $1 \le j \le m$. In particular, for $f \in \mathcal{E}(U)$, we have

$$df = \sum_{j=1}^{m} \frac{\partial f}{\partial x^j} \, dx^j \in \Omega^1(U) \,.$$

Proof (4.3) and (4.5) imply

$$\langle dx^j, \partial_k \rangle = \partial_k x^j = \varphi^* \partial_k (\varphi_* x^j) = \varphi^* \partial_k \big[(\mathrm{pr}_j \circ \varphi) \circ \varphi^{-1} \big] = \varphi^* \partial_k \, \mathrm{pr}_j = \delta_k^j$$

and hence (4.6). For v with the representation given in (4.7), we obtain

$$\langle dx^j(p), v(p) \rangle_p = \sum_{k=1}^{m} v^k(p) \Big\langle dx^j(p), \frac{\partial}{\partial x^k} \Big|_p \Big\rangle_p = \sum_{k=1}^{m} v^k(p) \delta_k^j = v^j(p) \tag{4.9}$$

for $p \in U$ and $1 \le j \le m$, because $dx^j(p)$ is a linear form on $T_p M = T_p U$. Therefore the first part of (4.7) and Remark 4.3(c) imply the first claim of (4.8).

For the push forward dx^j by φ, we find by applying Remarks VII.10.9(b) and 1.14(c) as well as (4.2) and (4.6) that

$$\begin{aligned}
\big\langle (\varphi_* dx^j)(\varphi(p)), (\varphi(p), e_k) \big\rangle_{\varphi(p)} &= \big\langle dx^j(p), (T_{\varphi(p)} \varphi^{-1})(\varphi(p), e_k) \big\rangle_p \\
&= \big\langle dx^j(p), (T_p \varphi)^{-1}(\varphi(p), e_k) \big\rangle_p \\
&= \Big\langle dx^j(p), \frac{\partial}{\partial x^k} \Big|_p \Big\rangle_p = \delta_k^j \,.
\end{aligned} \tag{4.10}$$

This shows that $(\varphi_* dx^1, \ldots, \varphi_* dx^m)$ is, at every point $\varphi(p) \in \varphi(U)$, the basis dual to the canonical basis of $T_{\varphi(p)} \varphi(U)$. In particular, the covector part of $\varphi_* dx^j$ is constant on $\varphi(U)$.

Remark 4.4(e) guarantees that φ_* is a vector space isomorphism from $\Omega^1(U)$ to $\Omega^1(\varphi(U))$. From this, (4.10), and Proposition 2.3, we conclude that every $\alpha \in \Omega^1(U)$ has a representation of the form given in (4.7) by real-valued functions a_j on U. Because of

$$\varphi_* \alpha = \sum_{j=1}^{m} (\varphi_* a_j) \, \varphi_* dx^j \tag{4.11}$$

and due to the constancy of the covector part of 1-forms $\varphi_* dx^j$ on $\varphi(U)$, we learn from Remark 3.1(e) that α belongs to $\Omega^1(U)$ if and only if $a_j \in \mathcal{E}(U)$ for $1 \le j \le m$. Finally $a_j = \langle \alpha, \partial_j \rangle$ follows by a calculation analogous to (4.9). $\blacksquare$

(c) For $r \in \mathbb{N}$, $\Omega^r(U)$ is a free $\mathcal{E}(U)$-module of dimension $\binom{m}{r}$, and

$$\big\{ \, dx^{(j)} = dx^{j_1} \wedge \cdots \wedge dx^{j_r} \; ; \; (j) = (j_1, \ldots, j_r) \in \mathbb{J}_r \, \big\} \tag{4.12}$$

is a basis. An r-form α on U has a uniquely determined **basis representation in local coordinates**

$$\alpha = \sum_{(j)\in \mathbb{J}_r} a_{(j)}\, dx^{(j)} \tag{4.13}$$

whose coefficients are

$$a_{(j)} = \alpha\left(\frac{\partial}{\partial x^{j_1}}, \ldots, \frac{\partial}{\partial x^{j_r}}\right) \quad \text{for } (j)\in \mathbb{J}_r . \tag{4.14}$$

If $k \in \mathbb{N} \cup \{\infty\}$, then α belongs to the class C^k on U if and only if $a_{(j)} \in C^k(U)$ for $(j) \in \mathbb{J}_r$.

Proof From (4.10) and the properties of the pull back $(\varphi^{-1})^* = \varphi_*$ given in Remark 4.4(e), it follows that

$$\varphi_*\, dx^{(j)} = \varepsilon^{(j)} \quad \text{for } (j)\in \mathbb{J}_r , \tag{4.15}$$

where $(\varepsilon^1, \ldots, \varepsilon^m)$ denotes the basis dual to the canonical basis of $T_{\varphi(p)}\varphi(U)$ for $p \in U$. Because φ_* is a vector space isomorphism from $\Omega^r(U)$ to $\Omega^r\big(\varphi(U)\big)$, we derive from Proposition 2.3 and (4.2) that every r-form α on U has a unique representation of the form (4.13), whose coefficients are given by (4.14). Because

$$\varphi_*\alpha = \sum_{(j)\in \mathbb{J}_r} (\varphi_* a_{(j)})\varphi_*\, dx^{(j)}$$

and by (4.15), the definition of the differentiability of an r-form of class C^k implies that α belongs to the class C^k if and only if the $a_{(j)}$ lie in $C^k(U)$. ∎

(d) Note that we have only shown that $\mathcal{V}(U)$ and $\Omega(U)$ are free modules, while we have made no such statements about $\mathcal{V}(M)$ and $\Omega(M)$. Indeed, corresponding statements are not generally true in the global case, that is, for manifolds that cannot be described by a single chart. For example, it is known[10] that the n-sphere does not support n (nontrivial) linearly independent vector fields (that is, $\mathcal{V}(S^n)$ is not a free module of dimension n) unless $n = 0, 1, 3$, or 7.

(e) (regularity) If $k \in \mathbb{N}$, then the statements of (c) remain true if M is a C^{k+1} manifold. ∎

The local coordinates $x^1, \ldots, x^m$ on U belonging to a chart φ are smooth functions on U; namely, they are the maps $\mathrm{pr}_j \circ \varphi \in \mathcal{E}(U)$ for $1 \leq j \leq m$. On the other hand, we also use $(x^1, \ldots, x^m)$ as the notation for a general point of $\varphi(U)$, that is, the coordinates of $\mathbb{R}^m$ are also called $x^1, \ldots, x^m$. This use of the same notation for two different things is deliberate. It simplifies calculations with (local) coordinates considerably, if it is clear from context which interpretation is correct. For example, the expression

$$\alpha = \sum_{(j)\in \mathbb{J}_r} a_{(j)}\, dx^{(j)} \tag{4.16}$$

---

[10] By work of Bott, Kervaire, and Milnor.

has two meanings if no other specification is made (which is usual in practice). First, we can regard (4.16) as the basis representation of an r-form on the open subset $X = \varphi(U)$ of $\overline{\mathbb{H}}^m$, as we have done in the previous sections. Or, we can interpret (4.16) as the basis representation of an r-form on U with respect to the local coordinates in the corresponding chart. This is the standpoint we have taken here. In the first case, the $a_{(j)}$ are functions on X, and the $dx^{(j)}$ are the constant basis forms of $\mathbb{R}^m$. In the second, the $a_{(j)}$ are functions on $U \subset M$, and the $dx^{(j)}$ are the position-dependent r-forms that "live" on U. *Because of (4.15), we must, in order to pass from second interpretation to the first, "pass down" the coefficient functions $a_{(j)} = a_{(j)}(p)$ to the parameter domain using φ. That is, $a_{(j)}$ must be interpreted as $\varphi_* a_{(j)} = a_{(j)} \circ \varphi^{-1}$, and we must think $a_{(j)} = a_{(j)}(x)$ for $x \in X$.*

4.6 Examples (a) Denote the upper and lower hemispheres of the m-sphere S^m in $\mathbb{R}^{m+1}$ by S_+^m and S_-, respectively. That is, let

$$S_\pm^m := \left\{ x \in \mathbb{R}^{m+1} \;;\; |x| = 1,\ \pm x^{m+1} > 0 \right\} .$$

Also let

$$\varphi_\pm : S_\pm^m \to \mathbb{B}^m , \qquad x \mapsto x' := (x^1, \ldots, x^m)$$

be the projection of $\mathbb{B}^m = \mathbb{B}^m \times \{0\}$ onto the hyperplane orthogonal to the x^{m+1}-axis. Then (φ_+, S_+^m) and (φ_-, S_-^m) are charts of S^m. For

$$\alpha := \sum_{j=1}^{m+1} (-1)^{j-1} x^j\, dx^1 \wedge \cdots \wedge \widehat{dx^j} \wedge \cdots \wedge dx^{m+1} \in \Omega^m(\mathbb{R}^{m+1}) ,$$

the restriction to S^m reads, in the local coordinates induced by $\varphi_\pm$, as

$$\alpha \,|\, S_\pm^m = \pm \frac{(-1)^m}{\sqrt{1 - |x'|^2}}\, dx^1 \wedge \cdots \wedge dx^m .$$

Proof Let $g_\pm(x') := \left(x', \pm\sqrt{1 - |x'|^2}\right)$ for $x' \in \mathbb{B}^m$. Then $g_\pm$ is smooth and is the parametrization belonging to $\varphi_\pm$ of the hemisphere $S_\pm^m$ as a graph over $\mathbb{B}^m$. Also $g_\pm = i \circ \varphi_\pm^{-1}$ with $i : S^m \hookrightarrow \mathbb{R}^{m+1}$. Therefore $(\varphi_\pm, S_\pm^m)$ are charts of S^m. For these we find

$$(\varphi_\pm)_*(\alpha \,|\, S_\pm^m) = (\varphi_\pm^{-1})^* \circ i^* \alpha = g_\pm^* \alpha$$

$$= \sum_{j=1}^{m+1} (-1)^{j-1} g_\pm^j\, dg_\pm^1 \wedge \cdots \wedge \widehat{dg_\pm^j} \wedge \cdots \wedge dg_\pm^{m+1}$$

$$= \sum_{j=1}^{m} (-1)^{j-1} x^j\, dx^1 \wedge \cdots \wedge \widehat{dx^j} \wedge \cdots \wedge dx^m \wedge \sum_{k=1}^{m} \frac{\mp x^k\, dx^k}{\sqrt{1 - |x'|^2}}$$

$$\pm (-1)^m \sqrt{1 - |x'|^2}\, dx^1 \wedge \cdots \wedge dx^m$$

$$= \pm \frac{(-1)^m}{\sqrt{1 - |x'|^2}} \left[-\sum_{j=1}^{m} (-1)^{m+j-1+m-j} (x^j)^2 + 1 - |x'|^2 \right] dx^1 \wedge \cdots \wedge dx^m .$$

The claim follows because the expression in the square brackets reduces to 1. ∎

(b) Let $\omega_{S^1} := (x\, dy - y\, dx)\,|\, S^1$, and make

$$g_1 : (0, 2\pi) \to S^1 \setminus \{(1,0)\}\ , \quad t \mapsto (\cos t, \sin t)$$

a parametrization of $S^1 \setminus \{(1,0)\}$. Then with respect to the local coordinates induced by the chart (φ, U) with $\varphi := g_1^{-1}$ and $U := S^1 \setminus \{(1,0)\}$, we have

$$\omega_{S^1}\,|\, U = dt\ .$$

Proof This follows from $\varphi_* \omega_{S^1} = (g_1^1 \dot{g}_1^2 - g_1^2 \dot{g}_1^1)\, dt$. ∎

(c) Let $U := S^2 \setminus H_3$ be the 2-sphere S^2 minus the half circle where it intersects the half plane $H_3 := \mathbb{R}^+ \times \{0\} \times \mathbb{R}$.[11] Also let

$$(0, 2\pi) \times (0, \pi) \to U\ , \quad (\varphi, \vartheta) \mapsto (\cos\varphi \sin\vartheta, \sin\varphi \sin\vartheta, \cos\vartheta)$$

be the parametrization of U by spherical coordinates. Finally, let

$$\alpha := x\, dy \wedge dz + y\, dz \wedge dx + z\, dx \wedge dy \in \Omega^2(\mathbb{R}^3)\ .$$

Then the form $\omega_{S^2} := \alpha\,|\, S^2 \in \Omega^2(S^2)$ has the representation

$$\omega_{S^2}\,|\, U = -\sin\vartheta\, d\varphi \wedge d\vartheta$$

with respect to the local coordinates (φ, ϑ).

Proof After a simple calculation,[12] we obtain this from Example 3.4(d). ∎

Coordinate transformations

To carry out concrete calculations efficiently, it is important to choose the coordinates best suited to the problem. So, for example, we use polar coordinates when we want to describe rotationally symmetric problems, as we have already done in our treatment of integration theory in Section X.8.

Because a given problem is usually already described in a coordinate system, we must be able to change to another coordinate system without undue trouble. This background frames the following transformation theorem for vector fields and Pfaff forms.

Let (φ, U) and (ψ, V) be charts of M with $U \cap V \neq \emptyset$. Let $\varphi = (x^1, \ldots, x^m)$ and $\psi = (y^1, \ldots, y^m)$. On $U \cap V$, we can regard the y^j as functions of local coordinates $x = (x^1, \ldots, x^m)$; we could also regard the x^j as functions of $y = (y^1, \ldots, y^m)$. Here is it usual and expedient not to introduce new symbols but rather to write simply $y = y(x)$ and $x = x(y)$. Clearly the map $y(\cdot)$ is a

---

[11] See Example VII.9.11(b).

[12] Note that (for example) $\partial(x, y)/\partial(\varphi, \vartheta)$ is the determinant of the matrix obtained from (VII.9.3) by removing the last row.

diffeomorphism from $U \cap V$ to itself, a **coordinate transformation**, which we also denote by $x \mapsto y$. The inverse map is $x(\cdot)$, that is, the coordinate transformation $y \mapsto x$. However, we can also regard x [or y] as a generic point in $X := \varphi(U \cap V)$ [or $Y := \psi(U \cap V)$] in $\overline{\mathbb{H}}^m$. Then the coordinate transformation $x \mapsto y$ is nothing but the transition function $\psi \circ \varphi^{-1} \in \mathrm{Diff}(X, Y)$. It will always be clear from context which of these two interpretations is to be chosen.

In the following formulas, we leave it to you to determine from context whether x^j means an independent variable or the function $x^j(\cdot)$. The double meaning, which is scarcely a problem in practice, is used on purpose since it helps to cast formulas into a form that is more intuitively understandable and easier to remember.

4.7 Proposition *For the coordinate transformation $x \mapsto y$, we have*

$$\frac{\partial}{\partial y^j} = \sum_{k=1}^{m} \frac{\partial x^k}{\partial y^j} \frac{\partial}{\partial x^k} \quad \text{and} \quad dy^j = \sum_{k=1}^{m} \frac{\partial y^j}{\partial x^k} dx^k$$

for $1 \le j \le m$.

Proof From Remark 4.5(c), it follows that

$$\frac{\partial}{\partial y^j} = \sum_{k=1}^{m} v_j^k \frac{\partial}{\partial x^k} \quad \text{and} \quad v_j^k = \left\langle dx^k, \frac{\partial}{\partial y^j} \right\rangle \quad \text{for } 1 \le j, k \le m .$$

With $x = f(y)$ and (4.5), we find

$$\left\langle dx^k, \frac{\partial}{\partial y^j} \right\rangle = \frac{\partial x^k}{\partial y^j} \quad \text{for } 1 \le j, k \le m , \tag{4.17}$$

which proves the first claim.

Analogously, we have

$$dy^j = \sum_{k=1}^{m} a_k \, dx^k \quad \text{and} \quad a_k = \left\langle dy^j, \frac{\partial}{\partial x^k} \right\rangle = \frac{\partial y^j}{\partial x^k}$$

for $1 \le j, k \le m$, which proves the second. ∎

4.8 Corollary (a) *The Jacobi matrix of the coordinate transformation $x \mapsto y$ satisfies*

$$\left[\frac{\partial y^j}{\partial x^k} \right] = \left[\frac{\partial x^j}{\partial y^k} \right]^{-1} .$$

(b)
$$dy^1 \wedge \cdots \wedge dy^m = \frac{\partial(y^1, \ldots, y^m)}{\partial(x^1, \ldots, x^m)} dx^1 \wedge \cdots \wedge dx^m .$$

Proof (a) Because

$$y(\,\cdot\,) = \psi \circ \varphi^{-1} \in \mathrm{Diff}(X, Y) \quad \text{and} \quad y(\,\cdot\,)^{-1} = x(\,\cdot\,) = \varphi \circ \psi^{-1} \in \mathrm{Diff}(Y, X) \;,$$

the claim is immediate.

(b) This is a consequence of Example 3.4(c), the considerations after Remark 4.5(e), and the fact that

$$\frac{\partial(y^1, \ldots, y^m)}{\partial(x^1, \ldots, x^m)}$$

is the Jacobian of the coordinate transformation $x \mapsto y$ (see Remark VII.7.9(a)). ∎

4.9 Examples (a) (plane polar coordinates) Using the polar coordinate transformation

$$V_2 \to \mathbb{R}^2 \;, \quad (r, \varphi) \mapsto (x, y) := (r \cos \varphi, r \sin \varphi)$$

with $V_2 := (0, \infty) \times (0, 2\pi)$, we have

$$\frac{\partial}{\partial r} = \frac{\partial x}{\partial r} \frac{\partial}{\partial x} + \frac{\partial y}{\partial r} \frac{\partial}{\partial y} = \cos \varphi \frac{\partial}{\partial x} + \sin \varphi \frac{\partial}{\partial y}$$

and

$$\frac{\partial}{\partial \varphi} = \frac{\partial x}{\partial \varphi} \frac{\partial}{\partial x} + \frac{\partial y}{\partial \varphi} \frac{\partial}{\partial y} = -r \sin \varphi \frac{\partial}{\partial x} + r \cos \varphi \frac{\partial}{\partial y} \;.$$

(b) (spherical coordinates) Let $V_3 := (0, \infty) \times (0, 2\pi) \times (0, \pi)$. Using the spherical coordinate transformation

$$V_3 \to \mathbb{R}^3 \;, \quad (r, \varphi, \vartheta) \mapsto (x, y, z) = (r \cos \varphi \sin \vartheta, r \sin \varphi \sin \vartheta, r \cos \vartheta) \;,$$

we find

$$\frac{\partial}{\partial r} = \quad \cos \varphi \sin \vartheta \frac{\partial}{\partial x} + \quad \sin \varphi \sin \vartheta \frac{\partial}{\partial y} + \quad \cos \vartheta \frac{\partial}{\partial z}$$

$$\frac{\partial}{\partial \varphi} = -r \sin \varphi \sin \vartheta \frac{\partial}{\partial x} + r \cos \varphi \sin \vartheta \frac{\partial}{\partial y}$$

$$\frac{\partial}{\partial \vartheta} = \quad r \cos \varphi \cos \vartheta \frac{\partial}{\partial x} + r \sin \varphi \cos \vartheta \frac{\partial}{\partial y} - r \sin \vartheta \frac{\partial}{\partial z} \;.$$

(c) (cylindrical coordinates) Let $X := (0, \infty) \times (0, 2\pi) \times \mathbb{R}$. For the cylindrical coordinate transformation

$$X \to \mathbb{R}^3 \;, \quad (r, \varphi, \zeta) \mapsto (x, y, z) := (r \cos \varphi, r \sin \varphi, \zeta) \;,$$

we find

$$\frac{\partial}{\partial r} = \cos \varphi \frac{\partial}{\partial x} + \sin \varphi \frac{\partial}{\partial y} \;, \quad \frac{\partial}{\partial \varphi} = -r \sin \varphi \frac{\partial}{\partial x} + r \cos \varphi \frac{\partial}{\partial y} \;, \quad \frac{\partial}{\partial \zeta} = \frac{\partial}{\partial z} \;. \quad ∎$$

The exterior derivative

The next theorem shows that the exterior derivative can be generalized so that it is defined globally on manifolds.

4.10 Theorem *There is exactly one map*

$$d : \Omega(M) \to \Omega(M) ,$$

the **exterior** *(or* **Cartan**) *derivative, with these four properties:*

(i) *d is $\mathbb{R}$-linear and maps $\Omega^r(M)$ to $\Omega^{r+1}(M)$.*

(ii) *d satisfies the* **product rule**

$$d(\alpha \wedge \beta) = d\alpha \wedge \beta + (-1)^r \alpha \wedge d\beta \quad \text{for } \alpha \in \Omega^r(M) \text{ and } \beta \in \Omega(M) .$$

(iii) *$d^2 = d \circ d = 0$.*

(iv) *The differential df of $f \in \mathcal{E}(M) = \Omega^0(M)$ is the same as the differential of f.*

Also

$$d \circ h^* = h^* \circ d \tag{4.18}$$

for $h \in C^\infty(M, N)$.

Proof (a) (existence) Let (φ, U) be a chart of M. According to Theorem 3.5, there is exactly one map $d : \Omega(\varphi(U)) \to \Omega(\varphi(U))$ with the properties (i)–(iv). We define $d_U : \Omega(U) \to \Omega(U)$ by requiring the commutativity of the diagram

$$
\begin{array}{ccc}
\Omega(U) & \xrightarrow{\ \ d_U\ \ } & \Omega(U) \\[2mm]
\varphi_* \downarrow & & \uparrow \varphi^* \\[2mm]
\Omega(\varphi(U)) & \xrightarrow{\ \ d\ \ } & \Omega(\varphi(U)) .
\end{array}
\tag{4.19}
$$

Equivalently, we set $d_U := \varphi^* \circ d \circ \varphi_*$. We learn from Remark 4.4(e) that φ_* is an algebra isomorphism with $(\varphi_*)^{-1} = (\varphi^{-1})_*$. With this and (4.19), we verify easily that d_U has the properties (i)–(iv) and is uniquely defined.

Let (ψ, V) be another chart of M such that $U \cap V \neq \emptyset$. Then it follows from $\varphi = (\varphi \circ \psi^{-1}) \circ \psi$, the properties of pull backs, and (3.12) that

$$
\begin{aligned}
d_U &= \varphi^* \circ d \circ \varphi_* = \psi^* \circ (\varphi \circ \psi^{-1})^* \circ d \circ (\varphi \circ \psi^{-1})_* \circ \psi_* \\
&= \psi^* \circ d \circ \psi_* = d_V
\end{aligned}
\tag{4.20}
$$

(of course, on $U \cap V$). Therefore d_U is independent of the special coordinates chosen.

Let $\{ (\varphi_\kappa, U_\kappa) \; ; \; \kappa \in \mathsf{K} \}$ be an atlas for M, and let $i_\kappa : U_\kappa \hookrightarrow M$ be the natural embedding. Then we define $d : \Omega(M) \to \Omega(M)$ by

$$d\alpha(p) := d_{U_\kappa} \big[(i_\kappa)^* \alpha \big](p) \quad \text{for } \alpha \in \Omega(M) \;,$$

where $\kappa \in \mathsf{K}$ is chosen so that p lies in U_κ. By (4.20) this definition is meaningful, and it is clear that d has the properties (i)–(iv).

(b) (uniqueness) Let $\alpha \in \Omega^r(M)$ and $p \in M$. Also let (φ, U) be a chart around p. According to Remark 4.5(c), $\alpha \,|\, U$ can be written in local coordinates as

$$\alpha \,|\, U = \sum_{(j) \in \mathbb{J}_r} a_{(j)} \, dx^{(j)}$$

with $a_{(j)} \in \mathcal{E}(U)$. Now it follows from (4.19) and Remarks 3.6(a) and 4.4(e) that

$$\begin{aligned} d_U(\alpha \,|\, U) &= \varphi^* \, d\varphi_*(\alpha \,|\, U) = \varphi^* \sum_{(j) \in \mathbb{J}_r} d(\varphi_* a_{(j)}) \wedge \varphi_* \, dx^{(j)} \\ &= \sum_{(j) \in \mathbb{J}_r} d_U a_{(j)} \wedge dx^{(j)} = \sum_{(j) \in \mathbb{J}_r} da_{(j)} \wedge dx^{(j)} \;, \end{aligned} \tag{4.21}$$

because $d_U a_{(j)}$ is the differential of $a_{(j)} \in \mathcal{E}(U)$.

Let V be an open neighborhood of p with $V \subset\subset U$. Then $\varphi(V) \subset\subset \varphi(U)$. Hence Remark 1.21(a) implies the existence of $\widetilde{\chi} \in \mathcal{D}\big(\varphi(U)\big)$ such that $\widetilde{\chi} \,|\, \varphi(V) = 1$. For

$$\chi := \begin{cases} \varphi^* \widetilde{\chi} & \text{on } U \;, \\ 0 & \text{on } M \backslash U \;, \end{cases}$$

we have $\chi \in \mathcal{E}(M)$ and $\chi \,|\, V = 1$. This implies that both

$$b_{(j)} := \chi a_{(j)} \text{ for } (j) \in \mathbb{J}_r \quad \text{and} \quad \xi^j := \chi x^j \text{ for } 1 \le j \le m$$

belong to $\mathcal{E}(M)$. Therefore the differentials $d\xi^j \in \Omega^1(M)$ are defined, which implies that

$$\beta := \sum_{(j) \in \mathbb{J}_r} b_{(j)} \, d\xi^{(j)}$$

is also defined and belongs to $\Omega(M)$.

Now suppose $\widetilde{d}$ is a map from $\Omega(M)$ to itself satisfying (i)–(iv). Then we find easily that

$$\widetilde{d}\beta = \sum_{(j) \in \mathbb{J}_r} db_{(j)} \wedge d\xi^{(j)} \;.$$

For $a \in \mathcal{E}(U)$, the product rule gives

$$d(\chi a) = a \, d\chi + \chi \, da$$

(see Corollary VII.3.8 and the definition of the tangential). Because $\chi \,|\, V = 1$, we may use the natural embedding $i \colon V \hookrightarrow M$ to conclude

$$\big\langle i^* d(\chi a)(q), v(q) \big\rangle_q = \big\langle d(\chi a)(q), v(q) \big\rangle_q = \big\langle da(q), v(q) \big\rangle_q \quad \text{for } q \in V \ ,$$

for $v \in \mathcal{V}(M)$. That is, $d(\chi a) \,|\, V = da \,|\, V$. This and (4.21) imply $\beta \,|\, V = \alpha \,|\, V$ and

$$\widetilde{d}\beta \,|\, V = d_V(\alpha \,|\, V) \ . \tag{4.22}$$

Because d_V is unique and every $p \in M$ has an open coordinate neighborhood V for which (4.22) holds, we see that $\widetilde{d} = d$.

(c) To prove (4.18), we can use our previous work to restrict to the local situation, that is, we can assume that $M = U$. Then the claim follows from (4.19), (3.12), and Theorem 3.5. ∎

4.11 Remarks (a) Let

$$\alpha \,|\, U = \sum_{(j) \in \mathbb{J}_r} a_{(j)} \, dx^{(j)}$$

be the representation of $\alpha \in \Omega^r(M)$ in the local coordinates of the chart (φ, U). Then

$$d(\alpha \,|\, U) = \sum_{(j) \in \mathbb{J}_r} da_{(j)} \wedge dx^{(j)} \ .$$

Proof This follows from (4.21). ∎

(b) (regularity) For $k \in \mathbb{N}$, the map

$$d \colon \Omega^r_{(k+1)}(M) \to \Omega^{r+1}_{(k)}(M) \quad \text{for } r \in \mathbb{N}$$

is defined and $\mathbb{R}$-linear. This remains true when M is a C^{k+2} manifold. ∎

Closed and exact forms

As in the local theory, we say $\alpha \in \Omega(M)$ is **closed** if $d\alpha = 0$. We say it is **exact** if there is a $\beta \in \Omega(M)$, an **antiderivative**, such that $d\beta = \alpha$.

4.12 Remarks and examples (a) Because $d^2 = 0$, every exact form is closed.

(b) Every m-form on M is closed.

Proof This is because $\Omega^{m+1}(M) = \{0\}$. ∎

(c) (Poincaré lemma) Let $r \in \mathbb{N}^\times$, and let $\alpha \in \Omega^r(M)$ be closed. Then α is **locally exact**, that is, every $p \in M$ has an open neighborhood U and a $\beta \in \Omega^{r-1}(U)$ such that $d\beta = \alpha \,|\, U$.

Proof Let (φ, U) be a chart around p, in which $\varphi(U)$ is star shaped. Because $d\alpha = 0$ and $d\varphi_* \alpha = \varphi_* \, d\alpha$, the form $\varphi_* \alpha \in \Omega^r(\varphi(U))$. Since $\varphi(U)$ is contractible, it follows from the Poincaré lemma (Theorem 3.11) that there exists a $\beta_0 \in \Omega^{r-1}(\varphi(U))$ such that $d\beta_0 = \varphi_* \alpha$. For $\beta := \varphi^* \beta_0 \in \Omega^{r-1}(U)$, we then have $d\beta = \varphi^* \, d\beta_0 = \varphi^* \varphi_* \alpha = \alpha \,|\, U$. ∎

Contractions

Let $\alpha \in \Omega^{r+1}(M)$ and $v \in \mathcal{V}(M)$. Then the **contraction** $v \lrcorner \alpha$ of α by v is defined by

$$v \lrcorner \alpha(v_1, \ldots, v_r) := \alpha(v, v_1, \ldots, v_r) \quad \text{for } v_j \in \mathcal{V}(M) \text{ and } 1 \leq j \leq r .$$

We sometimes write $v \lrcorner \cdot$ as i_v and call $i_v \alpha$ the **interior product of** v by α. We verify easily that $v \lrcorner \alpha$ belongs to $\Omega^r(M)$. For completeness and to avoid a bothersome special case, we simply set

$$v \lrcorner \alpha := 0 \quad \text{for } \alpha \in \Omega^0(M) .$$

4.13 Remarks and examples   **(a)** If $\varphi : M \to N$ is a diffeomorphism, then

$$v \lrcorner (\varphi^* \alpha) = \varphi^* (\varphi_* v \lrcorner \alpha)$$

for $\alpha \in \Omega(N)$ and $v \in \mathcal{V}(M)$. In particular, for every r the diagram

$$
\begin{array}{ccc}
\Omega^{r+1}(M) & \xleftarrow{\;\varphi^*\;} & \Omega^{r+1}(N) \\[2pt]
{\scriptstyle v \lrcorner} \big\downarrow & & \big\downarrow {\scriptstyle \varphi_* v \lrcorner} \\[2pt]
\Omega^r(M) & \xleftarrow{\;\varphi^*\;} & \Omega^r(N)
\end{array}
$$

commutes.

Proof   If α is a null form, then the claim is trivially true. Therefore we can assume $\alpha \in \Omega^{r+1}(N)$. Then we find for $p \in M$ and $v_1, \ldots, v_r \in T_p M$ that

$$
\begin{aligned}
v \lrcorner (\varphi^* \alpha)(p)(v_1, \ldots, v_r) &= (\varphi^* \alpha)(p)\big(v(p), v_1, \ldots, v_r\big) \\
&= \alpha\big(\varphi(p)\big)\big((T_p\varphi)v(p), (T_p\varphi)v_1, \ldots, (T_p\varphi)v_r\big) \\
&= \alpha\big(\varphi(p)\big)\big(\varphi_* v(\varphi(p)), (T_p\varphi)v_1, \ldots, (T_p\varphi)v_r\big) \\
&= (\varphi_* v \lrcorner \alpha)\big(\varphi(p)\big)\big((T_p\varphi)v_1, \ldots, (T_p\varphi)v_r\big) \\
&= \varphi^* (\varphi_* v \lrcorner \alpha)(p)(v_1, \ldots, v_r) ,
\end{aligned}
$$

which proves the claim. $\blacksquare$

(b) Suppose X is open in $\overline{\mathbb{H}}^m$ and $\omega := dx^1 \wedge \cdots \wedge dx^m$. For $v = \sum_{j=1}^m v^j \partial_j$, we have

$$v \lrcorner \omega = \sum_{j=1}^m (-1)^{j-1} v^j \, dx^1 \wedge \cdots \wedge \widehat{dx^j} \wedge \cdots \wedge dx^m .$$

Proof   We set $v_1 := v$. Then for $v_2, \ldots, v_m \in \mathcal{V}(X)$, we have

$$(v_1 \lrcorner \omega)(v_2, \ldots, v_m) = \omega(v_1, \ldots, v_m) = \det\big[\langle dx^j, v_k \rangle\big] .$$

By expanding this determinant in the first column, we find it has the value

$$\sum_{j=1}^{m}(-1)^{j+1}\langle dx^{j}, v_1\rangle \det(A_j) ,$$

where A_j is the matrix obtained by striking the first column and the j-th row from $[\langle dx^j, v_k\rangle]$. From this it follows that

$$\det(A_j) = dx^1 \wedge \cdots \wedge \widehat{dx^j} \wedge \cdots \wedge dx^m (v_2, \ldots, v_m) .$$

The claim now follows because

$$\langle dx^j, v_1\rangle = \sum_{k=1}^{m} v^k \langle dx^j, \partial_k\rangle = v^j . \quad \blacksquare$$

(c) Let

$$\rho\colon \mathbb{R}^{m+1}\setminus\{0\} \to S^m , \quad x \mapsto x/|x|$$

be the **radial retraction**[13] on the m-sphere in $\mathbb{R}^{m+1}$.
Also let

$$\alpha := \sum_{j=1}^{m+1}(-1)^{j-1}x^j\, dx^1 \wedge \cdots \wedge \widehat{dx^j} \wedge \cdots \wedge dx^{m+1}$$

and

$$\omega_{S^m} := \alpha\,|\, S^m .$$

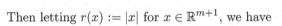

Then letting $r(x) := |x|$ for $x \in \mathbb{R}^{m+1}$, we have

$$\rho^*\omega_{S^m} = \frac{1}{r^{m+1}}\alpha = \sum_{j=1}^{m+1}(-1)^{j-1}\frac{x^j}{|x|^{m+1}}\, dx^1 \wedge \cdots \wedge \widehat{dx^j} \wedge \cdots \wedge dx^{m+1} ,$$

and $\rho^*\omega_{S^m}$ is closed.

Proof Because $\rho \in C^\infty(\mathbb{R}^{m+1}\setminus\{0\}, \mathbb{R}^{m+1})$ with $\mathrm{im}(\rho) = S^m$, we know ρ is a smooth map from $\mathbb{R}^{m+1}\setminus\{0\}$ to S^m. Therefore $\rho^*\omega_{S^m} \in \Omega^m(\mathbb{R}^{m+1}\setminus\{0\})$ is defined. It is closed by Remark 4.12(b) and because $d(\rho^*\omega_{S^m}) = \rho^*\, d\omega_{S^m} = 0$.

To show that $\rho^*\omega_{S^m} = r^{-(m+1)}\alpha$, we must verify that for every $p \in \mathbb{R}^{m+1}\setminus\{0\}$, both sides agree on every m-tuple from a system of basis vectors of $T_p\mathbb{R}^{m+1}$. Suppose therefore $p \in \mathbb{R}^{m+1}\setminus\{0\}$. A basis of $T_p\mathbb{R}^{m+1}$ is given by the vectors $\{(p)_p, (v_1)_p, \ldots, (v_m)_p\}$, where

---

[13]If X is topological space and A is a subset of X, then a continuous map $\rho\colon X \to A$ is called a **retraction** of X on A if $\rho(a) = a$ for $a \in A$. If there is a retraction of X on A, then A is a **retract** of X.

$\{(v_1)_p, \ldots, (v_m)_p\}$ is a basis of $T_p(r(p)S^m)$. If the m-tuple $(w_1)_p, \ldots, (w_m)_p$ contains the vector $(p)_p$, then by letting $\omega := dx^1 \wedge \cdots \wedge dx^{m+1}$, we can use (b) to find

$$\alpha(p)\big((w_1)_p, \ldots, (w_m)_p\big) = \big((p)_p \lrcorner \omega\big)\big((w_1)_p, \ldots, (w_m)_p\big)$$
$$= \omega\big((p)_p, (w_1)_p, \ldots, (w_m)_p\big) = 0 ,$$

because two entries are equal. Therefore $r^{-(m+1)}(p)\alpha(p)$ also vanishes on this m-tuple. We also find

$$\rho^*\omega_{S^m}(p)\big((w_1)_p, \ldots, (w_m)_p\big) = \omega_{S^m}\big(\rho(p)\big)\big((T_p\rho)(w_1)_p, \ldots, (T_p\rho)(w_m)_p\big)$$

with

$$(T_p\rho)(w_j)_p = \big(\rho(p), \partial\rho(p)w_j\big) ,$$

where, according to Proposition VII.2.5, we have

$$\partial\rho(p)w_j = \partial_t\rho(p + tw_j)\big|_{t=0} \quad \text{for } 1 \le j \le m .$$

Because $\rho(p+tp) = \rho(p)$ for $t \in (-1, 1)$, it follows in particular that $(T_p\rho)(p)_p = 0$. Therefore $\rho^*\omega_{S^m}(p)\big((w_1)_p, \ldots, (w_m)_p\big)$ also vanishes if the m-tuple $(w_1)_p, \ldots, (w_m)_p$ contains the vector $(p)_p$.

It remains to show

$$\rho^*\omega_{S^m}(p)\big((v_1)_p, \ldots, (v_m)_p\big) = \frac{1}{r(p)^{m+1}} \alpha(p)\big((v_1)_p, \ldots, (v_m)_p\big) . \tag{4.23}$$

For $(v)_p \in T_p\big(r(p)S^m\big)$, Theorem VII.10.6 gives an $\varepsilon > 0$ and a $\gamma \in C^1\big((-\varepsilon, \varepsilon), r(p)S^m\big)$ such that $\gamma(0) = p$ and $\dot\gamma(0) = v$. Now we use $\rho \circ \gamma(t) = \gamma(t)/r(p)$ to get

$$\partial\rho(p)v = (\rho \circ \gamma)^{\cdot}(0) = v/r(p) .$$

From this we derive

$$\rho^*\omega_{S^m}(p)\big((v_1)_p, \ldots, (v_m)_p\big) = r(p)^{-m}\alpha\big(\rho(p)\big)\big((v_1)_p, \ldots, (v_m)_p\big) ,$$

which implies (4.23), thus finishing the proof. ∎

(d) (regularity) Let $k \in \mathbb{N}$, and suppose M is a C^{k+1} manifold. For $\alpha \in \Omega^{r+1}_{(k)}(M)$ and $v \in \mathcal{V}^k(M)$, the contraction $v \lrcorner \alpha$ belongs to $\Omega^r_{(k)}(M)$. ∎

Orientability

As we learned in Section 2, T_pM can be oriented by choosing a volume form $\alpha(p) \in \bigwedge^m T_p^*M$. Thereby one gets an m-form α on M with $\alpha(p) \ne 0$ for $p \in M$. Conversely, every map $p \mapsto \alpha(p) \in \bigwedge^m T_p^*M$ such that $\alpha(p) \ne 0$ for $p \in M$ induces an orientation on every T_pM. However, such α will generally not be continuous. Intuitively, this means that the orientation of the tangent spaces is not "coherent", that is, the tangent spaces can "flip over" in moving from one point to the next. To avoid this, we also require that α be smooth (more precisely, as regular as permitted by the regularity of the manifold).

A manifold M is said to be **orientable** if there is an $\alpha \in \Omega^m(M)$ such that $\alpha(p) \neq 0$ for every $p \in M$; such an m-form α is called a **volume form on** M.

4.14 Remarks (a) If M is orientable, then $\Omega^m(M)$ is a one-dimensional $\mathcal{E}(M)$-module.

Proof Let α be a volume form on M, and let $\beta \in \Omega^m(M)$. Because $\dim \bigwedge^m T_p^* M = 1$ for $p \in M$, there is an $f : M \to \mathbb{R}$ such that $\beta = f\alpha$. We must show that f is smooth. In local coordinates, we have

$$\alpha \,|\, U = a\, dx^1 \wedge \cdots \wedge dx^m \quad \text{and} \quad \beta \,|\, U = b\, dx^1 \wedge \cdots \wedge dx^m$$

with $a, b \in \mathcal{E}(U)$ and $a(p) \neq 0$ for $p \in U$. From this we deduce that $\beta \,|\, U = f\alpha \,|\, U$, where $f := b/a$ belongs to $\mathcal{E}(U)$. ∎

(b) (regularity) Suppose $k \in \mathbb{N}$ and M is a C^{k+1} manifold. Then M is orientable if and only if there is an $\alpha \in \Omega^m_{(k)}(M)$ such that $\alpha(p) \neq 0$ for $p \in M$. This is the case if and only if the $C^k(M)$-module $\Omega^m_{(k)}(M)$ is one-dimensional. ∎

The next proposition shows that one can also characterize the orientability of a manifold by its charts.

If X and Y are open in $\overline{\mathbb{H}}^m$, then we say $\varphi \in \mathrm{Diff}(X, Y)$ is **orientation-preserving** [or **orientation-reversing**] if $\det \partial\varphi(x) > 0$ [or $\det \partial\varphi(x) < 0$] for every $x \in X$, that is, if $\partial\varphi(x) \in \mathcal{L}(\mathbb{R}^m)$ is an orientation-preserving [or orientation-reversing] automorphism for every $x \in X$. An atlas of M is said to be **oriented** if all of its transition functions are orientation-preserving.

4.15 Proposition *A manifold of dimension ≥ 2 is orientable if and only if it has an oriented atlas.*

Proof (a) Suppose M is orientable and $\alpha \in \Omega^m(M)$ is a volume form. In addition, let $\{ (\varphi_\kappa, U_\kappa) \,;\, \kappa \in \mathsf{K} \}$ be an atlas of M. Then $(\varphi_\kappa)_* \alpha = a_\kappa\, dx^1 \wedge \cdots \wedge dx^m$ on $X_\kappa := \varphi_\kappa(U_\kappa) \subset \overline{\mathbb{H}}^m$, with $a_\kappa(x) \neq 0$ for $x \in X_\kappa$. Because we can change coordinates (if necessary) as $x \mapsto (-x^1, x^2, \ldots, x^m)$, we can assume that $a_\kappa(x_\kappa)$ is strictly positive for some $x_\kappa \in X_\kappa$. Because we can assume that U_κ and therefore also X_κ are connected, it follows from the intermediate value theorem (Theorem III.4.7) that $a_\kappa(x) > 0$ for all $x \in X_\kappa$ and every $\kappa \in \mathsf{K}$.

Suppose now $(\varphi_\kappa, U_\kappa)$ and $(\varphi_\lambda, U_\lambda)$ are local charts with $U_\kappa \cap U_\lambda \neq \emptyset$. Also let $\varphi_\kappa = (x^1, \ldots, x^m)$ and $\varphi_\lambda = (y^1, \ldots, y^m)$. Then we find

$$(\varphi_\lambda \circ \varphi_\kappa^{-1})^* \big(a_\lambda\, dy^1 \wedge \cdots \wedge dy^m \,\big|\, \varphi_\lambda(U_\kappa \cap U_\lambda)\big)$$
$$= (\varphi_\kappa)_* \varphi_\lambda^* \big(a_\lambda\, dy^1 \wedge \cdots \wedge dy^m \,\big|\, \varphi_\lambda(U_\kappa \cap U_\lambda)\big) \tag{4.24}$$
$$= (\varphi_\kappa)_* \alpha \,\big|\, (U_\kappa \cap U_\lambda) = a_\kappa\, dx^1 \wedge \cdots \wedge dx^m .$$

By Example 3.4(c), we have

$$(\varphi_\lambda \circ \varphi_\kappa^{-1})^* dy^1 \wedge \cdots \wedge dy^m = \det \partial(\varphi_\lambda \circ \varphi_\kappa^{-1})\, dx^1 \wedge \cdots \wedge dx^m .$$

By comparing with (4.24), we see

$$(\varphi_\lambda \circ \varphi_\kappa^{-1})^* a_\lambda(x) \det \partial(\varphi_\lambda \circ \varphi_\kappa^{-1})(x) = a_\kappa(x) > 0 \quad \text{for } x \in \varphi_\kappa(U_\kappa \cap U_\lambda) \ .$$

Because a_λ is positive, it follows that M has an oriented atlas.

(b) Let $\{ (\varphi_\kappa, U_\kappa) \ ; \ \kappa \in \mathsf{K} \}$ be an oriented atlas. Proposition 1.20 guarantees the existence of a smooth partition of unity $\{ \pi_\kappa \ ; \ \kappa \in \mathsf{K} \}$ that is subordinate to the cover $\{ U_\kappa \ ; \ \kappa \in \mathsf{K} \}$ of M. For $\kappa \in \mathsf{K}$, define $\alpha_\kappa \in \Omega^m(U_\kappa)$ by

$$\alpha_\kappa := \begin{cases} \pi_\kappa \varphi_\kappa^* \, dx^1 \wedge \cdots \wedge dx^m & \text{in } U_\kappa \ , \\ 0 & \text{otherwise} \ . \end{cases}$$

We can verify easily that the definition

$$\alpha := \sum_{\kappa \in \mathsf{K}} \alpha_\kappa \in \Omega^m(M)$$

is meaningful. We must show that $\alpha(p) \neq 0$ for $p \in M$.

Let $p \in M$, and choose $\kappa \in \mathsf{K}$ so that $\pi_\kappa(p) > 0$. For $\lambda \in \mathsf{K}$ with $\lambda \neq \kappa$ and $U_\kappa \cap U_\lambda \neq \emptyset$, it follows, as in (a), that

$$\begin{aligned} \alpha_\lambda = \pi_\lambda \varphi_\lambda^* \, dy^1 \wedge \cdots \wedge dy^m &= \pi_\lambda \varphi_\kappa^* (\varphi_\lambda \circ \varphi_\kappa^{-1})^* \, dy^1 \wedge \cdots \wedge dy^m \\ &= \pi_\lambda \big(\varphi_\kappa^* \det \big(\partial(\varphi_\lambda \circ \varphi_\kappa^{-1}) \big) \big) \varphi_\kappa^* \, dx^1 \wedge \cdots \wedge dx^m \ . \end{aligned}$$

From this we obtain

$$\alpha(p) = \Big(\pi_\kappa(p) + \sum_{\substack{\lambda \in \mathsf{K} \\ \lambda \neq \kappa}} \pi_\lambda(p) \det \big(\partial(\varphi_\lambda \circ \varphi_\kappa^{-1}) \big) \big(\varphi_\kappa(p) \big) \Big) \varphi_\kappa^* \, dx^1 \wedge \cdots \wedge dx^m(p) \ ,$$

where only finitely many summands differ from zero. Because $\pi_\lambda(p) \geq 0$ and because the transition functions are orientation-preserving, we see that $\alpha(p) \neq 0$. Therefore α is a volume form, and M is orientable. $\blacksquare$

Suppose M is orientable. Then we say two volume forms $\alpha, \beta \in \Omega^m(M)$ are **equivalent** if there is an $f \in \mathcal{E}(M)$ such that $f(p) > 0$ for $p \in M$, and $\alpha = f\beta$. This is obviously an equivalence relation on the set of all volume forms on M. Every equivalence class with respect to this relation is called an **orientation** on M. Given $\mathcal{O}r := \mathcal{O}r(M)$ an orientation of M, then we call $(M, \mathcal{O}r)$ an **oriented manifold**. If the orientation of M is clear from context, we may write M for $(M, \mathcal{O}r)$.

If $\alpha \in \mathcal{O}r$, then $-\alpha$ is a volume form that does not belong to $\mathcal{O}r$. We denote the associated equivalence class by $-\mathcal{O}r$ and call it the **orientation opposite to** $\mathcal{O}r$. It is clear that $-\mathcal{O}r$ is independent of its particular representative.

4.16 Remarks (a) An orientable manifold is connected if and only if it has exactly two orientations.

Proof Suppose M is connected, and α and β are two volume forms. By Remark 4.14(a), there is an $f \in \mathcal{E}(M)$ such that $\alpha = f\beta$. Because α vanishes nowhere, we have $f(p) \neq 0$ for $p \in M$. Because M is connected, the intermediate value theorem (see Theorem III.4.7) implies that either $f(p) > 0$ or $f(p) < 0$ for every $p \in M$. Hence α is equivalent either to β or to $-\beta$. Therefore M has precisely two orientations.

Now suppose M is connected. Proposition III.4.2 guarantees the existence of a nonempty, open, and closed proper subset X of M. For α a volume form on M, we set

$$\beta(p) := \begin{cases} \alpha(p) & \text{if } p \in X , \\ -\alpha(p) & \text{if } p \in M \backslash X . \end{cases}$$

Then β is obviously a volume form with $\beta \notin \mathcal{O}r \cup (-\mathcal{O}r)$, where $\mathcal{O}r$ is the equivalence class of α. Therefore M has more than two orientations. ∎

(b) Let $M = (M, \mathcal{O}r)$ be an oriented manifold. A chart (φ, U) of M is said to be **positive(ly oriented)** if $\varphi_*(\alpha \,|\, U)$ for $\alpha \in \mathcal{O}r$ is equivalent to the m-form $dx^1 \wedge \cdots \wedge dx^m \,|\, \varphi(U)$. Otherwise it is **negative(ly oriented)**. M has an atlas consisting only of positive charts, an **oriented atlas**.

Proof For $\beta \in \mathcal{O}r$, we have $\alpha = f\beta$ with $f \in \mathcal{E}(M)$ and $f(p) > 0$ for $p \in M$. With $i : U \hookrightarrow M$, it follows from this that

$$\varphi_* \alpha \,|\, U = \varphi_* i^* \alpha = \varphi_* i^* (f\beta) = (\varphi_* i^* f)(\varphi_* i^* \beta) = g\varphi_*(\beta \,|\, U) ,$$

where $g := f \circ \varphi^{-1} \in \mathcal{E}(\varphi(U))$ and $g(x) > 0$ for $x \in \varphi(U)$. This shows that the definition does not depend on the chosen representative. That there is indeed an atlas with positive charts was shown in part (a) of the proof of Proposition 4.15. ∎

(c) Let M be oriented. Then (φ, U) is a positive chart if and only if $(\partial_1|_p, \ldots, \partial_m|_p)$ is a positive basis of $T_p M$ for $p \in U$.

Proof For $\alpha \in \Omega^m(M)$, Remark 4.5(c) says that the basis representation in local coordinates is

$$\alpha \,|\, U = a \, dx^1 \wedge \cdots \wedge dx^m$$

with $a(p) = \alpha(p)(\partial_1|_p, \ldots, \partial_m|_p)$ for $p \in U$. The claim is now clear. ∎

4.17 Examples (a) Every open subset U of an orientable manifold M is itself orientable.[14]

Proof For $\alpha \in \mathcal{O}r(M)$, the restriction $\alpha \,|\, U$ is a volume form on U. ∎

(b) If M and N are orientable and one of these manifolds is without boundary, then the product manifold[15] $M \times N$ is orientable.

---

[14] We stipulate that the empty set is orientable.
[15] See Exercise VII.9.4 and Exercise 3. Why do we assume that one of these two manifolds is without boundary?

Proof If $\{ (\varphi_\kappa, U_\kappa) \ ; \ \kappa \in K \}$ and $\{ (\psi_\lambda, V_\lambda) \ ; \ \lambda \in L \}$ are oriented atlases of M and N, respectively, then it is easy to see that $\{ \varphi_\kappa \times \psi_\lambda \ ; \ (\kappa, \lambda) \in K \times L \}$ with

$$\varphi_\kappa \times \psi_\lambda(p, q) := (\varphi_\kappa(p), \psi_\lambda(q)) \in \mathbb{R}^m \times \mathbb{R}^n \quad \text{for } (p, q) \in U_\kappa \times V_\lambda \ ,$$

is an oriented atlas of $M \times N$. Because we can assume without loss of generality that M and N are at least one-dimensional, the claim follows from Proposition 4.15. ∎

(c) Any manifold that can be described by a single chart (that is, one that has an atlas with only one chart) is orientable.

Proof This is trivial (see the first part of the proof of Proposition 4.15). ∎

(d) (graphs) Suppose X is open in $\mathbb{R}^m$ and $f \in C^\infty(X, \mathbb{R}^n)$. Then graph$(f)$ is an m-dimensional orientable submanifold of $\mathbb{R}^{m+n}$.

Proof For Proposition VII.9.2, we know that graph(f) is an m-dimensional submanifold of $\mathbb{R}^{m+n}$. The proof of that result shows that

$$\varphi : \text{graph}(f) \to X \ , \quad (x, f(x)) \mapsto x$$

is a chart that describes graph(f). Therefore the claim follows from (c). ∎

(e) (fibers of regular maps) Suppose X is open in $\mathbb{R}^m$ and $\ell \in \{0, \dots, m-1\}$. Also let q be regular value of $f \in C^\infty(X, \mathbb{R}^{m-\ell})$. Then the ℓ-dimensional submanifold $f^{-1}(q)$ of X is orientable.

Proof Let $\omega := dx^1 \wedge \cdots \wedge dx^m \,|\, X$ and

$$\nabla f^k := \sum_{j=1}^m \partial_j f^k \frac{\partial}{\partial x^j} \in \mathcal{V}(X) \quad \text{for } 1 \leq k \leq m - \ell \ .$$

With the notations of Remark VII.10.11(a), we have $\nabla f^k(p) = \nabla_p f^k$ for $p \in X$. We can assume that $L := f^{-1}(q)$ is not empty. Then

$$\alpha := \nabla f^1 \lrcorner \left(\nabla f^2 \lrcorner \left(\cdots \lrcorner (\nabla f^{m-\ell} \lrcorner \omega) \cdots \right) \right) \Big| L \in \Omega^\ell(L) \ .$$

Proposition VII.10.13 guarantees that $\nabla f^1(p), \dots, \nabla f^{m-\ell}(p)$ are linearly independent. Therefore

$$\alpha(p) = \omega\left(\nabla f^{m-\ell}(p), \dots, \nabla f^1(p), \dots\right) \neq 0 \quad \text{for } p \in L \ ,$$

that is, α is a volume form on L. ∎

(f) If M and N are diffeomorphic, then M is orientable if and only if N is orientable.

Proof Let $f \in \text{Diff}(M, N)$, and let (φ, U) be a chart of M. Then $\psi := \varphi \circ f^{-1}$ is a chart of N with $V := f(U) = \text{dom}(\psi)$. Because M and N are diffeomorphic, $m = n$. Suppose now $\beta \in \Omega^m(N)$ is a volume form on N. The local coordinates $(y^1, \dots, y^m) = \psi$ give β the representation $\beta \,|\, V = b \, dy^1 \wedge \cdots \wedge dy^m$ with $b(q) \neq 0$ for $q \in V$. From this it follows that

$$f^*(\beta \,|\, V) = (f^*b)f^*(dy^1 \wedge \cdots \wedge dy^m) = b \circ f \, dx^1 \wedge \cdots \wedge dx^m \ ,$$

because $f^* y^j = \text{pr}_j \circ \psi \circ f = \text{pr}_j \circ \varphi = x^j$ with $(x^1, \dots, x^m) = \varphi$. Because $b \circ f(p) \neq 0$ for $p \in U$, we see that $f^*\beta$ is a volume form on M. Now the claim is immediate. ∎

(g) Every one-dimensional manifold is orientable.

Proof We can assume that the manifold M is connected since it suffices to show that every connected component in orientable. Then by Theorem 1.18, M is diffeomorphic to an interval J or to S^1. Because J and S^1 are orientable (where the orientability of S^1 follows from (e), for example), the claim is implied by (f). ∎

(h) (hypersurfaces) A hypersurface M in $\mathbb{R}^{m+1}$ is orientable if and only if there is a smooth **unit normal field** on M, that is, a $\nu \in C^\infty(M, \mathbb{R}^{m+1})$ such that $|\nu(p)| = 1$ and $\boldsymbol{\nu}(p) = (p, \nu(p)) \in T_p^\perp M$ for $p \in M$.

Proof If ν is a unit normal field on M, then $(\boldsymbol{\nu} \lrcorner dx^1 \wedge \cdots \wedge dx^{m+1}) \mid M$ is a volume form on M. Therefore M is orientable.

Let M be orientable. If (φ, U) is a positive chart with $\varphi = (x^1, \ldots, x^m)$, then, because $\dim(T_p^\perp M) = 1$, there is for every $p \in U$ exactly one $\boldsymbol{\nu}(p) = (p, \nu(p)) \in T_p^\perp M$ with $|\nu(p)| = 1$ such that

$$\left(\boldsymbol{\nu}(p), \tfrac{\partial}{\partial x^1}\big|_p, \ldots, \tfrac{\partial}{\partial x^m}\big|_p\right)$$

is a positive basis of $T_p\mathbb{R}^{m+1}$. By shrinking U, we can assume that there are open sets $\widetilde{U}$ and $\widetilde{V}$ of $\mathbb{R}^{m+1}$, with $U = \widetilde{U} \cap M$, and a $\Phi \in \mathrm{Diff}(\widetilde{U}, \widetilde{V})$ such that $U = f^{-1}(0)$ for $f := \Phi^{m+1} \in \mathcal{E}(\widetilde{U})$. It follows because $\nabla f(p) \neq 0$ for $p \in \widetilde{U}$ that f is regular. Hence it follows from Proposition VII.10.13, that

$$\nu(p) = \varepsilon \nabla f(p) / |\nabla f(p)| \quad \text{for } p \in U$$

with $\varepsilon \in \{\pm 1\}$. This shows that ν is smooth.

Now let (ψ, V) be a second positive chart with $U \cap V \neq \emptyset$ and $\psi = (y^1, \ldots, y^m)$, and suppose $\boldsymbol{\mu}(q) = (q, \mu(q)) \in T_q^\perp M$ satisfies $\mu \in C^\infty(V, \mathbb{R}^{m+1})$ and $|\mu(q)| = 1$ for $q \in V$. Also suppose $\left(\boldsymbol{\mu}(q), \tfrac{\partial}{\partial y^1}\big|_q, \ldots, \tfrac{\partial}{\partial y^m}\big|_q\right)$ is a positive basis of $T_q\mathbb{R}^{m+1}$ for $q \in V$. Because the two bases

$$\left(\tfrac{\partial}{\partial x^1}\big|_p, \ldots, \tfrac{\partial}{\partial x^m}\big|_p\right) \quad \text{and} \quad \left(\tfrac{\partial}{\partial y^1}\big|_p, \ldots, \tfrac{\partial}{\partial y^m}\big|_p\right)$$

have the same orientation for $p \in U \cap V$, it follows that $\mu(p) = \nu(p)$ for $p \in U \cap V$. Now the existence of a unit normal field follows from the existence of an oriented atlas of M. ∎

(i) (Möbius strip) Suppose $R > 0$, and define

$$f : [-\pi, \pi) \times (-1, 1) \to \mathbb{R}^3$$

by

$$f(\theta, t) := \left(\left(R + t \cos \tfrac{\theta}{2}\right) \cos \theta, \left(R + t \cos \tfrac{\theta}{2}\right) \sin \theta, t \sin \tfrac{\theta}{2} \right).$$

Then the image M of f is a nonorientable surface, the **Möbius strip**. Visually, the map f works as follows: Because

$$f(\pm\pi, t) = (-R, 0, \pm t),$$

it twists the end $\{\pi\} \times (-1, 1)$ of the rectangle $[-\pi, \pi] \times (-1, 1)$ by 180 degrees relative to the start $\{-\pi\} \times (-1, 1)$. These two ends are then glued together.

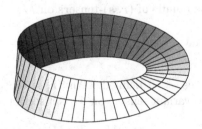

Representing f in the form

$$f(\theta, t) = R(\cos\theta, \sin\theta, 0) + tg(\theta)$$

with $g(\theta) := \big(\cos(\theta/2)\cos\theta, \sin(\theta/2)\sin\theta, \sin(\theta/2)\big)$, we obtain an interpretation of the parametrization of f: A point with angular velocity 1 traces a circle in the (x, y)-plane with center 0 and radius R; this describes the first summand. The midpoint of a rod of length 2 is affixed to this point (along its length) and is allowed to simultaneously rotate about its own midpoint with an angular velocity of $1/2$, so that its direction is reversed after one rotation; this is the second summand.

Proof   The proof that M is a smooth surface is left to you.

For $-\pi \leq \theta \leq \pi$, we have

$$v_1(\theta) := \partial_1 f(\theta, 0) = R(-\sin\theta, \cos\theta, 0) \ ,$$
$$v_2(\theta) := \partial_2 f(\theta, 0) = \big(\cos(\theta/2)\cos\theta, \cos(\theta/2)\sin\theta, \sin(\theta/2)\big) \ .$$

It follows that for every $\theta \in [-\pi, \pi)$ the vectors $v_1(\theta)$, $v_2(\theta)$ attached to $p(\theta) := f(\theta, 0)$ form a basis of $T_{p(\theta)}M$. Therefore the vector

$$n(\theta) := \big(-v_1(\theta) \times v_2(\theta)\big)/R = \big(-\cos\theta\sin(\theta/2), -\sin\theta\sin(\theta/2), \cos(\theta/2)\big)$$

attached to $p(\theta)$ is a unit normal vector for $-\pi \leq \theta < \pi$. In particular, $n(0) = e_3$.

We assume that $\nu \colon M \to \mathbb{R}^3$ is a unit normal field with $\nu\big(p(0)\big) = e_3$. Then because $T_{p(\theta)}^\perp M$ is continuous and one-dimensional, it follows that the vectors $\nu\big(p(\theta)\big)$ and $n(\theta)$ coincide in $-\pi \leq \theta < \pi$. From this and the relation $p(-\pi) = p(\pi)$, we find as $\theta \to \pi$ that

$$-e_1 = n(-\pi) = \nu\big(p(-\pi)\big) = \nu\big(p(\pi)\big) = n(\pi) = e_1 \ ,$$

which is not possible. Thus there is no smooth (or even continuous) unit normal field on M; this, by (h), shows that M is not orientable. ∎

(j) (regularity)   With obvious modifications, the statements above remain true for C^1 manifolds. ∎

Tensor fields

Let $r, s \in \mathbb{N}$. Then, according to Section 2, the vector space $T_s^r(T_pM)$, which consists of r-contravariant and s-covariant tensors, is well defined on T_pM. Therefore the **bundle of (r, s)-tensors** on M,

$$T_s^r(M) := \bigcup_{p \in M} T_s^r(T_pM) \ ,$$

is also well defined. An **(r, s)-tensor** (more precisely, an r-contravariant and s-covariant tensor) on M is a section of this bundle, that is, it is a map

$$\gamma \colon M \to T_s^r(M) \qquad \text{with } \gamma(p) \in T_s^r(T_pM) \text{ for } p \in M \ .$$

If γ and δ are (r,s)-tensors on M and $f \in \mathbb{R}^M$, then the **sum**, $\gamma + \delta$, the **product with functions**, $f\gamma$, and the **tensor product**, $\gamma \otimes \delta$, are again defined pointwise as

$$(\gamma + \delta)(p) := \gamma(p) + \delta(p) \ , \quad (f\gamma)(p) := f(p)\gamma(p) \ , \quad \gamma \otimes \delta(p) := \gamma(p) \otimes \delta(p)$$

for $p \in M$. Likewise, the effect of $\gamma \in T_s^r(M)$ on an r-tuple $\alpha_1, \ldots, \alpha_r$ of Pfaff forms and an s-tuple $v_1, \ldots, v_s$ of vector fields is defined pointwise by

$$\gamma(\alpha_1, \ldots, \alpha_r, v_1, \ldots, v_s)(p) := \gamma(p)(\alpha_1(p), \ldots, \alpha_r(p), v_1(p), \ldots, v_s(p)) \text{ for } p \in M \ .$$

Finally let $\varphi \in \mathrm{Diff}^1(M, N)$. Then we define the **push forward** by φ of $\gamma \in T_s^r(M)$ through

$$(\varphi_* \gamma)(\alpha_1, \ldots, \alpha_r, v_1, \ldots, v_s) := (\gamma \circ \varphi^{-1})(\varphi^* \alpha_1, \ldots, \varphi^* \alpha_r, \varphi^* v_1, \ldots, \varphi^* v_s) \ ,$$

where $\alpha_1, \ldots, \alpha_r$ are Pfaff forms and $v_1, \ldots, v_s$ are vector fields on N, and we have set

$$\varphi^* v := (\varphi^{-1})_* v \tag{4.25}$$

with v a vector field on N. Naturally, $\varphi^* \gamma := (\varphi^{-1})_* \gamma$ is then the **pull back** of $\gamma \in T_s^r(N)$.

Let $k \in \mathbb{N} \cup \{\infty\}$. Then an (r,s)-tensor γ belongs to the class C^k (or, is k-times **continuously differentiable** or **smooth** in the case $k = \infty$) if every point of M has a chart (φ, U) such that $\varphi_* \gamma$ is a (r,s)-tensor on $\varphi(U)$ of class C^k. We denote the set of all smooth (r,s)-tensors on M by

$$\mathcal{T}_s^r(M) \ .$$

The proofs of the following remarks are straightforwardly transferred from the corresponding proofs for differential forms in Sections 2 and 3.[16] Therefore we leave these proofs to you.

4.18 Remarks (a) The definition of differentiability is coordinate independent.

(b) $\mathcal{T}_s^r(M)$ is an $\mathcal{E}(M)$-module. The tensor product map

$$\otimes : \mathcal{T}_{s_1}^{r_1}(M) \times \mathcal{T}_{s_2}^{r_2}(M) \to \mathcal{T}_{s_1+s_2}^{r_1+r_2}(M) \ , \quad (\gamma, \delta) \mapsto \gamma \otimes \delta$$

is $\mathcal{E}(M)$-bilinear and associative.

(c) An (r,s)-tensor γ on M is smooth if and only if

$$\gamma(\alpha_1, \ldots, \alpha_r, v_1, \ldots, v_s) \in \mathcal{E}(M)$$

for all $v_1, \ldots, v_s \in \mathcal{V}(M)$ and $\alpha_1, \ldots, \alpha_r \in \Omega^1(M)$.

---

[16]This repetition can be avoided if one first develops the (elementary) theory of vector bundles.

(d) Let (φ, U) be a chart of M. Then

$$\left\{ \frac{\partial}{\partial x^{j_1}} \otimes \cdots \otimes \frac{\partial}{\partial x^{j_r}} \otimes dx^{k_1} \otimes \cdots \otimes dx^{k_s} \; ; \; j_i, k_i \in \{1, \ldots, m\} \right\} \qquad (4.26)$$

is a module basis of $T_s^r(M)$. Then $\gamma \in T_s^r(U)$ if and only if the coefficients of γ in the basis representation (4.26) are smooth.

(e) Let $\varphi \in \mathrm{Diff}(M, N)$. Then φ_* maps the module $T_s^r(M)$ to $T_s^r(N)$ and operates covariantly as

$$(\psi \circ \varphi)_* = \psi_* \circ \varphi_* \text{ and } (\mathrm{id}_M)_* = \mathrm{id}_{T_s^r(M)} \; .$$

Analogously, $\varphi^* \big(T_s^r(N) \big) = T_s^r(M)$, and φ^* operates contravariantly. Finally φ_* and therefore also φ^* is compatible with the tensor product map, that is,

$$\varphi_*(\gamma \otimes \delta) = \varphi_* \gamma \otimes \varphi_* \delta \; .$$

(f) For $f \in C^\infty(M, N)$ and $\gamma \in T_s^0(N)$, the **pull back** $f^* \gamma$ of γ by f is determined by

$$f^* \gamma(v_1, \ldots, v_s) := (\gamma \circ f) \big((Tf) v_1, \ldots, (Tf) v_s \big) \quad \text{for } v_1, \ldots, v_s \in \mathcal{V}(M)$$

with $((Tf)v)(p) := (T_p f) v(p)$ for $p \in M$. Then the map

$$f^* : T_s^0(N) \to T_s^0(M) \; , \quad \gamma \mapsto f^* \gamma$$

is well defined and $\mathbb{R}$-linear, operates contravariantly, and is compatible with the tensor product map. For $f \in \mathrm{Diff}(M, N)$, it is the same as the previously defined pull back.

(g) $T_0^1(M) = \mathcal{V}(M)$, $T_1^0(M) = \Omega^1(M)$, and dual pairing $\langle \cdot, \cdot \rangle$ is a $(1, 1)$-tensor on M.

(h) (regularity) Let $k \in \mathbb{N}$. Then the statements above hold analogously if M is a C^{k+1} manifold and C^∞ is replaced by C^k. ∎

Exercises

1 Let N be a submanifold of a manifold without boundary. Show that (with the canonical identification) $\mathcal{V}^k(N) \subset \mathcal{V}^k(M)$ for $k \in \mathbb{N} \cup \{\infty\}$.

2 Suppose $\alpha \in \Omega^r(M)$ and $\beta \in \Omega(M)$, and let $v \in \mathcal{V}(M)$. Show that

$$v \lrcorner (\alpha \wedge \beta) = (v \lrcorner \alpha) \wedge \beta + (-1)^r \alpha \wedge (v \lrcorner \beta) \; .$$

3 Verify the statements made in the proof of Example 4.17(b).

4 For $\alpha \in \Omega^1(M)$ and $v \in \mathcal{V}(M)$, calculate $d\langle \alpha, v \rangle$ in local coordinates.

5 Riemannian metrics

We already know from Section VII.10 that the Euclidean inner product $(\cdot\,|\,\cdot)$ on $\mathbb{R}^{\overline{m}}$ can be used to define another inner product by restricting it to the tangent space T_pM of a submanifold M. This gives a way to measure lengths and angles on T_pM. So, for example, we can determine if two curves Γ_1 and Γ_2 on M intersect orthogonally at a point p by verifying that the tangent spaces $T_p\Gamma_1$ and $T_p\Gamma_2$ are themselves orthogonal in T_pM.

That the Euclidean structure of $\mathbb{R}^{\overline{m}}$ induces one on M, or precisely on the tangent bundle of M, is the foundation for the theory of integration on manifolds, which we will treat in the next chapter. In this section, we explore a few consequences of the existence of a Euclidean structure on M, and we study several examples. We also introduce the Hodge star operator and the codifferential, which are of significance for a deeper incursion into the theory of differential forms — in particular, these concepts are important in (theoretical) physics.

To ease the introduction to the material, we consider first the case of the Euclidean structure on M induced by $(\cdot\,|\,\cdot)$. It will be apparent, however, that all abstract theorems remain true in an essentially more general framework, namely, that of Riemannian geometry. Because these facts are of great theoretical and practical importance, we will introduce the concept of a (pseudo) Riemannian metric, which forms the general framework for our subsequent considerations.

For the entire section, suppose the following:

- M is an m-dimensional submanifold of $\mathbb{R}^{\overline{m}}$; N is an n-dimensional submanifold of $\mathbb{R}^{\overline{n}}$.

- The indices i, j, k, l always range from 1 to m unless otherwise stated, and $\sum_j$ means that j is summed from 1 to m.

The volume element

Suppose M is oriented. Then $\mathcal{O}r$ induces an orientation on every tangent space T_pM. Also T_pM is an inner product space with inner product $(\cdot\,|\,\cdot)_p$ induced by the Euclidean scalar product of the surrounding space $\mathbb{R}^{\overline{m}}$. Thus, by Remark 2.12(b), there is a unique volume element ω_p on T_pM. Therefore

$$\omega_M(p) := \omega_p \quad \text{for } p \in M$$

defines an m-form on M, the **volume element** on M.

5.1 Proposition *Suppose M is oriented. Then ω_M belongs to $\mathcal{O}r(M)$. If (φ, U) is a positive chart with $\varphi = (x^1, \ldots, x^m)$, then*

$$\omega_M \,|\, U = \sqrt{G}\, dx^1 \wedge \cdots \wedge dx^m \ , \tag{5.1}$$

where $G := \det[g_{jk}] \in \mathcal{E}(U)$ *is the* **Gram determinant** *and*

$$g_{jk}(p) := \left(\partial_j|_p \mid \partial_k|_p\right)_p \quad \text{for } 1 \leq j, k \leq m \text{ and } p \in U .$$

Letting $g_\varphi := i \circ \varphi^{-1} \in C^\infty(\varphi(U), \mathbb{R}^{\overline{m}})$, *with* $i : M \hookrightarrow \mathbb{R}^{\overline{m}}$, *be the parametrization belonging to* φ, *we have*

$$\varphi_* g_{jk}(x) = \left(\partial_j g_\varphi(x) \mid \partial_k g_\varphi(x)\right) \quad \text{for } 1 \leq j, k \leq m \text{ and } x \in \varphi(U) . \tag{5.2}$$

Proof Because $(\partial_1|_p, \ldots, \partial_m|_p)$ is a positive basis of T_pM, it follows from Proposition 2.13 that

$$\omega_p = \sqrt{G(p)} \, dx^1 \wedge \cdots \wedge dx^m(p) \quad \text{for } p \in U .$$

Therefore (5.1) holds. Because

$$\varphi_*(\omega_M \mid U) = \varphi_* \sqrt{G} \, dx^1 \wedge \cdots \wedge dx^m \mid \varphi(U)$$

with $\varphi_* \sqrt{G} = \sqrt{G \circ \varphi^{-1}}$, it follows from Remark 4.3(a) that (5.2) is satisfied. Because the scalar product and the determinant function are smooth (see Proposition VII.4.6 and Exercise VII.4.2) and because $G(p) > 0$ for $p \in U$, the chain rule gives $\varphi_* \sqrt{G} \in \mathcal{E}(\varphi(U))$. Therefore $\omega_M \mid U$ is smooth, which proves $\omega_M \in \mathcal{O}r$. ∎

5.2 Remark (regularity) By modifying the statement of this proposition in the obvious way, we find it remains true when M is a C^1 manifold. ∎

5.3 Examples **(a)** (open sets in $\overline{\mathbb{H}}^m$) Let X be a nonempty open subset of $\overline{\mathbb{H}}^m$. Then X is endowed with a **natural orientation** with respect to which every tangent space $T_pX = T_p\mathbb{R}^m$ with $p \in X$ is naturally oriented, that is, this orientation makes the canonical basis $((e_1)_p, \ldots, (e_m)_p)$ positive. Then the volume element of X is given by

$$\omega_X = dx^1 \wedge \cdots \wedge dx^m \mid X .$$

The trivial chart (id_X, X) is positive.

(b) (fibers of regular maps) Suppose X is open in $\mathbb{R}^m$ and q is a regular value of $f \in \mathcal{E}(X)$ with $M := f^{-1}(q) \neq \emptyset$. We provide the hypersurface M with the **orientation** $\mathcal{O}r(M, \nabla f)$ **induced by** ∇f, as follows: For every $p \in M$, the basis $(v_1, \ldots, v_{m-1})$ of T_pM is positive if and only if the basis

$$\left(\nabla f(p), v_1, \ldots, v_{m-1}\right)$$

of $T_pX = T_p\mathbb{R}^m$ is positive with $\nabla f = \sum_k \partial_k f \, \partial/\partial x^k$. With

$$\nu := \nabla f / |\nabla f|$$

the unit normal field of M, the volume element of $\left(M, \mathcal{O}r(M, \nabla f)\right)$ is given by $\omega_M := (\nu \lrcorner \omega_X) \mid M$.

If $m = 3$, a basis $\big(v_1(p), v_2(p)\big)$ of $T_p M$ is positive if and only if the three vectors $\big(v_1, v_2, \nu(p)\big)$ form a "right handed basis" of $T_p \mathbb{R}^3$. Here (w_1, w_2, w_3) is a **right handed basis** if one can stretch out the thumb, first, and second fingers of one's right hand (with the middle finger bent palmward) so that these three fingers point in the direction (and the same order) of these three vectors. This is called the **right hand rule**.

Proof Because q is a regular point, $\nabla f(q) \neq 0$ for $q \in M$. By the regular value theorem, M is a smooth hypersurface in X. The proof of Example 4.17(e) shows that ω_M is a smooth volume form. Now all is clear. ∎

(c) (spheres) The m-sphere S^m in $\mathbb{R}^{m+1}$ for $m \in \mathbb{N}$ is **canonically** oriented by the outward unit normal field

$$\nu(x) := (x, x) \in T_x \mathbb{R}^{m+1} .$$

If $m = 0$, S^0 consists of the two points $\{\pm 1\} \subset \mathbb{R}$, and the outward unit normal field at 1 [or -1] is given by $(1, 1) \in T_1 \mathbb{R}$ [or $(-1, -1) \in T_{-1}\mathbb{R}$].[1]

When $m = 1$, the canonical orientation of S^1 is the same as the one given in Remark VIII.5.8. Therefore "one traverses S^1 is the positive direction" exactly when the traversal is counterclockwise. In this case, ν coincides with the negative unit normal vector $-\mathbf{n}$ in the sense of the Frenet two-frame.

The volume element of the canonically oriented m-sphere is the m-form[2]

$$\omega_{S^m} = \big(\nu \lrcorner \; \omega_{\mathbb{R}^{m+1}}\big)\big|\, S^m$$

$$= \sum_{j=1}^{m+1} (-1)^{j-1} x^j \, dx^1 \wedge \cdots \wedge \widehat{dx^j} \wedge \cdots \wedge dx^{m+1} \Big|\, S^m .$$

The chart $(\varphi_{\pm}, S^m_{\mp})$ describes the upper [or lower] hemisphere $S^m_{\mp}$ that is projected along the x^{m+1}-axis onto $\mathbb{B}^m \times \{0\}$; this chart is positively oriented when m is even [or odd] and is negatively oriented for odd [or even] m.

The spherical coordinate chart of S^1 is positive, whereas that of S^2 is negative.

Proof The formula for ω_{S^m} is a special case of (b). The statements about the various charts of S^m follow from Examples 4.6(a)–(b). ∎

(d) (graphs) Let X be open in $\overline{\mathbb{H}}^m$ and $f \in C^\infty(X, \mathbb{R}^n)$. Then the **natural orientation of the graph** $M := \text{graph}(f)$ is the one for which the natural chart (φ, M) with

$$\varphi : M \to \mathbb{R}^m , \quad \big(x, f(x)\big) \mapsto x$$

---

[1] See Example 1.17(a).
[2] This justifies the notations used in Examples 4.6 and 4.13(c).

is positive. In the case $n = 1$, the volume element ω_M has the local representation

$$\omega_M \mid M = \sqrt{1 + |\nabla f|^2} \, dx^1 \wedge \cdots \wedge dx^m$$

with $\nabla f = \sum_{j=1}^m \partial_j f \, \partial_j$.

Proof Because $g_\varphi(x) = (x, f(x))$ for $x \in X = \varphi(M)$, it follows from Remark 4.3(a) that

$$\partial_j|_p = \big(p, (e_j, \partial_j f(x))\big) \in T_p\mathbb{R}^{m+1} \quad \text{for } p = (x, f(x)) \in M \text{ and } 1 \le j \le m ,$$

where T_pM is identified canonically with the vector subspace $(T_p i_M)(T_pM)$ of $T_p\mathbb{R}^{m+1}$. Putting $d_j := \partial_j f$, we then get $g_{jk} = \delta_{jk} + d_j d_k$.

Let $D_m := [\varphi_* g_{jk}]$. Then it follows that

$$G = \det D_m = \det \begin{bmatrix} 1 + d_1^2 & d_1 d_2 & \cdots & d_1 d_m \\ d_2 d_1 & 1 + d_2^2 & \cdots & d_2 d_m \\ \vdots & \vdots & \ddots & \vdots \\ d_m d_1 & d_m d_2 & \cdots & 1 + d_m^2 \end{bmatrix}$$

$$= \det \left[\begin{array}{ccc:c} & & & 0 \\ & D_{m-1} & & \vdots \\ & & & 0 \\ \hdashline d_m d_1 & \cdots & d_m d_{m-1} & 1 \end{array} \right] + d_m^2 \det \left[\begin{array}{ccc:c} & & & d_1 \\ & D_{m-1} & & \vdots \\ & & & d_{m-1} \\ \hdashline d_1 & \cdots & d_{m-1} & 1 \end{array} \right] .$$

To compute the last determinant, we subtract d_j times the last column from the j-th column for $1 \le j \le m - 1$, and so find that its value is 1. This then gives the recursion formula

$$\det D_m = \det D_{m-1} + d_m^2 .$$

Because $\det D_1 = 1 + d_1^2$, the recursion yields

$$G = \det D_m = 1 + d_1^2 + \cdots + d_m^2 = 1 + |\nabla f|^2$$

and hence the claim. ∎

(e) (curves) Suppose J is a perfect interval in $\mathbb{R}$, and $\gamma: J \to \mathbb{R}^m$ is a smooth embedding. Then $M := \gamma(J)$ is an embedded curve in $\mathbb{R}^m$. Also let M be **oriented by** γ, that is, let $(\gamma(t), \dot\gamma(t))$ be a positive basis of $T_{\gamma(t)}M$ for $t \in J$. Finally, suppose $\varphi: M \to \mathbb{R}$, with $\gamma = i_M \circ \varphi^{-1}$ the chart of M belonging to γ. Then $\omega_M = |\dot\gamma| \, dt$.

Proof This is an immediate consequence of Proposition 5.1. ∎

(f) (parametrized surfaces) Let X be open in $\overline{\mathbb{H}}^2$, and $h: X \to \mathbb{R}^n$ let be a smooth embedding. Then $M := h(X)$ is a two-dimensional submanifold in $\mathbb{R}^n$, a **surface** in $\mathbb{R}^n$, which is described by a single chart. Therefore M is orientable. By the **orientation induced by the parametrization** h, we mean that orientation for which $(\partial_1 h(x), \partial_2 h(x))$ is a positive basis of $T_{h(x)}M$ for every $x \in X$.

Let $\varphi \colon M \to \mathbb{R}^2$ with $\varphi = (u, v)$ charge belonging to h, that is, $h = i_M \circ \varphi^{-1}$. With the classical notations

$$\mathsf{E} := |\partial_1 h|^2 \,, \quad \mathsf{F} := (\partial_1 h \,|\, \partial_2 h) \,, \quad \mathsf{G} := |\partial_2 h|^2 \,,$$

we have

$$\omega_M = \sqrt{\mathsf{EG} - \mathsf{F}^2} \, du \wedge dv \,.$$

Proof This follows from $\varphi_* G = \mathsf{EG} - \mathsf{F}^2$. ∎

(g) (boundaries) Suppose M is an oriented manifold with boundary and $\nu(p)$ is the outward (unit) normal vector ∂M at $p \in \partial M$. Then we say a basis $(v_1, \ldots, v_{m-1})$ of $T_p \partial M$ is positive if $(\nu(p), v_1, \ldots, v_{m-1})$ is a positive basis of $T_p M$. This is turn determines an orientation on ∂M, the **orientation induced by the outward normal**. The volume element $\omega_{\partial M}$ of ∂M satisfies

$$\omega_{\partial M} = (\nu \lrcorner \omega_M)\,|\,\partial M = i^*_{\partial M}(\nu \lrcorner \omega_M) \,,$$

where $i_{\partial M} \colon \partial M \hookrightarrow M$ is the natural embedding.

Obviously (c) is a special case of this situation. Note also that the orientation induced by the outward normal must not agree with that induced by ∇f if ∂M can be represented as in (b) as the fiber of a regular map.

Proof From Theorem 1.15 and Remark 1.16(a), we know that ν can be locally described in the form $\nu(p) = \nabla f(p)/|\nabla f(p)|$, where f is a smooth function satisfying $\nabla f(p) \neq 0$. This shows the unit normal vector field is smooth. From this it follows easily that $(\nu \lrcorner \omega_M)\,|\,\partial M$ belongs to $\Omega^{m-1}(\partial M)$. If $(v_1, \ldots, v_{m-1})$ is an ONB of $T_p \partial M$, then $(\nu(p), v_1, \ldots, v_{m-1})$ is an ONB of $T_p M$. When $(\nu(p), v_1, \ldots, v_{m-1})$ is a positive ONB of $T_p M$,

$$1 = \omega_M(\nu(p), v_1, \ldots, v_{m-1}) = (\nu \lrcorner \omega_M)(p)(v_1, \ldots, v_{m-1}) \,. \quad ∎$$

(h) Let $m \geq 2$. Then the orientation of $\partial \mathbb{H}^m = \mathbb{R}^{m-1}$ induced by the outward normal $\nu = -e_m$ coincides with the natural orientation of $\mathbb{R}^{m-1}$ if and only if m is even.

Proof This we may read off from

$$\det[\nu, e_1, \ldots, e_{m-1}] = (-1)^{m-1} \det[e_1, \ldots, e_{m-1}, -e_m] = (-1)^m \,. \quad ∎$$

Riemannian manifolds

The proof of Proposition 5.1 depends on the fact that every tangent space $T_p M$ is endowed naturally with an inner product that varies differentiably with $p \in M$. Such situations appear quite frequently, although the scalar product on TM is often generated in another way. Therefore it is useful to explore these issues somewhat more precisely.

A **Riemannian metric** on M is a tensor $g \in T_2^0(M)$ such that $g(p)$ is an inner product on T_pM for every $p \in M$. Then (M, g) is a **Riemannian manifold**.

Let (M, g) be a Riemannian manifold. We will often write $((x^1, \ldots, x^m), U)$ for the chart (φ, U) of M with $\varphi = (x^1, \ldots, x^m)$. Then we set

$$g_{jk} := g\left(\frac{\partial}{\partial x^j}, \frac{\partial}{\partial x^k}\right) \in \mathcal{E}(U) \tag{5.3}$$

and put

$$[g^{jk}] := [g_{jk}]^{-1} \in C^\infty(U, \mathbb{R}^{m\times m}_{\mathrm{sym}}) \quad \text{and} \quad G := \det[g_{jk}] \in \mathcal{E}(U) . \tag{5.4}$$

We also call g the **(first) fundamental tensor**. Here $[g_{jk}]$ is the representation matrix (or simply, the matrix) of g in the local coordinates $(x^1, \ldots, x^m)$; it is also called the **(first) fundamental matrix**.[3] As before, G is called the **Gram determinant**.

5.4 Remarks **(a)** If g is a Riemannian metric on M, the map

$$\mathcal{V}(M) \times \mathcal{V}(M) \to \mathcal{E}(M) , \quad (v, w) \mapsto g(v, w) \tag{5.5}$$

is well defined, bilinear, symmetric, and **positive** in the sense that

$$g(v, v) \geq 0 \quad \text{and} \quad g(v, v) = 0 \Longleftrightarrow v = 0 . \tag{5.6}$$

Proof That the map (5.5) is well defined follows immediately from Remark 4.18(c). The remaining claims are direct consequences of the properties of scalar products. ∎

(b) Let $((x^1, \ldots, x^m), U)$ be a chart of a Riemannian manifold (M, g). Then

$$g \mid U = \sum_{j,k} g_{jk} \, dx^j \otimes dx^k .$$

In this context, we usually write $dx^j dx^k$ for $dx^j \otimes dx^k$.

Proof According to (4.8), any $v \in \mathcal{V}(U)$ has a basis representation

$$v = \sum_j \langle dx^j, v\rangle \frac{\partial}{\partial x^j} .$$

The claim follows from this, the bilinearity of the map (5.5), and the definitions of $dx^j \otimes dx^k$ and g_{jk}. ∎

(c) Let $((x^1, \ldots, x^m), U)$ be a positive chart of an oriented Riemannian manifold (M, g). Then the volume element ω_M of M satisfies

$$\omega_M \mid U = \sqrt{G} \, dx^1 \wedge \cdots \wedge dx^m .$$

Proof This follows from Proposition 2.13. ∎

---

[3]See Remark VII.10.3(b).

(d) Let g be a Riemannian metric on M, and let $(x^1, \ldots, x^m)$ and $(y^1, \ldots, y^m)$ be local coordinates on an open set U of M. Then

$$g \,|\, U = \sum_{j,k} g_{jk} \, dx^j \otimes dx^k = \sum_{r,s} \overline{g}_{rs} \, dy^r \otimes dy^s$$

with

$$\overline{g}_{rs} = \sum_{j,k} \frac{\partial x^j}{\partial y^r} \frac{\partial x^k}{\partial y^s} g_{jk} \quad \text{for } 1 \le r, s \le m .$$

Proof Because

$$dx^j = \sum_r \frac{\partial x^j}{\partial y^r} \, dy^r \quad \text{for } 1 \le r \le m ,$$

this is a consequence of (b). ∎

(e) If we only require of $g \in \mathcal{T}_2^0(M)$ that the bilinear form $g(p)$ on $T_p M$ is symmetric and nondegenerate for every $p \in M$, then we call g an **indefinite Riemannian metric**, and (M, g) is a **pseudo-Riemannian manifold**. In this case, we again use the notations (5.3) and (5.4). Then (a), (b) and (d), with the exception of (5.6), remain true. Every Riemannian manifold is also pseudo-Riemannian.

(f) Let (M, g) be a (pseudo-)Riemannian manifold, and suppose W is open in M. If $v_1, \ldots, v_m \in \mathcal{V}(W)$ satisfy

$$g(v_j, v_k) = \pm \delta_{jk} \quad \text{for } 1 \le j, k \le m ,$$

we say $(v_1, \ldots, v_m)$ is an **orthonormal frame** on W. Of course, Riemannian manifolds have $g(v_j, v_j) = 1$ for $1 \le j \le m$. An orthonormal frame $(v_1, \ldots, v_m)$ on W is therefore an m-tuple of (smooth) vector fields W that form an ONB (with respect to the (indefinite) inner product $g(p)$ of $T_p M$) at every point $p \in W$. Such an orthonormal frame does not exist in general, because, according to Remark 4.5(d), one cannot generally find m vector fields that are everywhere linearly independent.

If (φ, U) is a chart of M, then there is an orthonormal frame on U.

Proof The basis vector fields $\partial_1, \ldots, \partial_m \in \mathcal{V}(U)$ are linearly independent at every point. Because g is nondegenerate, the Gram–Schmidt orthonormalization procedure (see for example [Art93, §§ 7.1 and 7.2]) then generates an orthonormal frame. The details are left to you. ∎

(g) Let (M, g) be an oriented pseudo-Riemannian manifold. If (φ, U) is a positive chart of M with $\varphi = (x^1, \ldots, x^m)$, we set

$$\omega_M \,|\, U := \sqrt{|G|} \, dx^1 \wedge \cdots \wedge dx^m .$$

This then defines a volume form $\omega_M \in \Omega^m(M)$ on M, which we call the **volume element** of M. Every positive orthonormal frame $(v_1, \ldots, v_m)$ on U satisfies

$$\omega_M(v_1, \ldots, v_m) = 1 .$$

Proof We show first that $\omega_M \in \Omega^m(M)$ is well defined. So let $(e_1, \ldots, e_m)$ be any orthonormal frame on U, with $(\varepsilon^1, \ldots, \varepsilon^m)$ its dual frame; that is, $\varepsilon^j \in \Omega^1(U)$ and $\langle \varepsilon^j, e_k \rangle = \delta_k^j$ for $1 \leq j, k \leq m$. Then it follows from Remark 2.18(d) that

$$\varepsilon^1 \wedge \cdots \wedge \varepsilon^m = \sqrt{|G|} \, dx^1 \wedge \cdots \wedge dx^m \ .$$

Because this is true for every positive coordinate system $(x^1, \ldots, x^m)$ on U, it follows that $\omega_M \mid U \in \Omega^m(U)$ is well defined and independent of the special choice of local coordinates. Suppose now $\{ (\varphi_\alpha, U_\alpha) \, ; \, \alpha \in A \}$ is a positive atlas of M and $(v_{\alpha,1}, \ldots, v_{\alpha,m})$ is a positive orthonormal frame on U_α with dual frame $(\varepsilon_\alpha^1, \ldots, \varepsilon_\alpha^m)$. Then we define ω_M on M by $\omega_M \mid U_\alpha := \varepsilon_\alpha^1 \wedge \cdots \wedge \varepsilon_\alpha^m$. From the previous considerations, it follows that ω_M is well defined and belongs to $\Omega^m(M)$. The last claim is now obvious. ∎

(h) (regularity) Let $k \in \mathbb{N}$, and let M be a C^{k+1} manifold. Then the definitions and statements above remain true if $\mathcal{V}(M)$ and $\mathcal{E}(M)$ are replaced everywhere by $\mathcal{V}^k(M)$ and $C^k(M)$, respectively. ∎

Suppose $(N, \bar{g})$ is a Riemannian manifold and $f : M \to N$ is an immersion. Then $f^* \bar{g}$ (the pull back of $\bar{g}$ by f) is a Riemannian metric on M. If M is a submanifold of N and $i : M \hookrightarrow N$ is the natural embedding, then $i^* \bar{g}$ is the **Riemannian metric induced by N** (more precisely, by $(N, \bar{g})$).

Let (M, g) and $(N, \bar{g})$ be Riemannian manifolds. An immersion $f : M \to N$ is said to be an **isometry** if $g = f^* \bar{g}$. If f is an isometric diffeomorphism, that is, both an isometry and a diffeomorphism, then M and N are **isometrically isomorphic**.

5.5 Examples **(a)** Suppose (M, g) is a Riemannian manifold and (φ, U) is a chart with $\varphi = (x^1, \ldots, x^m)$. Then (U, g) and $\big(\varphi(U), \varphi_* g \big)$ are isometrically isomorphic, and

$$\varphi_* g = \sum_{j,k} g_{jk} \, dx^j \, dx^k \ .$$

Proof This follows immediately from the definition of the fundamental matrix. ∎

(b) $\mathbb{R}^{\overline{m}}$ is a Riemannian manifold with the Euclidean metric $g_{\overline{m}} := (\cdot \mid \cdot)$, the standard metric. Therefore $\mathbb{R}^{\overline{m}}$ induces a Riemannian metric g on M, which we also call the **standard metric**. It is obviously independent of $\mathbb{R}^{\overline{m}}$ in the sense that $\mathbb{R}^n$ induces the same metric on M when M lies in $\mathbb{R}^n$. In particular, $g(p) = (\cdot \mid \cdot)_p$ for $p \in M$ (with the notation we have been using) for the scalar product induced by $(\cdot \mid \cdot)$ in $T_p M$.

If (φ, U) is a chart of M with $\varphi = (x^1, \ldots, x^m)$ and $h := i \circ \varphi^{-1}$ with $i : M \hookrightarrow \mathbb{R}^{\overline{m}}$ is the associated parametrization, then

$$\varphi_* g = \sum_{j,k} (\partial_j h \mid \partial_k h) \, dx^j \, dx^k \ .$$

In other words, the first fundamental matrix $[g_{jk}]$ is given in local coordinates $(x^1, \ldots, x^m)$ by

$$[(\partial_j h \mid \partial_k h)] \in C^\infty \big(\varphi(U), \mathbb{R}^{m \times m} \big) \ .$$

This is consistent with Remark 5.4(b) and also shows that Proposition 5.1 is a
special case of Remark 5.4(c).

Proof   From $g = i^* g_{\overline{m}}$, it follows that $\varphi_* g = (\varphi^{-1})^* i^* g_{\overline{m}} = h^* g_{\overline{m}}$. Let $(y^1, \ldots, y^{\overline{m}})$ be
Euclidean coordinates of $\mathbb{R}^{\overline{m}}$. Then

$$g_{\overline{m}} = \sum_{j=1}^{\overline{m}} (dy^j)^2 \quad \text{and} \quad h^* g_{\overline{m}} = \sum_{j=1}^{\overline{m}} h^*(dy^j \otimes dy^j) = \sum_{j=1}^{\overline{m}} dh^j \otimes dh^j ,$$

which follow easily from the definition of the pull back of $(0,2)$-tensors $dy^j \otimes dy^j$ and
from Example 3.4(a). Now the claim follows easily from the bilinearity of $(\alpha, \beta) \mapsto \alpha \otimes \beta$
for $\alpha, \beta \in \Omega^1(\varphi(U))$ and from $dh^j = \sum_k \partial_k h^j \, dx^k$. ∎

(c) (graphs)   Suppose X is open in $\overline{\mathbb{H}}^m$ and $f \in C^\infty(X, \mathbb{R}^n)$. Let M be the graph
of f, and let

$$\varphi: M \to \mathbb{R}^m , \quad (x, f(x)) \mapsto x$$

be the natural chart (φ, M). Then the standard metric g of M satisfies

$$g = \sum_j (dx^j)^2 + \sum_{j,k} (\partial_j f \mid \partial_k f) \, dx^j dx^k .$$

In particular, in the case of a surface $(m = 2)$,

$$g = (1 + |\partial_1 f|^2)(dx)^2 + 2(\partial_1 f \mid \partial_2 f) \, dxdy + (1 + |\partial_2 f|^2)(dy)^2 .$$

Proof   Because $g_{jk} = \delta_{jk} + (\partial_j f \mid \partial_k f)$ for $1 \leq j, k \leq m$, this follows from (b). ∎

(d) (parametrized surfaces)   Suppose X is open in $\overline{\mathbb{H}}^2$ and $h: X \to \mathbb{R}^n$ is an
embedding. Then the standard metric of the surface $M := h(X)$ is given by

$$g = \mathsf{E}(du)^2 + 2\mathsf{F} \, dudv + \mathsf{G}(dv)^2 ,$$

where we have used the notations of Example 5.3(f).

(e) (plane polar coordinates)   Let $V_2 := (0, \infty) \times (0, 2\pi)$. Then the polar coordinate
map

$$f_2: V_2 \to \mathbb{R}^2 , \quad (r, \varphi) \mapsto (x, y) := (r \cos \varphi, r \sin \varphi)$$

is an embedding with $M := f_2(V_2) = \mathbb{R}^2 \setminus (\mathbb{R}^+ \times \{0\})$, and

$$g_2 \mid M = (dx)^2 + (dy)^2 = (dr)^2 + r^2 (d\varphi)^2 .$$

Proof   This follows easily from (d). ∎

(f) (circular coordinates)   With respect to the parametrization

$$h: (0, 2\pi) \to \mathbb{R}^2 , \quad t \mapsto (\cos t, \sin t)$$

of $S^1 \setminus \{(1,0)\}$, the standard metric on the circle satisfies $g_{S^1} = (dt)^2$.

Proof Because $|\partial h| = 1$, this follows from (b). ∎

(g) (m-dimensional polar coordinates) With $m \geq 3$, let[4]

$$f_m : V_m \to \mathbb{R}^m , \quad (r, \varphi, \vartheta_1, \ldots, \vartheta_{m-2}) \mapsto (x^1, x^2, x^3, \ldots, x^m)$$

be the (restriction to V_m of the) polar coordinate map (X.8.17). Then f_m is a parametrization of $\mathbb{R}^m \backslash H_{m-1}$, and

$$\sum_{j=1}^{m} (dx^j)^2 = (dr)^2 + r^2 \left[a_{m,0} (d\varphi)^2 + \sum_{k=1}^{m-2} a_{m,k} (d\vartheta_k)^2 \right]$$

with

$$a_{m,k} := \prod_{i=k+1}^{m-2} \sin^2 \vartheta_i \quad \text{for } 0 \leq k \leq m - 3 \quad \text{and} \quad a_{m,m-2} := 1 .$$

In particular, spherical coordinates satisfy ($m = 3$)

$$(dx)^2 + (dy)^2 + (dz)^2 = (dr)^2 + r^2 \left[\sin^2 \vartheta (d\varphi)^2 + (d\vartheta)^2 \right] .$$

Proof With $y = (r, z) \in \mathbb{R} \times \mathbb{R}^{m-1}$, we read off from (X.8.14) that

$$\partial_1 f_m(y) = h_{m-1}(z) \quad \text{and} \quad \partial_j f_m(y) = r \partial_{j-1} h_{m-1}(z) \quad \text{for } 2 \leq j \leq m . \tag{5.7}$$

Therefore (X.8.13) implies

$$|\partial_1 f_m|^2 = 1 . \tag{5.8}$$

Differentiation of $|h_{m-1}|^2 = 1$ gives $(h_{m-1} | \partial_k h_{m-1}) = 0$ for $1 \leq k \leq m - 1$. Then it follows from (5.7) that

$$\left(\partial_1 f_m(y) | \partial_k f_m(y) \right) = r \left(h_{m-1}(z) | \partial_{k-1} h_{m-1}(z) \right) = 0 \quad \text{for } 2 \leq k \leq m . \tag{5.9}$$

From (5.7) we also get

$$\left(\partial_j f_m(y) | \partial_k f_m(y) \right) = r^2 \left(\partial_{j-1} h_{m-1}(z) | \partial_{k-1} h_{m-1}(z) \right) \quad \text{for } 2 \leq j, k \leq m . \tag{5.10}$$

The recursion formula (X.8.12) with $z = (z', z_{m-1}) \in \mathbb{R}^{m-2} \times \mathbb{R}$ leads to

$$\partial_j h_{m-1}(z) = \left(\partial_j h_{m-2}(z') \sin z_{m-1}, 0 \right) \quad \text{for } 1 \leq j \leq m - 2 , \tag{5.11}$$

and

$$\partial_{m-1} h_{m-1}(z) = \left(h_{m-2}(z') \cos z_{m-1}, - \sin z_{m-1} \right) .$$

From this and (X.8.13), it follows that

$$|\partial_{m-1} h_{m-1}(z)|^2 = |h_{m-2}(z')|^2 \cos^2 z_{m-1} + \sin^2 z_{m-1} = 1 ,$$

and, in analogy to the above, we have

$$\left(\partial_j h_{m-1}(z) | \partial_{m-1} h_{m-1}(z) \right) = \sin z_{m-1} \cos z_{m-1} \left(h_{m-2}(z') | \partial_j h_{m-2}(z') \right) = 0$$

---

[4]We use the notations of (X.8.11)–(X.8.24).

for $1 \leq j \leq m - 2$. With (5.7), this proves

$$|\partial_m f_m(z)|^2 = r^2 \quad \text{and} \quad (\partial_j f_m \,|\, \partial_m f_m) = 0 \quad \text{for } 2 \leq j \leq m - 1 . \tag{5.12}$$

Finally (5.11) implies

$$(\partial_j h_{m-1}(z) \,|\, \partial_k h_{m-1}(z)) = \sin^2 z_{m-1}(\partial_j h_{m-2}(z') \,|\, \partial_k h_{m-2}(z')) \quad \text{for } 1 \leq j \leq m - 2 .$$

Thus (5.7) and (5.10) give the recursion formula

$$(\partial_j f_m(y) \,|\, \partial_k f_m(y)) = \sin^2 z_{m-1}(\partial_j f_{m-1}(y') \,|\, \partial_k f_{m-1}(y')) \tag{5.13}$$

for $2 \leq j, k \leq m - 1$, with $y = (y', y_m) \in \mathbb{R}^{m-1} \times \mathbb{R}$. Because (5.8) and (5.12) are true for all $m \geq 3$, induction on (5.13) gives

$$|\partial_j f_m|^2 = r^2 a_{m,j-2} \quad \text{for } 2 \leq j \leq m - 1 , \tag{5.14}$$

and

$$(\partial_j f_m \,|\, \partial_k f_m) = 0 \quad \text{for } 2 \leq j, k \leq m - 1 \text{ and } j \neq k . \tag{5.15}$$

Now the claim follows from (5.8), (5.9), (5.12), (5.14), (5.15), and (b). ∎

(h) (m-dimensional spherical coordinates) For $m \geq 2$, let

$$h_m : W_m \to \mathbb{R}^{m+1} , \quad (\varphi, \vartheta_1, \ldots, \vartheta_{m-1}) \mapsto (y^1, y^2, \ldots, y^{m+1}) ,$$

where $W_m := (0, 2\pi) \times (0, \pi)^{m-1}$ and

$$
\begin{aligned}
y^1 &= \cos\varphi \sin\vartheta_1 \sin\vartheta_2 \cdots \sin\vartheta_{m-1} , \\
y^2 &= \sin\varphi \sin\vartheta_1 \sin\vartheta_2 \cdots \sin\vartheta_{m-1} , \\
y^3 &= \cos\vartheta_1 \sin\vartheta_2 \cdots \sin\vartheta_{m-1} , \\
&\;\;\vdots \\
y^m &= \cos\vartheta_{m-2} \sin\vartheta_{m-1} , \\
y^{m+1} &= \cos\vartheta_{m-1}
\end{aligned}
$$

are (**m-dimensional**) **spherical coordinates**.[5] Then h_m is a parametrization of the open subset $U_m := S^m \setminus H_m$ of the m-sphere. The standard metric g_{S^m} of S^m satisfies

$$g_{S^m} = a_{m+1,0}(d\varphi)^2 + \sum_{k=1}^{m-1} a_{m+1,k}(d\vartheta_k)^2 .$$

In the case of the 2-sphere (with $\vartheta := \vartheta_1$), this becomes

$$g_{S^2} = \sin^2\vartheta(d\varphi)^2 + (d\vartheta)^2 .$$

Proof Because $h_m = f_{m+1}(1, \cdot)$, the claim is a simple consequence of (g). ∎

---

[5] See Example VII.9.11(b).

(i) (Minkowski metric)   We denote the Euclidean coordinates of $\mathbb{R}^4$ by (t, x, y, z) or (x^0, x^1, x^2, x^3) and set $\mathbb{R}^4_{1,3} := \left(\mathbb{R}^4, (\cdot \,|\, \cdot)_{1,3} \right)$ with the Minkowski metric

$$(\cdot \,|\, \cdot)_{1,3} = (dt)^2 - (dx)^2 - (dy)^2 - (dz)^2 = (dx^0)^2 - \sum_{j=1}^{3} (dx^j)^2 \ .$$

Then $\mathbb{R}^4_{1,3}$ is a pseudo-Riemannian manifold, the **spacetime** or **Minkowski space** of (special) relativity theory.

For $v = (v^0, \dots, v^3) \subset \mathbb{R}^4_{1,3}$, we call

$$|v|^2_{1,3} := (v \,|\, v)_{1,3} = (v^0)^2 - \sum_{j=1}^{3} (v^j)^2$$

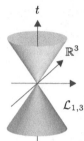

the **Minkowski norm** of the vector v. Vectors with positive Minkowski norm are said to be timelike; those with negative norm are **spacelike**. Those whose Minkowski norm is zero are lightlike; in $\mathbb{R}^4_{1,3}$, the lightlike vectors form a (double) cone, the **light cone** $\mathcal{L}_{1,3}$.

(j) (pseudospherical coordinates)   Let $V_{1,3} := \mathbb{R} \times V_3$ and

$$f_{1,3} : V_{1,3} \to \mathbb{R}^4 \ , \quad (\rho, \chi, \varphi, \vartheta) \mapsto (x^0, x^1, x^2, x^3)$$

with

$$x^0 = \rho \cosh \chi \ ,$$
$$x^1 = \rho \sinh \chi \cos \varphi \sin \vartheta \ ,$$
$$x^2 = \rho \sinh \chi \sin \varphi \sin \vartheta \ ,$$
$$x^3 = \rho \sinh \chi \cos \vartheta \ ;$$

this is the **pseudospherical coordinate map**. Then $f_{1,3}$ is a smooth diffeomorphism from $V_{1,3} \backslash \{0\}$ to the **interior**

$$\mathring{\mathcal{L}}_{1,3} := \left\{ x \in \mathbb{R}^4 \ ; \ |x|^2_{1,3} > 0 \right\}$$

of the **light cone**, and

$$(\cdot \,|\, \cdot)_{1,3} = (d\rho)^2 - \rho^2 \left[(d\chi)^2 + \sinh^2 \chi \sin^2 \vartheta (d\varphi)^2 + \sinh^2 \chi (d\vartheta)^2 \right] \ .$$

Proof   This follows easily from the properties of sinh and cosh (see Exercises III.6.5 and IV.2.5), and from Remark 5.4(d). ∎

(k) (hyperbolic spaces)   To generalize the Minkowski space, we set

$$(\cdot \,|\, \cdot)_{1,m} := (dx^0)^2 - \sum_{j=1}^{m} (dx^j)^2$$

for $n \in \mathbb{N}^\times$. Then $\mathbb{R}^{m+1}_{1,m} := \left(\mathbb{R}^{m+1}, (\cdot \,|\, \cdot)_{1,m} \right)$ is an m-dimensional pseudo-Riemannian manifold.

Let

$$M^m := \left\{ (x^0, x) \in \mathbb{R} \times \mathbb{R}^m \ ; \ (x^0)^2 - |x|^2 = 1, \ x^0 > 0 \right\},$$

that is, M^m is the upper connected component of the m-dimensional two-shelled hyperboloid

$$K_1 := \left\{ x \in \mathbb{R}^{m+1} \ ; \ (Ax \,|\, x) = 1 \right\}, \quad \text{where} \quad A := \mathrm{diag}(1, -1, \ldots, -1)$$

(see Example 1.17(b)). Also let $i \colon M^m \hookrightarrow \mathbb{R}^{m+1}$ be the canonical embedding, and let

$$g_{H^m} := -i^* (\,\cdot \,|\, \cdot\,)_{1,m} \,.$$

Then

$$H^m := (M^m, g_{H^m})$$

is an m-dimensional Riemannian manifold, the m-dimensional **hyperbolic space**. If $N := (N, g)$ is isometrically isomorphic to H^m, we say N is a **model** of H^m. In particular, if we provide $\mathbb{R}^m$ with the metric

$$\frac{(dr)^2}{1 + r^2} + r^2 g_{S^{m-1}} \,,$$

written in the "polar coordinates" $(r, \sigma) \in \mathbb{R}^+ \times S^{m-1}$, then $\mathbb{R}^m$ is a model of H^m.

Proof For $u \colon \mathbb{R}^m \to \mathbb{R}^{m+1}$, $x \mapsto \sqrt{1 + |x|^2}$, we have $M^m = \mathrm{graph}(u)$. Therefore

$$\varphi \colon M^m \to \mathbb{R}^m \ , \quad \big(h(x), x\big) \mapsto x$$

is a diffeomorphism from the hypersurface M^m in $\mathbb{R}^{m+1}$ to $\mathbb{R}^m$. Hence we have only to show that the bilinear form $g_{H^m}(0)$ induced on M by $-(\,\cdot \,|\, \cdot\,)_{1,m}$ is positive definite and that $\varphi_* g_{H^m}$ has the form indicated, because one could read off from this that $g_{H^m}(p)$ is positive definite for every $p \in M \backslash \{\varphi^{-1}(0)\}$.

With $h(x) := \big(u(x), x\big)$ for $x \in \mathbb{R}^m$, we have $h = i \circ \varphi^{-1}$ and

$$\partial_j h = (\partial_j u, e_j) \quad \text{for } 1 \leq j \leq m \,,$$

where e_j is the j-th standard basis vector of $\mathbb{R}^m$. Because $\partial_j u(x) = x^j / u(x)$, it follows that

$$(\varphi_* g_{H^m})_{jk}(x) = (\partial_j h \,|\, \partial_k h)_{1,m}(x) = (\delta_{jk} - x^j x^k) / u^2(x) \quad \text{for } x \in \mathbb{R}^m \,;$$

in particular $(\varphi_* g_{H^m})(0) = \sum_j (dx^j)^2$.

As in (g), let

$$f_m \colon (0, \infty) \times W_{m-1} \to \mathbb{R}^m \ , \quad (r, \vartheta) \mapsto r h_{m-1}(\vartheta)$$

be the m-dimensional polar coordinate map. Then $\psi := f_m^{-1} \circ \varphi$ is a local chart of M, and $a := i \circ \psi^{-1} = h \circ f_m = f_m^* h$ is the associated parametrization. This has

$$a(r, \vartheta) = \big(\sqrt{1 + r^2}, r h_{m-1}(\vartheta)\big) \,,$$

and therefore

$$\partial_r a(r,\vartheta) = \left(\frac{r}{\sqrt{1+r^2}}, h_{m-1}(\vartheta) \right) , \quad \partial_{\vartheta^j} a(r,\vartheta) = (0, r\partial_j h_{m-1}(r,\vartheta))$$

for $(r,\vartheta) \in (0,\infty) \times W_{m-1}$. Because $|h_{m-1}| = 1$, we derive

$$\psi_* g_{H^m} = -a^*(\cdot\,|\,\cdot)_{1,m}$$

$$= r^2 \sum_{j,k} (\partial_j h_{m-1} \,|\, \partial_k h_{m-1}) \, dx^j \, dx^k + \left(1 - \frac{r^2}{\sqrt{1+r^2}} \right) (dr)^2 .$$

The claim now follows from this because of (h) and because the part still missing from $M^m \backslash \{\varphi^{-1}(0)\}$ can be analogously parametrized by rotating M^m around the x^0-axis. ∎

(l) (the Poincaré model) In analogy to
the stereographic projection of the sphere onto
the plane, consider the stereographic projection of the **pseudosphere**

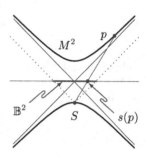

$$S^2_{1,3} := \left\{ (t,x,y) \in \mathbb{R}^3 \,;\, t^2 - x^2 - y^2 = 1 \right\} .$$

We set $N := (1,0,0)$, the **north pole** of $S^2_{1,3}$,
and define the **south pole** as $S := (-1,0,0)$.
Then the value $s(p)$ of the point $p \in M^2$ of
the **stereographic projection** $s \colon M^2 \to \mathbb{R}^2$ is
defined as the point where the line from S to p intersects the plane $\mathbb{R}^2 \times \{0\}$ in $\mathbb{R}^3$.
If the (Euclidean) coordinates of $p \in M^2$ are (t,x,y), and those of $s(p)$ are (u,v),
we learn from the figure above that

$$\frac{x}{u} = \frac{t+1}{1} \quad \text{and} \quad \frac{y}{v} = \frac{t+1}{1} .$$

Because $t^2 - x^2 - y^2 = 1$, it follows that $t^2 - (u^2 + v^2)(t+1)^2 = 1$. From this we
calculate

$$t = \frac{1+u^2+v^2}{1-u^2-v^2} , \quad x = \frac{2u}{1-u^2-v^2} , \quad y = \frac{2v}{1-u^2-v^2} .$$

This shows that

$$\pi \colon \mathbb{B}^2 \to M^2 , \quad (u,v) \mapsto \left(\frac{1+u^2+v^2}{1-u^2-v^2}, \frac{2u}{1-u^2-v^2}, \frac{2v}{1-u^2-v^2} \right)$$

is a parametrization of M^2 over $\mathbb{B}^2$. It satisfies

$$\pi^* g_{H^2} = 4 \frac{(du)^2 + (dv)^2}{(1-u^2-v^2)^2} .$$

Therefore

$$\left(\mathbb{B}^2, 4 \frac{(dx)^2 + (dy)^2}{(1-x^2-y^2)^2} \right)$$

is a model of the hyperbolic plane, the **Poincaré model**.

Proof The proof that $\pi^* g_{H^2}$ has the form given is left to you as an exercise. ∎

(m) (the Lobachevsky model) Following what we did in (h) and (j), we can parametrize M^2 by the **pseudospherical coordinates**

$$h_{1,2} : \mathbb{R}^+ \times [0, 2\pi) \to \mathbb{R}^3 , \quad (\chi, \varphi) \mapsto (t, x, y)$$

with

$$t = \cosh \chi , \quad x = \sinh \chi \cos \varphi , \quad y = \sinh \chi \sin \varphi .$$

These satisfy $h_{1,2}^* g_{H^3} = (d\chi)^2 + \sinh^2 \chi \, (d\varphi)^2$. Therefore

$$\left(\mathbb{R}^+ \times [0, 2\pi), (d\chi)^2 + \sinh^2 \chi \, (d\varphi)^2 \right)$$

is a model of the hyperbolic plane H^2, the **Lobachevsky model.**

Proof The verification of the given formulas is again left to you as an exercise. ∎

(n) (general pseudo-Riemannian metrics) Let X be open in $\overline{\mathbb{H}}^m$, and suppose $g_{jk} = g_{kj} \in \mathcal{E}(X)$ for $1 \le j, k \le m$ and $\det[g_{jk}(x)] \ne 0$ for $x \in X$. Then

$$g := \sum_{j,k} g_{jk} \, dx^j dx^k$$

defines a pseudo-Riemannian metric on X. If the matrix $[g_{jk}(x)]$ is positive definite for every $x \in X$, then g is a Riemannian metric on X.

Now suppose $\{ (\varphi_\alpha, U_\alpha) ; \alpha \in \mathsf{A} \}$ is an atlas for M,

$$g_{\alpha,jk} = g_{\alpha,kj} \in \mathcal{E}(\varphi_\alpha(U_\alpha)) \quad \text{for } 1 \le j, k \le m ,$$

and $\det[g_{\alpha,jk}(x)] \ne 0$ for $x \in \varphi_\alpha(U_\alpha)$ and $\alpha \in \mathsf{A}$. Then there is exactly one pseudo-Riemannian metric g on M such that

$$g \,|\, U_\alpha = g_\alpha := \sum_{j,k} g_{\alpha,jk} \, dx^j dx^k$$

if the transition function $h := \varphi_\beta^{-1} \circ \varphi_\alpha$ satisfies

$$g_{\beta,rs} = \sum_{j,k} \frac{\partial h^j}{\partial x^r} \frac{\partial h^k}{\partial x^s} g_{\alpha,jk}$$

for $\alpha, \beta \in \mathsf{A}$ with $U_\alpha \cap U_\beta \ne \emptyset$.

Proof This is a consequence of Remark 5.4(d). ∎

A Riemannian manifold has a Euclidean structure on every tangent space, which allows lengths and angles to be measured. This allows many concepts from Euclidean geometry to be extended. For example, we presented in Section VIII.1

a formula for the length of a curve. It can now be naturally generalized: a curve $\gamma \colon I \to M$ on a Riemannian manifold M has length

$$\int_I \sqrt{g\big(\dot{\gamma}(t), \dot{\gamma}(t)\big)}\, dt \ ,$$

where $\dot{\gamma}(t) \in T_{\gamma(t)}M$ is the "velocity vector"

$$\dot{\gamma}(t) = (T_t\gamma)(t,1) \quad \text{for } t \in I$$

at the point $\gamma(t)$. We will not expand here on this subject, as the questions raised are best treated in the framework of *Riemannian geometry* (see however Exercise 5).

The Hodge star[6]

Suppose (M,g) is an Riemannian manifold and ω_M is the volume element of M.

For $0 \le r \le m$, we define bilinear maps

$$(\,\cdot\,|\,\cdot\,)_{g,r} : \Omega^r(M) \times \Omega^r(M) \to \mathcal{E}(M) \tag{5.16}$$

by

$$(\alpha\,|\,\beta)_{g,r}(p) := \big(\alpha(p)\,|\,\beta(p)\big)_{g(p),r} \quad \text{for } p \in M \text{ and } \alpha, \beta \in \Omega^r(M) \ , \tag{5.17}$$

where $(\,\cdot\,|\,\cdot\,)_{g(p),r}$ denotes the scalar product on $\bigwedge^r T_p^* M$ introduced in (2.14) and (2.15). The **Hodge star operator** (or simply Hodge star)

$$* : \Omega^r(M) \to \Omega^{m-r}(M) \ , \quad \alpha \mapsto *\alpha \tag{5.18}$$

is also defined pointwise:

$$(*\alpha)(p) := *\alpha(p) \quad \text{for } p \in M \text{ and } \alpha \in \Omega(M) \ .$$

5.6 Remarks (a) The map (5.16) is well defined, bilinear, symmetric, and positive.

Proof We need only show that $(\alpha\,|\,\beta)_{g,r}$ belongs to $\mathcal{E}(M)$ for $\alpha, \beta \in \Omega^r(M)$, because the other statements follow from the properties of $(\,\cdot\,|\,\cdot\,)_{g(p),r}$. Suppose therefore (φ, U) is a positive chart of M. According to Remark 5.4(f), we can choose an oriented orthonormal frame $(v_1, \ldots, v_m)$ on U. Let $(\eta^1, \ldots, \eta^m)$ be its dual frame. Then Remark 3.1(e) implies

$$\alpha\,|\,U = \sum_{(j) \in \mathbb{J}_r} \alpha_{(j)}\, \eta^{j_1} \wedge \cdots \wedge \eta^{j_r} \tag{5.19}$$

with

$$\alpha_{(j)} = \alpha(v_{j_1}, \ldots, v_{j_r}) \in \mathcal{E}(U) \quad \text{for } (j) \in \mathbb{J}_r \ . \tag{5.20}$$

Now the claim follows from (2.14), (2.15), and Remark 2.15(c). ∎

---

[6]The rest of this chapter can be skipped on first reading.

(b) The star operator is a well-defined $\mathcal{E}(M)$-module isomorphism with

$$**\alpha = (-1)^{r(m-r)}\alpha \quad \text{for } \alpha \in \Omega^r(M) . \tag{5.21}$$

Proof Because (5.21) follows from Example 2.17(e) and the pointwise definition (5.18), and because (5.21) also shows that the star operator is bijective, it only remains to show that $*\alpha$ is smooth. So let (φ, U) be a positive chart of M. As in the proof of (a), let $(v_1, \ldots, v_m)$ be an orthonormal frame on U, and let $(\eta^1, \ldots, \eta^m)$ be its dual frame. Then $*\alpha \,|\, U \in \mathcal{E}(U)$ follows from (5.19), (5.20), and the explicit representation of $*\alpha$ in Example 2.17(d). ∎

(c) For $\alpha, \beta \in \Omega^r(M)$, we have

$$\alpha \wedge *\beta = \beta \wedge *\alpha = (\alpha \,|\, \beta)_{g,r}\omega_M . \tag{5.22}$$

Proof This follows immediately from Example 2.17(f) and the pointwise definition of all operations involved. ∎

(d) $*1 = \omega_M$ and $*\omega_M = 1$.

(e) (regularity) It is clear that the statements above are still true when M is a C^{k+1} manifold and $\Omega(M)$ is replaced by $\Omega_{(k)}(M)$. ∎

Using the pointwise definition of the star operator, we can transfer the other formulas of Example 2.17 to the present case. The following examples gather several rules so obtained.

Let $((x^1, \ldots, x^m), U)$ be a chart of M. When $(\partial_1, \ldots, \partial_m)$ is an orthonormal frame on U, we say $(x^1, \ldots, x^m)$ are **orthonormal coordinates** on U. If the ∂_j are not necessarily normalized, that is, we only know $g(\partial_j, \partial_k) = 0$ for $j \neq k$, then the coordinates are **orthogonal**.

5.7 Examples In these examples, $(x^1, \ldots, x^m)$ are orthonormal coordinates on $U \subset M$, and $\alpha \in \Omega(U)$.

(a) Euclidean coordinates are orthonormal coordinates. Polar, spherical, and pseudospherical coordinates are orthogonal.

Proof This follows from Examples 5.5. ∎

(b) $* \sum_j a_j \, dx^j = \sum_{j=1}^m (-1)^{j-1} a_j \, dx^1 \wedge \cdots \wedge \widehat{dx^j} \wedge \cdots \wedge dx^m.$

(c) $* \sum_j (-1)^{j-1} a_j \, dx^1 \wedge \cdots \wedge \widehat{dx^j} \wedge \cdots \wedge dx^m = (-1)^{m-1} \sum_j a_j \, dx^j.$

(d) For $m = 3$, we have

$$*d\left(\sum_j a_j \, dx^j\right) = (\partial_2 a_3 - \partial_3 a_2) \, dx^1 + (\partial_3 a_1 - \partial_1 a_3) \, dx^2 + (\partial_1 a_2 - \partial_2 a_1) \, dx^3 .$$

Proof This follows from Example 3.7(a) (and the remarks following (4.16)), because (2.20) implies the relations

$$*(dx^2 \wedge dx^3) = dx^1 , \quad *(dx^3 \wedge dx^1) = dx^2 , \quad *(dx^1 \wedge dx^2) = dx^3 . \blacksquare$$

Of course, we can also explicitly calculate $* \sum_{(j) \in \mathbb{J}_r} a_{(j)} \, dx^{(j)}$ even if we are not using orthonormal coordinates. For simplicity, we only consider the case of 1-forms.

5.8 Proposition Let $((x^1, \dots, x^m), U)$ be a positive chart of M. Then

$$*dx^j = \sum_k (-1)^{k-1} g^{jk} \sqrt{G} \, dx^1 \wedge \cdots \wedge \widehat{dx^k} \wedge \cdots \wedge dx^m .$$

Proof Because $*dx^j \in \Omega^{m-1}(U)$, Example 3.2(b) guarantees that there are $a_{j\ell}$ in $\mathcal{E}(U)$ such that

$$*dx^j = \sum_\ell (-1)^{\ell-1} a_{j\ell} \, dx^1 \wedge \cdots \wedge \widehat{dx^\ell} \wedge \cdots \wedge dx^m .$$

This gives

$$\begin{aligned}
dx^k \wedge *dx^j &= \sum_\ell (-1)^{\ell-1} a_{j\ell} \, dx^k \wedge dx^1 \wedge \cdots \wedge \widehat{dx^\ell} \wedge \cdots \wedge dx^m \\
&= a_{jk} \, dx^1 \wedge \cdots \wedge dx^m .
\end{aligned} \tag{5.23}$$

From Remark 2.14(b) we get $(dx^j \mid dx^k)_{g,1} = g^{jk}$. Thus Remark 5.6(c) gives

$$dx^k \wedge *dx^j = g^{kj} \omega_M = g^{jk} \sqrt{G} \, dx^1 \wedge \cdots \wedge dx^m , \tag{5.24}$$

where the last equality follows from Remark 5.4(c). Now the claim follows from (5.23) and (5.24). $\blacksquare$

The codifferential

Let (M, g) be an oriented Riemannian manifold. To avoid an exceptional case, we set $\Omega^{-1}(M) := \{0\}$ so that, because $\Omega^{m+1}(M) = \{0\}$, we can also define the star operation $* : \Omega^{m+1}(M) \to \Omega^{-1}(M)$. With help of the (so extended) star operator and the exterior derivative, we define for $0 \le r \le m$ the **codifferential**

$$\delta : \Omega^r(M) \to \Omega^{r-1}(M)$$

by[7]

$$\delta \alpha := (-1)^{m(r+1)} * d * \alpha \quad \text{for } \alpha \in \Omega^r(M) .$$

---

[7]The normalization factor $(-1)^{m(r+1)+1}$ is often used instead of $(-1)^{m(r+1)}$, particularly in differential geometry. The reason for our choice will be made clear in Remark 6.23(c).

In other words, we require that the diagram

$$
\begin{array}{ccc}
\Omega^r(M) & \xrightarrow{\quad * \quad} & \Omega^{m-r}(M) \\
{\scriptstyle (-1)^{m(r+1)}\delta} \downarrow & & \downarrow {\scriptstyle d} \\
\Omega^{r-1}(M) & \xleftarrow{\quad * \quad} & \Omega^{m-r+1}(M)
\end{array}
$$

commutes.

The following remarks list several properties of the codifferential.

5.9 Remarks (a) $\delta^2 = 0$.

Proof Because $**\alpha = (-1)^{r(m-r)}\alpha$, we have $\delta\delta\alpha = \pm *d**d*\alpha = \pm *d^2*\alpha = 0$ because $d^2 = 0$. ∎

(b) $*\delta d = d\delta *$ and $*d\delta = \delta d *$.

Proof If $\alpha \in \Omega^r(M)$, then $d\alpha$ belongs to $\Omega^{r+1}(M)$. Therefore

$$*\delta d\alpha = (-1)^{m(r+2)} **d*d\alpha = (-1)^{mr} **d*d\alpha .$$

Because $d*d\alpha \in \Omega^{m-r}(M)$, we thus find $*\delta d\alpha = (-1)^{-r^2} d*d\alpha$. Analogously,

$$d\delta *\alpha = (-1)^{m(m-r+1)} d*d**\alpha = (-1)^{m(m+1)-r^2} d*d\alpha .$$

Because $m(m+1)$ is even, this proves the first claim. The second follows analogously. ∎

(c) $d*\delta = \delta*d = 0$.

Proof We leave the simple proof to you. ∎

(d) $*\delta\alpha = (-1)^{r+1}d*\alpha$ and $\delta(*\alpha) = (-1)^r *d\alpha$ for $\alpha \in \Omega^r(M)$.

Proof The first statement follows from

$$*\delta\alpha = (-1)^{m(r+1)} **d*\alpha = (-1)^{mr+m}(-1)^{(m-r+1)(r-1)} d*\alpha = (-1)^{r+1}d*\alpha .$$

The second follows from an analogous calculation. ∎

(e) (regularity) From the definition of δ and Remarks 4.11(b) and 5.6(e), it follows immediately that δ is an $\mathbb{R}$-linear map from $\Omega^r_{(k)}$ to $\Omega^{r-1}_{(k-1)}$ for $1 \le r \le m$ and $k \in \mathbb{N}^\times$. This remains true for C^{k+1} manifolds. ∎

5.10 Examples Let $((x^1, \ldots, x^m), U)$ be a positive chart on M.

(a) For $\alpha = \sum_j a_j \, dx^j \in \Omega^1(U)$, we have

$$\delta\alpha = \frac{1}{\sqrt{G}} \sum_{j,k} \frac{\partial}{\partial x^j} \left(g^{jk} a_k \sqrt{G} \right) \in \mathcal{E}(U) .$$

Proof It follows from Proposition 5.8 that

$$*\alpha = \sum_j a_j \sum_k (-1)^{k-1} g^{jk} \sqrt{G} \, dx^1 \wedge \cdots \wedge \widehat{dx^k} \wedge \cdots \wedge dx^m .$$

From this we derive (because $r = 1$) that

$$\delta\alpha = *d*\alpha = *\sum_j \sum_k \sum_\ell (-1)^{k-1} \frac{\partial}{\partial x^\ell} \left(a_j g^{jk} \sqrt{G}\right) dx^\ell \wedge dx^1 \wedge \cdots \wedge \widehat{dx^k} \wedge \cdots \wedge dx^m$$

$$= *\sum_{j,k} \frac{\partial}{\partial x^j} \left(g^{jk} a_k \sqrt{G}\right) dx^1 \wedge \cdots \wedge dx^m \ .$$

The claim now follows from

$$dx^1 \wedge \cdots \wedge dx^m = \frac{1}{\sqrt{G}}\, \omega_M \tag{5.25}$$

and from Remark 5.6(d). ∎

(b) For orthonormal coordinates $(x^1, \ldots, x^m)$, it follows from (a) that

$$\delta\left(\sum_j a_j\, dx^j\right) = \sum_j \partial_j a_j \ .$$

(c) $\delta a = 0$ for $a \in \mathcal{E}(M)$.

(d) $\delta(a\, dx^1 \wedge \cdots \wedge dx^m)$

$$= \sum_{j,k} (-1)^{k-1} \frac{\partial}{\partial x^j}\left(\frac{a}{\sqrt{G}}\right) g^{jk} \sqrt{G}\, dx^1 \wedge \cdots \wedge \widehat{dx^k} \wedge \cdots \wedge dx^m \ .$$

Proof   Using (5.25) and Remark 5.6(d), we get

$$*(a\, dx^1 \wedge \cdots \wedge dx^m) = a/\sqrt{G} \ .$$

Therefore

$$*d*(a\, dx^1 \wedge \cdots \wedge dx^m) = *\sum_j \frac{\partial}{\partial x^j}\left(\frac{a}{\sqrt{G}}\right) dx^j \ .$$

Now the claim follows from Proposition 5.8. ∎

(e) With orthonormal coordinates, we have

$$\delta \sum_{(j) \in \mathbb{J}_r} a_{(j)}\, dx^{(j)} = \sum_{(j) \in \mathbb{J}_r} \sum_{k=1}^r (-1)^{k-1} \partial_{j_k} a_{(j)}\, dx^1 \wedge \cdots \wedge \widehat{dx^{j_k}} \wedge \cdots \wedge dx^{j_r} \ .$$

Proof   Because of the linearity, it suffices to consider $\alpha = a\, dx^{(j)}$ with $(j) \in \mathbb{J}_r$. From (2.20) and Theorem 4.10(ii), we obtain

$$d*\alpha = s(j)\, da \wedge dx^{(j^c)} = s(j) \sum_{k=1}^r \partial_{j_k} a\, dx^{j_k} \wedge dx^{(j^c)} \ .$$

Therefore Example 2.17(d) implies

$$*d*\alpha = s(j) \sum_{k=1}^r s(j_k, (j^c)) \partial_{j_k} a\, dx^{j_1} \wedge \cdots \wedge \widehat{dx^{j_k}} \wedge \cdots \wedge dx^{j_r} .$$

with $s\big(j_k,(j^c)\big) := \mathrm{sign}\big(j_k,(j^c),j_1,\ldots,\widehat{j_k},\ldots,j_r\big)$. Because (j^c) consists of $m-r$ elements, it follows that

$$
\begin{aligned}
s\big(j_k,(j^c)\big) &= (-1)^{(m-r)(r-1)}\,\mathrm{sign}\big(j_k,j_1,\ldots,\widehat{j_k},\ldots,j_r,(j^c)\big)\\
&= (-1)^{(m-r)(r-1)+k-1}s(j)\;.
\end{aligned}
$$

Due to the (mod 2) congruences

$$
(m-r)(r-1)+k-1+m(r+1) \equiv k-r(r+1)-1 \equiv k-1\;,
$$

the claim then follows from the definition of δ. ∎

5.11 Remarks (a) By making appropriate modifications, the above properties of the star operator and the codifferential can be extended to the case of pseudo-Riemannian manifolds.

More precisely, suppose (M,g) is an oriented pseudo-Riemannian manifold. We can provide T_pM with the inner product induced by that of $\mathbb{R}^{\overline{m}}$. By Remark 2.18(a), it follows that the representation matrix g of $g(p)$ at every $p \in M$ is diagonal in an appropriately chosen basis, and its diagonal entries are ± 1. Now $(-1)^s = \mathrm{sign}\,g(p)$ is uniquely determined by $g(p)$, where s denotes the number of negative elements. We now assume that $\mathrm{sign}(g) = \mathrm{sign}\,g(p)$ is constant on M, that is, it is independent of p. From (the proof of) Remark 5.4(f), it follows that this assumption is satisfied if M can be described by a single chart.

Under this assumption the star operator, as defined through (2.25), can also be defined pointwise. Then (5.21) and (5.22) must be replaced by

$$
**\alpha = \mathrm{sign}(g)(-1)^{r(m-r)}\alpha \quad \text{for } \alpha \in \Omega^r(M)\;,
$$

and

$$
\alpha \wedge *\beta = \beta \wedge *\alpha = \mathrm{sign}(g)(\alpha\,|\,\beta)_{g,r}\omega_M \quad \text{for } \alpha,\beta \in \Omega^r(M)\;,
$$

as we learn from Remarks 2.19(d) and (e), respectively. Here ω_M is the volume element of M defined in Remark 5.4(g). Remark 2.19(c) also implies

$$
*1 = \mathrm{sign}(g)\omega_M \text{ for } *\omega_M = 1\;.
$$

The codifferential is defined in this case by

$$
\delta\alpha := \mathrm{sign}(g)(-1)^{m(r+1)}*d*\alpha \quad \text{for } \alpha \in \Omega^r(M)\;. \tag{5.26}
$$

We verify easily that with these modifications, the statements of Remark 5.6 hold as written.

Proof The claims follows from Remarks 2.19. ∎

(b) Let $\big((x^1,\ldots,x^m),U\big)$ be a positive chart of M. Then

$$
\delta\sum_j a_j\,dx^j = \frac{1}{\sqrt{|G|}}\sum_{j,k}\frac{\partial}{\partial x^j}\big(g^{jk}a_k\sqrt{|G|}\big) \in \mathcal{E}(U)\;.
$$

Proof This follows, in analogy to the proof of Example 5.10(a), from Remark 5.4(g). ∎

5.12 Examples We consider now the Minkowski space $\mathbb{R}_{1,3}^4$ with the metric $(\,\cdot\mid\cdot\,)_{1,3}$
and therefore with $g := (dt)^2 - (dx)^2 - (dy)^2 - (dz)^2$.

(a) If (i, j, k) is a cyclic permutation of $(1, 2, 3)$, then with $(x^1, x^2, x^3) := (x, y, z)$,
we have

$$*(dx^i \wedge dt) = dx^j \wedge dx^k \quad \text{and} \quad *(dx^i \wedge dx^j) = -dx^k \wedge dt \ .$$

Proof Let (e_0, e_1, e_2, e_3) be the canonical basis of $\mathbb{R}_{1,3}^4$. Then

$$g(e_0, e_0) = 1 \quad \text{and} \quad g(e_j, e_j) = -1 \quad \text{for } 1 \leq j \leq 3 \ .$$

This implies

$$(dt \mid dt)_{g,1} = 1 \quad \text{and} \quad (dx^j \mid dx^j)_{g,1} = -1 \ .$$

Therefore

$$(dt \wedge dx^j \mid dt \wedge dx^j)_{g,2} = -1 \quad \text{and} \quad (dx^j \wedge dx^k \mid dx^j \wedge dx^k)_{g,2} = 1 \quad \text{for } 1 \leq j < k \leq 3 \ .$$

Now the claim follows from Remark 2.19(c). ∎

(b) Let $E_j, H_j \in \mathcal{E}(\mathbb{R}_{1,3}^4)$ and

$$\begin{aligned}
\alpha := \ &(E_1 dx^1 + E_2 dx^2 + E_3 dx^3) \wedge dt \\
&+ H_1 dx^2 \wedge dx^3 + H_2 dx^3 \wedge dx^1 + H_3 dx^1 \wedge dx^2 \ .
\end{aligned}$$

Then

$$\begin{aligned}
*\alpha = \ &-(H_1 dx^1 + H_2 dx^2 + H_3 dx^3) \wedge dt \\
&+ E_1 dx^2 \wedge dx^3 + E_2 dx^3 \wedge dx^1 + E_3 dx^1 \wedge dx^2 \ .
\end{aligned}$$

Proof This is an immediate consequence of (a) and the $\mathcal{E}(\mathbb{R}_{1,3}^4)$-linearity of the star
operator. ∎

(c) The α from (b) satisfies

$$\delta\alpha = \sum_{j=1}^{3} \frac{\partial E_j}{\partial t} dx^j - \sum_{k=1}^{3} \frac{\partial E_k}{\partial x^k} dt + \sum_{(i,j,k)} \left(\frac{\partial H_i}{\partial x^j} - \frac{\partial H_j}{\partial x^i} \right) dx^k \ ,$$

where the last term is summed over all cyclic permutations of $(1, 2, 3)$.

Proof From (b), we know that

$$*\alpha = -\sum_{i=1}^{3} H_i \, dx^i \wedge dt + \sum_{(i,j,k)} E_i \, dx^j \wedge dx^k \ .$$

This implies

$$\begin{aligned}
d*\alpha = \ &-\sum_{i=1}^{3} \sum_{\substack{j=1 \\ j\neq i}}^{3} \frac{\partial H_i}{\partial x^j} dx^j \wedge dx^i \wedge dt \\
&+ \sum_{(i,j,k)} \left(\frac{\partial E_i}{\partial t} dt \wedge dx^j \wedge dx^k + \frac{\partial E_i}{\partial x^i} dx^i \wedge dx^j \wedge dx^k \right) \ .
\end{aligned}$$

From Remark 2.19(c), we derive

$$*(dt \wedge dx^i \wedge dx^j) = -dx^k \quad \text{and} \quad *(dx^i \wedge dx^j \wedge dx^k) = dt \ .$$

With this we get

$$*d*\alpha = - \sum_{(i,j,k)} \left(\frac{\partial H_i}{\partial x^j} - \frac{\partial H_j}{\partial x^i} \right) dx^k - \sum_{i=1}^{3} \frac{\partial E_i}{\partial t} dx^i + \sum_{k=1}^{3} \frac{\partial E_k}{\partial x^k} dt \ .$$

Now the claim follows because $m = 4$ and $\text{sign}(g) = -1$. ∎

Exercises

1 Let (M_j, g_j) for $j = 1, 2$ be pseudo-Riemannian manifolds with $\partial M_1 = \emptyset$, and denote by $\pi_j : M_1 \times M_2 \to M_j$ the canonical projection onto M_j. Prove these statements:

(i) $(M_1 \times M_2, \pi_1^* g_1 + \pi_2^* g_2)$ is a Riemannian manifold, the **product** of M_1 and M_2.

(ii) Two points (p_1, p_2) yield submanifolds $M_1 \times \{p_2\}$ and $\{p_1\} \times M_2$ of $M_1 \times M_2$.

(iii) $T_{(p_1,p_2)}(M_1 \times M_2) = T_{(p_1,p_2)}(M_1 \times \{p_2\}) \oplus T_{(p_1,p_2)}(\{p_1\} \times M_2)$.

(iv) $\omega_{M_1 \times M_2} = \pi_1^* \omega_{M_1} \wedge \pi_2^* \omega_{M_2}$.

2 Let M be an oriented hypersurface in $\mathbb{R}^{m+1}$. We call $\nu : M \to T\mathbb{R}^{m+1}$ a **positive unit normal field** when ν is a unit normal of M such that, for every $p \in M$ and every positive basis $(v_1, \ldots, v_m)$ of $T_p M$, the $(m+1)$-tuple $(\nu(p), v_1, \ldots, v_m)$ is a positive basis of $T_p\mathbb{R}^{m+1}$.

(a) Show that ν is well defined and unique.

(b) Determine the unit normal on these surfaces in $\mathbb{R}^3$:

(i) graph f, with X open in $\mathbb{R}^2$ and $f \in \mathcal{E}(X)$,   (ii) $\mathbb{R} \times S^1$,   (iii) S^2 ,   (iv) $\mathsf{T}_{a,r}^2$.

(Hint: (iv) Exercise VII.10.10 and Example VII.9.11(f).)

3 Let M be an oriented hypersurface in $\mathbb{R}^{m+1}$, provided with the standard metric. Denote by ν the positive unit normal of M. Show these facts:

(i) ν defines a smooth map from M to S^m, the **Gauss map** (which is also denoted by ν).

(ii) For $p \in M$ and $v \in T_p M$, we have $((T_p\nu)v \,|\, \nu(p))_{\mathbb{R}^{m+1}} = 0$. Therefore $(T_p\nu)v$ belongs to $T_p M$.

(iii) The map

$$L : M \to \bigcup_{p \in M} \mathcal{L}(T_p M) \in \mathcal{L}(T_p M) \ , \quad p \mapsto T_p \nu$$

is well defined. This is called the **Weingarten map** of M.

(iv) For $p \in M$ and $v, w \in T_p M$, we have

$$g(p)(L(p)v, w) = g(p)(v, L(p)w) \ ,$$

that is, $L(p)$ is symmetric on the inner product space $(T_p M, g(p))$. The tensor $h \in T_2^0(M)$ defined by

$$h(p)(v, w) := g(p)(L(p)v, w) \quad \text{for } p \in M \text{ and } v, w \in T_p M$$

is called the **second fundamental tensor** of M.

(v) In local coordinates (U, φ), with the natural embedding $i : M \hookrightarrow \mathbb{R}^{m+1}$, and with $f := i \circ \varphi^{-1}$, we have

$$h_{jk} = (\partial_j \nu \,|\, \partial_k f) = -(\nu \,|\, \partial_j \partial_k f) ,$$

where $h_{jk} := h(\partial/\partial x^j, \partial/\partial x^k)$.

4 Calculate the second fundamental forms of $\mathbb{R}^2$, S^2, $\mathbb{R} \times S^1$, and $\mathsf{T}^2_{a,r}$ as submanifolds of $\mathbb{R}^3$.

5 Suppose I is a compact interval in $\mathbb{R}$ and M is a Riemannian manifold. Also let $\gamma \in C^1(I, M)$. Let $i : M \hookrightarrow \mathbb{R}^{\overline{m}}$ be the natural embedding, and put $\widetilde{\gamma} := i \circ \gamma$. Then the length $L(\widetilde{\gamma})$ of $\widetilde{\gamma}$ is defined as in Section VIII.1. Show that if $\dot{\gamma}(t) := (T_t \gamma)(t, 1)$ for $t \in I$, then

$$L(\widetilde{\gamma}) = \int_I \sqrt{g(\dot{\gamma}(t), \dot{\gamma}(t))} \, dt .$$

When $L(\widetilde{\gamma}) = L(I)$, we say γ is **parametrized by arc length**.

6 Suppose M is an oriented surface in $\mathbb{R}^3$ and $\gamma \in C^2(I, M)$ is parametrized by arc length. Also denote by ν the positive unit normal bundle of M. Then we call

$$\kappa_g(\gamma) := \det[\dot{\gamma}, \ddot{\gamma}, \nu]$$

the **curvature of γ in M** or the **geodesic curvature** of γ.

(a) Verify in the Euclidean case $M = \mathbb{R}^2$ that the geodesic curvature is the same as the (usual) curvature from Section VIII.2.

(b) Suppose $M = S^2$ and (x, y, z) are the Euclidean coordinates in $\mathbb{R}^3$. Also let γ_z for $z \in (-1, 1)$ be a parametrization by arc length of $L_z := \sqrt{1 - |z|^2} \, S^1 \times \{z\}$ (see Example 1.5(a)). Show then that

$$\kappa_g(\gamma_z) = \frac{z}{\sqrt{1 - |z|^2}} .$$

Therefore the geodesic curvature is constant on the circle L_z and vanishes at the equator.

7 Prove the equality $\rho^* \omega_{S^m} = r^{-(m+1)} \alpha$ from Example 4.13(c) by direct calculation for $m = 2$ and 3.

8 Prove the statements made in the proof of Example 5.5(b).

9 Show that
$$\{z \in \mathbb{C} \,;\, \operatorname{Im} z > 0\} \to \mathbb{D} , \quad z \mapsto (1 + iz)/(1 - iz)$$
gives a diffeomorphism from the "upper half of complex half plane" to the unit disc. Then use this map to show that

$$\left(H^2, \frac{(dx)^2 + (dy)^2}{y^2} \right)$$

is a model of the hyperbolic plane, the **Klein model**.

10 Show that the Lobachevsky plane from Example 5.5(m) is a model of H^2.

11   For $\alpha \in \Omega^{r-1}(M)$ and $\beta \in \Omega^r(M)$, show

$$d(\alpha \wedge *\beta) = d\alpha \wedge *\beta + \alpha \wedge *\delta\beta .$$

12   Show that the codifferential does not depend on the orientation of the underlying Riemannian manifold.

13   Suppose M is oriented, $(N, \bar{g})$ is another oriented m-dimensional Riemannian manifold, and $f : M \to N$ is an isometric diffeomorphism. Show that $f^*\omega_N = \pm\omega_M$. Also show that $f^*\omega_N = \omega_M$ if and only if f is orientation-preserving.

14   Suppose M and N are as in Exercise 13 and $f : M \to N$ is an orientation-preserving isometric diffeomorphism. Show that the diagram

$$
\begin{array}{ccc}
\Omega^r(M) & \xrightarrow{\;\;*\;\;} & \Omega^{m-r}(M) \\[4pt]
{\scriptstyle f^*} \Big\uparrow & & \Big\uparrow {\scriptstyle f^*} \\[4pt]
\Omega^r(N) & \xrightarrow[\;\;*\;\;]{} & \Omega^{m-r}(N)
\end{array}
$$

is commutative for $0 \le r \le m$.

15   Suppose M and N are as in Exercise 13, and $f : M \to N$ is an isometric diffeomorphism.[8] Show that the diagram

$$
\begin{array}{ccc}
\Omega^r(M) & \xrightarrow{\;\;\delta\;\;} & \Omega^{r-1}(M) \\[4pt]
{\scriptstyle f^*} \Big\uparrow & & \Big\uparrow {\scriptstyle f^*} \\[4pt]
\Omega^r(N) & \xrightarrow[\;\;\delta\;\;]{} & \Omega^{r-1}(N)
\end{array}
$$

commutes for $0 \le r \le m$.

---

[8]Note Exercise 12.

6   Vector analysis

Vector fields and Pfaff forms can be interchanged using the Riesz isomorphism. While vector fields have an immediate geometrical interpretation, the calculus of differential forms is of great value in calculations. The exterior product and derivative obey relatively simple rules, which themselves stand for a more complicated set of prescriptions for how to change from one system of local coordinates to another. In this section, we will use the Riesz isomorphism to translate some of the concepts and theorems of differential forms into the language of classical vector analysis. In so doing, we will learn about the divergence and curl of vector fields, which are of fundamental significance in physics and the theory of partial differential equations.

For the entire section suppose the following:

- M is an m-dimensional manifold; N is an n-dimensional manifold.
- The indices i, j, k, ℓ always range from 1 to m unless stated otherwise, and $\sum_j$ means that j is summed from 1 to m.

The Riesz isomorphism

Let g be a pseudo-Riemannian metric on M. Then we define the **Riesz isomorphism**, Θ_g, by

$$\Theta_g : \mathcal{V}(M) \to \Omega^1(M) , \quad v \mapsto \Theta_g v \tag{6.1}$$

and

$$(\Theta_g v)(p) := \Theta_{g(p)} v(p) \quad \text{for } p \in M ,$$

where $\Theta_{g(p)} : T_p M \to T_p^* M$ is the Riesz isomorphism of (2.12) (or Remark 2.18(b)) and is defined by

$$\langle \Theta_{g(p)} u, w \rangle = g(p)(u, w) \quad \text{for } u, w \in T_p M .$$

When no confusion is expected, we may write Θ instead of Θ_g.

6.1 Remarks   (a) The map (6.1) is well defined.

Proof   We must show that Θv belongs to $\Omega^1(M)$ for $v \in \mathcal{V}(M)$. In local coordinates, we have

$$v \,|\, U = \sum_j v^j \frac{\partial}{\partial x^j}$$

with $v^j \in \mathcal{E}(U)$. From this and from Remarks 2.14(a) and 2.18(b), it follows that

$$\Theta v(p) = \Theta_{g(p)} \sum_j v^j(p) \frac{\partial}{\partial x^j}\Big|_p = \sum_j v^j(p) \Theta_{g(p)} \frac{\partial}{\partial x^j}\Big|_p$$

$$= \sum_j v^j(p) \sum_k g_{jk}(p) \, dx^k(p) = \sum_j a_j(p) \, dx^j(p) ,$$

where

$$a_k := \sum_j g_{kj} v^j \in \mathcal{E}(U) \ .$$

Now the claim follows from Remarks 4.5(c) and 5.4(e). ∎

(b) In local coordinates,

$$\Theta\Big(\sum_j v^j \frac{\partial}{\partial x^j}\Big) = \sum_j a_j \, dx^j \qquad \text{with} \quad a_j := \sum_k g_{jk} v^k \ . \qquad (6.2)$$

Instead of Θv, we often write $v^\flat$ or $g^\flat v$, because, as seen in (6.2), Θ effects a "lowering of indices" (see Remark 2.14(d)).

Proof This was shown in the proof of (a). ∎

(c) The map $\Theta : \mathcal{V}(M) \to \Omega^1(M)$ is an $\mathcal{E}(M)$-module isomorphism.

Proof Let $\alpha \in \Omega^1(M)$. Then $\alpha(p) \in T_p^* M$ for $p \in M$. From Section 2, we know that $\Theta_{g(p)}$ is a vector space isomorphism. Therefore $\Theta_{g(p)}^{-1}\alpha(p) \in T_p M$ is well defined. We set

$$(\overline{\Theta}_g \alpha)(p) := \Theta_{g(p)}^{-1}\alpha(p) \quad \text{for } p \in M \text{ and } \alpha \in \Omega^1(M) \ .$$

In local coordinates, we know from Remarks 2.14(a) and 2.18(b) that

$$\overline{\Theta}_g \alpha(p) = \Theta_{g(p)}^{-1} \sum_j a_j(p) \, dx^j(p) = \sum_j a_j(p) \Theta_{g(p)}^{-1} dx^j(p)$$

$$= \sum_j a_j(p) \sum_k g^{jk}(p) \frac{\partial}{\partial x^k}\Big|_p = \sum_j v^j(p) \frac{\partial}{\partial x^j}\Big|_p \ ,$$

where

$$v^j := \sum_k g^{jk} a_k \in \mathcal{E}(U) \ .$$

Thus it follows from Remark 4.3(c) that $\overline{\Theta}_g \alpha$ belongs to $\mathcal{V}(M)$. From the definitions of Θ_g and $\overline{\Theta}_g$, it follows immediately that $\Theta_g \overline{\Theta}_g = \mathrm{id}_{\Omega^1(M)}$ and $\overline{\Theta}_g \Theta_g = \mathrm{id}_{\mathcal{V}(M)}$. Therefore Θ_g is bijective, and $\Theta_g^{-1} = \overline{\Theta}_g$.

Finally we see that for $a \in \mathcal{E}(M)$ and $v \in \mathcal{V}(M)$, we have

$$\Theta_g(av)(p) = \Theta_{g(p)} a(p) v(p) = a(p) \Theta_{g(p)} v(p) = (a\Theta_g v)(p) \quad \text{for } p \in M \ .$$

Therefore Θ is an $\mathcal{E}(M)$-Module isomorphism. ∎

(d) In local coordinates,

$$\Theta^{-1}\Big(\sum_j a_j \, dx^j\Big) = \sum_j v^j \frac{\partial}{\partial x^j} \qquad \text{with} \quad v^j := \sum_k g^{jk} a_k \ . \qquad (6.3)$$

Instead of $\Theta^{-1}\alpha$, we often write $\alpha^\sharp$ or $g^\sharp \alpha$, because Θ^{-1} "raises indices".

Proof This was shown in the proof of (c). ∎

(e) (orthogonal coordinates) If $(x^1, \ldots, x^m)$ are orthogonal coordinates, that is, if

$$g\Big(\frac{\partial}{\partial x^j}, \frac{\partial}{\partial x^k}\Big) = 0 \quad \text{for } j \neq k \ ,$$

then (6.2) and (6.3) simplify respectively to

$$\Theta v = \sum_j g_{jj} v^j \, dx^j \quad \text{and} \quad \Theta^{-1}\alpha = \sum_j g^{jj} a_j \frac{\partial}{\partial x^j}$$

for $v = \sum_j v^j \, \partial/\partial x^j$ and $\alpha = \sum_j a_j \, dx^j$.

(f) Let $(N, \bar{g})$ be a pseudo-Riemannian manifold, and let $\varphi \in \mathrm{Diff}(M, N)$ with $\varphi_* g = \lambda \bar{g}$ for some $\lambda \neq 0$. Then the diagram

$$
\begin{array}{ccc}
\mathcal{V}(M) & \xleftarrow[\cong]{\varphi^*} & \mathcal{V}(N) \\[2mm]
{\scriptstyle \Theta_M} \Big\downarrow {\scriptstyle \cong} & & {\scriptstyle \cong} \Big\downarrow {\scriptstyle \lambda\Theta_N} \\[2mm]
\Omega^1(M) & \xleftarrow[\cong]{\varphi^*} & \Omega^1(N)
\end{array}
$$

commutes. Therefore $\Theta_M \varphi^* = \lambda \varphi^* \Theta_N$.

Proof Using the definition and properties of the push forward and the pull back of vector fields and forms (see in particular (4.25)), we find for $v, w \in \mathcal{V}(N)$ that

$$\lambda \bar{g}(v, w) = \varphi_* g(v, w) = g(\varphi^* v, \varphi^* w) = \langle \Theta_M \varphi^* v, \varphi^* w \rangle_M = \langle \varphi_* \Theta_M \varphi^* v, w \rangle_N$$
$$= \bar{g}(\Theta_N^{-1} \varphi_* \Theta_M \varphi^* v, w) \ .$$

Because $\bar{g}$ is nondegenerate and $\mathbb{R}$-linear, it follows that

$$\lambda v = \Theta_N^{-1} \varphi_* \Theta_M \varphi^* v \quad \text{for } v \in \mathcal{V}(M) \ ,$$

which proves the claim. ∎

(g) (regularity) Suppose $k \in \mathbb{N}$ and M is a C^{k+1} manifold. Then the definitions and statements above remain true when smooth vector fields, differential forms, and functions are replaced by C^k vector fields, C^k differential forms, and C^k functions. ∎

6.2 Examples **(a)** (Euclidean coordinates) Let M be open in $\mathbb{R}^m$. We denote Euclidean coordinates by $(x^1, \ldots, x^m)$, that is, $(\cdot \mid \cdot) = \sum_j (dx^j)^2$. Then

$$\Theta\Big(\sum_j v^j \frac{\partial}{\partial x^j}\Big) = \sum_j a_j \, dx^j \quad \text{for } a_j := v^j \ .$$

The last assignment means that in the Euclidean case we do not need to introduce new notation; instead we normally write $\sum_j v^j \, dx^j$ for the image of $\sum_j v^j \, \partial/\partial x^j$ under Θ. That is, Θ allows us to regard the components v^j of the vector field $\sum_j v^j \, \partial/\partial x^j$ as those of the Pfaff form $\sum_j v^j \, dx^j$.

Proof Because $g_{jk} = \delta_{jk}$, this follows from Remark 6.1(b). ∎

(b) (spherical coordinates) Let $V_3 := (0, \infty) \times (0, 2\pi) \times (0, \pi)$, and let

$$f : V_3 \to \mathbb{R}^3 \ , \quad (r, \varphi, \vartheta) \mapsto (x, y, z)$$

be the spherical coordinate transformation of Example VII.9.11(a). Then with respect to the standard metric, we have

$$\Theta\left(v^1\frac{\partial}{\partial r} + v^2\frac{\partial}{\partial\varphi} + v^3\frac{\partial}{\partial\vartheta}\right) = v^1\,dr + r^2\sin^2(\vartheta)v^2\,d\varphi + r^2v^3\,d\vartheta .$$

Proof This follows immediately from Remark 6.1(b). ∎

(c) (Minkowski metric) On $\mathbb{R}^4_{1,3}$, we have

$$\Theta_g\left(\sum_{\mu=0}^{3} v^\mu\frac{\partial}{\partial x^\mu}\right) = v^0\,dx^0 - \sum_{j=1}^{3} v^j\,dx^j$$

for $g := (\cdot\,|\,\cdot)_{1,3}$. ∎

The gradient

If $f \in \mathcal{E}(M)$, then df belongs to $\Omega^1(M)$. Therefore

$$\operatorname{grad}_g f := \Theta_g^{-1}\,df \in \mathcal{V}(M)$$

is a well-defined vector field on M, the **gradient** of f on the (pseudo-)Riemannian manifold (M,g) (or with respect to g). We may also write it as $\operatorname{grad}_M f$ or $\operatorname{grad} f$ if no misunderstanding is expected. Therefore $\operatorname{grad} f$ is defined by the commutativity of the diagram

$$\mathcal{E}(M) = \Omega^0(M)$$

$$\begin{array}{ccc} & & \\ \operatorname{grad}\nearrow & & \searrow d \\ \mathcal{V}(M) & \xrightarrow[\cong]{\Theta} & \Omega^1(M) . \end{array} \qquad (6.4)$$

6.3 Remarks **(a)** The map $\operatorname{grad}: \mathcal{E}(M) \to \mathcal{V}(M)$, $f \mapsto \operatorname{grad} f$ is $\mathbb{R}$-linear.

(b) For $f \in \mathcal{E}(M)$, the vector field $\operatorname{grad} f$ is characterized by the relation

$$g(\operatorname{grad} f, w) = \langle df, w\rangle \quad \text{for } w \in \mathcal{V}(M) .$$

(c) In local coordinates, we have

$$\operatorname{grad} f = \sum_j\left(\sum_k g^{jk}\frac{\partial f}{\partial x^k}\right)\frac{\partial}{\partial x^j} . \qquad (6.5)$$

Proof Because we know from (4.5) and (4.8) that $df = \sum_j \partial f/\partial x^j\,dx^j$, the claim follows from Remark 6.1(d). ∎

(d) (orthogonal coordinates) In orthogonal coordinates, (6.5) simplifies to

$$\operatorname{grad} f = \sum_j g^{jj} \frac{\partial f}{\partial x^j} \frac{\partial}{\partial x^j} .$$

Because in this case g has the form

$$g = \sum_j g_{jj} (dx^j)^2 , \tag{6.6}$$

that is, because the fundamental matrix is diagonal, we have $g^{jj} = 1/g_{jj}$. Thus the coefficients g^{jj} can be read directly from the representation (6.6).

(e) Suppose $(N, \bar{g})$ is a pseudo-Riemannian manifold and $\varphi \in \operatorname{Diff}(M, N)$ with $\varphi_* g = \lambda \bar{g}$ for some $\lambda \neq 0$. Then the diagram

$$
\begin{array}{ccc}
 & \lambda \operatorname{grad}_M & \\
\mathcal{E}(M) & \xrightarrow{\hspace{2cm}} & \mathcal{V}(M) \\[4pt]
\varphi^* \Big\uparrow & & \Big\uparrow \varphi^* \\[4pt]
\mathcal{E}(N) & \xrightarrow[\operatorname{grad}_N]{\hspace{2cm}} & \mathcal{V}(N)
\end{array}
$$

commutes. Therefore $\operatorname{grad}_M \circ \varphi^* = \lambda^{-1} \varphi^* \circ \operatorname{grad}_N$.

Proof   Because the relation $\lambda \Theta_M^{-1} \varphi^* = \varphi^* \Theta_N^{-1}$ follows from Remark 6.1(f), we find for $f \in \mathcal{E}(N)$ that

$$\lambda \operatorname{grad}_M (\varphi^* f) = \lambda \Theta_M^{-1} d(\varphi^* f) = \lambda \Theta_M^{-1} \varphi^* df = \varphi^* \Theta_N^{-1} df = \varphi^* \operatorname{grad}_N f ,$$

where we have used (4.19). ∎

(f) (regularity) Let $k \in \mathbb{N}$. For $f \in C^{k+1}(M)$, we have $\operatorname{grad} f \in \mathcal{V}^k(M)$. Here it suffices to assume that M is a C^{k+1} manifold. ∎

6.4 Examples   **(a)** (Euclidean coordinates) Let M be open in $\mathbb{R}^m$. Denoting Euclidean coordinates by $(x^1, \ldots, x^m)$, we have $g_{jk} = \delta_{jk}$ and therefore

$$\operatorname{grad} f = \sum_j \frac{\partial f}{\partial x^j} \frac{\partial}{\partial x^j} .$$

This representation obviously coincides with that of Proposition VII.2.16. For an arbitrary locally Riemannian metric, we have already confirmed (6.4) in Remark VII.2.17(c).

(b) (spherical coordinates) Let $V_3 \to \mathbb{R}^3$, $(r, \varphi, \vartheta) \mapsto (x, y, z)$ be the spherical coordinate map. In these coordinates, the gradient with respect to the standard metric reads

$$\operatorname{grad} f = \frac{\partial f}{\partial r} \frac{\partial}{\partial r} + \frac{1}{r^2 \sin^2 \vartheta} \frac{\partial f}{\partial \varphi} \frac{\partial}{\partial \varphi} + \frac{1}{r^2} \frac{\partial f}{\partial \vartheta} \frac{\partial}{\partial \vartheta} .$$

Proof   Because spherical coordinates are orthogonal, this follows from Example 5.5(g). ∎

(c) (spherical coordinates) Suppose $h_2 : W_2 \to \mathbb{R}^3$, $(\varphi, \vartheta) \mapsto (x, y, z)$ is the parametrization of the open subset $U_2 := S^2 \setminus H_2$ of the 2-sphere. Then for $f \in C^1(U_2, \mathbb{R})$, we have

$$\operatorname{grad}_{S^2} f = \frac{1}{\sin^2 \vartheta} \frac{\partial f}{\partial \varphi} \frac{\partial}{\partial \varphi} + \frac{\partial f}{\partial \vartheta} \frac{\partial}{\partial \vartheta} .$$

Proof This can be read from the representation of g_{S^2} in Example 5.5(h). ∎

(d) (Minkowski metric) Suppose X is open in $\mathbb{R}^4_{1,3}$ and $f \in C^1(X, \mathbb{R})$. Then, with respect to the Minkowski metric, we have

$$\operatorname{grad} f = \frac{\partial f}{\partial t} \frac{\partial}{\partial t} - \frac{\partial f}{\partial x} \frac{\partial}{\partial x} - \frac{\partial f}{\partial y} \frac{\partial}{\partial y} - \frac{\partial f}{\partial z} \frac{\partial}{\partial z} ,$$

as we see immediately from the definition of $(\cdot | \cdot)_{1,3}$. ∎

The divergence

Now suppose M is oriented and that ω_M denotes the volume element of (M, g). Then the maps

$$\bullet \, \omega_M : \mathcal{E}(M) \to \Omega^m(M) , \qquad a \mapsto a\omega_M \tag{6.7}$$

and

$$\lrcorner \, \omega_M : \mathcal{V}(M) \to \Omega^{m-1}(M) , \qquad v \mapsto v \lrcorner \omega_M \tag{6.8}$$

are defined pointwise.

6.5 Lemma *The maps (6.7) and (6.8) are well-defined $\mathcal{E}(M)$-module isomorphisms. If $\big((x^1, \ldots, x^m), U\big)$ is a chart of M, then*

$$a\omega_M \,|\, U = \pm a \sqrt{|G|} \, dx^1 \wedge \cdots \wedge dx^m \tag{6.9}$$

and

$$\Big(\sum_j v^j \frac{\partial}{\partial x^j}\Big) \lrcorner \, \omega_M$$
$$= \sum_j (-1)^{j-1} v^j \sqrt{|G|} \, dx^1 \wedge \cdots \wedge \widehat{dx^j} \wedge \cdots \wedge dx^m , \tag{6.10}$$

where the positive sign is used in (6.9) when the chart is positively oriented, and the negative is used otherwise.

Proof (i) From the pointwise definition of $\bullet \, \omega_M$ and from Remarks 5.4(c) and (g), the truth of (6.9) follows immediately. From this and Remark 4.5(c), we conclude that $a\omega_M$ belongs to $\Omega^m(M)$ for $a \in \mathcal{E}(M)$. Therefore the map (6.7) is well defined.

It is clearly $\mathcal{E}(M)$-linear. By Remark 4.14(a), every $\alpha \in \Omega^m(M)$ has exactly one $a \in \mathcal{E}(M)$ such that $\alpha = a\omega_M$. Thus (6.7) is also bijective.

(ii) The validity of (6.10) follows from Remark 4.13(b) if the chart is positive. Otherwise we replace[1] x^1 by $-x^1$. Then v^1 is substituted by $-v^1$. This shows that (6.10) is independent of the chart's orientation.

Because $\sqrt{|G|} \in \mathcal{E}(U)$, (6.10) and Remark 4.5(c) show that $v \lrcorner \omega_M$ belongs to $\Omega^{m-1}(M)$ for $v \in \mathcal{V}(M)$. Thus the map (6.8) is well defined and clearly $\mathcal{E}(M)$-linear.

Let $\alpha \in \Omega^{m-1}(M)$. Then it follows from Example 3.2(b) and Remark 4.5(c) that there is a unique $a_j \in \mathcal{E}(U)$ such that

$$\alpha \,|\, U = \sum_j (-1)^{j-1} a_j \, dx^1 \wedge \cdots \wedge \widehat{dx^j} \wedge \cdots \wedge dx^m \ .$$

Then $v^j := a_j / \sqrt{|G|}$ belongs to $\mathcal{E}(U)$. Therefore

$$v := \sum_j v^j \frac{\partial}{\partial x^j} \in \mathcal{V}(U) \ ,$$

and (6.10) shows $(v \lrcorner \omega_M) \,|\, U = \alpha \,|\, U$. This implies that the map $\lrcorner \omega_M$ is surjective. Because its injectivity is clear, we see that it is an isomorphism from $\mathcal{V}(M)$ to $\Omega^{m-1}(M)$. ∎

6.6 Remarks   (a) Let $(N, \bar{g})$ be an oriented pseudo-Riemannian manifold, and suppose $\varphi \in C^\infty(M, N)$ with $\varphi^* \omega_N = \mu \omega_M$ for some $\mu \neq 0$. Then the diagram

$$
\begin{array}{ccc}
\mathcal{E}(M) & \xrightarrow{\ \mu(\bullet\, \omega_M)\ } & \Omega^m(M) \\[2mm]
\varphi^* \uparrow & & \uparrow \varphi^* \\[2mm]
\mathcal{E}(N) & \xrightarrow[\ \bullet\, \omega_N\ ]{} & \Omega^n(N)
\end{array}
$$

commutes, that is,

$$\mu(\varphi^* a) \bullet \omega_M = \varphi^*(a \bullet \omega_N) \quad \text{for } a \in \mathcal{E}(N) \ .$$

Proof   This follows immediately from the behavior of (exterior) products under pull backs. ∎

---

[1] Consider how this proof should be modified for the case of a one-dimensional manifold with boundary.

(b) Let $(N, \bar{g})$ be an oriented pseudo-Riemannian manifold, and suppose φ belongs to $\mathrm{Diff}(M, N)$ and satisfies $\varphi^* \omega_N = \mu \omega_M$ for some $\mu \neq 0$. Then

$$
\begin{array}{ccc}
\mathcal{V}(M) & \xrightarrow[\cong]{\mu(\lrcorner\, \omega_M)} & \Omega^{m-1}(M) \\
\varphi^* \Big\uparrow & & \Big\uparrow \varphi^* \\
\mathcal{V}(N) & \xrightarrow[\cong]{\lrcorner\, \omega_N} & \Omega^{m-1}(N)
\end{array}
$$

is a commutative diagram, that is, $\mu\big((\varphi^* v) \lrcorner\, \omega_M\big) = \varphi^*(v \lrcorner\, \omega_N)$ for $v \in \mathcal{V}(N)$.

Proof We derive from Remark 4.13(a) that

$$
\mu\big((\varphi^* v) \lrcorner\, \omega_M\big) = \varphi^* v \lrcorner\, (\mu \omega_M) = \varphi^* v \lrcorner\, \varphi^* \omega_N = \varphi^*(\varphi_* \varphi^* v \lrcorner\, \omega_N) = \varphi^*(v \lrcorner\, \omega_N)
$$

for $v \in \mathcal{V}(N)$. ∎

(c) (regularity) Let $k \in \mathbb{N}$. Clearly then

$$
\bullet\, \omega_M : C^k(M) \to \Omega^m_{(k)}(M)
$$

and

$$
\lrcorner\, \omega_M : \mathcal{V}^k(M) \to \Omega^{m-1}_{(k)}(M) ,
$$

and these maps are $C^k(M)$-module isomorphisms. Thus it suffices to assume that M is a C^{k+1} manifold. ∎

With help of the isomorphisms (6.7) and (6.8), we define a map

$$
\mathrm{div}_g : \mathcal{V}(M) \to \mathcal{E}(M) , \quad v \mapsto \mathrm{div}_g v \tag{6.11}
$$

by demanding that the diagram

$$
\begin{array}{ccc}
\mathcal{V}(M) & \xrightarrow{\mathrm{div}_g} & \mathcal{E}(M) \\
\lrcorner\, \omega_M \Big\downarrow \cong & & \cong \Big\downarrow \bullet\, \omega_M \\
\Omega^{m-1}(M) & \xrightarrow{d} & \Omega^m(M)
\end{array} \tag{6.12}
$$

commutes. In other words, for $v \in \mathcal{V}(M)$, the **divergence** $\mathrm{div}_g v$ of a vector field v on an oriented pseudo-Riemannian manifold (M, g) (or, with respect to g) is defined by the relation

$$
(\mathrm{div}_g v) \omega_M = d(v \lrcorner\, \omega_M) . \tag{6.13}
$$

Instead of div_g, we may also write div_M or, if no confusion is anticipated, simply div.

6.7 Remarks (a) The map (6.11) is $\mathbb{R}$-linear.

(b) Let $((x^1,\ldots,x^m),U)$ be a chart of M. For $v := \sum_j v^j\,\partial/\partial x^j \in \mathcal{V}(U)$, we have

$$\operatorname{div} v = \frac{1}{\sqrt{|G|}}\sum_j \frac{\partial}{\partial x^j}\left(\sqrt{|G|}\,v^j\right). \tag{6.14}$$

In orthogonal coordinates, we also have $\sqrt{|G|} = \sqrt{|g_{11}\cdot g_{22}\cdot\ \cdots\ \cdot g_{mm}|}$.

Proof Let $\varepsilon := 1$ if the chart is positive; use $\varepsilon := -1$ if it is negative. From (6.9), (6.10), and (6.13), we obtain (on U) that

$$\operatorname{div}(v)\omega_M = d(v \lrcorner \omega_M) = \varepsilon d\left(\sum_j (-1)^{j-1} v^j\sqrt{|G|}\,dx^1 \wedge\cdots\wedge \widehat{dx^j}\wedge\cdots\wedge dx^m\right)$$

$$= \varepsilon\sum_{j,k}(-1)^{j-1}\frac{\partial\big(v^j\sqrt{|G|}\big)}{\partial x^k}\,dx^k\wedge dx^1\wedge\cdots\wedge\widehat{dx^j}\wedge\cdots\wedge dx^m$$

$$= \varepsilon\left(\sum_j \frac{\partial\big(v^j\sqrt{|G|}\big)}{\partial x^j}\right)dx^1\wedge\cdots\wedge dx^m$$

$$= \left(\frac{1}{\sqrt{|G|}}\sum_j \frac{\partial\big(v^j\sqrt{|G|}\big)}{\partial x^j}\right)\omega_M$$

for $v\in\mathcal{V}(M)$. $\blacksquare$

(c) Suppose $(N,\bar{g})$ is an oriented pseudo-Riemannian manifold and a map $\varphi \in \operatorname{Diff}(M,N)$ satisfies $\varphi^*\omega_N = \mu\omega_M$ for some $\mu \neq 0$. Then

$$
\begin{array}{ccc}
\mathcal{V}(M) & \xrightarrow{\ \operatorname{div}_M\ } & \mathcal{E}(M)\\[4pt]
{\scriptstyle\varphi^*}\big\uparrow & & \big\uparrow{\scriptstyle\varphi^*}\\[4pt]
\mathcal{V}(N) & \xrightarrow{\ \operatorname{div}_N\ } & \mathcal{E}(N)
\end{array}
$$

is a commutative diagram, that is, $\operatorname{div}_M\circ\varphi^* = \varphi^*\circ\operatorname{div}_N$.

Proof From Remark 6.6(b) and from (6.13) we obtain, by using $d\circ\varphi^* = \varphi^*\circ d$, that

$$\mu\operatorname{div}_M(\varphi^*v)\omega_M = \mu\,d(\varphi^*v \lrcorner \omega_M) = d\varphi^*(v\lrcorner\omega_N) = \varphi^*d(v\lrcorner\omega_N)$$
$$= \varphi^*\big[(\operatorname{div}_N v)\omega_N\big] = \varphi^*(\operatorname{div}_N v)\varphi^*\omega_N = \mu\varphi^*(\operatorname{div}_N v)\omega_M$$

for $v\in\mathcal{V}(N)$. Now the claim follows from Lemma 6.5. $\blacksquare$

(d) (regularity) Let $k\in\mathbb{N}$. Then $\operatorname{div} v$ belongs to $C^k(M)$ for $v\in\mathcal{V}^{k+1}(M)$, and the map

$$\operatorname{div}: \mathcal{V}^{k+1}(M)\to C^k(M), \quad v\mapsto \operatorname{div} v$$

is $\mathbb{R}$-linear. So it suffices to assume that M is a C^{k+2} manifold.

Proof This is a consequence of Remarks 4.11(b) and 6.6(c). $\blacksquare$

As we shall see in the next chapter, the divergence of a vector field has interesting geometric and physical interpretations.

6.8 Examples **(a)** (Euclidean coordinates) Suppose U is open in $\mathbb{R}^m$. Denoting Euclidean coordinates by $(x^1, \ldots, x^m)$, we have

$$\operatorname{div} v = \sum_j \frac{\partial v^j}{\partial x^j}$$

for $v = \sum_j v^j \, \partial/\partial x^j$. This formula also holds when $((x^1, \ldots, x^m), U)$ are any other orthonormal coordinates on (M, g).

(b) (plane polar coordinates) Let $V_2 := (0, \infty) \times (0, 2\pi)$, and let

$$f_2 : V_2 \to \mathbb{R}^2 , \quad (r, \varphi) \mapsto (x, y) := (r \cos \varphi, r \sin \varphi)$$

be the plane polar coordinate map. Then with respect to the standard metric, we have

$$\operatorname{div}\left(v^1 \frac{\partial}{\partial r} + v^2 \frac{\partial}{\partial \varphi}\right) = \frac{1}{r} \frac{\partial(r v^1)}{\partial r} + \frac{\partial v^2}{\partial \varphi} = \frac{v^1}{r} + \frac{\partial v^1}{\partial r} + \frac{\partial v^2}{\partial \varphi} .$$

Proof This follows from $\sqrt{G} = r$, as can be read off the representation of g_2 given in Example 5.5(e). ∎

(c) (spherical coordinates) Let $V_3 := (0, \infty) \times (0, 2\pi) \times (0, \pi)$, and let

$$f_3 : V_3 \to \mathbb{R}^3 , \quad (r, \varphi, \vartheta) \mapsto (x, y, z)$$

be the spherical coordinate map of Example 5.5(g). With respect to the standard metric $g_3 := (dx)^2 + (dy)^2 + (dz)^2$, we have

$$\operatorname{div}\left(v^1 \frac{\partial}{\partial r} + v^2 \frac{\partial}{\partial \varphi} + v^3 \frac{\partial}{\partial \vartheta}\right) = \frac{1}{r^2} \frac{\partial(r^2 v^1)}{\partial r} + \frac{\partial v^2}{\partial \varphi} + \frac{1}{\sin \vartheta} \frac{\partial(v^3 \sin \vartheta)}{\partial \vartheta}$$

$$= \frac{2}{r} v^1 + \frac{\partial v^1}{\partial r} + \frac{\partial v^2}{\partial \varphi} + \cot(\vartheta) v^3 + \frac{\partial v^3}{\partial \vartheta} .$$

Proof Example 5.5(g) gives $\sqrt{|G|} = r^2 \sin \vartheta$, as the claim requires. ∎

(d) (Minkowski metric) Let $M := \mathbb{R}^4_{1,3}$ and $g := (dt)^2 - (dx)^2 - (dy)^2 - (dz)^2$. Then

$$\operatorname{div}\left(v^0 \frac{\partial}{\partial t} + v^1 \frac{\partial}{\partial x} + v^2 \frac{\partial}{\partial y} + v^3 \frac{\partial}{\partial z}\right) = \frac{\partial v^0}{\partial t} - \frac{\partial v^1}{\partial x} - \frac{\partial v^2}{\partial y} - \frac{\partial v^3}{\partial z}$$

for $v^j \in \mathcal{E}(\mathbb{R}^4_{1,3})$ with $0 \leq j \leq 3$. ∎

The Laplace–Beltrami operator

By combining the two first order differential operators grad and div, we obtain the most important second order differential operator, the **Laplace–Beltrami operator** Δ_g.

Let (M, g) be an oriented pseudo-Riemannian manifold. Then we define Δ_g by

$$\Delta_g := \operatorname{div}_g \operatorname{grad}_g$$

or, equivalently, by requiring that the diagram

$$
\begin{array}{ccc}
\mathcal{E}(M) & \xrightarrow{\ \Delta_g\ } & \mathcal{E}(M) \\
& \operatorname{grad}_g \searrow \quad \nearrow \operatorname{div}_g & \\
& \mathcal{V}(M) &
\end{array}
$$

commutes. Instead of Δ_g, we may also write Δ_M or simply Δ if g is clear from context.

6.9 Remarks (a) The map $\Delta_M : \mathcal{E}(M) \to \mathcal{E}(M)$ is $\mathbb{R}$-linear.

(b) If $((x^1, \ldots, x^m), U)$ is a chart of M, then

$$\Delta_M f = \frac{1}{\sqrt{|G|}} \sum_{j,k} \frac{\partial}{\partial x^j}\left(\sqrt{|G|}\, g^{jk} \frac{\partial f}{\partial x^k}\right) \quad \text{for } f \in \mathcal{E}(U) \ . \tag{6.15}$$

In orthogonal coordinates, (6.15) simplifies to

$$\Delta_M f = \frac{1}{\sqrt{|G|}} \sum_{j} \frac{\partial}{\partial x^j}\left(\sqrt{|G|}\, g^{jj} \frac{\partial f}{\partial x^j}\right) \quad \text{for } f \in \mathcal{E}(U) \ , \tag{6.16}$$

where $\sqrt{|G|} = \sqrt{|g_{11} \cdot g_{22} \cdot \cdots \cdot g_{mm}|}$.

Proof This follows from Remarks 6.3(c) and (d) and Remark 6.7(b). ∎

(c) Suppose $(N, \bar{g})$ is an oriented pseudo-Riemannian manifold. Also let $\varphi \in \operatorname{Diff}(M, N)$, and suppose there are $\lambda \neq 0$ and $\mu \neq 0$ such that $\varphi_* g = \lambda \bar{g}$ and $\varphi^* \omega_N = \mu \omega_M$. Then the diagram

$$
\begin{array}{ccc}
\mathcal{E}(M) & \xrightarrow{\ \lambda\Delta_M\ } & \mathcal{E}(M) \\
\varphi^* \big\uparrow \cong & & \cong \big\uparrow \varphi^* \\
\mathcal{E}(N) & \xrightarrow[\ \Delta_N\ ]{} & \mathcal{E}(N)
\end{array}
$$

commutes: $\lambda\Delta_M \circ \varphi^* = \varphi^* \circ \Delta_N$.

Proof This is a consequence of Remarks 6.3(e) and 6.7(c). ∎

(d) (regularity) Let $k \in \mathbb{N}$. Then obviously

$$\Delta_M : C^{k+2}(M) \to C^k(M) \ ,$$

and this map is $\mathbb{R}$-linear. Here it suffices to assume that M is a C^{k+2} manifold. ∎

6.10 Examples **(a)** (Euclidean coordinates) Suppose M is open in $\mathbb{R}^m$, with Euclidean coordinates $\big((x^1,\ldots,x^m),M\big)$. Then Δ_M is the same as the (usual) m-**dimensional Laplace operator**

$$\Delta_m := \sum_j \partial_j^2 .$$

See Exercise VII.5.3.

(b) (circular coordinates) With respect to the parametrization

$$h \colon (0,2\pi) \to \mathbb{R}^2 , \quad \varphi \mapsto (\cos\varphi, \sin\varphi)$$

of $S^1 \backslash \{(1,0)\}$ (and the standard metric), we have $\Delta_{S^1} = \partial_\varphi^2$.
Proof Remark 6.9(b) and Example 5.5(f). ∎

(c) (plane polar coordinates) In plane polar coordinates

$$(0,\infty) \times (0,2\pi) \to \mathbb{R}^2 , \quad (r,\varphi) \mapsto (r\cos\varphi, r\sin\varphi) ,$$

the Laplace–Beltrami operator (with respect to the standard metric $\mathbb{R}^2$) is

$$\Delta_2 = \frac{1}{r}\partial_r(r\partial_r \cdot) + \frac{1}{r^2}\partial_\varphi^2 = \partial_r^2 + \frac{1}{r}\partial_r + \frac{1}{r^2}\partial_\varphi^2 = \frac{1}{r^2}\big[(r\partial_r)^2 + \Delta_{S^1}\big] .$$

Proof This follows from Remark 6.9(b), Example 5.5(e), and (b). ∎

(d) (m-dimensional spherical coordinates) For $m \geq 2$, the Laplace–Beltrami operator of S^m (with respect to the standard metric) in the spherical coordinates of Example 5.5(h) assumes the form

$$\Delta_{S^m} = \frac{1}{\sin^2\vartheta_1 \cdot \;\cdots\; \cdot \sin^2\vartheta_{m-1}}\frac{\partial^2}{\partial\varphi^2}$$

$$+ \sum_{k=1}^{m-1} \frac{1}{\sin^k\vartheta_k \sin^2\vartheta_{k+1} \cdot \;\cdots\; \cdot \sin^2\vartheta_{m-1}}\frac{\partial}{\partial\vartheta_k}\Big(\sin^k\vartheta_k \frac{\partial}{\partial\vartheta_k}\Big) .$$

In particular,

$$\Delta_{S^2} = \frac{1}{\sin^2\vartheta}\partial_\varphi^2 + \frac{1}{\sin\vartheta}\partial_\vartheta(\sin\vartheta\,\partial_\vartheta \cdot) = \frac{1}{\sin^2\vartheta}\partial_\varphi^2 + \partial_\vartheta^2 + \cot\vartheta\,\partial_\vartheta .$$

Proof From Examples 5.5(g) and (h), it follows

$$G = \prod_{k=0}^{m-1} a_{m+1,k} = \prod_{k=0}^{m-2} \prod_{i=k+1}^{m-1} \sin^2\vartheta_i .$$

Exchanging the order of the two products gives

$$G = \prod_{i=1}^{m-1} \sin^{2i}\vartheta_i = \big[w_{m+1}(\vartheta)\big]^2 , \qquad (6.17)$$

where we use the abbreviated notation introduced in Proposition X.8.9.

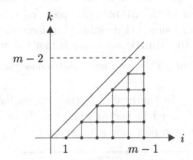

From the orthogonality of the spherical coordinates, it also follows from the given examples that

$$g^{jj} = \frac{1}{a_{m+1,j-1}} = \frac{1}{\prod_{i=j}^{m-1} \sin^2 \vartheta_i} \quad \text{for } 1 \le j \le m .$$

From this we read off

$$\sqrt{G} \, g^{jj} = \Big(\prod_{\substack{i=1 \\ i \ne j-1}}^{m-1} \sin^i \vartheta_i \prod_{k=j}^{m-1} \frac{1}{\sin^2 \vartheta_k} \Big) \sin^{j-1} \vartheta_{j-1}$$

for $2 \le j \le m$. Thus we find

$$\frac{1}{\sqrt{G}} \frac{\partial}{\partial \vartheta_{j-1}} \Big(\sqrt{G} \, g^{jj} \frac{\partial}{\partial \vartheta_{j-1}} \Big)$$

$$= \frac{1}{\sin^{j-1} \vartheta_{j-1} \prod_{i=j}^{m-1} \sin^2 \vartheta_i} \frac{\partial}{\partial \vartheta_{j-1}} \Big(\sin^{j-1} \vartheta_{j-1} \frac{\partial}{\partial \vartheta_{j-1}} \Big)$$

for $2 \le j \le m$. Now the claim is clear. ∎

(e) (*m*-dimensional polar coordinates) In *m*-dimensional polar coordinates with $m \ge 2$, the *m*-dimensional Laplace operator reads

$$\Delta_m = \frac{1}{r^{m-1}} \partial_r (r^{m-1} \partial_r \cdot) + \frac{1}{r^2} \Delta_{S^{m-1}} = \partial_r^2 + \frac{m-1}{r} \partial_r + \frac{1}{r^2} \Delta_{S^{m-1}}$$

$$= \frac{1}{r^2} \big[(r \partial_r)^2 + (m-2) r \partial_r + \Delta_{S^{m-1}} \big] .$$

Proof From Examples 5.5(g) and (h), we read off $g_m = (dr)^2 + r^2 g_{S^{m-1}}$. This then implies $G = r^{2(m-1)} G_{S^{m-1}}$. It also implies $g^{11} = 1$ and

$$g^{jj} = \frac{1}{r^2} g_{S^{m-1}}^{(j-1)(j-1)} \quad \text{for } 2 \le j \le m .$$

Now the claim follows from (6.16) because of the orthogonality of the coordinates. ∎

(f) (Minkowski metric) In orthonormal coordinates, the Laplace–Beltrami operator of the Minkowski space $\mathbb{R}_{1,3}^4$ has the form $\partial_t^2 - \Delta_3$, where Δ_3 is the three-dimensional (Euclidian) Laplace operator. That is, the Laplace–Beltrami operator in the Minkowski space is just the wave operator.[2]

Proof This is an immediate consequence of (6.16). ∎

In the next proposition, we list some basic properties of differential operators used in vector analysis. Here and in the following, we denote the pseudo-Riemannian metric of M by $(\cdot | \cdot)_M$.

---

[2]See Exercise VII.5.10.

6.11 Proposition *Suppose* $(M, (\cdot\,|\,\cdot)_M)$ *is an oriented pseudo-Riemannian mani-*
fold, $f, g \in \mathcal{E}(M)$, *and* $v, w \in \mathcal{V}(M)$. *Then*

(i) $\operatorname{grad}(fg) = f \operatorname{grad} g + g \operatorname{grad} f$;

(ii) $\operatorname{div}(fv) = f \operatorname{div} v + (\operatorname{grad} f \,|\, v)_M$;

(iii) $\Delta(fg) = f\Delta g + 2(\operatorname{grad} f \,|\, \operatorname{grad} g)_M + g\Delta f$;

(iv) $f\Delta g - g\Delta f = \operatorname{div}(f \operatorname{grad} g) - \operatorname{div}(g \operatorname{grad} f)$.

Proof (i) Because Θ is a module isomorphism, it follows from (6.4) that (i) is
equivalent to

$$d(fg) = f\,dg + g\,df\ . \tag{6.18}$$

Because (6.18) is a local statement, it suffices to prove this formula in local coordinates. In this case, it is an immediate consequence of the product rule.

(ii) From $(fv) \lrcorner\, \omega_M = f(v \lrcorner\, \omega_M) = f \wedge (v \lrcorner\, \omega_M)$ and the product rule of
Theorem 4.10, it follows that

$$d\big((fv) \lrcorner\, \omega_M\big) = d\big(f \wedge (v \lrcorner\, \omega_M)\big) = df \wedge (v \lrcorner\, \omega_M) + f\,d(v \lrcorner\, \omega_M)\ . \tag{6.19}$$

Because this is also a local statement, we can use local representations. Then,
with $v = \sum_j v^j\, \partial/\partial x^j$ and a positive chart, we obtain from (6.9) and (6.10) that

$$
\begin{aligned}
& df \wedge (v \lrcorner\, \omega_M) \\
&= \Big(\sum_j \frac{\partial f}{\partial x^j}\,dx^j\Big) \wedge \sum_k (-1)^{k-1} v^k \sqrt{|G|}\,dx^1 \wedge \cdots \wedge \widehat{dx^k} \wedge \cdots \wedge dx^m \tag{6.20} \\
&= \Big(\sum_j \frac{\partial f}{\partial x^j}\,v^j\Big)\sqrt{|G|}\,dx^1 \wedge \cdots \wedge dx^m = \sum_j \frac{\partial f}{\partial x^j}\,v^j \omega_M\ .
\end{aligned}
$$

We then read from Remark 6.3(b) and (4.4) that

$$(\operatorname{grad} f \,|\, v)_M = \langle df, v\rangle = \sum_j \Big\langle df, \frac{\partial}{\partial x^j}\Big\rangle v^j = \sum_j \frac{\partial f}{\partial x^j}\,v^j\ . \tag{6.21}$$

Therefore it follows from (6.19)–(6.21) and the definition (6.13) that

$$\operatorname{div}(fv)\omega_M = d\big((fv) \lrcorner\, \omega_M\big) = (\operatorname{grad} f \,|\, v)_M\,\omega_M + f \operatorname{div} v\,\omega_M\ ,$$

which implies the claim.

(iii) This we get immediately from $\Delta = \operatorname{div} \operatorname{grad}$ and (i), (ii).

(iv) From (ii), it follows that

$$\operatorname{div}(f \operatorname{grad} g) = f\Delta g + (\operatorname{grad} f \,|\, \operatorname{grad} g)_M\ . \tag{6.22}$$

Exchanging f and g and subtracting the result from (6.22) then yields (iv). ∎

The curl

Suppose now (M, g) is a 3-dimensional oriented pseudo-Riemannian manifold. Then we define the **curl**[3] $\operatorname{curl} v$ of the vector field $v \in \mathcal{V}(M)$ by requiring that the diagram

$$
\begin{array}{ccc}
\mathcal{V}(M) & \xrightarrow{\ \Theta\ } & \Omega^1(M) \\[2pt]
{\scriptstyle\operatorname{curl}}\Big\downarrow & {\scriptstyle\cong} & \Big\downarrow{\scriptstyle d} \\[2pt]
\mathcal{V}(M) & \xrightarrow[\ \cong\ ]{\ \lrcorner\,\omega_M\ } & \Omega^2(M)
\end{array}
\qquad (6.23)
$$

commutes, that is, by requiring

$$
(\operatorname{curl} v) \lrcorner\, \omega_M = d(\Theta v) \quad \text{for } v \in \mathcal{V}(M) . \qquad (6.24)
$$

The definition is clearly only possible in the case $m = 3$.

6.12 Remarks (a) The map $\operatorname{curl} \colon \mathcal{V}(M) \to \mathcal{V}(M)$, $v \mapsto \operatorname{curl} v$ is $\mathbb{R}$-linear.

(b) Let $((x^1, x^2, x^3), U)$ be a chart of M. Then

$$
\operatorname{curl} v = \frac{1}{\sqrt{|G|}} \sum_{i=1}^{3} \sum_{(j,k,\ell) \in S_3} \operatorname{sign}(j, k, \ell) \, \frac{\partial}{\partial x^j} (g_{ki} v^i) \frac{\partial}{\partial x^\ell}
$$

for $v = \sum_{j=1}^{3} v^j \, \partial / \partial x^j$. If the coordinates are orthogonal, this expression simplifies to

$$
\operatorname{curl} v = \frac{1}{\sqrt{|G|}} \sum_{(j,k,\ell) \in S_3} \operatorname{sign}(j, k, \ell) \, \frac{\partial}{\partial x^j} (g_{kk} v^k) \frac{\partial}{\partial x^\ell}
$$

$$
= \frac{1}{\sqrt{|G|}} \Big[(\partial_2(g_{33} v^3) - \partial_3(g_{22} v^2)) \frac{\partial}{\partial x^1} + (\partial_3(g_{11} v^1) - \partial_1(g_{33} v^3)) \frac{\partial}{\partial x^2}
$$

$$
+ (\partial_1(g_{22} v^2) - \partial_2(g_{11} v^1)) \frac{\partial}{\partial x^3} \Big]
$$

with $\sqrt{|G|} = \sqrt{|g_{11} g_{22} g_{33}|}$. If the coordinates are orthonormal, this becomes

$$
\operatorname{curl} v = (\partial_2 v^3 - \partial_3 v^2) \frac{\partial}{\partial x^1} + (\partial_3 v^1 - \partial_1 v^3) \frac{\partial}{\partial x^2} + (\partial_1 v^2 - \partial_2 v^1) \frac{\partial}{\partial x^3} .
$$

Proof Remark 6.1(b) and the properties of the exterior derivative give

$$
d(\Theta v) = d \sum_{k} \Big(\sum_{i} g_{ki} v^i \Big) dx^k = \sum_{k} \sum_{j \neq k} \frac{\partial}{\partial x^j} \Big(\sum_{i} g_{ki} v^i \Big) dx^j \wedge dx^k .
$$

---

[3] Sometimes written rot, short for "rotation".

From (6.10), we read off

$$\operatorname{curl} v \; \lrcorner \; \omega_M = \sqrt{|G|} \left((\operatorname{curl} v)^1 \, dx^2 \wedge dx^3 + (\operatorname{curl} v)^2 \, dx^3 \wedge dx^1 + (\operatorname{curl} v)^3 \, dx^1 \wedge dx^2 \right) . \quad (6.25)$$

Therefore the claim follows from (6.24). ∎

(c) (regularity) Let $k \in \mathbb{N}$. Then $\operatorname{curl} v \in \mathcal{V}^k(M)$ for $v \in \mathcal{V}^{k+1}(M)$. So it suffices here to assume that M is a C^{k+2} manifold. ∎

In the case $m = 3$, there are important relations between the operators grad, div, and curl. These are summarized diagrammatically in the following theorem.

6.13 Theorem Let (M, g) be a three-dimensional oriented (pseudo-)Riemannian manifold.

(i) The diagram

$$
\begin{array}{ccccccc}
& \xrightarrow{\text{grad}} & & \xrightarrow{\text{curl}} & & \xrightarrow{\text{div}} & \\
\mathcal{E}(M) & & \mathcal{V}(M) & & \mathcal{V}(M) & & \mathcal{E}(M) \\
\Big\| & & \cong \Big\downarrow \Theta_M & & \cong \Big\downarrow \lrcorner \, \omega_M & & \cong \Big\downarrow \bullet \, \omega_M \\
\Omega^0(M) & \xrightarrow{d} & \Omega^1(M) & \xrightarrow{d} & \Omega^2(M) & \xrightarrow{d} & \Omega^3(M)
\end{array}
\quad (6.26)
$$

commutes.

(ii) $\operatorname{curl} \circ \operatorname{grad} = 0$.

(iii) $\operatorname{div} \circ \operatorname{curl} = 0$.

Proof (i) follows immediately from the commutativity of the diagrams (6.4), (6.12), and (6.23).

(ii) and (iii) are now direct consequence of $d^2 = 0$. ∎

6.14 Corollary Let X be open and contractible in $\mathbb{R}^3$. Also let v be a smooth vector field on X.

(i) If $\operatorname{curl} v = 0$, then there is an $f \in \mathcal{E}(X)$ such that $v = \operatorname{grad} f$, a **potential** for v.

(ii) If $\operatorname{div} v = 0$, then there is a $w \in \mathcal{V}(X)$ with $v = \operatorname{curl} w$, a **vector potential** for v.

Proof (i) From (6.26) we learn that $\operatorname{curl} v = 0$ is equivalent to $d(\Theta_X v) = 0$. Therefore the 1-form $\Theta_M v$ is closed, and the Poincaré lemma (Theorem 3.11) guarantees the existence of an $f \in \Omega^0(X) = \mathcal{E}(X)$ such that $\Theta_X v = df$. From this it follows that $v = \Theta_X^{-1} df = \operatorname{grad} f$.

(ii) Analogously to (i), it follows from $\operatorname{div} v = 0$ that the 2-form $v \; \lrcorner \; \omega_X$ is closed and therefore exact, again by the Poincaré lemma. Thus there is an $\alpha \in \Omega^1(X)$ with $d\alpha = v \; \lrcorner \; \omega_X$. Then $w := \Theta_X^{-1} \alpha \in \mathcal{V}(X)$, by the commutativity of the middle "loop" of (6.26), satisfies $\operatorname{curl} w = v$. ∎

6.15 Remarks Suppose X is open in $\mathbb{R}^3$.

(a) In Euclidean coordinates, the equality $\operatorname{curl} v = 0$ is equivalent to the integrability conditions

$$\partial_j v^k = \partial_k v^j \quad \text{for } 1 \le j, k \le 3 ,$$

which can be seen from Remark 6.12(b). Therefore Corollary 6.14(i) is a special case of Remark VIII.4.10(a).

(b) (classical notation) In Euclidean coordinates, we know from Example 6.4(a) that $\operatorname{grad} f$ agrees with the ∇f from Proposition VII.2.16. The physics and engineering literatures, and many mathematical texts, use the formal **nabla vector**

$$\nabla := \left(\frac{\partial}{\partial x}, \frac{\partial}{\partial y}, \frac{\partial}{\partial z} \right) .$$

With the notation $x \cdot y$ for the Euclidean scalar product in $\mathbb{R}^3$ and $x \times y$ for the vector product, the nabla vector notation leads to the (formal) relations

$$\operatorname{div} v = \nabla \cdot v , \quad \operatorname{curl} v = \nabla \times v , \quad \Delta v = (\nabla \cdot \nabla) v =: \nabla^2 v .$$

These follow easily from the corresponding local representations of these operators and from Remark VIII.2.14(d). In particular, the components of the vector $\operatorname{curl} v$ can be found by expanding the (formal) determinant

$$\begin{vmatrix} \vec{e}_1 & \vec{e}_2 & \vec{e}_3 \\ \partial/\partial x & \partial/\partial y & \partial/\partial z \\ v^1 & v^2 & v^3 \end{vmatrix}$$

in its first row. Here $\vec{e}_1$, $\vec{e}_2$, $\vec{e}_3$ are the standard basis vectors of $\mathbb{R}^3$, and $\partial/\partial x$, $\partial/\partial y$, $\partial/\partial z$ are *not* interpreted as tangent vectors, but as differential operators.

Because the symbol ∇ has another meaning in the context of Riemannian geometry, we will rarely use the nabla vector in the rest of this book.

(c) (the physical meaning of the curl[4]) We consider a rigid body rotating at constant (angular) velocity about a fixed axis. We then choose an orthonormal basis $(\vec{e}_1, \vec{e}_2, \vec{e}_3)$ and the coordinate origin so that $\vec{e}_3$ points along the rotation axis. Also let ω be the **angular velocity**, that is, ω is the speed of any point P fixed in the rotating body at unit distance from the axis of rotation. If $\vec{r}$ is the radius vector of the point P, that

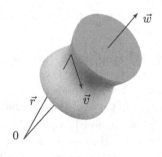

---

[4]A deeper interpretation of the curl of a vector field is given in Section XII.3.

is, the position vector of the point P in the coordinate system $(O; \vec{e}_1, \vec{e}_2, \vec{e}_3)$ (see the statements after Remarks I.12.6) and if θ is the angle between $\vec{e}_3$ and $\vec{r}$ (in the plane spanned by $\vec{e}_3$ and $\vec{r}$), then the distance a from P to the rotation axis satisfies $a = |\vec{r}| \sin \theta$. Therefore the modulus of the velocity vector $\vec{v}$ of the point P is given by

$$|\vec{v}| = \omega a = \omega |\vec{r}| \sin \theta .$$

Denote by $\vec{w} := \omega \vec{e}_3$ the "angular velocity vector" and orient it so that the body rotates clockwise about it. Then it follows from the properties of the vector product that

$$\vec{v} = \vec{w} \times \vec{r} , \tag{6.27}$$

since the point P moves with constant speed ω in a circle centered at and in a plane orthogonal to the $\vec{e}_3$-axis.[5]

Let (x, y, z) be the coordinates of P with respect to $(O; \vec{e}_1, \vec{e}_2, \vec{e}_3)$. Then

$$\vec{r} = x \frac{\partial}{\partial x} + y \frac{\partial}{\partial y} + z \frac{\partial}{\partial z} \quad \text{and} \quad \vec{w} = \omega \frac{\partial}{\partial z} .$$

Therefore

$$\vec{v} = \vec{w} \times \vec{r} = -\omega y \frac{\partial}{\partial x} + \omega x \frac{\partial}{\partial y} .$$

For the curl of the vector field $\vec{v}$, we find $\operatorname{curl} \vec{v} = 2\omega \, \partial/\partial z = 2\vec{w}$. In words, for a rigid body rotating about a fixed axis, the curl of the velocity vector is a vector field whose elements are parallel to the rotation axis and have absolute value twice the angular velocity.

(d) (regularity) The statements of Theorem 6.13 and Corollary 6.14 can be proved with weaker differentiability assumptions that are easily derived from earlier remarks about regularity. ∎

The Lie derivative

Now suppose M is again an arbitrary manifold. For $f \in \mathcal{E}(M)$ and $v \in \mathcal{V}(M)$, we set

$$L_v f := \langle df, v \rangle \in \mathcal{E}(M)$$

and call $L_v f$ the **Lie derivative** of f with respect to v.

6.16 Proposition

(i) *The map $L_v : \mathcal{E}(M) \to \mathcal{E}(M)$, the **Lie derivative** with respect to v, has the properties that*

    (α) *L_v is $\mathbb{R}$-linear;*

    (β) *$L_v(fg) = L_v(f)g + f L_v g$ for $f, g \in \mathcal{E}(M)$.*

---

[5] We leave the formal proof of (6.27) to you.

(ii) *In local coordinates,*

$$L_v f = \sum_j v^j \frac{\partial f}{\partial x^j} \quad \text{and} \quad v = \sum_j v^j \frac{\partial}{\partial x^j} \ .$$

Proof   (i) follows immediately from the properties of d (see (6.18)).

(ii) is a consequence of (4.4). ∎

6.17 Remarks   (a) Proposition 6.16(ii) makes it clear that the Lie derivative generalizes the directional derivative of Section VII.2.

(b) Let A be an $\mathbb{R}$-algebra. A map $D: A \to A$ is said to be a **derivation** (of A) if D is $\mathbb{R}$-linear and satisfies the product rule

$$D(ab) = (Da)b + a(Db) \quad \text{for } a, b \in A \ .$$

Therefore the Lie derivative with respect to $v \in \mathcal{V}(M)$ is a derivation of the algebra $\mathcal{E}(M)$.

(c) If A is an algebra with unity e and D is a derivation of A, then $De = 0$.
Proof   The product rule gives

$$De = D(ee) = (De)e + e(De) = De + De = 2De$$

and hence the claim. ∎

The next theorem shows that every derivation of $\mathcal{E}(M)$ is given by a Lie derivative.

6.18 Theorem   *Let D be a derivation of $\mathcal{E}(M)$. Then there is exactly one $v \in \mathcal{V}(M)$ such that $D = L_v$.*

Proof   (i) We show first that D is a "local operator". Let U be an open and K a compact neighborhood of $p \in M$ with $K \subset\subset U$. Remark 1.21(a) guarantees the existence of a $\chi \in \mathcal{E}(M)$ with $\chi \,|\, K = 1$ and $\operatorname{supp}(\chi) \subset\subset U$.

Let $f \in \mathcal{E}(M)$ with $f \,|\, U = 0$. Then $f = f\chi + f(1 - \chi) = f(1 - \chi)$, and thus

$$Df(p) = Df(p)\big(1 - \chi(p)\big) + f(p)D(1 - \chi)(p) = 0 \ .$$

Because this is true for every $p \in U$, it follows that $D(f)\,|\,U = 0$. If $\chi_1 \in \mathcal{E}(M)$ is another function with $\operatorname{supp}(\chi_1) \subset\subset U$ and which is identically equal to 1 in a neighborhood of p, then $f\chi - f\chi_1 \in \mathcal{E}(M)$ for $f \in \mathcal{E}(U)$ vanishes in a neighborhood of p. Then it follows from the above that $D(f\chi) = D(f\chi_1)$ for $f \in \mathcal{E}(U)$. Hence the "restriction of D to U" is well defined by

$$D_U f := D(f\chi) \quad \text{for } f \in \mathcal{E}(U)$$

and is independent of the special choice of χ.

(ii) Suppose now (φ, U) is a chart with $\varphi = (x^1, \ldots, x^m)$. We can assume that $X := \varphi(U)$ is convex. For every fixed $p \in U$, it follows from the mean value theorem in integral form (Theorem VII.3.10) with $a := \varphi(p)$ that

$$(\varphi_* f)(x) = (\varphi_* f)(a) + \sum_j (x^j - a^j) \widetilde{f}_j(x) \quad \text{for } x \in X ,$$

where we have set

$$\widetilde{f}_j(x) := \int_0^1 \partial_j f\big(a + t(x - a)\big) \, dt \quad \text{for } x \in X .$$

Therefore

$$f_j := \varphi^* \widetilde{f}_j \in \mathcal{E}(U) , \qquad f_j(p) = \frac{\partial f}{\partial x^j}(p) ,$$

and

$$f(q) = f(p) + \sum_j \big(\varphi^j(q) - \varphi^j(p)\big) f_j(q) \quad \text{for } q \in U .$$

From this, the properties of D, and Remark 6.17(c), it follows that

$$Df(p) = \sum_j D\varphi^j(p) \frac{\partial f}{\partial x^j}(p) \quad \text{for } p \in U , \tag{6.28}$$

where we have written D instead of D_U.

(iii) Let (ψ, V) be a second chart around p with $\psi = (y^1, \ldots, y^m)$. Now define the transition function $k := \psi \circ \varphi^{-1}$. Then, in analogy to (ii) and because we can assume $U = V$, we have

$$k^j(x) = k^j(a) + \sum_\ell (x^\ell - a^\ell) k_\ell^j(x) \quad \text{for } x \in X , \tag{6.29}$$

with $k_\ell^j \in \mathcal{E}(X)$ and $k_\ell^j(a) = \partial_\ell k^j(a)$. Because $\varphi^* k = \psi$, applying φ^* to (6.29) gives

$$\psi^j(q) = \psi^j(p) + \sum_\ell \big(\varphi^\ell(q) - \varphi^\ell(p)\big) h_\ell^j(q) \quad \text{for } q \in U , \tag{6.30}$$

with $h_\ell^j := \varphi^* k_\ell^j \in \mathcal{E}(U)$ and

$$h_\ell^j(p) = (\varphi^* \partial_\ell k^j)(p) = \frac{\partial y^j}{\partial x^\ell}(p) .$$

This and (6.30) imply

$$D\psi^j(p) = \sum_k D\varphi^k(p) \frac{\partial y^j}{\partial x^k}(p) \quad \text{for } p \in U \text{ and } 1 \le j, k \le m . \tag{6.31}$$

Now we set

$$v_\varphi := \sum_j D\varphi^j \frac{\partial}{\partial x^j} \quad \text{and} \quad v_\psi := \sum_j D\psi^j \frac{\partial}{\partial y^j} . \tag{6.32}$$

Then it follows from (6.31) and Proposition 4.7 that

$$v_\psi = \sum_j \sum_k D\varphi^k \frac{\partial y^j}{\partial x^k} \frac{\partial}{\partial y^j} = \sum_k D\varphi^k \frac{\partial}{\partial x^k} = v_\varphi \ .$$

This shows that (6.32) defines a vector field $v_U \in \mathcal{V}(U)$ on U that is independent of the coordinates chosen. From (6.28), (6.31), and Proposition 6.16(ii), we read off $D_U f = L_{v_U} f$ for $f \in \mathcal{E}(U)$.

(iv) Suppose now $\{(\varphi_\alpha, U_\alpha) \ ; \ \alpha \in \mathsf{A}\}$ is an atlas for M. Then it follows from (iii) that for every $\alpha \in \mathsf{A}$ there is an $v_\alpha \in \mathcal{V}(U_\alpha)$ such that $D_{U_\alpha} f = L_{v_\alpha} f$ for $f \in \mathcal{E}(U_\alpha)$. Moreover, the observations in (iii) show that there is exactly one $v \in \mathcal{V}(M)$ such that $v \,|\, U_\alpha = v_\alpha$ for $\alpha \in \mathsf{A}$. Now (i) and Proposition 6.16(ii) give $D = L_v$.

(v) Suppose $v, w \in \mathcal{V}(M)$ with $D = L_v$ and $D = L_w$. Then $L_v f = L_w f$ for every $f \in \mathcal{E}(M)$. In an arbitrary local chart $\big((x^1, \ldots, x^m), U\big)$, we then have

$$\sum_j (v^j - w^j) \frac{\partial f}{\partial x^j} = 0 \quad \text{for } f \in \mathcal{E}(U) \ .$$

Choosing $f := x^k$, we find $\partial f / \partial x^j = \delta_j^k$ and therefore $v^k - w^k = 0$. That this is true for $1 \le k \le m$ implies $v \,|\, U = w \,|\, U$ and hence $v = w$. Thus we are done. ∎

6.19 Lemma   For $L_v L_w - L_w L_v$ is a derivation of $\mathcal{E}(M)$ for $v, w \in \mathcal{V}(M)$.

Proof   Clearly $L_v L_w - L_w L_v$ is an $\mathbb{R}$-linear map of $\mathcal{E}(M)$ to itself. For $f, g \in \mathcal{E}(M)$, we know because $\mathcal{E}(M)$ is commutative that

$$L_v L_w(fg) = L_v\big(L_w(f)g + fL_w g\big)$$
$$= gL_v L_w f + L_v f L_w g + L_v g L_w f + f L_v L_w g \ .$$

The claim is now obvious. ∎

Let $v, w \in \mathcal{V}(M)$. Then it follows from Theorem 6.18 and Lemma 6.19 that there is exactly one smooth vector field $[v, w]$ on M such that

$$L_{[v,w]} = L_v L_w - L_w L_v \ . \tag{6.33}$$

We call $[v, w]$ the **Lie bracket** or the **commutator** of v and w.

6.20 Proposition

(i) *The map* $\mathcal{V}(M) \times \mathcal{V}(M) \to \mathcal{V}(M)$, $(v, w) \mapsto [v, w]$ *has the properties that*

    (α) *(bilinearity)* $[\,\cdot\,,\,\cdot\,]$ *is $\mathbb{R}$-bilinear.*

    (β) *(skew-symmetry)* $[v, w] = -[w, v]$ *for $v, w \in \mathcal{V}(M)$.*

(γ) (Jacobi identity) $u, v, w \in \mathcal{V}(M)$ *satisfy the relation*

$$\big[u, [v, w]\big] + \big[v, [w, u]\big] + \big[w, [u, v]\big] = 0 \ .$$

(ii) *In local coordinates,*

$$[v, w] = \sum\nolimits_{j,k} \left(v^k \frac{\partial w^j}{\partial x^k} - w^k \frac{\partial v^j}{\partial x^k} \right) \frac{\partial}{\partial x^j} \tag{6.34}$$

for $v = \sum_j v^j \, \partial/\partial x^j$ *and* $w = \sum_j w^j \, \partial/\partial x^j$.

Proof The simple proofs are left to you. ∎

6.21 Remarks (a) Suppose M is open in $\mathbb{R}^m$ and $(x^1, \dots, x^m)$ are Euclidean coordinates on M. Using the nabla vector ∇, (6.34) can be written symbolically in the intuitive form

$$[v, w] = (v \cdot \nabla)w - (w \cdot \nabla)v \ .$$

(b) Suppose V is a vector space and $[\cdot, \cdot]\colon V \times V \to V$ is a map with the properties (α)–(γ) of Proposition 6.20(i). Then $\big(V, [\cdot, \cdot]\big)$ is called a **Lie algebra**. Because of (β), the "multiplication" $[\cdot, \cdot]$ is generally not commutative. It follows from (β) and (γ) that

$$\big[a, [b, c]\big] - \big[[a, b], c\big] = \big[[c, a], b\big] \quad \text{for } a, b, c \in V \ .$$

So the multiplication is generally not associative either. Thus a Lie algebra is generally a noncommutative, nonassociative algebra.[6] Therefore $\big(\mathcal{V}(M), [\cdot, \cdot]\big)$ is a Lie algebra.

(c) (regularity) Let $k \in \mathbb{N}$ and $v, w \in \mathcal{V}^k(M)$. Let M be a manifold of class C^{k+1}. Then L_v is not a derivation on $\mathcal{E}^{k+1}(M)$, because $L_v f$ for $f \in \mathcal{E}^{k+1}(M)$ generally only belongs to $\mathcal{E}^k(M)$. Therefore the Lie bracket cannot be defined through (6.33) either. We are left to choose local coordinates and to define $[v, w]$ for $v, w \in \mathcal{V}^k(M)$ through (6.34).[7] Then $[v, w] \in \mathcal{V}^{k-1}(M)$. ∎

The Hodge–Laplace operator

In the rest of this section, we use the codifferential and the star operator to derive some more important relations from vector analysis.

Let $\big(M, (\cdot \mid \cdot)_M\big)$ be an oriented pseudo-Riemannian manifold. First we write the divergence in terms of the codifferential.

---

[6] In the trivial commutative case in which $[a, b] = 0$ for $a, b \in V$, it is of course commutative and associative.

[7] You may want to consider why $[v, w]$ so defined is well defined on all of M.

6.22 Proposition   *The diagram*

$$\mathcal{V}(M) \xrightarrow{\quad\Theta\quad} \Omega^1(M)$$

$$\mathrm{div} \searrow \qquad \swarrow \delta$$

$$\mathcal{E}(M) = \Omega^0(M)$$

is commutative, that is, $\mathrm{div} = \delta \circ \Theta$.

Proof   If suffices to prove this equation locally. So let $\big((x^1, \ldots, x^m), U\big)$ be local coordinates. Then for $v = \sum_j v^j \, \partial/\partial x^j \in \mathcal{V}(U)$, Remarks 6.1(b) and 5.11(b) imply

$$\delta\Theta v = \delta \sum_j \left(\sum_k g_{jk} v^k \right) dx^j = \frac{1}{\sqrt{|G|}} \sum_j \frac{\partial}{\partial x^j} \left(\sqrt{|G|}\, v^j \right) \, .$$

Therefore the claim follows from (6.14). ∎

Using the exterior derivative and the codifferential, we define for $0 \le r \le m$ an $\mathbb{R}$-linear map on $\Omega^r(M)$ by

$$\Delta_M := d\delta + \delta d : \Omega^r(M) \to \Omega^r(M) \, . \tag{6.35}$$

This is the **Hodge–Laplace operator**. For $a \in \mathcal{E}(M)$, it follows from (6.4) and Proposition 6.22 that

$$(d\delta + \delta d)a = \delta da = \delta\Theta(\Theta^{-1}da) = \mathrm{div}\,\mathrm{grad}\,a \, .$$

Therefore the Hodge–Laplace operator on $\Omega^0(M) = \mathcal{E}(M)$ is the same as the Laplace–Beltrami operator, which justifies the notation. When M is clear from context, we write Δ for Δ_M. ∎

6.23 Remarks   **(a)** $*\Delta = \Delta*$.

Proof   Remarks 5.9(b) and 5.11(a) give

$$*\Delta = *d\delta + *\delta d = \delta d* + d\delta* = \Delta*$$

and therefore the claim. ∎

(b) $d\Delta = \Delta d = d\delta d$ and $\delta\Delta = \Delta\delta = \delta d\delta$.

Proof   From $d^2 = 0$ we get

$$d\Delta = dd\delta + d\delta d = d\delta d = d\delta d + \delta dd = \Delta d \, .$$

The second claim follows analogously. ∎

(c) Suppose M is open in $\mathbb{R}^m$ and $(x^1, \ldots, x^m)$ are Euclidean coordinates. Then

$$\Delta\left(\sum\nolimits_{(j)\in\mathbb{J}_r} a_{(j)}\, dx^{(j)}\right) = \sum\nolimits_{(j)\in\mathbb{J}_r} \Delta a_{(j)}\, dx^{(j)}$$

for $1 \leq r \leq m$.

Proof Because of the linearity, it suffices to show the statement for $\alpha := a\, dx^{(j)}$ with $(j) \in \mathbb{J}_r$. Using Example 5.10(e), we find

$$d\delta\alpha = d\left(\sum_{k=1}^{r} (-1)^{k-1} \partial_{j_k} a\, dx^{j_1} \wedge \cdots \wedge \widehat{dx^{j_k}} \wedge \cdots \wedge dx^{j_r}\right)$$

$$= \sum_{k=1}^{r} (-1)^{k-1} \sum_{\ell=1}^{m} \partial_\ell \partial_{j_k} a\, dx^\ell \wedge dx^{j_1} \wedge \cdots \wedge \widehat{dx^{j_k}} \wedge \cdots \wedge dx^{j_r}$$

$$= \sum_{k=1}^{r} \partial_{j_k}^2 a\, dx^{(j)} + \sum_{k=1}^{r} (-1)^{k-1} \sum_{\substack{\ell=1 \\ \ell \notin \{j_1,\ldots,j_r\}}}^{m} \partial_\ell \partial_{j_k} a\, dx^\ell \wedge dx^{j_1} \wedge \cdots \wedge \widehat{dx^{j_k}} \wedge \cdots \wedge dx^{j_r} \,.$$

Analogously, we get

$$\delta d\alpha = \delta \sum_{\substack{\ell=1 \\ \ell \notin \{j_1,\ldots,j_r\}}}^{m} \partial_\ell a\, dx^\ell \wedge dx^{(j)} = \sum_{\substack{\ell=1 \\ \ell \notin \{j_1,\ldots,j_r\}}}^{m} \partial_\ell^2 a\, dx^{(j)}$$

$$- \sum_{\substack{\ell=1 \\ \ell \notin \{j_1,\ldots,j_r\}}}^{m} \sum_{k=1}^{r} (-1)^{k-1} \partial_{j_k} \partial_\ell a\, dx^\ell \wedge dx^{j_1} \wedge \cdots \wedge \widehat{dx^{j_k}} \wedge \cdots \wedge dx^{j_r} \,.$$

This implies[8]

$$\Delta_M \alpha = (d\delta + \delta d)\alpha = \left(\sum_k \partial_k^2 a\right) dx^{(j)} = (\Delta a)\, dx^{(j)}$$

and therefore the claim. ∎

(d) (regularity) Clearly Δ_M is an $\mathbb{R}$-linear map from $\Omega^r_{(k)}(M)$ to $\Omega^r_{(k-2)}(M)$ when $0 \leq r \leq m$ and $k \in \mathbb{N}$ with $k \geq 2$. In this case, it suffices to assume that M is a C^{k+2} manifold. ∎

Finally, we define the **Laplace operator for vector fields**, namely, $\vec{\Delta}$, by

$$\vec{\Delta} := \vec{\Delta}_M := \Theta_M^{-1} \circ \dot{\Delta}_M \circ \Theta_M : \mathcal{V}(M) \to \mathcal{V}(M)$$

and therefore by the commutativity of the diagram

$$
\begin{array}{ccc}
\mathcal{V}(M) & \xrightarrow{\ \vec{\Delta}\ } & \mathcal{V}(M) \\[2pt]
\Theta \downarrow & & \downarrow \Theta \\[2pt]
\Omega^1(M) & \xrightarrow{\ \Delta\ } & \Omega^1(M)\,.
\end{array}
$$

---

[8]We chose the sign in the definition of δ to be $(-1)^{m(r+1)}$ so that this formula would take this form. For the sign convention typically used in geometry, the formula is $(d\delta + \delta d)\alpha = -(\Delta a)\, dx^{(j)}$.

6.24 Remarks (a) Proposition 6.22 and (6.4) give $\vec{\Delta} = \text{grad div} + \Theta^{-1}\delta d\Theta$.

(b) Suppose M is open in $\mathbb{R}^m$, and $(x^1, \ldots, x^m)$ are Euclidean coordinates on M. Then

$$\vec{\Delta}\left(\sum_j v^j \frac{\partial}{\partial x^j}\right) = \sum_j \Delta v^j \frac{\partial}{\partial x^j} \ .$$

If we identify as usual the vector field $v = \sum_j v^j \, \partial/\partial x^j$ with $(v^1, \ldots, v^m)$, then $\vec{\Delta}v$ means that the Laplace operator can be applied componentwise

$$\vec{\Delta}v = (\Delta v^1, \ldots, \Delta v^m) \ .$$

In this case, we usually write Δ, not $\vec{\Delta}$.

Proof This follows from Example 6.2(a) and Remark 6.23(c). ∎

(c) (regularity) Let $k \in \mathbb{N}$. Then $\vec{\Delta}$ maps the $\mathbb{R}$-vector space $\mathcal{V}^{k+2}(M)$ linearly into $\mathcal{V}^k(M)$. So here it suffices to assume that M is a C^{k+2} manifold. ∎

The vector product and the curl

In this last section, we derive the most important properties of the curl operator.

Let $(M, (\cdot\,|\,\cdot)_M)$ be a three-dimensional oriented Riemannian[9] manifold with volume element ω_M.

On $\mathcal{V}(M)$, we define the **vector product** or **cross product**,

$$\times : \mathcal{V}(M) \times \mathcal{V}(M) \to \mathcal{V}(M) \ , \quad (v, w) \mapsto v \times w \ , \tag{6.36}$$

by

$$v \times w := \Theta_M^{-1} \omega_M(v, w, \cdot) \ . \tag{6.37}$$

Clearly this map is well defined.

6.25 Remarks (a) Suppose $(M, (\cdot\,|\,\cdot)_M) = (\mathbb{R}^3, (\cdot\,|\,\cdot))$. Then in the case of a constant vector field, (6.37) agrees with the definition of Section VIII.2.

(b) The vector product is bilinear, alternating (skew symmetric), and satisfies

$$(u \,|\, v \times w)_M = \omega_M(u, v, w) \quad \text{for } u, v, w \in \mathcal{V}(M) \ . \tag{6.38}$$

For $p \in M$, the vector product $(v \times w)(p)$ is orthogonal to $v(p)$ and $w(p)$ with respect to the inner product $(\cdot\,|\,\cdot)_M(p)$ of T_pM. Letting $|v|_M := \sqrt{(v\,|\,v)_M}$, we have

$$|v \times w|_M = \sqrt{|v|_M^2 \, |w|_M^2 - (v\,|\,w)_M^2} = |v|_M \, |w|_M \sin\varphi \ ,$$

where $\varphi(p) \in [0, \pi]$ is the angle between the vectors $v(p)$ and $w(p)$ for $p \in M$ and $v, w \in \mathcal{V}(M)$.

---

[9]For simplicity, we will restrict here to Riemannian metrics, as they are the most important in applications.

The vector product satisfies the **Grassmann identity**

$$v_1 \times (v_2 \times v_3) = (v_1 \,|\, v_3)_M v_2 - (v_1 \,|\, v_2)_M v_3 \, ,$$

the **Jacobi identity**

$$v_1 \times (v_2 \times v_3) + v_2 \times (v_3 \times v_1) + v_3 \times (v_1 \times v_2) = 0$$

and the relation

$$(v_1 \times v_2) \times (v_3 \times v_4) = \omega_M(v_1, v_2, v_4) v_3 - \omega_M(v_1, v_2, v_3) v_4$$

for $v_1, v_2, v_3, v_4 \in \mathcal{V}(M)$. In particular, $(\mathcal{V}(M), \times)$ is a Lie algebra.

Proof   All of these reduce easily to pointwise statements already proved in Exercise 2.3. ∎

(c) Suppose $\big((x^1, x^2, x^3), U\big)$ are positive orthonormal coordinates[10] on M. Then the cross product of vector fields $v = \sum_j v^j \, \partial/\partial x^j$ and $w = \sum_j w^j \, \partial/\partial x^j$ takes the form

$$v \times w = (v^2 w^3 - v^3 w^2) \frac{\partial}{\partial x^1} + (v^3 w^1 - v^1 w^3) \frac{\partial}{\partial x^2} + (v^1 w^2 - v^2 w^1) \frac{\partial}{\partial x^3} \, .$$

Proof   Exercise 2.3. ∎

(d) (regularity)   For $k \in \mathbb{N}$, the statements above remain true for C^k vector fields, and it suffices to assume that M is a C^{k+1} manifold. ∎

The next theorem shows how the vector product is related to the exterior product of 1-forms.

6.26 Proposition   For $v, w \in \mathcal{V}(M)$, we have $v \times w = \Theta^{-1} * (\Theta v \wedge \Theta w)$, that is, the diagram

$$
\begin{array}{ccc}
& \Theta \times \Theta & \\
\mathcal{V}(M) \times \mathcal{V}(M) & \longrightarrow & \Omega^1(M) \times \Omega^1(M) \\[2mm]
\Big\downarrow{\scriptstyle \times} & & \Big\downarrow{\scriptstyle \wedge} \\[2mm]
\mathcal{V}(M) \quad\longleftarrow\quad \Omega^1(M) & \longleftarrow & \Omega^2(M) \\
\quad\quad \Theta^{-1} & * &
\end{array}
$$

commutes.

Proof   It suffices to prove the equality locally, where we can choose positive orthonormal coordinates $\big((x^1, x^2, x^3), U\big)$. If (v^1, v^2, v^3) and (w^1, w^2, w^3) are the components of vector fields $v, w \in \mathcal{V}(M)$, then it follows from Remark 6.1(e) that

$$
\begin{aligned}
\Theta v \wedge \Theta w &= \sum_j v^j \, dx^j \wedge \sum_k w^k \, dx^k \\
&= (v^2 w^3 - v^3 w^2) \, dx^2 \wedge dx^3 + (v^3 w^1 - v^1 w^3) \, dx^3 \wedge dx^1 \\
&\qquad\qquad + (v^1 w^2 - v^2 w^1) \, dx^1 \wedge dx^2 \, .
\end{aligned}
$$

---

[10]That is, $(\partial/\partial x^1, \partial/\partial x^2, \partial/\partial x^3)$ is a positive orthonormal frame.

From the proof of Example 5.7(d), we know that

$$*(dx^2 \wedge dx^3) = dx^1 , \quad *(dx^3 \wedge dx^1) = dx^2 , \quad *(dx^1 \wedge dx^2) = dx^3 . \qquad (6.39)$$

Now the claim follows from Remarks 6.1(e) and 6.25(c). ∎

We next derive a representation of the curl operator.

6.27 Proposition   *The diagram*

$$
\begin{array}{ccccc}
\mathcal{V}(M) & \xrightarrow{\;\Theta\;} & \Omega^1(M) & \xrightarrow{\;d\;} & \Omega^2(M) \\
& \text{curl} \searrow & & \swarrow * & \\
& & \mathcal{V}(M) \xleftarrow{\;\Theta^{-1}\;} \Omega^1(M) & &
\end{array}
$$

commutes, that is, curl $= \Theta^{-1}{*}d\Theta$.

Proof   It again suffices to prove the equality locally in positive orthonormal coordinates $\big((x^1, x^2, x^3), U\big)$. Then for $v = \sum_{j=1}^{3} v^j\, \partial/\partial x^j$, we find using Remark 6.1(e) that

$$
\begin{aligned}
d(\Theta v) = d\Big(\sum_j v^j\, dx^j\Big) &= \sum_{j,k} \frac{\partial v^j}{\partial x^k}\, dx^k \wedge dx^j \\
&= \Big(\frac{\partial v^3}{\partial x^2} - \frac{\partial v^2}{\partial x^3}\Big) dx^2 \wedge dx^3 + \Big(\frac{\partial v^1}{\partial x^3} - \frac{\partial v^3}{\partial x^1}\Big) dx^3 \wedge dx^1 \\
&\quad + \Big(\frac{\partial v^2}{\partial x^1} - \frac{\partial v^1}{\partial x^2}\Big) dx^1 \wedge dx^2 .
\end{aligned}
$$

Then the claim follows from (6.39) and Remarks 6.1(e) and 6.12(b). ∎

We are now ready to deduce several important differential identities involving three-dimensional vector fields.

6.28 Proposition   *For $f \in \mathcal{E}(M)$ and $v, w \in \mathcal{V}(M)$,*

   (i) $\operatorname{div}(v \times w) = (\operatorname{curl} v \,|\, w)_M - (v \,|\, \operatorname{curl} w)_M$;

   (ii) $\operatorname{curl}(fv) = f \operatorname{curl} v + \operatorname{grad} f \times v$;

   (iii) $\operatorname{curl}(v \times w) = (\operatorname{div} w)v - (\operatorname{div} v)w - [v, w]$;

   (iv) $\operatorname{curl}(\operatorname{curl} v) = \operatorname{grad} \operatorname{div} v - \vec{\Delta} v$.

Proof   (i) Putting $m = 3$ in Remark 5.9(d), we obtain

$$*\delta \alpha = (-1)^{m(r+1)} {*}{*}d{*}\alpha = d{*}\alpha \quad \text{for } \alpha \in \Omega^2(M) .$$

Now we use Propositions 6.22 and 6.26 to deduce

$$\operatorname{div}(v \times w) = \delta\Theta\big(\Theta^{-1}*(\Theta v \wedge \Theta w)\big) = \delta*(\Theta v \wedge \Theta w)$$
$$= *d(\Theta v \wedge \Theta w) = *(d\Theta v \wedge \Theta w - \Theta v \wedge d\Theta w) \ .$$

From Proposition 6.27, it follows that $\Theta \operatorname{curl} = *d\Theta$. Now Remark 2.19(d) implies that $d\Theta = *\Theta \operatorname{curl}$, because $m = 3$ and $r = 2$. Hence we get

$$\operatorname{div}(v \times w) = *\big((*\Theta \operatorname{curl} v) \wedge \Theta w - \Theta v \wedge *\Theta \operatorname{curl} w\big)$$
$$= *(\Theta w \wedge *\Theta \operatorname{curl} v - \Theta v \wedge *\Theta \operatorname{curl} w) \ ,$$

where we have used $*\Theta \operatorname{curl} v \in \Omega^2(M)$. Now (2.22), with $r = 1$, and (2.13) give

$$\operatorname{div}(v \times w) = *\big[(w \,|\, \operatorname{curl} v)_M - (v \,|\, \operatorname{curl} w)\big]\omega_M \ .$$

The claim now follows from $*\omega_M = 1$.

(ii) Proposition 6.27 gives

$$\operatorname{curl}(fv) = \Theta^{-1}*d\Theta(fv) = \Theta^{-1}*d(f\Theta v)$$
$$= \Theta^{-1}*(df \wedge \Theta v + f d\Theta v)$$
$$= \Theta^{-1}*(\Theta \operatorname{grad} f \wedge \Theta v) + f\Theta^{-1}*d\Theta v$$
$$= \operatorname{grad} f \times v + f \operatorname{curl} v \ .$$

Here we have also made use of Proposition 6.26 and properties of d.

(iii) It suffices to prove the statement locally. We can use positive orthonormal coordinates. Then the claim follows from the local representations of Remarks 6.12(b) and 6.25(c) and from Proposition 6.20 after a simple calculation, which we leave to you.

(iv) It follows from Proposition 6.27 and the definition of δ that

$$\operatorname{curl} \operatorname{curl} v = \Theta^{-1}*d\Theta\Theta^{-1}*d\Theta v = \Theta^{-1}*d*d\Theta v$$
$$= (-1)^{3(2+1)}\Theta^{-1}\delta d\Theta v = -\Theta^{-1}\delta d\Theta v \ .$$

Now the claim follows from Remark 6.24(a). ∎

To demonstrate the power of the new calculus, we proved part (i) using the properties of the codifferential and the star operator. Of course, we could also have worked in the orthonormal coordinates of a positive chart. In other words, we can assume that M is open in $\mathbb{R}^3$ and $(\cdot \,|\, \cdot)_M$ is the standard metric $(\cdot \,|\, \cdot)$. Using the (formal) nabla operator in (6.38), we obtain

$$\nabla \cdot (v \times w) = \det[\nabla, v, w]$$
$$= \partial_1(v^2w^3 - v^3w^2) + \partial_2(v^3w^1 - v^1w^3) + \partial_3(v^1w^2 - v^2w^1)$$

by expanding the (formal) determinant in its first row. By using the product rule, we see easily that the last row agrees with the expression $w \cdot \operatorname{curl} v - v \cdot \operatorname{curl} w$, as claimed in (i).

However, the formal calculus with the nabla operator must be used with caution. For example, if we formally calculate $\mathrm{curl}(v \times w) = \nabla \times (v \times w)$ using the Grassmann identity, we find the *false* statement

$$\nabla \times (v \times w) = (\nabla \cdot w)v - (\nabla \cdot v)w .$$

Where was the mistake?

Exercises

1 Find the representation of the Laplace–Beltrami operator with respect to
  (i) the cylindrical coordinates $(0, 2\pi) \times \mathbb{R} \to \mathbb{R}^3$, $(\varphi, z) \mapsto (\cos\varphi, \sin\varphi, z)$;
  (ii) the parametrization

$$(0, 2\pi)^2 \to \mathbb{R}^3 , \quad (\alpha, \beta) \mapsto \big((2 + \cos\alpha)\cos\beta, (2 + \cos\alpha)\sin\beta, \sin\alpha\big)$$

  of the 2-torus $\mathsf{T}^2_{2.1}$ of Example VII.9.11(f);
  (iii) the parametrization $X \to \mathbb{R}^3$, $x \mapsto (x, f(x))$ of the graph of $f \in \mathcal{E}(X)$, when X is open in $\mathbb{R}^2$.

2 Let (M_j, g_j) for $j = 1, 2$ be Riemannian manifolds with $\partial M_1 = \emptyset$, and let π_j denote the canonical projection $M_1 \times M_2 \to M_j$. Show that

$$\Delta_{M_1 \times M_2} = \pi_1^* \Delta_{M_1} + \pi_2^* \Delta_{M_2} .$$

3 Suppose M and N are Riemannian manifolds and $f : M \to N$ is an isometric diffeomorphism. Then for $0 \le r \le m$, show that the diagram

$$
\begin{array}{ccc}
\Omega^r(M) & \xrightarrow{\ \Delta_M\ } & \Omega^r(M) \\[4pt]
f^* \uparrow & & \uparrow f^* \\[4pt]
\Omega^r(N) & \xrightarrow{\ \Delta_N\ } & \Omega^r(N)
\end{array}
$$

commutes.

4 Let (M, g) be a pseudo-Riemannian manifold. Show the commutativity of the diagram

$$
\begin{array}{ccc}
\Omega^1(M) & \xrightarrow{\quad * \quad} & \Omega^{m-1}(M) \\
 & \Theta \searrow \quad \nearrow \lrcorner & \\
 & \mathcal{V}(M) &
\end{array}
$$

and derive the relations
  (i) $\mathrm{div} = *d*\Theta$;
  (ii) $\mathrm{curl} = \Theta^{-1}*d\Theta$ $(m = 3)$;

(iii) $\Delta_M = *d*d$,

where Δ_M is the Laplace–Beltrami operator of M.

5 Let Ω be open in $\mathbb{R}^3$. For $E, B, j \in C^\infty(\mathbb{R} \times \Omega, \mathbb{R}^3)$, $\rho \in C^\infty(\mathbb{R} \times \Omega, \mathbb{R})$, and $c > 0$, set

$$F := \Theta_e E \wedge (c\,dt) + *(\Theta_e B \wedge (c\,dt)) , \quad J := \Theta_e j - \rho\,dt \in \Omega(\mathbb{R}^4_{1,3}) ,$$

where E, B, and j are seen as time-dependent vector fields, ρ is seen as a time-dependent function on Ω, and $\Theta_e : \mathcal{V}(\mathbb{R}^3) \to \Omega^1(\mathbb{R}^3)$ denotes the (Euclidean) Riesz isomorphism. Also let dt be the first standard basis vector in $\Omega^1(\mathbb{R}^4_{1,3})$. Now show these facts:

(a) The statements

   (i) $dF = 0$;

   (ii) $\partial B/\partial t + c\operatorname{curl} E = 0$ and $\operatorname{div} B = 0$

are equivalent. (These are the homogeneous Maxwell's equations.) That is, the 2-form F is closed if and only if the vector fields E and B satisfy these two of Maxwell's equations.

(b) The statements

   (i) $dF = 4\pi J$;

   (ii) $\partial E/\partial t - c\operatorname{curl} B = 4\pi j$ and $\operatorname{div} E = 4\pi \rho$

are equivalent. (These are the Maxwell's equations with sources.)

(c) The statments

   (i) $\Delta_{\mathbb{R}^4_{1,3}} F = 0$;

   (ii) $\partial B/\partial t + c\operatorname{curl} E = 0$, $\partial E/\partial t - c\operatorname{curl} B = 0$, $\operatorname{div} E = 0$, $\operatorname{div} B = 0$

are equivalent. Therefore the 2-form F is harmonic if and only if the vector fields E and B satisfy the homogeneous Maxwell's equations.

(d) If $dF = 0$, then j and ρ satisfy the **continuity equation**

$$\partial \rho/\partial t + \operatorname{div} j = 0 ,$$

which can also be written $\operatorname{grad}_{\mathbb{R}^4_{1,3}} J = 0$.

(e) These two statements are equivalent:

   (i) F is exact;

   (ii) There are an $A \in C^\infty(\mathbb{R} \times \Omega, \mathbb{R})$, a **vector potential**, and $\Phi \in C^\infty(\mathbb{R} \times \Omega, \mathbb{R})$, a **scalar potential**, with
   $$\operatorname{curl} A = B \quad \text{and} \quad -\partial A/\partial t - \operatorname{grad} \Phi = E .$$

6 Suppose X is open in $\mathbb{R}^3$ and contractible. Also suppose $f, g \in \mathcal{E}(X)$. Show:

   (i) There is a $v \in \mathcal{V}(X)$ such that $\operatorname{grad} f \times \operatorname{grad} g = \operatorname{curl} v$.

   (ii) If $f(x) \neq 0$ for $x \in X$, then there is an $h \in \mathcal{E}(X)$ with $(\operatorname{grad} f)/f = \operatorname{grad} h$.

7 Verify that

$$\vec{\Delta}_M(f \operatorname{grad} f) = \operatorname{grad} \operatorname{div}(f \operatorname{grad} f) = \Delta_M f \operatorname{grad} f + \operatorname{grad} |\operatorname{grad} f|_M^2 + f \operatorname{grad} \Delta_M f$$

for $f \in \mathcal{E}(M)$, where $|v|_M^2 := (v \mid v)_M$ for $v \in \mathcal{V}(M)$.

8 Show that $\alpha \in \Omega^1(M)$ and $v, w \in \mathcal{V}(M)$ satisfy

$$d\alpha(v, w) = \mathcal{L}_v \langle \alpha, w \rangle - \mathcal{L}_w \langle \alpha, v \rangle - \langle \alpha, [v, w] \rangle \ .$$

9 Let M and N be m-dimensional manifolds, and let $\varphi \in \mathrm{Diff}(M, N)$. Show that

$$\varphi_*[v, w] = [\varphi_* v, \varphi_* w] \quad \text{for } v, w \in \mathcal{V}(M) \ .$$

10 Let $T^2 := S^1 \times S^2 \subset \mathbb{R}^4$, and let $\alpha, \beta \in \Omega^1(T^2)$ with

$$\alpha := -x^2 \, dx^1 + x^1 \, dx^2 \ , \quad \beta := -x^4 \, dx^3 + x^3 \, dx^4 \ .$$

Show that $\Delta\alpha = \Delta\beta = 0$.

11 Show that for $H \in \mathcal{E}(\mathbb{R}^{2m})$ the vector field sgrad $H \in \mathcal{V}(\mathbb{R}^{2m})$ is divergence free.

Chapter XII

Integration on manifolds

In the first two chapters of this book, we developed the basics of measure and integration theory, and, in the third, we deepened our knowledge of manifolds and introduced the theory of differential forms. We are now ready to extend integration theory to manifolds, which means we will be able to integrate over "curved spaces".

In Section 1, we introduce the Riemann–Lebesgue measure on manifolds. Its construction is based entirely on the properties of the Lebesgue measure, which we "lift up" from $\mathbb{R}^m$ to the manifold using local charts. The transformation theorem plays a principal role, because it guarantees that the construction is independent of the local coordinates. We show that Riemann–Lebesgue volume measure is a complete Radon measure, which makes available the entire integration theory developed in the second chapter. As a first application of the general theory, we calculate the volumes of several manifolds.

In Section 2, we generalize the theory of line integrals which tells how to integrate 1-forms over curves and therefore over 1-dimensional manifolds. We now show how to integrate m-forms over m-dimensional manifolds. To make the calculus more powerful, we extend the transformation theorem and Fubini's theorem so that they can be used with integrals of differential forms. Then we will be ready to treat more complicated integration problems, as we demonstrate in a series of examples.

We also give physical and geometric interpretations to integrals of differential forms, and we present the basic idea of the flux of a vector field. As an application, we prove the transport theorem and derive some of its consequences.

The high point of the differential and integral calculus of manifolds is undoubtedly Stokes's theorem, to which the last section, Section 3, is devoted. We prove a version for manifolds with singularities; this version will suffice for most cases seen in practice. Of course, we show several classical application of Stokes's theorem and also give a first glimpse at its topological consequences. However, in the framework of this introduction, we must forgo any deeper explorations. It is the goal of this work to give you the basics for your further progress into this fascinating branch of mathematics. Should you choose to take further courses in analysis or delve deeper into its literature, you will learn a great number of applications and generalizations of the theory developed here.

1 Volume measure

In Section VIII.1, we learned how to calculate the length of a curve. We also know how to find the area under graphs and the volume of simple shapes. Now we turn to the problem of determining the area of curve surfaces and the content of general manifolds.

In this section, we introduce the Riemann–Lebesgue volume measure of a pseudo-Riemannian manifold and show that it is a complete massive Radon measure. Then the entire theory of integration developed in Chapter X will also be available for manifolds. With the help of local representations, we can explicitly calculate integrals on manifolds in many cases, as our examples will show.

In the entire section

- M is an m-dimensional manifold with $m \in \mathbb{N}^\times$.

The Lebesgue σ-algebra of M

Because a manifold locally "looks" like an open subset of $\overline{\mathbb{H}}^m$, it is clear that measurability can be "lifted up" from $\overline{\mathbb{H}}^m$ to M using local charts.

A subset A of M is said to be (**Lebesgue**) **measurable** if around every $p \in A$ there is a chart (φ, U) such that $\varphi(A \cap U)$ belongs to $\mathcal{L}(m)$, that is, $\varphi(A \cap U)$ is λ_m-measurable. We set

$$\mathcal{L}_M := \{ A \subset M \ ; \ A \text{ is measurable} \} .$$

The following remarks shows that this definition is meaningful.

1.1 Remarks   (a) The definition is coordinate-independent.

Proof   Suppose $A \subset M$ and (φ_p, U_p) is a chart around $p \in A$ with $\varphi_p(A \cap U_p) \in \mathcal{L}(m)$. Further let (ψ, V) be a chart of M, and let $q \in A \cap V$. Because the set $\varphi_p(U_q \cap V)$ is open in $\overline{\mathbb{H}}^m$, it is λ_m-measurable and therefore so is

$$\varphi_q(A \cap V \cap U_q) = \varphi_q(A \cap U_q) \cap \varphi_q(V \cap U_q) .$$

By Corollary IX.5.13, it now follows from

$$\psi(A \cap V \cap U_q) = \psi \circ \varphi_q^{-1}\big(\varphi_q(A \cap V \cap U_q)\big)$$

that $\psi(A \cap V \cap U_q)$ belongs to $\mathcal{L}(m)$. This holds for every $q \in A \cap V$, and according to Remark IX.5.14(c), measurability is a local property. Hence $\psi(A \cap V) \in \mathcal{L}(m)$, and we are done. ∎

(b) If M is open in $\mathbb{R}^m$, then $\mathcal{L}_M = \mathcal{L}(m) \,|\, M$. Therefore the notation introduced here is consistent with that of Section X.5.

Proof   This follows by using the trivial chart (id, M). ∎

1.2 Proposition $\mathcal{L}_M$ is a σ-algebra over M, the **Lebesgue σ-algebra of M**. It contains the Borel σ-algebra $\mathcal{B}(M)$.

Proof   Let (φ, U) be a chart of M. If $A \in \mathcal{L}_M$, then $\varphi(A \cap U)$ belongs to $\mathcal{L}(m)$. Because $\mathcal{L}(m)$ is a σ-algebra, we have $\varphi(A^c \cap U) = \varphi(U)\backslash\varphi(A \cap U) \in \mathcal{L}(m)$. Because this is true for every chart, it follows that $A^c \in \mathcal{L}_M$.

If (A_j) is a sequence in $\mathcal{L}_M$, then we analogously find

$$\varphi\Big(\Big(\bigcup_j A_j\Big) \cap U\Big) = \varphi\Big(\bigcup_j (A_j \cap U)\Big) = \bigcup_j \varphi(A_j \cap U) \in \mathcal{L}(m) ,$$

which implies $\bigcup_j A_j \in \mathcal{L}_M$.

Finally, it is obvious that M belongs to $\mathcal{L}_M$. Thus we have shown that $\mathcal{L}_M$ is a σ-algebra.

If O is open in M, then $\varphi(O \cap U)$ is open in $\overline{\mathbb{H}}^m$ and therefore belongs to $\mathcal{L}(m)$. Therefore O belongs to $\mathcal{L}_M$, which implies $\mathcal{L}_M \supset \mathcal{B}(M)$. ∎

The definition of the volume measure

Suppose now g is a pseudo-Riemannian metric on M and (φ, U) is a chart of M with $\varphi = (x^1, \dots, x^m)$. Then the Gram determinant $G = \det[g_{jk}]$ with $g_{jk} = g(\partial/\partial x^j, \partial/\partial x^k)$ is well defined, and $\sqrt{|G|} \in \mathcal{E}(U)$. For $A \in \mathcal{L}_M$ with $A \subset U$, we set

$$\mathrm{vol}_{g,U}(A) := \int_{\varphi(A)} \varphi_* \sqrt{|G|} \, d\lambda_m = \int_{\varphi(A)} \varphi_* \sqrt{|G|} \, dx . \tag{1.1}$$

1.3 Lemma   For $A \in \mathcal{L}_M$, the volume $\mathrm{vol}_{g,U}(A)$ is independent of the chart (φ, U) when $A \subset U$.

Proof   Suppose (ψ, V) is another chart with $A \subset V$ and $\psi = (y^1, \dots, y^m)$. We can assume $V = U$. Now we regard U as an oriented manifold with positive atlas $\{(\varphi, U)\}$. Then, according to Remark XI.5.4(g),

$$\omega_U := \sqrt{|G|} \, dx^1 \wedge \cdots \wedge dx^m \in \Omega^m(U)$$

is the volume element of U. Also

$$\omega_U = \pm\sqrt{|G|} \, dy^1 \wedge \cdots \wedge dy^m ,$$

where the positive sign is chosen if ψ positively oriented and the negative is chosen otherwise. Because $f := \psi \circ \varphi^{-1} \in \mathrm{Diff}\big(\varphi(U), \psi(U)\big)$ and $\varphi = f^{-1} \circ \psi$, it follows from Example XI.3.4(c) that

$$
\begin{aligned}
\varphi_* \sqrt{|G|}\, dx^1 \wedge \cdots \wedge dx^m &= \varphi_* \omega_U = (f^{-1})_* \psi_* \omega_U = f^* \psi_* \omega_U \\
&= \pm f^* \big(\psi_* \sqrt{|G|}\, dy^1 \wedge \cdots \wedge dy^m \big) \\
&= \pm f^* \big(\psi_* \sqrt{|G|} \big) \det(\partial f)\, dx^1 \wedge \cdots \wedge dx^m \\
&= f^* \big(\psi_* \sqrt{|G|} \big) \, |\det(\partial f)|\, dx^1 \wedge \cdots \wedge dx^m ,
\end{aligned}
$$

because f is orientation-preserving if and only if ψ is positively oriented, and f is orientation-reversing otherwise. Therefore

$$
\varphi_* \sqrt{|G|} = \big(\psi_* \sqrt{|G|} \big) \circ f \, |\det(\partial f)| \ .
$$

Because $\varphi(U \cap \partial M) = \varphi(U) \cap \partial \mathbb{H}^m$ is a λ_m-null set and $\varphi(U \backslash \partial M) = \varphi(U) \backslash \partial \mathbb{H}^m$ is open in $\mathbb{R}^m$, and because $f\big(\varphi(U)\big) = \psi(U)$, it follows from the transformation theorem in the version of Corollary X.8.5 that

$$
\int_{\varphi(U)} \varphi_* \sqrt{|G|}\, dx = \int_{\psi(U)} \psi_* \sqrt{|G|}\, dy \ .
$$

This proves the claim. ∎

By Remark XI.1.21(b) and Proposition IX.1.8, M is a Lindelöf space. Therefore M has a countable atlas $\mathfrak{A} := \big\{ (\varphi_j, U_j) \; ; \; j \in \mathbb{N} \big\}$. For $A \in \mathcal{L}_M$, we set

$$
A_0 := A \cap U_0 \quad \text{and} \quad A_{n+1} := (A \cap U_{n+1}) \backslash \bigcup_{k=0}^{n} A_k \quad \text{for } n \in \mathbb{N} \ .
$$

Then (A_j) is a disjoint sequence in $\mathcal{L}_M$ with $A_j \subset U_j$, and $A = \bigcup_j A_j$. Therefore

$$
\mathrm{vol}_g(A) := \sum_{j=0}^{\infty} \mathrm{vol}_{g,U_j}(A_j) \tag{1.2}
$$

is a well-defined element of $\overline{\mathbb{R}}^+$.

1.4 Lemma *The definition (1.2) is independent of the special choice of an atlas.*

Proof Let $\widetilde{\mathfrak{A}} := \big\{ (\widetilde{\varphi}_j, \widetilde{U}_j) \; ; \; j \in \mathbb{N} \big\}$ be another countable atlas, and let $\widetilde{A}_j$ be defined in analogy to A_j under the application of $\widetilde{U}_j$ rather than U_j. Now vol_{g,U_j} and $\mathrm{vol}_{g,\widetilde{U}_k}$ are measures on U_j and $\widetilde{U}_k$, respectively, and they are therefore σ-additive. Then it follows from $\bigcup_j A_j = \bigcup_k \widetilde{A}_k = A$ that

$$
\begin{aligned}
\mathrm{vol}_{g,U_j}(A_j) &= \mathrm{vol}_{g,U_j}(A_j \cap A) = \mathrm{vol}_{g,U_j}\Big(A_j \cap \bigcup_k \widetilde{A}_k \Big) \\
&= \mathrm{vol}_{g,U_j}\Big(\bigcup_k (A_j \cap \widetilde{A}_k) \Big) = \sum_k \mathrm{vol}_{g,U_j}(A_j \cap \widetilde{A}_k) \ .
\end{aligned}
$$

By exchanging the roles of $\mathfrak{A}$ and $\widetilde{\mathfrak{A}}$, we find analogously that

$$\text{vol}_{g,\tilde{U}U_k}(\tilde{A}_k) = \sum_j \text{vol}_{g,\tilde{U}U_k}(\tilde{A}_k \cap A_j) \ .$$

Because $A_j \cap \tilde{A}_k \subset U_j \cap \tilde{U}_k$, Lemma 1.3 results in

$$\text{vol}_{g,U_j}(A_j \cap \tilde{A}_k) = \text{vol}_{g,\tilde{U}U_k}(\tilde{A}_k \cap A_j) \ .$$

Then from Remark X.3.6(b), we obtain

$$\sum_j \text{vol}_{g,U_j}(A_j) = \sum_j \sum_k \text{vol}_{g,U_j}(A_j \cap \tilde{A}_k) = \sum_j \sum_k \text{vol}_{g,\tilde{U}U_k}(\tilde{A}_k \cap A_j)$$
$$= \sum_k \sum_j \text{vol}_{g,\tilde{U}U_k}(\tilde{A}_k \cap A_j) = \sum_k \text{vol}_{g,\tilde{U}U_k}(\tilde{A}_k) \ ,$$

as desired. ∎

Clearly, $\text{vol}_g(A)$ should represent the "volume" of $A \subset M$, measured with respect to the measure tensor g. We will now think about how this is to be understood.

Definition (1.2) shows that an arbitrary measurable subset A of M is decomposed into countably many pairwise disjoint parts A_j and that the "volume" of A is just the "sum" of the volumes of these parts. The definition requires that every A_j is contained in a chart of an atlas. Therefore it suffices to understand how $\text{vol}_g(A)$ is to be interpreted when A is contained in the domain U of a chart (φ, U).

So let $\bar{x} \in \varphi(U)$ and $\bar{p} := \varphi^{-1}(\bar{x})$, and let $\ell^1, \ldots, \ell^m$ be positive numbers such that

$$Q := \prod_{j=1}^m [\bar{x}^j, \bar{x}^j + \ell^j] \ ,$$

the rectangle with "lower left corner" $\bar{x}$ and volume $\lambda_m(Q) = \ell^1 \cdots \cdot \ell^m$, still lies entirely in $\varphi(U)$. If the side lengths ℓ^j are sufficiently small, the parallelepiped $\hat{Q} := T_{\bar{x}}\varphi^{-1}(Q)$ spanned by the vectors

$$\ell^1 \frac{\partial}{\partial x^1}\Big|_{\bar{p}}, \ldots, \ell^m \frac{\partial}{\partial x^m}\Big|_{\bar{p}}$$

in $T_{\bar{p}}M$ will approximate well the image $A := \varphi^{-1}(Q)$ of Q in M.

We can assume that (φ, U) is a positive chart of U. Then it follows from Remark XI.2.12(c) that the volume of $\hat{Q}$ satisfies

$$\omega(\bar{p})\Big(\ell^1 \frac{\partial}{\partial x^1}\Big|_{\bar{p}}, \ldots, \ell^m \frac{\partial}{\partial x^m}\Big|_{\bar{p}}\Big) = \omega(\bar{p})\Big(\frac{\partial}{\partial x^1}\Big|_{\bar{p}}, \ldots, \frac{\partial}{\partial x^m}\Big|_{\bar{p}}\Big)\ell^1 \cdots \cdot \ell^m$$
$$= \sqrt{|G|}\,(\bar{p})\lambda_m(Q) = \varphi_* \sqrt{|G|}\,(\bar{x})\lambda_m(Q) \ .$$

Now we decompose Q into finitely many rectangles Q_j (with edges along the Cartesian axes) that intersect at most along their common sides.

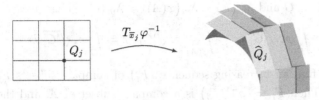

We denote by $\bar{x}_j$ the lower left corner of Q_j and set $\bar{p}_j := \varphi^{-1}(\bar{x}_j)$. We let $\widehat{Q}_j$ be the image of Q_j in $T_{\bar{p}_j}M$ under the tangential map $T_{\bar{x}_j}\varphi^{-1}$. Then it follows from the considerations above that

$$\mathrm{vol}_g(A) = \int_Q \varphi_* \sqrt{|G|}\, dx \approx \sum_j \varphi_* \sqrt{|G|}\, (\bar{x}_j) \lambda_m(Q_j) \ .$$

This shows that $\mathrm{vol}_g\big(\varphi^{-1}(Q)\big)$ is approximately equal to the sum of the volumes of the parallelepipeds that approximate $A \subset M$. By "unlimited refinement", these sums approach the integral and therefore $\mathrm{vol}_g(A)$, while at the same time $\widehat{Q}_j$ gets arbitrarily close to the A. This shows that $\mathrm{vol}_g(A)$ does agree with our usual understanding of the volume of A.

In the following, we set $\lambda_{(M,g)} := \mathrm{vol}_g$. We may also write this as λ_M or λ_g if the context allows it. Also

$$\mathrm{vol}(M) := \lambda_M(M)$$

is the **volume of M** or, in the case $m = 2$, the **area of M**.

1.5 Proposition λ_M is a Radon measure of M, the **Riemann–Lebesgue volume measure** of M.

Proof (a) It is clear that λ_M maps the σ-algebra $\mathcal{L}_M$ to $[0, \infty]$ and assigns the empty set the value 0.

Suppose $\big\{ (\varphi_j, U_j) \ ; \ j \in \mathbb{N} \big\}$ is a countable atlas and (A_k) is a disjoint sequence in $\mathcal{L}_M$. Then $(A_k \cap U_j)_{k \in \mathbb{N}}$ is disjoint sequence in U_j. Therefore Proposition 1.2 with $A := \bigcup_k A_k$ and $\lambda_{M,U_j} := \mathrm{vol}_{g,U_j}$ gives

$$\lambda_{M,U_j}(A \cap U_j) = \sum_k \lambda_{M,U_j}(A_k \cap U_j) \ .$$

Because $\lambda_M(A_k) = \sum_j \lambda_{M,U_j}(A_k \cap U_j)$, we find

$$\lambda_M(A) = \sum_j \lambda_{M,U_j}(A \cap U_j) = \sum_j \sum_k \lambda_{M,U_j}(A_k \cap U_j)$$

$$= \sum_k \sum_j \lambda_{M,U_j}(A_k \cap U_j) = \sum_k \lambda_M(A_k) \ ,$$

where we have again referred to Remark X.3.6(b). Thus λ_M is a σ-additive function and therefore a measure.

(b) Let $A \in \mathcal{L}_M$, and suppose (φ, U) is a chart with $A \subset U$. Then it follows from Corollary IX.5.5 that there are a G_δ-set $\widetilde{G}$ and an F_σ-set $\widetilde{F}$ in $\varphi(U)$ such that $\widetilde{F} \subset \varphi(A) \subset \widetilde{G}$ and $\lambda_m(\widetilde{F}) = \lambda_m(\varphi(A)) = \lambda_m(\widetilde{G})$. This gives

$$\int_{\widetilde{F}} \varphi_* \sqrt{|G|}\, dx = \int_{\varphi(A)} \varphi_* \sqrt{|G|}\, dx = \int_{\widetilde{G}} \varphi_* \sqrt{|G|}\, dx \ .$$

Therefore we find an increasing sequence $(\widetilde{F}_j)$ of compact subsets $\widetilde{F}_j$ of $\widetilde{F}$ with $\bigcup_j \widetilde{F}_j = \widetilde{F}$. Then $F_j := \varphi^{-1}(\widetilde{F}_j)$ is a compact subset of A, and the monotone convergence theorem gives

$$\lambda_M(F_j) = \int_{\widetilde{F}_j} \varphi_* \sqrt{|G|}\, dx \uparrow \int_{\widetilde{F}} \varphi_* \sqrt{|G|}\, dx = \int_{\varphi(A)} \varphi_* \sqrt{|G|}\, dx = \lambda_M(A) \ .$$

Therefore

$$\lambda_M(A) = \sup\{\, \lambda_M(K)\ ;\ K \subset A,\ K \text{ is compact in } M \,\} \ . \tag{1.3}$$

Analogously, we show

$$\lambda_M(A) = \inf\{\, \lambda_M(O)\ ;\ O \supset A,\ O \text{ is open in } M \,\} \ . \tag{1.4}$$

(c) Suppose now A is an arbitrary set in $\mathcal{L}_M$. Then the proof of Lemma 1.4 shows that there is a disjoint sequence (A_j) in $\mathcal{L}_M$ with $A_j \subset U_j$ and $\bigcup_j A_j = A$. If $\lambda_M(A_{j_0}) = \infty$ for some $j_0 \in \mathbb{N}$, then $\lambda_M(A) = \infty$. Also, (b) then shows that to every $\alpha > 0$ there is a compact set K such that $K \subset A_{j_0} \subset A$ and $\lambda_M(K) > \alpha$. From this it follows that (1.3) is true in this case as well. Therefore suppose $\lambda_M(A_j) < \infty$ for $j \in \mathbb{N}$, and let $\alpha < \beta < \lambda_M(A)$. Then there is an N with $\sum_{j=0}^{N} \lambda_M(A_j) > \beta > \alpha$. According to (b), we find for every j a compact subset K_j of A_j with $\lambda_M(K_j) > \lambda_M(A_j) - (\beta - \alpha)2^{-j-1}$. Then $K := \bigcup_{j=0}^{N} K_j$ is a compact subset of A, and

$$\lambda_M(K) = \sum_{j=0}^{N} \lambda_M(K_j) > \sum_{j=0}^{N} \lambda_M(A_j) - (\beta - \alpha) \sum_{j=0}^{N} 2^{-j-1} > \beta - (\beta - \alpha) = \alpha \ .$$

Because this it true for every $\alpha < \lambda_M(A)$, we see that (1.3) also holds in this case.

From (b) it also follows that to every $\varepsilon > 0$ and $j \in \mathbb{N}$ there is an open set O_j with $A \subset O_j \subset U_j$ and $\lambda_M(O_j) < \lambda_M(A_j) + \varepsilon\, 2^{-j-1}$. Then $O := \bigcup_j O_j$ is open in M and satisfies $O \supset A$ and

$$\lambda_M(O) \le \sum_j \lambda_M(O_j) < \sum_j \lambda_M(A_j) + \varepsilon = \lambda_M(A) + \varepsilon \ .$$

This shows that (1.4) is also true.

(d) Let $p \in M$, and let (φ, U) be a chart around p. Then there is a compact neighborhood K of p in M with $K \subset U$. Because $\varphi(K)$ is compact in $\overline{\mathbb{H}}^m$ and therefore in $\mathbb{R}^m$, it follows from (1.1), the continuity of $\varphi_* \sqrt{|G|}$, Theorem X.5.1(i), and Corollary X.3.15(iii) that $\varphi_* \sqrt{|G|}$ is integrable over $\varphi(K)$. Therefore $\lambda_M(K)$ is finite, which shows that λ_M is locally finite. Thus it is a Radon measure. ∎

Properties

The next proposition characterizes λ_M-null sets.

1.6 Proposition   Let $A \in \mathcal{L}_M$. These statements are equivalent:
 (i) $\lambda_M(A) = 0$;
 (ii) $\lambda_m\big(\varphi(A \cap U)\big) = 0$ for every chart (φ, U);
 (iii) $\lambda_m\big(\varphi(A \cap U)\big) = 0$ for every chart of a countable atlas of M.

Proof   "(i)$\Rightarrow$(ii)" Because $A \cap U \in \mathcal{L}_M$ and $A \cap U \subset A$, it follows that

$$0 = \lambda_M(A) \geq \lambda_M(A \cap U) = \int_{\varphi(A\cap U)} \varphi_* \sqrt{|G|}\, dx \ .$$

Because $\varphi_* \sqrt{|G|}$ is continuous and pointwise strictly positive, Remark X.3.3(c) gives $\lambda_m\big(\varphi(A \cap U)\big) = 0$.
 "(ii)$\Rightarrow$(iii)" This is clear.
 "(iii)$\Rightarrow$(i)" From $\lambda_m\big(\varphi(A\cap U)\big) = 0$ and (1.1), it follows that $\lambda_M(A\cap U) = 0$. Let $\big\{ (\varphi_j, U_j) \ ; \ j \in \mathbb{N} \big\}$ be a countable atlas. Then $A = A \cap \bigcup_j U_j = \bigcup_j (A \cap U_j)$, and the claim follows from the σ-subadditivity of λ_M. ∎

This proposition shows that the concept of a λ_M-null set is independent of the particular pseudo-Riemannian metric. Therefore we may simply call λ_M-null sets **null sets** or **Lebesgue null sets of M**.

The next theorem lists the basic properties of Riemann–Lebesgue measures.

1.7 Theorem   Let (M, g) be a pseudo-Riemannian manifold.
 (i) $(M, \mathcal{L}_M, \lambda_M)$ is a σ-compact complete measure space.
 (ii) λ_M is a massive Radon measure.
 (iii) All n-dimensional submanifolds of M with $n < m$ are null sets in M, and ∂M is also a null set in M.
 (iv) If M is an m-dimensional submanifold of $\mathbb{R}^m$, then $\lambda_M = \lambda_m$.

Proof   (i) Since M is a locally compact metric and a Lindelöf space, $(M, \mathcal{L}_M, \lambda_M)$ is σ-compact by Remark X.1.16(e). Its completeness follows from Propositions 1.5 and 1.6 and the completeness of the Lebesgue measure.
 (ii) Let O open in M and nonempty, and let (φ, U) be a chart with $U \cap O \neq \emptyset$. Then $\varphi(O \cap U)$ has positive Lebesgue measure. Because $\varphi_* \sqrt{|G|}$ is continuous and pointwise strictly positive, Remark X.3.3(c) gives

$$\lambda_M(O) \geq \lambda_M(O \cap U) = \int_{\varphi(O\cap U)} \varphi_* \sqrt{|G|}\, dx > 0 \ .$$

Thus we get (ii) from (i) and Proposition 1.5.

(iii) This follows from Proposition 1.6 and Example IX.5.2.

(iv) Using the trivial chart, we get

$$\lambda_M(A) = \int_A dx = \lambda_m(A) \quad \text{for } A \in \mathcal{L}_M \ ,$$

as desired. ∎

Integrability

Let $E := (E, |\cdot|)$ be a Banach space. By Theorem 1.7, we can use the entire integration theory developed in Sections X.1–X.4. In particular, the spaces $\mathcal{L}_p(M, \lambda_M, E)$ and $L_p(M, \lambda_M, E)$ are defined for $p \in [1, \infty] \cup \{0\}$.

1.8 Proposition *For $f : M \to E$, these statements are equivalent:*

(i) $f \in \mathcal{L}_0(M, \lambda_M, E)$;

(ii) $\varphi_* f \in \mathcal{L}_0(\varphi(U), E)$ *for every chart (φ, U) of M;*

(iii) $\varphi_* f \in \mathcal{L}_0(\varphi(U), E)$ *for every chart of a countable atlas of M.*

Proof "(i)⇒(ii)" Suppose f is λ_M-measurable. Then by Theorem X.1.4, f is λ_M-almost separable valued and $\mathcal{L}_M$-measurable. From Proposition 1.6 and the definition of $\mathcal{L}_M$, it now follows easily that $\varphi_* f : \varphi(U) \to E$ is λ_m-almost separable valued and $\mathcal{L}_{\varphi(U)}$-measurable. Therefore Theorem X.1.4 implies that $\varphi_* f$ belongs to $\mathcal{L}_0(\varphi(U), E)$.

"(ii)⇒(iii)" This is trivial.

"(iii)⇒(i)" Let $\{ (\varphi_j, U_j) \ ; \ j \in \mathbb{N} \}$ be a countable atlas of M, and suppose $\varphi_{j*} f \in \mathcal{L}_0(\varphi_j(U_j), E)$ for $j \in \mathbb{N}$. Proposition XI.1.20 guarantees the existence of a smooth partition of unity $\{ \pi_j \ ; \ j \in \mathbb{N} \}$ subordinate to the cover $\{ U_j \ ; \ j \in \mathbb{N} \}$ of M. Then $\varphi_{j*}\pi_j$ belongs to $C^\infty(\varphi_j(U_j))$ for $j \in \mathbb{N}$. Hence it follows from Remark X.1.2(d) that

$$\varphi_{j*}(\pi_j f) = (\varphi_{j*}\pi_j)(\varphi_{j*} f) \in \mathcal{L}_0(\varphi_j(U_j), E) \quad \text{for } j \in \mathbb{N} \ .$$

As in the first part of the proof, we can deduce $\pi_j f \in \mathcal{L}_0(M, \lambda_M, E)$ for $j \in \mathbb{N}$. Because $f = \left(\sum_{j=1}^\infty \pi_j \right) f = \sum_{j=1}^\infty \pi_j f$, the λ_M-measurability of f now follows from Theorem X.1.14. ∎

Because λ_M is a massive Radon measure, we can use the convention we set after Proposition X.4.17. This is to be noted in the next proposition.

1.9 Proposition

(a) *In the sense of vector subspaces, the following are true:*

(i) $C(M, E) \subset L_0(M, \lambda_M, E)$.

(ii) $C_c(M, E)$ is dense in $L_p(M, \lambda_M, E)$ for $1 \leq p < \infty$.

(b) If K is a compact subset of M, then

$$\left| \int_K f \, d\lambda_M \right| \leq \int_K |f| \, d\lambda_M \leq \|f\|_{C(K,E)} \lambda_M(K) \quad \text{for } f \in C(M, E) .$$

Proof Because of Theorem 1.7, the proposition follows from Theorem X.4.18(i) and Corollary X.3.15(iii). ∎

We now show how the calculation of $\int_M f \, d\lambda_M$ can be reduced to integration in local coordinates. So first we consider the local case.

1.10 Theorem Suppose (φ, U) is a chart of M and $f \in L_0(U, \lambda_M, E)$. Then f belongs to $\mathcal{L}_1(U, \lambda_M, E)$ if and only if $(\varphi_* f)\varphi_* \sqrt{|G|}$ lies in $\mathcal{L}_1(\varphi(u), \lambda_m, E)$. In that case, we have

$$\int_U f \, d\lambda_M = \int_{\varphi(U)} (\varphi_* f)\varphi_* \sqrt{|G|} \, dx . \tag{1.5}$$

Proof (i) Let $f = \chi_A$ for some $A \in \mathcal{L}_M$ with $A \subset U$. Then

$$\int_U f \, d\lambda_M = \int_A d\lambda_M = \int_{\varphi(A)} \varphi_* \sqrt{|G|} \, dx = \int_{\varphi(U)} (\varphi_* f)\varphi_* \sqrt{|G|} \, dx$$

because $\varphi_* f = \varphi_* \chi_A = \chi_{\varphi(A)}$. Now it follows that (1.5) holds for simple functions.

(ii) Let $f \in \mathcal{L}_1(U, \lambda_M, E)$. Then there is an $\mathcal{L}_1$-Cauchy sequence (f_j) of simple functions such that $f_j \to f$ λ_M-a.e. and

$$\int_U f_j \, d\lambda_M \to \int_U f \, d\lambda_M . \tag{1.6}$$

Further $(\varphi_* f_j)$ is a sequence in $\mathcal{S}(\varphi(U), E)$, and

$$(\varphi_* f_j)\varphi_* \sqrt{|G|} \to (\varphi_* f)\varphi_* \sqrt{|G|} \qquad \lambda_m\text{-a.e. in } \varphi(U) .$$

In addition, it follows from the validity of (1.5) for simple functions that

$$\int_U |f_j - f_k| \, d\lambda_M = \int_{\varphi(U)} |\varphi_* f_j - \varphi_* f_k| \, \varphi_* \sqrt{|G|} \, dx = \int_{\varphi(U)} |h_j - h_k| \, dx ,$$

where $h_j := \varphi_*(f_j \sqrt{|G|})$. Therefore (h_j) is a $\mathcal{L}_1$-Cauchy sequence in $F := \mathcal{L}_1(\varphi(U), E)$. By Theorem X.2.10(ii), we find an $h \in F$ such that $h_j \to h$ in F. Then it follows from Theorem X.2.18(i) that there is a subsequence $(h_{j_k})_{k \in \mathbb{N}}$ of

(h_j) such that $h_{j_k} \to h$ λ_m-a.e. as $k \to \infty$. Because $f_j \to f$ λ_M-a.e. implies that $h_j \to \varphi_*\big(f\sqrt{|G|}\big)$ λ_m-a.e., we find that h agrees λ_m-a.e. with $\varphi_*\big(f\sqrt{|G|}\big)$. Then Theorem X.2.18(ii) implies

$$\int_U f_j \, d\lambda_M = \int_{\varphi(U)} h_j \, dx \to \int_{\varphi(U)} \varphi_*\big(f\sqrt{|G|}\big) \, dx \ .$$

Hence the claim follows from (1.6). ∎

We now treat the general case. We first introduce a useful abbreviation: Suppose $\big\{ (\varphi_j, U_j) \ ; \ j \in \mathbb{N} \big\}$ is a countable atlas for M, and let $\{ \pi_j \ ; \ j \in \mathbb{N} \}$ be a smooth partition of unity subordinate to the cover $\{ U_j \ ; \ j \in \mathbb{N} \}$ of M. Then we call $\big\{ (\varphi_j, U_j, \pi_j) \ ; \ j \in \mathbb{N} \big\}$ a **local system** for M. The existence of such systems is secured by Proposition XI.1.20.

1.11 Proposition Let $\big\{ (\varphi_j, U_j, \pi_j) \ ; \ j \in \mathbb{N} \big\}$ be a local system for M. Then $f \in \mathcal{L}_0(M, \lambda_M, E)$ belongs to $\mathcal{L}_1(M, \lambda_M, E)$ if and only if $\pi_j f$ lies in $\mathcal{L}_1(U_j, \lambda_M, E)$ for every $j \in \mathbb{N}$ and

$$\sum_{j=0}^{\infty} \int_{U_j} \pi_j \, |f| \, d\lambda_M < \infty \ . \tag{1.7}$$

In this case, we have

$$\int_M f \, d\lambda_M = \sum_{j=0}^{\infty} \int_{U_j} \pi_j f \, d\lambda_M \ . \tag{1.8}$$

Proof (i) Let $f \in \mathcal{L}_1(M, \lambda_M, E)$. Then $f = \big(\sum_{j=0}^{\infty} \pi_j\big) f = \sum_{j=0}^{\infty} \pi_j f$ holds with pointwise convergence, and

$$|\pi_k f| \le \sum_{j=0}^{n} |\pi_j f| = \sum_{j=0}^{n} \pi_j \, |f| \le |f| \quad \text{for } 0 \le k \le n < \infty \ .$$

Therefore $\pi_k f \in \mathcal{L}_1(U_k, \lambda_M, E)$ follows because $\mathrm{supp}(\pi_k) \subset\subset U_k$. Theorems X.3.9 and 1.10 and Lebesgue's theorem now imply (1.7) and (1.8).

(ii) Let $\pi_j f \in \mathcal{L}_1(U_j, \lambda_M, E)$ for $j \in \mathbb{N}$, and suppose (1.7) holds. Letting $h_n := \sum_{j=0}^{n} \pi_j f$ for $n \in \mathbb{N}$, it follows that

$$\int_M |h_j - h_k| \, d\lambda_M \le \sum_{i=j+1}^{k} \int_{U_i} \pi_i \, |f| \, d\lambda_M \quad \text{for } 0 \le j < k < \infty \ .$$

Therefore (h_j) is an $\mathcal{L}_1$-Cauchy sequence in $\mathcal{L}_1(M, \lambda_M, E)$. Because (h_j) converges pointwise to f, it follows from the completeness of $\mathcal{L}_1(M, \lambda_M, E)$ and from Theorem X.2.18 that f is integrable with respect to the measure λ_M. ∎

1.12 Remark (regularity) Obviously all the definitions and propositions above remain true when M is a C^1 manifold. ∎

Calculation of several volumes

Of course, Proposition 1.11 is chiefly of theoretical importance. In practical cases, one is often in the comfortable situation that M can be described, except for perhaps a null set, by a single chart. In this case, Theorem 1.10 can be used. In the following examples, we consider such cases in the special case $f = 1$.

1.13 Examples Unless we say otherwise, we use the standard metric on M.

(a) (curves) Let $\gamma : J \to \mathbb{R}^m$ be an embedding of a perfect interval $J \subset \mathbb{R}$. Then $M := \gamma(J)$ is an embedded curve $\mathbb{R}^m$, and $\mathrm{vol}(M) = L(M)$, where L denotes the arc length. More generally, $f \in \mathcal{L}_1(M, \lambda_M, E)$ satisfies

$$\int_M f \, d\lambda_M = \int_J (f \circ \gamma) \, |\dot{\gamma}| \, dt \; .$$

In this case, we set $ds := \sqrt{G} \, dt$ and call ds the **arc length element**, as motivated by Theorem VIII.1.7.

Proof This follows from Example XI.5.3(e) and Theorem VIII.1.7. ∎

(b) (graphs) Let X be open in $\mathbb{R}^m$ and $f \in C^\infty(X, \mathbb{R})$. Then[1]

$$\mathrm{vol}\big(\mathrm{graph}(f)\big) = \int_X \sqrt{1 + |\nabla f|^2} \, dx \; .$$

Proof Example XI.5.3(d). ∎

(c) (spheres) Let $m \in \mathbb{N}^\times$ and $r > 0$. Then

$$r \, \mathrm{vol}(rS^m) = (m + 1) \, \mathrm{vol}(r\mathbb{B}^{m+1}) \; .$$

In particular, $\mathrm{vol}(rS^1) = 2\pi r$, $\mathrm{vol}(rS^2) = 4\pi r^2$, and $\mathrm{vol}(rS^3) = 2\pi^2 r^3$.

Proof From Remark IX.5.26(b), we know that

$$\mathrm{vol}(r\mathbb{B}^{m+1}) = r^{m+1} \, \mathrm{vol}(\mathbb{B}^{m+1}) \; . \tag{1.9}$$

Then by Example VIII.1.9(c), we can assume $m \geq 2$. Let

$$h_\pm : \mathbb{B}_r^m \to \mathbb{R} \;, \quad x \mapsto \pm\sqrt{r^2 - |x|^2}$$

be the parametrizations of the upper and lower half spheres $rS_\pm^m$. For these we find $\nabla h_\pm(x) = -x/h_\pm(x)$ and therefore $|\nabla h_\pm(x)|^2 = |x|^2/(r^2 - |x|^2)$ for $x \in r\mathbb{B}^m$. Therefore

---

[1]See Example VIII.1.9(a).

it follows from (b) and the transformation theorem that

$$\text{vol}(rS^m_\pm) = \int_{r\mathbb{B}^m} \sqrt{r^2/(r^2 - |x|^2)}\, dx = r^m \int_{\mathbb{B}^m} \sqrt{1/(1 - |y|^2)}\, dy = r^m\, \text{vol}(S^m_\pm)\ .$$

Because $S^m = S^m_+ \cup S^m_- \cup S^{m-1}$, where S^{m-1} is identified with the "equatorial sphere" $S^{m-1} \times \{0\}$ of S^m, and because $\lambda_{S^m}(S^{m-1}) = 0$, we find

$$\text{vol}(rS^m) = r^m\, \text{vol}(S^m)\ . \tag{1.10}$$

Due to (1.9) and (1.10), it suffices to show

$$\text{vol}(S^m) = (m+1)\, \text{vol}(\mathbb{B}^{m+1})\ .$$

We apply the parametrization $h_m : W_m \to \mathbb{R}^{m+1}$ of $S^m \backslash H_m$ that we considered in Example XI.5.5(h) and denote its chart by ψ. Then we may read from the formula given there for g_{S^m} (because the fundamental matrix is diagonal) that

$$\psi_* G = \prod_{k=0}^{m-1} a_{m+1,k} = \prod_{k=0}^{m-1} \prod_{i=k+1}^{m-1} \sin^2 \vartheta_i = w_{m+1}^2(\vartheta)\ ,$$

where

$$w_{m+1}(\vartheta) := \sin\vartheta_1 \sin^2\vartheta_2 \cdot \cdots \cdot \sin^{m-1}\vartheta_{m-1} \quad \text{for } \vartheta = (\vartheta_1, \ldots, \vartheta_{m-1}) \in [0, \pi]^{m-1}\ .$$

Because of $W_m = (0, 2\pi) \times (0, \pi)^{m-1}$, Example X.8.10(c), and Fubini's theorem, we get

$$\lambda_{S^m}(S^m \backslash H_m) = \int_{W_m} \psi_* \sqrt{G}\, d(\varphi, \vartheta) = 2\pi \int_{(0,\pi)^{m-1}} w_{m+1}(\vartheta)\, d\vartheta = (m+1)\omega_{m+1}\ ,$$

where $\omega_{m+1} := \text{vol}(\mathbb{B}^{m+1})$. Because $S^m \cap H_m$ is a null set of S^m, the claim follows. ∎

(d) (helicoids) Let $0 \le \alpha < \beta < \infty$ and $a \ge 0$.
Let $T > 0$, and define

$$h : (\alpha, \beta) \times (0, T) \to \mathbb{R}^3$$

by

$$h(s, t) := (s\cos t, s\sin t, at)\ .$$

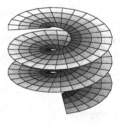

Then h is a parametrization of a helicoid F. It is
generated by beginning at $t = 0$ with the "rod"
consisting of the interval (α, β) lying on the x-axis and then rotating it with angular velocity 1 around the z-axis, while simultaneously raising it with velocity a. It satisfies

$$\text{vol}(F) = \frac{T}{2} \left[s\sqrt{s^2 + a^2} + a^2 \log\left(s + \sqrt{s^2 + a^2}\right) \right]\Big|_\alpha^\beta\ .$$

In particular, for $\alpha = 0$, $\beta = 1$, and $T = 2\pi$ — that is, for the complete rotation of a rod of length 1 about its endpoint — we have

$$
\begin{aligned}
\mathrm{vol}(F) &= \pi\left[\sqrt{1+a^2} + a^2 \log\left(1 + \sqrt{1+a^2}\right) - a^2 \log a\right] \\
&= \pi\left[\sqrt{1+a^2} + a^2 \log\left(\sqrt{1+1/a^2} + 1/a\right)\right] .
\end{aligned}
$$

This formula gives what we would expect in several special cases: When $a = 0$, we obtain the unit disc of area π, whereas $\mathrm{vol}(F) > \pi$ for $a > 0$.

Proof   A simple calculation shows that $\varphi_* \sqrt{G} = \sqrt{s^2 + a^2}$, where φ is the chart belonging to h, that is, $h = i \circ \varphi^{-1}$. Because

$$
\left[s\sqrt{s^2 + a^2} + a^2 \log\left(s + \sqrt{s^2 + a^2}\right)\right]^{\cdot} = 2\sqrt{s^2 + a^2} ,
$$

the claim follows from Fubini's theorem. ∎

(e) The volume of the disc $R\mathbb{B}^2$ in the hyperbolic plane H^2 satisfies

$$
\mathrm{vol}_{g_{H^2}}(R\mathbb{B}^2) = 2\pi\left(\sqrt{1 + R^2} - 1\right) .
$$

As R gets larger, this expression behaves approximately as $2\pi R$, while in the Euclidean case the volume grows as R^2.

Proof   Using polar coordinates, we recall Example XI.5.5(k) that

$$
g_{H^2} = \frac{(dr)^2}{1 + r^2} + r^2(d\varphi)^2 .
$$

From this we read

$$
\omega_{H^2} = \frac{r}{\sqrt{1 + r^2}}\, dr \wedge d\varphi .
$$

Because a single point set of H^2 is a null set, it follows that

$$
\mathrm{vol}_{g_{H^2}}(R\mathbb{B}^2) = \int_0^R \int_0^{2\pi} \frac{r}{\sqrt{1 + r^2}}\, dr\, d\varphi = 2\pi \int_0^R \frac{r\, dr}{\sqrt{1 + r^2}} ,
$$

as desired. ∎

Exercises

1   Let N be an m-dimensional manifold and $f \in \mathrm{Diff}^1(M, N)$. Show that $A \in \mathcal{L}_M \Longleftrightarrow f(A) \in \mathcal{L}_N$.

2   Determine ω_M for the hyperboloid $H := \{(x, y, z) \in \mathbb{R}^3 \ ; \ x^2 + y^2 - z^2 = 1\}$ with respect to the standard metric and the parametrization

$$
(t, \varphi) \mapsto (\cosh t \cos\varphi, \cosh t \sin\varphi, \sinh t) .
$$

Also calculate the volume of the part of H satisfying $0 < z < 1$.

3   Let $Z := S^1 \times (-1, 1)$ be a cylinder in $\mathbb{R}^3$. Calculate $\int_Z |x|^2\, d\lambda_Z$.

4 Calculate the integral $\int_{S^2} x^2 y^2 z^2 \, d\lambda_{S^2}$.

5 The solid cut out from the ball $R\bar{\mathbb{B}}^3$ by the cylinder

$$Z_R := \left\{ (x, y, z) \in \mathbb{R}^3 \; ; \; x^2 + y^2 = Rx \right\}$$

is called the **Viviani solid** V_R.
Show

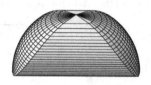

(i) $\operatorname{vol}(V_R) = 2(\pi - 4/3)R^3/3$;

(ii) the area of the intersection D_R of Z_R with RS^2 (the left and right "caps" of V_R in the figure) equals $4R^2(\pi/2 - 1)$.

6 Show that $\operatorname{vol}(M \times N) = \operatorname{vol}(M)\operatorname{vol}(N)$ if N is an n-dimensional manifold without boundary.

7 Suppose $\gamma : [0, 1] \to (0, \infty) \times \mathbb{R}, \; t \mapsto (\rho(t), \sigma(t))$ is a smooth embedding, $i : S^1 \hookrightarrow \mathbb{R}^2$, and

$$f : S^1 \times [0, 1] \to \mathbb{R}^3 , \quad (q, t) \mapsto (\rho(t)i(q), \sigma(t)) .$$

Finally denote by $Z_\gamma^2 := f(S^1 \times [0, 1])$ the surface of revolution in $\mathbb{R}^3$ generated by γ. Show that

$$\operatorname{vol}(Z_\gamma^2) = 2\pi \int_0^1 \rho(t) \, |\gamma'(t)| \, dt .$$

8 Let $E_{a,b}$ be an ellipsoid of revolution $\mathbb{R}^3$ with semiaxes $a \geq b > 0$. Also let $k := \sqrt{a^2 - b^2}/a$. Then[2]

$$\operatorname{vol}(E_{a,b}) = 4\pi ab \int_0^{\pi/2} \sqrt{1 - k^2 \sin t} \, dt .$$

9 Show that a torus $\mathsf{T}_{a,1}^2$ with $a > 1$ satisfies $\operatorname{vol}(\mathsf{T}_{a,1}^2) = 4\pi^2 a$.

10 Let $\alpha \in (1/2, 1]$ and $r(x) := x^{-\alpha}$ for $x \geq 1$.
Show that the set

$$\left\{ (x, y, z) \in [1, \infty) \times \mathbb{R}^2 \; ; \; y^2 + z^2 \leq r^2(x) \right\}$$

has finite volume, even though its surface area is infinite.

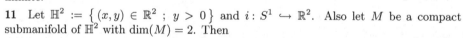

11 Let $\mathbb{H}^2 := \left\{ (x, y) \in \mathbb{R}^2 \; ; \; y > 0 \right\}$ and $i : S^1 \hookrightarrow \mathbb{R}^2$. Also let M be a compact submanifold of $\mathbb{H}^2$ with $\dim(M) = 2$. Then

$$R_M := \left\{ (x, u) \in \mathbb{R} \times \mathbb{R}^2 \; ; \; (x, |u|) \in M \right\} \subset \mathbb{R}^3$$

is the rotationally symmetric solid generated by rotating M about the x-axis.
Show these facts:

(a) $R_M \to M \times S^1, \; (x, u) \mapsto ((x, |u|), u/|u|)$ is a diffeomorphism, and

$$M \times S^1 \to R_M , \quad ((x, y), \sigma) \mapsto (x, y\, i(\sigma))$$

is its inverse map ("cylindrical coordinates");

---

[2]See Remark VIII.2.3(b).

(b) $\mathrm{vol}(R_M) = 2\pi \int_M y \, d\lambda_M$;

(c) **Guldin's first rule** says

$$\mathrm{vol}(R_M) = 2\pi S_2(M) \, \mathrm{vol}(M) \ ,$$

where $S_2(M)$ denotes the second coordinate of the centroid $S(M)$ of M and therefore the distance of this point from the rotation axis (see Exercise X.6.4). That is, *the volume of a solid of revolution is equal to the product of the area of a meridional slice and the circumference of the circle whose radius is the distance of the centroid of that slice from the axis of rotation.*

12 Use the notations of Exercise 11, but let $\dim(M) = 1$. Show:

(a) Statements (a) and (b) of Exercise 11 also hold in this case.[3]

(b) With the **centroid**

$$S(M) := \frac{1}{\mathrm{vol}(M)} \int_M i \, d_{\mathbb{R}^3} \, d\lambda_M \in \mathbb{R}^3 \ ,$$

we have **Guldin's second rule**

$$\mathrm{vol}(R_M) = 2\pi S_2(M) \, \mathrm{vol}(M) \ ,$$

that is, *the volume of the surface of revolution is equal to the product of the arc length of a meridional slice and the circumference of the circle whose radius is the distance of the centroid of that slice from the axis of rotation.*

13 Determine the centroid of the half disc $R\overline{\mathbb{B}}^2 \cap \mathbb{H}^2$. (Hint: Exercise 12.)

14 Suppose N is a pseudo-Riemannian manifold and $\varphi \in \mathrm{Diff}(M, N)$. Let $p \in [1, \infty]$, and let E be a Banach space. For $f \in E^N$, define $\mathcal{J}_\varphi f \in E^M$ by

$$\mathcal{J}_\varphi f(s) := \varphi^* f(s) \det T_s \varphi \quad \text{for } s \in M \ .$$

Prove these claims:

(a) $\mathcal{J}_\varphi$ maps $\mathcal{L}_p(N, d\lambda_N, E)$ linearly and continuously to $\mathcal{L}_p(M, d\lambda_M, E)$.

(b) For $f \in \mathcal{L}_0(N, d\lambda_N, E)$, the statements

  (i) $f = 0 \ \lambda_N$-a.e. and

  (ii) $\mathcal{J}_\varphi f = 0 \ \lambda_M$-a.e.

are equivalent.

(c) Suppose $J_\varphi[f] := [\mathcal{J}_\varphi f]$ for $[f] \in L_p(N, d\lambda_N, E)$. Then $J_\varphi[f]$ is well defined in $L_p(M, d\lambda_M, E)$, and J_φ itself is an isometric isomorphic map from $L_p(N, d\lambda_N, E)$ to $L_p(M, d\lambda_M, E)$.

15 For $p \in [1, \infty]$, give an isometric isomorphism from $L_p\big((0, \infty) \times S^{n-1}\big)$ to $L_p(\mathbb{R}^n)$.

16 For $f \in \mathcal{L}_0(\mathbb{R}^n)$ and $s \in S^{n-1}$, define a function $T_0 f(s) : (0, \infty) \to \mathbb{R}$ by

$$T_0 f(s)(r) := f(rs) r^{n-1} \quad \text{for } r \in (0, \infty) \ .$$

Also let $p \in [1, \infty)$.

---

[3] See Exercise 7.

Show:

(a) If $f \in \mathcal{E}_c(\mathbb{R}^n)$, then $T_0 f$ belongs to $\mathcal{L}_0\big(S^{n-1}, L_p(0, \infty)\big)$;

(b) Let $f \in \mathcal{E}_c(\mathbb{R}^n)$. Then the class $[T_0 f]$ of $T_0 f$ belongs to $L_p\big(S^{n-1}, L_p(0, \infty)\big)$.

(c) There is a unique extension

$$T \in \mathcal{L}\mathrm{is}\big(L_p(\mathbb{R}^n), L_p\big(S^{n-1}, L_p(0, \infty)\big)\big)$$

of $\mathcal{E}_c(\mathbb{R}^n) \to L_p\big(S^{n-1}, L_p(0, \infty)\big)$, $f \mapsto [T_0 f]$, and T is an isometry.
(Hint: Study the proofs of Lemma X.6.20 and Theorem X.6.22.)

2 Integration of differential forms

In Section VIII.4, we introduced the important concept of the line integral, an integral of a 1-form over an oriented curve and therefore over a 1-dimensional manifold. In this section, we introduce higher-dimensional analogues, namely, integrals of m-forms over oriented m-dimensional manifolds.

After we have introduced the basics, we prove generalizations of the transformation theorem and Fubini's theorem for differential forms. Through examples, we show how these ideas can be used in concrete cases.

Finally, we discuss fluxes on manifolds and prove the important transport theorem. The latter is not only significant in continuum mechanics, but also gives us a geometric interpretation of the divergence of a vector field.

In this section, let

- M be an m- and N be an n-dimensional oriented manifold with $m, n \in \mathbb{N}^{\times}$.

Integrals of m-forms

Let g be a pseudo-Riemannian metric on M, let ω_M be the volume element, and let λ_M be the associated Riemann-Lebesgue volume measure. If ω is an m-form of M, then, because $\dim(\bigwedge^m T_p^* M) = 1$ for $p \in M$, there is exactly one $f \in \mathbb{R}^M$ such that $\omega = f \omega_M$. Then the m-form ω on (M, g) is said to be **integrable** if f is λ_M-integrable, that is, if f belongs to $\mathcal{L}_1(M, \lambda_M)$. In this case, we set

$$\int_M \omega := \int_M f \, d\lambda_M \qquad (2.1)$$

and call $\int_M \omega$ the **integral of the m-form ω over M**.

2.1 Remarks (a) (local representation) Suppose (φ, U) is a positive chart of M such that $\varphi = (x^1, \ldots, x^m)$. The m-form $\omega := a \, dx^1 \wedge \cdots \wedge dx^m$ is integrable over U if and only if $\varphi_* a$ belongs to $\mathcal{L}_1(\varphi(U))$. If this is the case, then

$$\int_U \omega = \int_U a \, dx^1 \wedge \cdots \wedge dx^m = \int_{\varphi(U)} \varphi_* a \, dx = \int_{\varphi(U)} \varphi_* \omega . \qquad (2.2)$$

Proof It follows from $\omega_U = \sqrt{|G|} \, dx^1 \wedge \cdots \wedge dx^m$ that $\omega = (a/\sqrt{|G|}) \omega_U$. Therefore ω is integrable if and only if $a/\sqrt{|G|}$ belongs to $\mathcal{L}_1(U, \lambda_M)$. By Theorem 1.10, this is the case if and only if

$$\varphi_* a = \varphi_* (a/\sqrt{|G|}) \varphi_* \sqrt{|G|} \in \mathcal{L}_1(\varphi(U)) .$$

Now the claim follows from (1.5) and (2.1). ∎

(b) (reduction to local representations) Suppose ω is an m-form on M, and suppose that $\big\{\,(\varphi_j, U_j, \pi_j)\ ;\ j \in \mathbb{N}\,\big\}$ is a local system for M whose charts are all positive, a **positive local system**. For $j \in \mathbb{N}$, let $a_j\, dx_j^1 \wedge \cdots \wedge dx_j^m$ be the local representation of $\omega \,|\, U_j$ with respect to the chart (φ_j, U_j), and let $\omega_j := \pi_j \omega \,|\, U_j$. Then ω is integrable over M if and only if

$$\sum_{j=0}^{\infty} \int_{U_j} \pi_j \, |a_j| \, dx_j^1 \wedge \cdots \wedge dx_j^m = \sum_{j=0}^{\infty} \int_{\varphi_j(U_j)} \varphi_{j*}(\pi_j \, |a_j|) \, dx < \infty \ . \tag{2.3}$$

If (2.3) is satisfied, then

$$\int_M \omega = \sum_{j=0}^{\infty} \int_{U_j} \omega_j \ . \tag{2.4}$$

Proof This follows from (a) and (the proof of) Proposition 1.11. ∎

(c) Every continuous m-form on M with compact support is integrable over M.

Proof Suppose ω is a continuous m-form with compact support. Then there is a positive local system $\big\{\,(\varphi_j, U_j, \pi_j)\ ;\ j \in \mathbb{N}\,\big\}$ and a $k \in \mathbb{N}$ such that

$$\operatorname{supp}(\pi_j) \cap \operatorname{supp}(\omega) = \emptyset \quad \text{for } j > k \ .$$

Therefore the series in (2.3) reduces to a finite sum, and $\pi_j \, |a_j| \in C_c(U_j)$ for $j \in \mathbb{N}$. Proposition 1.9(b) implies the integrability of these functions. ∎

(d) From (a) and (b), we see that the integral of an m-form on M is independent of the special pseudo-Riemannian metric. We therefore lose no generality by always taking the standard metric. Indeed, we require no metric for the integration of differential forms. That is, we can *define* $\int_M \omega$ by the formulas (2.2)–(2.4), and thereby only use the Lebesgue integral in $\mathbb{R}^m$. From the considerations above, for which we can take some Riemannian metric, for example, the standard metric, it follows that these definitions are meaningful, that is, independent of the chosen local system. In particular, the integral of a continuous m-form with compact support is defined over an arbitrary[1] oriented m-dimensional manifold.

(e) (linearity) Let $\Omega_c^m(M)$ be the set of all smooth **m-forms on M with compact support**. In particular, $\mathcal{E}_c(M) := \Omega_c^0(M)$. Then $\Omega_c^m(M)$ is an $\mathcal{E}(M)$-submodule of $\Omega^m(M)$, and

$$\int_M : \Omega_c^m(M) \to \mathbb{R} \ , \quad \omega \mapsto \int_M \omega$$

is well defined and $\mathbb{R}$-linear.

Proof The first statement is obvious. Because of (c), the given map is well defined, and its linearity follows from the linearity of integrals with respect to λ_M. ∎

---

[1]This fact is significant when one considers abstract manifolds. One can show that Riemannian metrics always exist on such manifolds.

(f) (orientability) The integral of m-forms is **oriented**, that is, if the orientation of M is reversed, then the sign changes:

$$\int_{(M,-\mathcal{O}r)} \omega = -\int_M \omega .$$

Proof This follows from $\omega_{(M,-\mathcal{O}r)} = -\omega_M$. ∎

(g) An m-form ω on M is integrable if and only if $\chi_A \omega$ is integrable for every $A \in \mathcal{L}_M$. In this case, we set

$$\int_A \omega := \int_M \chi_A \omega \quad \text{for } A \in \mathcal{L}_M .$$

Proof This follows easily from (a) and (b). ∎

(h) (regularity) It suffices to assume that M is a C^1 manifold. ∎

Suppose (M, g) is a pseudo-Riemannian manifold. Then $A \in \mathcal{L}_M$ satisfies $\lambda_M(A) < \infty$. Thus $\chi_A \omega_M$ is an integrable m-form on M, and

$$\lambda_M(A) = \int_A \omega_M . \tag{2.5}$$

If $\lambda_M(A) = \infty$, we set $\int_A \omega_M := \infty$. This definition makes (2.5) hold for every $A \in \mathcal{L}_M$.

Restrictions to submanifolds

Suppose M is either a submanifold of N or the boundary ∂N of N; let $i : M \hookrightarrow N$ denote the natural embedding. Also suppose ω is an m-form on N and $\omega \,|\, M = i^* \omega$ is integrable over M. Then we set

$$\int_M \omega := \int_M i^* \omega . \tag{2.6}$$

2.2 Remarks Let (φ, U) with $\varphi = (x^1, \dots, x^m)$ be a positive chart of M, and put $h := i \circ \varphi^{-1}$.

(a) An m-form ω on N satisfies

$$\int_U \omega = \int_{\varphi(U)} h^* \omega$$

if $\omega \,|\, U$ is integrable.

Proof From (2.6) and Remark 2.1(a), it follows that

$$\int_U \omega = \int_U i^* \omega = \int_{\varphi(U)} \varphi_* i^* \omega = \int_{\varphi(U)} (\varphi^{-1})^* i^* \omega = \int_{\varphi(U)} (i \circ \varphi^{-1})^* \omega = \int_{\varphi(U)} h^* \omega .$$

The claim follows. ∎

(b) (line integrals) Let $X := N$ be open in $\mathbb{R}^n$, and let $\omega := \sum_{j=1}^n a_j \, dx^j \in \Omega^1(X)$. Also let $m = 1$. Then

$$\int_M \omega = \int_M a_1 \, dx^1 + \cdots + a_n \, dx^n$$

is a line integral.

Proof This follows from Section VIII.4 (by an obvious extension of its results about noncompact curves) and Theorem XI.1.18. ∎

(c) (vector line elements) Suppose X is open in $\mathbb{R}^n$ and $m = 1$. In physics, $\omega := \sum_{j=1}^n a_j \, dx^j \in \Omega^1(X)$ is often written as the formal inner product $\vec{a} \cdot d\vec{s}$ of the vector field $\vec{a} := (a_1, \ldots, a_n)$ and the **vector line element**[2] $d\vec{s} := (dx^1, \ldots, dx^n)$. Then

$$\int_M \omega = \int_M \vec{a} \cdot d\vec{s} .$$

Let $J := \varphi(U)$. Then J is an open interval in $\mathbb{R}$. Because M is one-dimensional and oriented, there is for $p \in M$ exactly one **positive unit tangent vector** $\mathfrak{t}(p)$ of M at p, that is, there is exactly one $\mathfrak{t}(p) = \sum_j \mathfrak{t}^j e_j \in T_p M$ such that $|\mathfrak{t}(p)| = 1$ and $\omega_M(p)\big(\mathfrak{t}(p)\big) = 1$. Then $t := \varphi(p) \in J$ satisfies $(\varphi_* \mathfrak{t}^j)(t) = \dot{h}^j(t)/|\dot{h}(t)|$. Therefore it follows from (a), Example 1.13(a), and Theorem 1.10 that

$$\int_U \omega = \int_{\varphi(U)} h^* \omega = \int_J \sum_{j=1}^n (a_j \circ \varphi^{-1}) \dot{h}^j \, dt = \int_{\varphi(U)} \sum_{j=1}^n \varphi_* a_j \, \varphi_* \mathfrak{t}^j \, |\dot{h}| \, dt$$

$$= \int_{\varphi(U)} \varphi_*(\Theta^{-1} \omega \,|\, \mathfrak{t}) \varphi_* \sqrt{G} \, dt = \int_U (\Theta^{-1} \omega \,|\, \mathfrak{t}) \, ds .$$

Therefore

$$\int_M \vec{a} \cdot d\vec{s} = \int_M (\vec{a} \,|\, \mathfrak{t}) \, ds ,$$

which is also expressed by $d\vec{s} = \mathfrak{t} \, ds$.

(d) (surfaces in space) Let $m = 2$, and let N be open in $\mathbb{R}^3$. Also let (x, y, z) be the Euclidean coordinates of $\mathbb{R}^3$, and put

$$\omega := a \, dy \wedge dz + b \, dz \wedge dx + c \, dx \wedge dy .$$

---

[2]See Remark VIII.4.10(b).

Denoting by (u, v) the local coordinates of U, we have

$$\int_U \omega = \int_U a\, dy \wedge dz + b\, dz \wedge dx + c\, dx \wedge dy$$

$$= \int_{\varphi(U)} \left[h^* a\, \frac{\partial(y, z)}{\partial(u, v)} + h^* b\, \frac{\partial(z, x)}{\partial(u, v)} + h^* c\, \frac{\partial(x, y)}{\partial(u, v)} \right] d(u, v)$$

$$= \int_{\varphi(U)} (h^* \vec{a}) \cdot (\vec{X}_u \times \vec{X}_v)\, d(u, v)$$

with $\vec{a} := (a, b, c)$ and $\vec{X} := h$.

Proof This is a consequence of Example XI.3.4(d) and Remark XI.6.25(c). ∎

(e) (**vector area element**) We use the assumptions and notations of (d). We define the **positive** (**unit**) **normal** $\nu := \nu_M$ of M by demanding that $\big(\nu(p), v_1, v_2 \big)$ is a positive ONB of $T_p \mathbb{R}^3$ for every $p \in M$ and every positive ONB (v_1, v_2) of $T_p M$. If (u, v) are positive coordinates on U, then

$$\nu \,|\, U = \Big(\frac{\partial}{\partial u} \times \frac{\partial}{\partial v} \Big) \Big/ \Big| \frac{\partial}{\partial u} \times \frac{\partial}{\partial v} \Big| . \qquad (2.7)$$

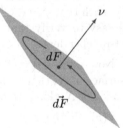

In physics, the triple $(dy \wedge dz, dz \wedge dx, dx \wedge dy)$ is often called the **oriented vector area element**. It is written $d\vec{F}$ and understood as an "infinitesimal area element" together with its orientation, as specified by the normal ν.

We also set

$$\int_M \vec{a} \cdot d\vec{F} := \int_M a\, dy \wedge dz + b\, dz \wedge dx + c\, dx \wedge dy .$$

Then, by using the **scalar area element** $dF := \sqrt{G}\, dx \wedge dy \wedge dz$, we find

$$\int_M \vec{a} \cdot d\vec{F} = \int_M \vec{a} \cdot \nu\, dF , \qquad (2.8)$$

which is also expressed by $d\vec{F} = \nu\, dF$.

Proof As in the proof of Example XI.5.3(g), we see that ν is well defined. We set $w_1 := \partial/\partial u$ and $w_2 := \partial/\partial v$. Then $(w_1(p), w_2(p))$ is a positive basis of $T_p M \subset T_p \mathbb{R}^3$. Therefore $w_1(p) \times w_2(p) \neq 0$. We define $\tilde{\nu} \,|\, U$ as the right side of (2.7). Then $\tilde{\nu} \,|\, U$ belongs to $C^\infty(U, \mathbb{R}^3)$. Remark XI.6.25(b) implies

$$\omega_{\mathbb{R}^3}(\tilde{\nu}, w_1, w_2) = \omega_{\mathbb{R}^3}(w_1, w_2, \tilde{\nu}) = (\tilde{\nu} \,|\, w_1 \times w_2) = |w_1 \times w_2| . \qquad (2.9)$$

Therefore $(w_1(p), w_2(p), \tilde{\nu}(p))$ is a positive basis of $T_p \mathbb{R}^3$, and $\tilde{\nu}(p)$ is orthogonal to $w_1(p)$ and $w_2(p)$. This implies that the positive normal on U is given by (2.7).

Remark XI.6.25(b) and Example XI.5.3(f) imply

$$|w_1 \times w_2| = \sqrt{|w_1|^2\,|w_2|^2 - (w_1\,|\,w_2)^2} = \sqrt{\mathsf{EG} - \mathsf{F}^2}\ .$$

Therefore $w_1 \times w_2 = \nu\sqrt{\mathsf{EG} - \mathsf{F}^2}$ on U, which again follows from Example XI.5.3(f). Now (2.8) is a consequence of (d) and the definition of λ_M. ∎

(f) Suppose either M is a submanifold of N or $M := \partial N$. By a **vector field of N along M**, we mean a map

$$v\colon M \to TN \qquad \text{with} \quad v(p) \in T_pN \text{ and } p \in M\ .$$

Note that $v(p)$ is not necessarily a tangent vector of M, that is, v is generally not a vector field on M. If $k \in \mathbb{N} \cup \{\infty\}$, we say v is a C^k vector field (a **smooth** vector field in the case $k = \infty$) of N along M if every $p \in M$ has a submanifold chart (φ, U) of N for M such that $\varphi_* v \in C^k\big(\varphi(U \cap M), \mathbb{R}^n\big)$. It is easy to verify that this definition is chart-independent.[3]

Now suppose $(\cdot\,|\,\cdot)$ is a Riemannian metric on N and M is an oriented hypersurface in N, that is, $m = n - 1$. Then there is exactly one smooth vector field, $\nu := \nu_M$, of N along M with the properties

(i) $\nu(p)\perp T_pM$ for $p \in M$;

(ii) $|\nu(p)| = 1$ for $p \in M$; and

(iii) if $(v_1, \ldots, v_m)$ is a positive basis of T_pM, then $\big(\nu(p), v_1, \ldots, v_m\big)$ is a positive basis of T_pN.

We call ν a **positive unit normal field** along M or, for short, a **positive normal** of M.

Proof The claim follows an obvious modification of the proof of Example XI.5.3(g). ∎

(g) Suppose $\big(N, (\cdot\,|\,\cdot)\big)$ is a Riemannian manifold and either M is an oriented hypersurface in N or $M := \partial N$. Then every vector field v of N along M satisfies

$$v \lrcorner \omega_N = (v\,|\,\nu)\omega_M\ .$$

Proof Let $p \in M$, and let $(v_1, \ldots, v_m)$ be a positive ONB of T_pM. Then the vectors $\big(\nu(p), v_1, \ldots, v_m\big)$ are a positive ONB of T_pN. Therefore $v(p) \in T_pN$ has the basis representation

$$v(p) = \big(v(p)\,|\,\nu(p)\big)\nu(p) + \sum_{j=1}^{m}(v(p)\,|\,v_j)v_j\ .$$

---

[3]See Remark XI.4.2(a).

Therefore we get from the alternating property of ω_N that

$$
\begin{aligned}
v(p) \mathbin{\lrcorner} \omega_N(p)(v_1,\dots,v_m) &= \omega_N(p)\big(v(p),v_1,\dots,v_m\big) \\
&= \big(v(p)\,|\,\nu(p)\big)\omega_N(p)\big(\nu(p),v_1,\dots,v_m\big) \\
&= \big(v(p)\,|\,\nu(p)\big)\big(\nu(p) \mathbin{\lrcorner} \omega_N(p)(v_1,\dots,v_m)\big) \ .
\end{aligned}
$$

Now

$$
\nu(p) \mathbin{\lrcorner} \omega_N(p)(v_1,\dots,v_m) = \omega_N(p)\big(\nu(p),v_1,\dots,v_m\big) = 1
$$

and $\nu \mathbin{\lrcorner} \omega_N \in \Omega^m(M)$ imply $\nu \mathbin{\lrcorner} \omega_N = \omega_M$, and we are done. ∎

(h) Suppose $\big(N,(\,\cdot\,|\,\cdot\,)\big)$ is a Riemannian manifold and either M is an oriented hypersurface in N or $M := \partial N$. Also let v be a vector field of N along M with $(v\,|\,\nu) \in \mathcal{L}(M,d\lambda_M)$. Then it follows from (g) that

$$
\int_M v \mathbin{\lrcorner} \omega_N = \int_M (v\,|\,\nu)\, d\lambda_M \ .
$$

This integral is called the **flux of the vector field v through M** (in the direction of the positive normal). In the situation of (e),

$$
\int_M \vec{a} \cdot \vec{dF}
$$

is the flux of the vector field $\vec{a}$ through M.

To motivate this idea, we consider the situation of (e) with $N = X$ and assume that X is filled with a (fictitious) flowing fluid (or, more generally, a continuously deformable medium). We consider an (infinitesimal) fluid element, which at time $t = 0$, goes "through the point x". We denote by $\chi^t(x) := \chi(x,t)$ its position at time t. Therefore $t \mapsto \chi^t(x)$ is the trajectory of the element located at x at time $t = 0$.

Now suppose $v(y,t)$ is the velocity vector of the element located at y at time t. Then

$$
\frac{d\chi^t(x)}{dt} = v\big(\chi^t(x),t\big) \ ,
$$

where $d\chi^t(x)/dt$ is the derivative of $s \mapsto \chi^s(x)$ at $s = t$.

We assume that the flow has a well-defined (smooth) mass (or charge) density $\rho(x,t) > 0$ at every time. This means that the total mass (or charge) of the fluid contained in the (measurable) subset A of X is given at time t by

$$
\rho(A,t) := \int_A \rho(x,t)\, dx \ .
$$

We now consider a vector area element $d\vec{F}_x$ attached to the point x. Then the fluid that flows outward (and thus in the direction of the positive normal) through

dF_x in the time interval $[t, t + \Delta t]$ approximately fills an inclined cylinder of base area
dF_x and height $\Delta t\big(v(x,t) \mid \nu_F(x)\big)$.

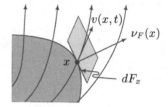

   Thus

$$\Delta t \rho(x,t)\big(v(x,t) \mid \nu(x)\big)\, dF_x$$

is approximately the fluid mass (or charge)
transported outward through $d\vec{F}_x$ in the interval $[t, t + \Delta t]$. (See Exercise X.6.1.)
These considerations show that the flux of the vector field $(\rho v)(\,\cdot\,, t)$ through M,
that is,

$$\int_M \big((\rho v)(\,\cdot\,, t) \mid \nu\big)\, d\lambda_M$$

specifies the rate at which total mass (or charge) flows out of M at time t.

(i) Suppose (N, g) is a pseudo-Riemannian manifold and either M is a hypersurface
in N or $M := \partial N$. Also let v be a vector field of N along M. Then we call

$$\int_M v \lrcorner\, \omega_N$$

the **flux of v through M**, if $v \lrcorner\, \omega_M$ is integrable over M. We have

$$\int_M v \lrcorner\, \omega_N = \int_M *\Theta v \ .$$

Proof   This follows from Exercise XI.6.4. ∎

The transformation theorem

The transformation theorem for the Lebesgue integral, one of the most important
aids both for concrete calculations and for theoretical purposes, has a globalization:

2.3 Theorem (transformation theorem)   *Suppose $f \in \mathrm{Diff}(M, N)$ is orientation-
preserving. Then an n-form ω is integrable on N if and only if $f^*\omega$ is integrable
on M.[4] Then $\int_N \omega = \int_M f^*\omega$.*

Proof   (i) Suppose ω is integrable on M and (φ, U) is a positive chart of M. Then
$(\psi, V) := \big(\varphi \circ f^{-1}, f(U)\big)$ is a positive chart of N, and f is an orientation-preserving
diffeomorphism from U to V. Also $\psi(V) = \varphi(U)$ and $\psi_* = (\varphi \circ f^{-1})_* = \varphi_* f^*$.
For an integrable differential form $\omega \in \Omega^m(V)$, we get from Remark 2.1(a) that

$$\int_V \omega = \int_{\psi(V)} \psi_* \omega = \int_{\varphi(U)} \varphi_*(f^*\omega) = \int_U f^*\omega \ .$$

Therefore the m-form $f^*\omega$ is integrable on U.

---

[4]Note that $M \cong N$ also implies $m = n$.

(ii) Let $\left\{\, (\varphi_j, U_j, \pi_j) \; ; \; j \in \mathbb{N} \,\right\}$ be a positive local system for M. Then $\left\{\, (\psi_j, V_j, \rho_j) \; ; \; j \in \mathbb{N} \,\right\}$ with

$$(\psi_j, V_j, \rho_j) := \left(\varphi_j \circ f^{-1}, f(U_j), \pi_j \circ f^{-1} \right)$$

is a positive local system for N. If ω is an integrable m-form of N and $j \in \mathbb{N}$, then $\rho_j \omega \,|\, V_j$ is an integrable m-form on V_j. It follows from (i) that

$$\int_{V_j} \rho_j \omega = \int_{U_j} f^*(\rho_j \omega) = \int_{U_j} (f^* \rho_j) f^* \omega = \int_{U_j} \pi_j f^* \omega \quad \text{for } j \in \mathbb{N} \,.$$

Thus we get from Remark 2.1(b) that

$$\int_N \omega = \sum_{j=0}^{\infty} \int_{V_j} \rho_j \omega = \sum_{j=0}^{\infty} \int_{U_j} \pi_j f^* \omega = \int_M f^* \omega \,.$$

In particular, $f^* \omega$ is integrable over M.

(iii) Now the claim follows by applying (ii) to f^{-1}. ∎

Fubini's theorem

Next we prove a global version of Fubini's theorem for the case of a product manifold. More precisely, we now assume that M or N is without boundary and set $L := M \times N$ and $\ell := m + n$. In addition, we provide L with the product orientation (see Example XI.4.17(b) and Exercise XI.4.3).

For every $(p, q) \in M \times N$ the product manifolds $\{p\} \times N$ and $M \times \{q\}$ are oriented, n- and m-dimensional, respectively, and submanifolds of L. Clearly the natural diffeomorphisms

$$\{p\} \times N \to N \,, \quad (p, q) \mapsto q \quad \text{and} \quad M \times \{q\} \to M \,, \quad (p, q) \mapsto p \qquad (2.10)$$

preserve orientation.

Because

$$T_{(p,q)} L = T_{(p,q)} \big(M \times \{q\} \big) \oplus T_{(p,q)} \big(\{p\} \times N \big)$$

(see Exercise XI.5.1), the diffeomorphisms (2.10) induce natural vector space isomorphisms

$$T_{(p,q)} L \to T_p M \times T_q N \quad \text{for } (p, q) \in M \times N \,.$$

In the following, we identify $\{p\} \times N$ with N and $M \times \{q\}$ with M by virtue of (2.10); thus $T_{(p,q)} L$ is identified with $T_p M \times T_q N$, so that we can write

$$T_{(p,q)} L = T_p M \oplus T_q N \quad \text{for } (p, q) \in M \times N \,, \qquad (2.11)$$

if the meaning is clear from context.

Let ω be an ℓ-form on L, and let $(p, q) \in M \times N$. For $a_1, \ldots, a_m \in T_p M$,

$$(b_1, \ldots, b_n) \mapsto \omega(p, q)(a_1, \ldots, a_m, b_1, \ldots, b_n)$$

belongs to $\bigwedge^n T_q^* N$. Therefore

$$\widehat{\omega}(p, \cdot)(a_1, \ldots, a_m) := \omega(p, \cdot)(a_1, \ldots, a_m, \cdot, \ldots, \cdot)$$

is an n-form on N. The map

$$\widehat{\omega}(p, \cdot) := \big((a_1, \ldots, a_m) \mapsto \widehat{\omega}(p, \cdot)(a_1, \ldots, a_m)\big)$$

is m-linear, alternating on $(T_p M)^m$, and assumes its values in the vector space of n-forms on N. Therefore it is an **n-form-valued m-form** on M. In addition, we say $\widehat{\omega}(p, \cdot)$ is **integrable** over N if the n-form $\widehat{\omega}(p, \cdot)(a_1, \ldots, a_m)$ is integrable over N for every m-tuple $(a_1, \ldots, a_m) \in (T_p M)^m$. Then it is clear that

$$\int_N \widehat{\omega}(p, \cdot) := \Big((a_1, \ldots, a_m) \mapsto \int_N \widehat{\omega}(p, \cdot)(a_1, \ldots, a_m)\Big)$$

belongs to $\bigwedge^m T_p^* M$.

Analogously,

$$\widehat{\omega}(\cdot, q)(b_1, \ldots, b_n) := \omega(\cdot, q)(\cdot, \ldots, \cdot, b_1, \ldots, b_n) \quad \text{for } b_1, \ldots, b_n \in T_q N$$

defines an m-form-valued n-form $\widehat{\omega}(\cdot, q)$ on N, and $\widehat{\omega}(\cdot, q)$ is integrable over M if the m-form $\widehat{\omega}(\cdot, q)(b_1, \ldots, b_n)$ is integrable over M for every n-tuple $(b_1, \ldots, b_n)$ of vectors in $T_q N$. In this case,

$$\int_M \widehat{\omega}(\cdot, q) := \Big((b_1, \ldots, b_n) \mapsto \int_M \widehat{\omega}(\cdot, q)(b_1, \ldots, b_n)\Big)$$

belongs to $\bigwedge^n T_q^* N$.

We can now prove a useful analogue of Fubini's theorem for differential forms.

2.4 Theorem (Fubini) *Suppose ω is an integrable ℓ-form on L. Then these three statements are valid:*

(i) *$\widehat{\omega}(p, \cdot)$ is integrable over N for λ_M-almost every $p \in M$, and $\widehat{\omega}(\cdot, q)$ is integrable over M for λ_N-almost every $q \in N$.*

(ii) *The m-form*

$$\int_N \omega := \Big(p \mapsto \int_N \widehat{\omega}(p, \cdot)\Big),$$

which is defined λ_M-a.e., is integrable over M, and the n-form

$$\int_M \omega := \Big(q \mapsto \int_M \widehat{\omega}(\cdot, q)\Big),$$

which is defined λ_N-a.e., is integrable over N.

(iii)
$$\int_L \omega = \int_M \left(\int_N \omega \right) = \int_N \left(\int_M \omega \right) .$$

Proof (a) Let (φ, U) be a positive chart of M with $\varphi = (x^1, \ldots, x^m)$. Let (ψ, V) be a positive chart of N with $\psi = (y^1, \ldots, y^n)$. Then $(\chi, W) := (\varphi \times \psi, U \times V)$ is a positive product chart of L. Finally, suppose

$$\omega := a \, dx^1 \wedge \cdots \wedge dx^m \wedge dy^1 \wedge \cdots \wedge dy^n$$

is integrable over W. Then $\chi_* a$ belongs to $\mathcal{L}_1\big(\varphi(U) \times \psi(V), \lambda_{m+n}\big)$. It therefore follows from Theorem X.6.9 that

$$\chi_* a(x, \cdot) \in \mathcal{L}_1\big(\psi(V), \lambda_n\big) \quad \text{for } \lambda_m\text{-almost every } x \in \varphi(U) , \tag{2.12}$$

that

$$\int_{\psi(V)} \chi_* a(\cdot, y) \, dy \in \mathcal{L}_1\big(\varphi(U), \lambda_m\big) \tag{2.13}$$

(where this function is only defined λ_m-a.e.), and that

$$\int_{\chi(W)} \chi_* a \, d\lambda_{m+n} = \int_{\varphi(U)} \left(\int_{\psi(V)} \chi_* a(x, y) \, dy \right) dx . \tag{2.14}$$

For $(p, q) \in U \times V$ and $v_1, \ldots, v_m \in T_p U$, it follows that

$$\widehat{\omega}(p, \cdot)(v_1, \ldots, v_m) = a(p, \cdot) \, dx^1 \wedge \cdots \wedge dx^m \wedge dy^1 \wedge \cdots \wedge dy^n (v_1, \ldots, v_m, \cdot, \ldots, \cdot)$$
$$= \alpha(p) a(p, \cdot) \, dy^1 \wedge \cdots \wedge dy^n$$

with

$$\alpha(p) := dx^1 \wedge \cdots \wedge dx^m (p)(v_1, \ldots, v_m) ,$$

as, because of (2.11), one may read from (XI.2.3). Therefore we get

$$\psi_* \widehat{\omega}(p, \cdot)(v_1, \ldots, v_m) = \alpha(p) \psi_* \big(a(p, \cdot)\big) \, dy^1 \wedge \cdots \wedge dy^n .$$

Now we note the relation

$$\psi_* \big(a(p, \cdot)\big)(y) = a\big(p, \psi^{-1}(y)\big) = a\big(\varphi^{-1}(x), \psi^{-1}(y)\big) = \chi_* a(x, y)$$

for $x = \varphi(p)$ and $y \in \psi(V)$. With this and by means of Remark 2.1(a) we find because of (2.12) that

$$\int_V \widehat{\omega}(p, \cdot) = \int_{\psi(V)} a\big(p, \psi^{-1}(y)\big) \, dy \, dx^1 \wedge \cdots \wedge dx^m (p) \in \bigwedge^m T_p^* U$$

is well defined for λ_M-almost every $p \in U$. Also, it follows from (2.13) that

$$\varphi_* \int_V \widehat{\omega}(p, \cdot)(x) = \int_{\psi(V)} a\big(\varphi^{-1}(x), \psi^{-1}(y)\big) \, dy \, dx^1 \wedge \cdots \wedge dx^m$$
$$= \int_{\psi(V)} \chi_* a(x, y) \, dy \, dx^1 \wedge \cdots \wedge dx^m$$

for λ_M-almost every $p \in U$ with $x = \varphi(p)$. Hence (2.14) and Remark 2.1(a) imply

$$\int_U \int_V \omega = \int_U \int_V \widehat{\omega}(p, \cdot) = \int_{\varphi(U)} \int_{\psi(V)} \chi_* a(x, y) \, dy \, dx$$

$$= \int_{\chi(W)} \chi_* a \, d\lambda_{m+n} = \int_W \omega \ .$$

We get the remaining statements in this case by exchanging the roles of U and V,

(b) Let $\{ (\varphi_j, U_j, \pi_j) \ ; \ j \in \mathbb{N} \}$ and $\{ (\psi_j, V_j, \rho_j) \ ; \ j \in \mathbb{N} \}$ be positive local systems for M and N, respectively. Then, with $\pi_j \otimes \rho_k(p, q) := \pi_j(p)\rho_k(q)$ for $(p, q) \in M \times N$,

$$\{ (\varphi_j \times \psi_k, U_j \times V_k, \pi_j \otimes \rho_k) \ ; \ (j, k) \in \mathbb{N}^2 \}$$

is a positive local system for L. We get from (a) that

$$\int_{U_j \times V_k} \pi_j \otimes \rho_k \omega = \int_{U_j} \pi_j \int_{V_k} \rho_k \omega \quad \text{for } (j, k) \in \mathbb{N}^2 \ .$$

Now the claim follows from Remark 2.1(b). ∎

2.5 Corollary   Let $L := M \times N$, and let $\pi_1 \colon L \to M$ and $\pi_2 \colon L \to N$ be the canonical projections. If α is an integrable m-form on M and β is an integrable n-form on N, then $\gamma := \pi_1^* \alpha \wedge \pi_2^* \beta$ is integrable over L, and $\int_L \gamma = \int_M \alpha \int_N \beta$.

2.6 Remark   (regularity)   It suffices that M and N are C^1 manifolds and that the f in Theorem 2.3 is a C^1 diffeomorphism. ∎

Calculations of several integrals

Formula (2.5) and the theorems of this section lay the foundation for calculating volumes. We illustrate this with some examples, in which we always use the standard metric.

2.7 Examples   **(a)** $(A, B) \in \mathcal{L}_M \times \mathcal{L}_N$ satisfy $\lambda_{M \times N}(A \times B) = \lambda_M(A)\lambda_N(B)$. In particular, it follows that

$$\text{vol}(M \times N) = \text{vol}(M) \, \text{vol}(N) \ .$$

Proof   We consider the $(m+n)$-form $\omega := \chi_{A \times B}\omega_L = \chi_A \omega_M \wedge \chi_B \omega_N$ on the product manifold $L := M \times N$ (see Exercise XI.5.1). Then the claim follows (in consideration of the convention $0 \cdot \infty := \infty \cdot 0 := 0$) from Corollary 2.5. ∎

(b) (spheres)   For $m \geq 1$, the canonically oriented m-sphere S^m in $\mathbb{R}^{m+1}$ satisfies

$$\int_{S^m} \sum_{j=1}^{m+1} (-1)^{j-1} x^j \, dx^1 \wedge \cdots \wedge \widehat{dx^j} \wedge \cdots \wedge dx^{m+1} = \text{vol}(S^m) \ .$$

In particular,[5]

$$\int_{S^1} x\,dy - y\,dx = 2\pi \tag{2.15}$$

and

$$\int_{S^2} x\,dy \wedge dz + y\,dz \wedge dx + z\,dx \wedge dy = 4\pi \ .$$

Proof This follows from Examples XI.5.3(c) and 1.13(c). ∎

(c) (star-shaped domains) Let $m \geq 1$, and define f as the "**polar coordinate diffeomorphism**"

$$(0,\infty) \times S^m \to \mathbb{R}^{m+1}\backslash\{0\} \ , \quad (r,\sigma) \mapsto r\sigma := ri(\sigma)$$

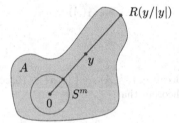

with the canonical embedding $i: S^m \hookrightarrow \mathbb{R}^{m+1}$.
Then

$$f^*(dy^1 \wedge \cdots \wedge dy^{m+1}) = r^m\,dr \wedge \omega_{S^m} \ . \tag{2.16}$$

Also let $R \in \mathcal{L}_1(S^m, \mathbb{R}^+)$ and

$$A := \{\, y \in \mathbb{R}^{m+1}\backslash\{0\} \ ; \ |y| \leq R(y/|y|) \,\} \ .$$

Therefore A is star shaped (with respect to 0), and its "outer boundary" is parametrized over the m-sphere. It satisfies

$$\lambda_{m+1}(A) = \frac{1}{m+1}\int_{S^m} R^{m+1}\,d\lambda_{S^m} \ . \tag{2.17}$$

Proof Let (φ, U) be a positive chart of S^m with $h := i \circ \varphi^{-1}: \varphi(U) \to \mathbb{R}^{m+1}$ the associated parametrization. Then f has the local representation

$$f_\psi : (0,\infty) \times \varphi(U) \to \mathbb{R}^{m+1} \ , \quad (r,x) \mapsto rh(x)$$

with respect to the positive chart

$$(\psi, V) := \big(\mathrm{id} \times \varphi, (0,\infty) \times U\big)$$

of $M := (0,\infty) \times S^m$ and the trivial chart of $\mathbb{R}^{m+1}\backslash\{0\}$. Therefore

$$df_\psi^j(r,x) = h^j(x)\,dr + r\,dh^j(x) \quad \text{for } 1 \leq j \leq m+1 \ .$$

By an easy induction, this gives

$$f_\psi^*(dy^1 \wedge \cdots \wedge dy^{m+1}) = df_\psi^1 \wedge \cdots \wedge df_\psi^{m+1}$$

$$= r^m \sum_{j=1}^{m+1} (-1)^{j-1} h^j\,dr \wedge dh^1 \wedge \cdots \wedge \widehat{dh^j} \wedge \cdots \wedge dh^{m+1}$$

$$+ r^{m+1} dh^1 \wedge \cdots \wedge dh^{m+1} \ .$$

---

[5]We already calculated the line integral (2.15) in Example VIII.4.2(a). Now we *understand* this formula.

From $|h|^2 = 1$, we get $\sum_{j=1}^{m+1} h^j \, dh^j = 0$, that is, the covectors $dh^1(x), \ldots, dh^{m+1}(x)$ are linearly independent for every $x \in \varphi(U)$. Therefore $dh^1 \wedge \cdots \wedge dh^{m+1} = 0$. From Example XI.5.3(c), it follows that

$$f_\psi^*(dy^1 \wedge \cdots \wedge dy^{m+1}) = r^m \, dr \wedge \sum_{j=1}^{m+1} (-1)^{j-1} h^j \, dh^1 \wedge \cdots \wedge \widehat{dh^j} \wedge \cdots \wedge dh^{m+1}$$

$$= r^m \, dr \wedge h^* \omega_{S^m} = \psi_*(r^m \, dr \wedge \omega_{S^m}) \ .$$

Because this is true for every positive chart of S^m, we get (2.16). This also implies that f is an orientation-preserving diffeomorphism from M to $N := \mathbb{R}^{m+1} \setminus \{0\}$.

We set $\omega := \chi_A \omega_N = \chi_A \, dy^1 \wedge \cdots \wedge dy^{m+1}$. Then the transformation theorem gives

$$\lambda_{m+1}(A) = \lambda_N(A) = \int_N \omega = \int_M f^* \omega = \int_M f^* \chi_A r^m \, dr \wedge \omega_{S^m}$$

$$= \int_{f^{-1}(A)} r^m \, dr \wedge \omega_{S^m} \ .$$

Because $f^{-1}(A) = \{ (r, \sigma) \in (0, \infty) \times S^m \ ; \ 0 < r \le R(\sigma) \}$, it follows from Fubini's theorem that

$$\int_{f^{-1}(A)} r^m \, dr \wedge \omega_{S^m} = \int_{S^m} \Big(\int_0^{R(\sigma)} r^m \, dr \Big) \omega_{S^m}(\sigma) = \frac{1}{m+1} \int_{S^m} R^{m+1} \omega_{S^m}$$

$$= \frac{1}{m+1} \int_{S^m} R^{m+1} \, d\lambda_{S^m} \ . \ \blacksquare$$

(d) (conical slices of spheres) Let B be a measurable subset of S^m for $m \ge 1$. Then

$$K(rB) := \{ t\sigma \ ; \ 0 \le t \le 1, \ \sigma \in rB \}$$

is a cone whose tip is at the origin and whose base is the subset of rB of the sphere rS^m. We have[6]

$$(m+1) \operatorname{vol}\big(K(rB)\big) = r \operatorname{vol}(rB) \ .$$

For $B := S^m$, we again find the formula $r \operatorname{vol}(rS^m) = (m+1) \operatorname{vol}(r\mathbb{B}^{m+1})$ of Example 1.13(c).

Proof With $R := r\chi_B$, this follows from (c) because of (1.10). $\blacksquare$

(e) (integration using polar coordinates) Let $g \in \mathcal{L}_0(\mathbb{R}^n)$. Then g is integrable if and only if

$$\big(r \mapsto g(r\sigma) r^{n-1} \big) \in \mathcal{L}_1\big((0, \infty)\big) \quad \text{for almost every } \sigma \in S^{n-1}$$

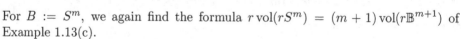

[6]The formula is the analogue of the statement of Exercise X.6.1 for cones with "flat bases".

and
$$\left(\sigma \mapsto \int_0^\infty g(r\sigma)r^{n-1}\,dr\right) \in \mathcal{L}_1(S^{n-1}) \ .$$

Then we have the formula[7]
$$\int_{\mathbb{R}^n} g\,dx = \int_{S^{n-1}} \int_0^\infty g(r\sigma)r^{n-1}\,dr\,d\lambda_{S^{n-1}} \ . \tag{2.18}$$

If g is rotationally symmetric with $g(x) = \overset{\bullet}{g}(|x|)$ for $x \in \mathbb{R}^n$, then g is integrable if and only if $\overset{\bullet}{g}(r)r^{n-1}$ is integrable over $(0, \infty)$. In this case, (2.18) reduces to

$$\int_{\mathbb{R}^n} g\,dx = \mathrm{vol}(S^{n-1}) \int_0^\infty \overset{\bullet}{g}(r)r^{n-1}\,dr \ ,$$

as we already know from Theorem X.8.11.

Proof We consider the n-form $\omega := g\,dy^1 \wedge \cdots \wedge dy^n$ on $N := \mathbb{R}^n \backslash \{0\}$. Then ω is integrable on N if and only if g belongs to $\mathcal{L}_1(\mathbb{R}^n)$. With the polar coordinate diffeomorphism of (c), we get from (2.16) that

$$f^*\omega = (f^*g)r^{n-1}\,dr \wedge \omega_{S^{n-1}} \ .$$

Now the claim follows from Theorem 2.3 and Corollary 2.5. ∎

Flows of vector fields

By a (global) **flow** on M, we mean a smooth map
$$\chi: M \times \mathbb{R} \to M \ , \quad (p,t) \mapsto \chi^t(p) := \chi(p,t)$$

for which
$$\chi^0 = \mathrm{id}_M \quad \text{and} \quad \chi^{s+t} = \chi^s \circ \chi^t \quad \text{for } s,t \in \mathbb{R} \ . \tag{2.19}$$

From (2.19), it follows that
$$\chi^t \in \mathrm{Diff}(M,M) \quad \text{and} \quad (\chi^t)^{-1} = \chi^{-t} \quad \text{for } t \in \mathbb{R} \ .$$

Because $\chi(p,\cdot) \in C^\infty(\mathbb{R},M)$ for $p \in M$, the vector $v(p) := T_0\chi(p,\cdot)(0,1) \in T_pM$ is well defined, and v is a smooth vector field on M. It also follows from (2.19) that
$$\frac{d\chi^t(p)}{dt} = v(\chi^t(p)) \quad \text{for } p \in M \text{ and } t \in \mathbb{R} \ . \tag{2.20}$$

This means that the **trajectory** $\chi(p,\cdot)$ is a global solution of the initial value problem (in $\mathbb{R}^{\overline{m}}$)
$$\dot{y} = v(y) \quad \text{and} \quad y(0) = p \tag{2.21}$$

for every $p \in M$.

---

[7]See Proposition X.8.9.

In the following, we denote by $\mathcal{V}_c^k(M)$ the $C^k(M)$ submodule of $\mathcal{V}^k(M)$ of C^k **vector fields with compact support**. We put $\mathcal{V}_c(M) := \mathcal{V}_c^\infty(M)$.

Conversely, suppose $v \in \mathcal{V}_c(M)$ and M is without boundary. Then it is shown in the theory of ordinary differential equations[8] that there is exactly one flow χ on M, the **flow generated by** v, that satisfies (2.20). Regarding $\chi(p, \cdot)$ as the trajectory of a "fluid element" that is found at the point p at time $t = 0$, we see that $v(p)$ is the velocity[9] at which this element passes through p. In this interpretation, χ^t is a snapshot of the entire flow field at time t.

We will now derive a connection between the divergence of a vector field and the flow it generates. So we first prove a result about linear, nonautonomous differential equations.

2.8 Proposition (Liouville)   *Suppose $A \in C\big(\mathbb{R}, \mathcal{L}(\mathbb{R}^m)\big)$ and $X \in C^1\big(\mathbb{R}, \mathcal{L}(\mathbb{R}^m)\big)$ is a solution of the homogeneous linear differential equation*

$$\dot{Y} = A(t)Y \quad \text{for } t \in \mathbb{R}$$

in $\mathcal{L}(\mathbb{R}^m)$. Then $W := \det(X)$ is a solution of the scalar equation

$$\dot{y} = \mathrm{tr}\big(A(t)\big)y \quad \text{for } t \in \mathbb{R} . \tag{2.22}$$

Therefore

$$W(t) = W(t_0)e^{\int_{t_0}^t \mathrm{tr}(A(s))\,ds} \quad \text{for } t, t_0 \in \mathbb{R} .$$

Proof   Let $r \in \mathbb{R}$. For $\eta \in \mathbb{R}^m$, we consider the initial value problem

$$\dot{y} = A(t)y \quad \text{for } t \in \mathbb{R} , \qquad y(r) = \eta . \tag{2.23}_{r,\eta}$$

From the Picard–Lindelöf theorem (Theorem VII.8.14), it follows easily that this problem has a unique global solution $u(\,\cdot\,, r, \eta) \in C^1(\mathbb{R}, \mathbb{R}^m)$.[10] Then

$$u\big(\,\cdot\,, s, u(s, r, \eta)\big) \quad \text{for } r, s \in \mathbb{R}$$

is the unique solution of $(2.23)_{s,u(s,r,\eta)}$. In other words, we follow the solution $u(\,\cdot\,, r, \eta)$ of $(2.23)_{r,\eta}$ until time s. Then we "start again", that is, we solve the differential equation $\dot{y} = A(t)y$ anew by now taking the value $u(s, r, \eta)$ as the starting value

$$r \bullet \eta$$
$$s \diagup$$
$$u(s, r, \eta)$$
$$t \quad u(t, r, \eta) = u\big(t, s, u(s, r, \eta)\big)$$

---

[8]For example [Ama95], if M is open in $\mathbb{R}^m$. This carries over to the general case by means of local charts (see[Con93], [Lan95]).

[9]Note that, in contrast to Remark 2.2(h), we consider here "stationary", that is, time-independent vector fields.

[10]See Exercise VII.8.13.

at the time point s. However, we can also follow the solution $u(\,\cdot\,,r,\eta)$ until time t. Then that $(2.23)_{r,\eta}$ has a unique solution for every $(r,\eta) \in \mathbb{R} \times \mathbb{R}^m$ implies

$$u(t,r,\eta) = u\big(t,s,u(s,r,\eta)\big) \quad \text{for } r,s,t \in \mathbb{R} \text{ and } \eta \in \mathbb{R}^m \ . \tag{2.24}$$

The linearity of the differential equation $\dot{y} = A(t)y$ and that it has a unique solution imply easily that $\eta \mapsto u(t,r,\eta)$ is a linear function for any pair $(t,r) \in \mathbb{R}^2$. Therefore

$$u(t,r,\eta) = U(t,r)\eta \quad \text{for } t,r \in \mathbb{R} \text{ and } \eta \in \mathbb{R}^m \ , \tag{2.25}$$

with $U(t,r) \in \mathcal{L}(\mathbb{R}^m) = \mathbb{R}^{m \times m}$. Hence we read from (2.24) that

$$U(t,r) = U(t,s)U(s,r) \ , \quad U(t,t) = 1_m \quad \text{for } r,s,t \in \mathbb{R} \ . \tag{2.26}$$

Finally, it follows from (2.25) and because $u(\,\cdot\,,r,\eta)$ is the solution of $(2.23)_{r,\eta}$ for every $\eta \in \mathbb{R}^m$ that

$$\partial_1 U(t,r) = A(t)U(t,r) \quad \text{for } t \in \mathbb{R} \ , \qquad U(r,r) = 1_m \ . \tag{2.27}$$

This shows that $U(\,\cdot\,,r)$ is the unique global solution to the initial value problem in $E := \mathcal{L}(\mathbb{R}^m)$ given by[11]

$$\dot{Y} = A(t)Y \quad \text{for } t \in \mathbb{R} \ , \qquad Y(r) = 1_m \ .$$

Suppose now $B = [b_1, \ldots, b_m] \in \mathbb{R}^{m \times m}$. Then these considerations imply that

$$U(\,\cdot\,,r)B = \big[U(\,\cdot\,,r)b_1, \ldots, U(\,\cdot\,,r)b_m\big]$$

is the unique global solution of

$$\dot{Y} = A(t)Y \quad \text{for } t \in \mathbb{R} \ , \qquad Y(r) = B$$

in E. If X is some solution of $\dot{Y} = A(t)Y$, then it follows with $B := X(r)$ that

$$X(t) = U(t,r)X(r) \quad \text{for } r,t \in \mathbb{R} \ . \tag{2.28}$$

We fix r and set $\alpha(t) := \det\big(U(t,r)\big)$. Then Example VII.4.8(a), together with (2.27) and the relation $U(t,r) = [u_1(t), \ldots, u_m(t)]$, give

$$\dot{\alpha}(t) = \sum_{j=1}^{m} \det\big[u_1(t), \ldots, u_{j-1}(t), \dot{u}_j(t), u_{j+1}(t), \ldots, u_m(t)\big]$$
$$= \sum_{j=1}^{m} \det\big[u_1(t), \ldots, u_{j-1}(t), A(t)u_j(t), u_{j+1}(t), \ldots, u_m(t)\big] \ . \tag{2.29}$$

---

[11]On the other hand, see Exercise VII.8.13.

For $t = r$, we read from $U(r, r) = 1_m$ that $u_j(r) = e_j$ for $1 \leq j \leq m$, where $(e_1, \ldots, e_m)$ denotes the standard basis of $\mathbb{R}^m$. Thus (2.29) gives

$$
\begin{aligned}
\dot{\alpha}(r) &= \sum_{j=1}^{m} \det\big[e_1, \ldots, e_{j-1}, a_j(r), e_{j+1}, \ldots, e_m\big] \\
&= \sum_{j=1}^{m} a_j^j(r) = \mathrm{tr}\big(A(r)\big)
\end{aligned}
\tag{2.30}
$$

with $A(r) = [a_1(r), \ldots, a_m(r)]$. Finally, $\dot{X}(t) = \partial_1 U(t, r) X(r)$ follows from (2.28), and therefore

$$
\dot{W}(t) = \dot{\alpha}(t) W(r) \quad \text{for } t \in \mathbb{R} .
$$

From this and (2.30), we get

$$
\dot{W}(r) = \mathrm{tr}\big(A(r)\big) W(r) \quad \text{for } r \in \mathbb{R} ,
$$

which shows that W satisfies Equation (2.22). The claim is now a consequence of Example VII.8.11(e). ∎

2.9 Remarks   (a) The proof of Proposition 2.8 has generated results for the initial value problem $(2.23)_{r,\eta}$ which generalize the corresponding results from Section VII.1 for the linear differential equation $\dot{x} = Ax$ with constant matrix $A \in \mathcal{L}(\mathbb{R}^n)$. In particular, this case has

$$
e^{(t-s)A} = U(t, s) \quad \text{for } s, t \in \mathbb{R} .
$$

In the time-independent case, Theorem VII.1.11(ii) says that $t \mapsto e^{tA}$ is a group homomorphism; in the nonautonomous case, that is, in the case of "time-dependent coefficients", this statement is replaced by (2.26).

(b) If X is a matrix solving the differential equation $\dot{Y} = A(t)Y$ in $\mathcal{L}(\mathbb{R}^m)$, then $W := \det(X)$ is called the **Wronskian** or **Wronski determinant**. From the explicit form of W given in Proposition 2.8, we learn that $W(t)$ is distinct from zero for every $t \in \mathbb{R}$ if and only if $W(t_0) \neq 0$ for some $t_0 \in \mathbb{R}$. In this case, the columns $x_1, \ldots, x_m$ of X form a **fundamental system** of the differential equation $\dot{y} = A(t)y$ in $\mathbb{R}^m$, since we can easily check that every solution of this equation can be written as a linear combination of $x_1, \ldots, x_m$. ∎

Now we can make the aforementioned connection between the flow of a vector field and its divergence.

2.10 Proposition   *Let M be without boundary and pseudo-Riemannian. Also suppose $v \in \mathcal{V}_c(M)$ and that χ is the flow on M generated by v. Then*

$$
\mathrm{div}(v)\omega_M = (\chi^t)_* \frac{d}{ds}\big[(\chi^s)^* \omega_M\big]\Big|_{s=t} \quad \text{for } t \in \mathbb{R} ,
$$

and therefore in particular

$$\mathrm{div}(v)\omega_M = \frac{d}{dt}\big[(\chi^t)^*\omega_M\big]\Big|_{t=0}.$$

Proof Because the claim is a local statement, it suffices to prove the claim in local orthonormal coordinates. Therefore we can assume that M is open in $\mathbb{R}^m$ and define $\omega := \omega_M = dx^1 \wedge \cdots \wedge dx^m$.

We take the derivative of (2.20) with respect to p. The chain rule then gives

$$[\partial\chi^t]^{\cdot} = \big((\chi^t)^*\partial v\big)\partial\chi^t \quad \text{for } t \in \mathbb{R},$$

with $\partial := \partial_p$. Thus Liouville's theorem (applied to every fixed $p \in M$) gives

$$\big[\det(\partial\chi^t)\big]^{\cdot} = \mathrm{tr}\big[(\chi^t)^*\partial v\big]\det(\partial\chi^t) \quad \text{for } t \in \mathbb{R}. \tag{2.31}$$

Then Example XI.6.8(a) gives

$$\mathrm{tr}\big[(\chi^t)^*\partial v\big] = (\chi^t)^* \sum_{j=1}^m \partial_j v^j = (\chi^t)^* \,\mathrm{div}\, v. \tag{2.32}$$

By Example XI.3.4(c), we know that

$$(\chi^t)^*\omega = \det(\partial\chi^t)\omega \quad \text{for } t \in \mathbb{R}.$$

From this, it follows from (2.31) and (2.32) that

$$\begin{aligned}
\big[(\chi^t)^*\omega\big]^{\cdot} &= \big[\det(\partial\chi^t)\big]^{\cdot}\omega = \big((\chi^t)^* \,\mathrm{div}\, v\big)\det(\partial\chi^t)\omega \\
&= (\chi^t)^* \,\mathrm{div}\, v\,(\chi^t)^*\omega = (\chi^t)^*\big(\mathrm{div}(v)\omega\big)
\end{aligned}$$

for $t \in \mathbb{R}$. $\blacksquare$

The transport theorem

Proposition 2.10 makes possible a geometric interpretation of the divergence of a vector field. To explain this, we first prove the important transport theorem, which is particularly useful in continuum mechanics.

2.11 Theorem (transport theorem) *Suppose M is without boundary and pseudo-Riemannian. Also suppose $v \in \mathcal{V}_c(M)$ and χ is the flow on M generated by v. Finally let $f \in \mathcal{E}(M \times \mathbb{R})$. Then for every relatively compact set $A \in \mathcal{L}_M$ we have*

$$\frac{d}{dt}\int_{A^t} f(\cdot, t)\, d\lambda_M = \int_{A^t} \big[\partial_2 f(\cdot, t) + \mathrm{div}\big(f(\cdot, t)v\big)\big]\, d\lambda_M \quad \text{for } t \in \mathbb{R},$$

with $A^t := \chi^t(A)$.

Proof Because $\chi^t \in \mathrm{Diff}(M, M)$, it follows that $A^t \in \mathcal{L}_M$ whenever $A \in \mathcal{L}_M$ (see Exercise 1.1), and also $\overline{A^t} = (\overline{A})^t$. Therefore A^t is relatively compact, and Proposition 1.9(b) implies that $f(\cdot, t)$ is integrable over A^t. Thus the m-form[12] $\omega_t := \chi_{A^t} f(\cdot, t)\omega_M$ is integrable over M, and the transformation theorem gives

$$\int_{A^t} f(\cdot, t)\, d\lambda_M = \int_M \omega_t = \int_M (\chi^t)^* \omega_t = \int_A (\chi^t)^* \big(f(\cdot, t)\omega_M\big)\ .$$

The theorem about the differentiability of parameter-dependent integrals (Theorem X.3.18) then gives

$$\frac{d}{dt} \int_{A^t} f(\cdot, t)\, d\lambda_M = \int_A \frac{d}{dt}(\chi^t)^*\big(f(\cdot, t)\omega_M\big)\ . \tag{2.33}$$

Because $(\chi^t)^*\big(f(\cdot, t)\omega_M\big) = f(\chi^t, t)(\chi^t)^*\omega_M$, Proposition 2.10 says

$$\begin{aligned}
\frac{d}{dt}(\chi^t)^*\big(f(\cdot, t)\omega_M\big) &= \Big(\frac{d}{dt}f(\chi^t, t)\Big)(\chi^t)^*\omega_M + f(\chi^t, t)\frac{d}{dt}\big((\chi^t)^*\omega_M\big) \\
&= \Big(\big\langle (\chi^t)^*\, df(\cdot, t), \frac{d}{dt}\chi^t\big\rangle + (\chi^t)^*\partial_2 f(\cdot, t)\Big)(\chi^t)^*\omega_M \\
&\quad + (\chi^t)^* f(\cdot, t)(\chi^t)^*\big(\mathrm{div}(v)\omega_M\big)\ .
\end{aligned}$$

Considering (2.20), we now find

$$\begin{aligned}
\frac{d}{dt}(\chi^t)^*\big(f(\cdot, t)\omega_M\big) &= (\chi^t)^*\big[\big(\langle df(\cdot, t), v\rangle + \partial_2 f(\cdot, t) + f(\cdot, t)\,\mathrm{div}\,v\big)\omega_M\big] \\
&= (\chi^t)^*\big[\big(\partial_2 f(\cdot, t) + (\mathrm{grad}\,f(\cdot, t)\mid v)_M + f(\cdot, t)\,\mathrm{div}\,v\big)\omega_M\big] \\
&= (\chi^t)^*\big[\big(\partial_2 f(\cdot, t) + \mathrm{div}(f(\cdot, t)v)\big)\omega_M\big]\ ,
\end{aligned}$$

where we have used Proposition XI.6.11(ii) in the last step. Now the claim follows from (2.33) and the transformation theorem. ∎

2.12 Corollary Let $t \in \mathbb{R}$. Then for every relatively compact set $A \in \mathcal{L}_M$ and every $v \in \mathcal{V}_c(M)$, we have

$$\frac{d}{dt}\,\mathrm{vol}_M(A^t) = \int_{A^t} \mathrm{div}\,v\, d\lambda_M \quad \text{for } v \in \mathcal{V}_c(M)\ .$$

A vector field $v \in \mathcal{V}(M)$ is said to be **divergence free** if $\mathrm{div}\,v = 0$. If v has compact support, then the flow χ generated by v is said to be **volume preserving** if

$$\mathrm{vol}_M(A^t) = \mathrm{vol}_M(A) \quad \text{for } t \in \mathbb{R} \text{ and } A \in \mathcal{L}_M\ .$$

From Corollary 2.12, we find that for relatively compact measurable subsets A of M, the volume of the transported set A^t increases with time if $\mathrm{div}\,v \geq 0$; when $\mathrm{div}\,v \leq 0$, this volume decreases.

---

[12]χ_{A^t} is the characteristic function of A^t.

2.13 Proposition *Suppose M is without boundary and pseudo-Riemannian. Also let $v \in \mathcal{V}_c(M)$. Then v is divergence free if and only if the flow generated by v is volume preserving.*

Proof "$\Rightarrow$" Suppose $\operatorname{div} v = 0$. Then it follows from Corollary 2.12 that

$$\operatorname{vol}(A^t) = \operatorname{vol}(A) \quad \text{for } t \in \mathbb{R} \tag{2.34}$$

for every compact subset of A of M. Because M is σ-compact and λ_M regular, we find that (2.34) holds for every $A \in \mathcal{L}_M$.

"$\Leftarrow$" Suppose χ is volume preserving. Then Corollary 2.12 shows in particular that

$$\int_A \operatorname{div} v \, d\lambda_M = 0 \quad \text{for } A \in \mathcal{L}_M \text{ with } \overline{A} \subset\subset M . \tag{2.35}$$

Because $\operatorname{div} v$ belongs to $C_c(M)$, (2.35) implies

$$\int_M f \operatorname{div} v \, d\lambda_M = 0 \quad \text{for } f \in \mathcal{S}(M, \lambda_M) . \tag{2.36}$$

Because $\operatorname{div} v \in \mathcal{L}_2(M, \lambda_M)$ and since $\mathcal{S}(M, \lambda_M)$ is dense in $\mathcal{L}_2(M, \lambda_M)$ according to Proposition X.4.8, there is a sequence (f_j) in $\mathcal{S}(M, \lambda_M)$ such that $f_j \to \operatorname{div} v$ in $\mathcal{L}_2(M, \lambda_M)$. Therefore it follows from the continuity of the scalar product in $\mathcal{L}_2(M, \lambda_M)$ (that is, the Cauchy–Schwarz inequality) that

$$\int_M (\operatorname{div} v)^2 \, d\lambda_M = \lim_{j \to \infty} \int_M f_j \operatorname{div} v \, d\lambda_M = 0 .$$

Now the claim follows from Remark X.3.3(c). ∎

2.14 Example (continuity equation) Suppose X is open and bounded in $\mathbb{R}^3$ and $v \in \mathcal{V}_c(X)$. Also let χ be the flow generated by v. As in Remark 2.2(h), we interpret χ as a flowing fluid in X with (smooth) mass density ρ. In fluid mechanics, it is frequently assumed that mass is neither created nor destroyed in X and hence it obeys the **law of mass conservation**: The mass contained in the domain A at time $t = 0$ stays constant as it is transported by the flow. This means that $\rho(A^t, t) = \rho(A, 0)$ for $t \in \mathbb{R}$ and therefore

$$\frac{d}{dt} \int_{A^t} \rho(\cdot, t) \, dx = 0 \quad \text{for } t \in \mathbb{R} \text{ and } A \in \mathcal{L}_X .$$

The transport theorem shows that this is equivalent to

$$\int_{A^t} \big(\partial_t \rho + \operatorname{div}(\rho v) \big) \, dx = 0 \quad \text{for } t \in \mathbb{R} \text{ and } A \in \mathcal{L}_X . \tag{2.37}$$

Therefore law of mass conservation is equivalent to the **continuity equation**

$$\partial_t \rho + \operatorname{div}(\rho v) = 0 \quad \text{in } X . \tag{2.38}$$

(Here and in analogous formulas, in which we treat "time-dependent" vector fields, divergence operators only operate on the "position variables".)

In the special case of a constant density $\rho > 0$, that is, an **incompressible fluid**, the law of mass conservation is equivalent to div $v = 0$, that is, to the vanishing divergence of the velocity field. For this reason, we also call divergence-free vector fields **incompressible**.

Proof The equivalence of (2.37) and (2.38) follows as in the proof of Proposition 2.13. ∎

2.15 Remarks (a) For simplicity, we have concentrated on the case of global fluids. If one relaxes the assumption that the vector fields have compact support, then $v \in \mathcal{V}(M)$ generates a *local* flow. Then corresponding local versions of Theorems 2.10, 2.11, and 2.13 and of Corollary 2.12 remain true.

Proof Compare this, for example, to Section 10 and Theorem 11.8 in [Ama95]. ∎

(b) (regularity) The statements about the flows generated by vector fields and their associated theorems remain true when M is a C^2 manifold and v is a C^1 vector field. Of course, we then only have $\chi \in C^1(M \times \mathbb{R}, M)$. In this case, the transport theorem requires the assumption that $f \in C^1(M \times \mathbb{R})$. ∎

Exercises

1 Prove the following form of **Lebesgue's theorem for differential forms**: Suppose ω is an integrable m-form on M and $f, f_j \in \mathcal{L}_\infty(M, \lambda_M)$ with $f_j \to f$ λ_M-a.e. and $\sup_j \|f_j\|_\infty < \infty$. Also let $\omega_j := f_j \omega$ for $j \in \mathbb{N}$. Show that $\int_M \omega_j \to \int_M \omega$ as $j \to \infty$.

2 Let $a \in \mathbb{R}^3$ and $R > 0$, and define $M := a + RS^2$. Calculate $\int_M \omega$ for

$$\omega := x^2 \, dy \wedge dz + y^2 \, dz \wedge dx + z^2 \, dx \wedge dy .$$

3 Suppose $v \in \mathcal{V}_c(M)$, and denote by χ the flow on M generated by v. A function $f \in \mathcal{E}(M)$ is said to be a **first integral** of v if

$$\frac{d}{dt} (\chi^t)^* f = 0 \qquad \text{on } M \times \mathbb{R} .$$

Show that these statements are equivalent:

(i) f is a first integral for v.

(ii) $(\chi^t)^* f = f$ for $t \in \mathbb{R}$.

(iii) $L_v f = 0$.

4 Let $H \in \mathcal{E}_c(\mathbb{R}^{2n})$. Show that the statements

(i) f is a first integral of sgrad H and

(ii) $\{f, H\} = 0$

are equivalent. In particular, H is a first integral of sgrad H.

5 Let $v \in \mathcal{V}_c(M)$, and let χ be the flow v generates. For $\alpha \in \Omega(M)$,

$$L_v(\alpha) := \frac{d}{dt}\left[(\chi^t)^*\alpha\right]\Big|_{t=0}$$

is called the **Lie derivative** of α.

Prove the following.

(i) If $\alpha \in \Omega^r(M)$ then $L_v(\alpha)$ belongs to $\Omega^r(M)$, and in the case $r = 0$, the definition above agrees with that of Section XI.6.

(ii) $L_v \circ d = d \circ L_v$.

(iii) The forms $\alpha, \beta \in \Omega(M)$ satisfy the **product rule** $L_v(\alpha \wedge \beta) = L_v(\alpha) \wedge \beta + \alpha \wedge L_v(\beta)$.

(iv) $L_v(\alpha) = d(v \lrcorner \alpha) + v \lrcorner d\alpha$ for $\alpha \in \Omega(M)$.

(v) $L_v(\omega_M) = \operatorname{div}(v)\omega_M$.

(vi) If $w \in \mathcal{V}_c(M)$ then $\Theta_M[v, w] = L_v(\Theta_M w)$.

6 Describe the flows generated by the vector fields

(i) $x\,\partial/\partial x + y\,\partial/\partial y \in \mathcal{V}(\mathbb{R}^2)$;

(ii) $-y\,\partial/\partial x + x\,\partial/\partial y \in \mathcal{V}(\mathbb{R}^2)$;

(iii) $-y\,\partial/\partial x + x\,\partial/\partial y \in \mathcal{V}(\mathbb{R}^3)$;

(iv) $(x - y)\,\partial/\partial x + (x + y)\,\partial/\partial y \in \mathcal{V}(\mathbb{R}^2)$;

(v) $(x + y)\,\partial/\partial x + x^2\,\partial/\partial y \in \mathcal{V}(\mathbb{R}^2)$;

(vi) $(x + y)\,\partial/\partial x + (x - y)\,\partial/\partial y + z\,\partial/\partial z \in \mathcal{V}(\mathbb{R}^3)$.

3 Stokes's theorem

In this section, we combine the differential and integral calculus on manifolds and prove the general Stokes's theorem. It is a higher-dimensional generalization of the fundamental theorem of calculus and has numerous applications in mathematics and theoretical physics. In particular, it forms the basis for theoretical explorations in topology and geometry, but we shall not go into these subjects here.

We show how Stokes's theorem can be used to calculate volume and that it provides physical interpretations to the operators div and curl. As a topological application, we prove the Brouwer fixed point theorem.

We close this section by making a connection between the exterior product and the coderivative, though we leave the full significance of this connection to further courses on global analysis.

In the entire section suppose

- $m \geq 2$;
- M is an m-dimensional oriented manifold.

If M is with boundary, then ∂M will be oriented by the outward normal (with respect to the metric induced by the surrounding space $\mathbb{R}^{\overline{m}}$).

Stokes's theorem for smooth manifolds

We denote by $i \colon \partial M \hookrightarrow M$ the natural embedding and recall the definition

$$\int_{\partial M} \omega := \int_{\partial M} i^* \omega = \int_{\partial M} \omega \,|\, \partial M \quad \text{for } \omega \in \Omega^{m-1}(M) \;,$$

which is meaningful if $\omega \,|\, \partial M$ is integrable. In addition, we set $\int_\emptyset \omega := 0$.

3.1 Theorem (Stokes) *Any $\omega \in \Omega_c^{m-1}(M)$ satisfies $\int_M d\omega = \int_{\partial M} \omega$.*

Proof (i) We consider first the case $M = \mathbb{R}^m$. Then $\partial M = \emptyset$, and ω has the representation

$$\omega = \sum_{j=1}^m (-1)^{j-1} a_j \, dx^1 \wedge \cdots \wedge \widehat{dx^j} \wedge \cdots \wedge dx^m \tag{3.1}$$

with $a_j \in \mathcal{D}(\mathbb{R}^m)$, as we know from Example XI.3.2(b). Example XI.3.7(b) implies

$$d\omega = \Big(\sum_{j=1}^m \partial_j a_j \Big) dx^1 \wedge \cdots \wedge dx^m \;. \tag{3.2}$$

Hence it follows from Remark 2.1(a) that

$$\int_M d\omega = \int_{\mathbb{R}^m} \sum_{j=1}^m \partial_j a_j \, dx = \sum_{j=1}^m \int_{\mathbb{R}^m} \partial_j a_j \, dx = 0 \ ,$$

where, since $\partial_j a_j = 1\partial_j a_j$, the last equality follows from integration by parts, according to Proposition X.7.22. Therefore the claim is correct in this case because $\int_{\partial M} \omega = \int_\emptyset \omega = 0$.

(ii) Let $M = \overline{\mathbb{H}}^m$. Then (3.1) and (3.2) likewise hold, were now the a_j belong to $C_c^\infty(\overline{\mathbb{H}}^m)$. From Fubini's theorem and Proposition X.7.22, we obtain with $x' = (x^1, \dots, x^{m-1})$ that

$$\int_{\overline{\mathbb{H}}^m} \partial_j a_j \, dx = \int_0^\infty \left(\int_{\mathbb{R}^{m-1}} \partial_j a_j \, dx' \right) dx^m = 0 \quad \text{for } 1 \le j \le m-1 \ . \tag{3.3}$$

Fubini's theorem and the fundamental theorem of calculus imply

$$\int_{\overline{\mathbb{H}}^m} \partial_m a_m \, dx = \int_{\mathbb{R}^{m-1}} \left(\int_0^\infty \partial_m a_m \, dx^m \right) dx' = - \int_{\mathbb{R}^{m-1}} a_m(x', 0) \, dx' \ . \tag{3.4}$$

Because $i(x') = (x', 0)$, it follows from Example XI.3.4(j) that

$$i^*\omega = (-1)^{m-1} i^* a_m \, dx^1 \wedge \dots \wedge dx^{m-1} \ . \tag{3.5}$$

Because of the standard orientation of $\partial \overline{\mathbb{H}}^m$ (see Example XI.5.3(h)) and by (3.2), (3.3), and (3.5), we can write (3.4) in the form

$$\int_M d\omega = \int_{\overline{\mathbb{H}}^m} \partial_m a_m \, dx = - \int_{\mathbb{R}^{m-1}} i^* a_m \, dx'$$

$$= (-1)^{m-1} \int_{\partial \overline{\mathbb{H}}^m} (i^* a_m) \, dx^1 \wedge \dots \wedge dx^{m-1} = \int_{\partial M} i^*\omega = \int_{\partial M} \omega \ .$$

Thus the claim holds in this case also.

(iii) Suppose now $\partial M = \emptyset$ and M is described by a single (positive) chart (φ, M). Because ω has compact support, $\varphi_* \omega$ belongs to $\Omega_c^{m-1}(\mathbb{R}^m)$. From $\varphi_* = (\varphi^{-1})^*$ and Theorem XI.4.10, it follows that $\varphi_* \circ d = d \circ \varphi_*$. Therefore, (i) gives

$$\int_M d\omega = \int_{\varphi(M)} \varphi_* \, d\omega = \int_{\varphi(M)} d(\varphi_* \omega) = \int_{\mathbb{R}^m} d(\varphi_* \omega) = 0 = \int_\emptyset \omega = \int_{\partial M} \omega \ .$$

(iv) Suppose $\partial M \ne \emptyset$ and M is described by a single (positive) chart (φ, M). Then $\{(\varphi_\partial, \partial M)\}$ with $\varphi_\partial := \varphi \,|\, \partial M$ is a positive atlas of ∂M. Also $\varphi_* \omega$ has compact support in $\overline{\mathbb{H}}^m$, and $i \circ \varphi_\partial^{-1} = \varphi^{-1} \circ i_{\partial \overline{\mathbb{H}}^m}$. Therefore

$$(\varphi_\partial)_* i^* = (i_{\partial \overline{\mathbb{H}}^m})^* \varphi_* \ .$$

Thus it follows from (ii), in analogy to step (iii), that

$$
\int_M d\omega = \int_{\varphi(M)} \varphi_* \, d\omega = \int_{\varphi(M)} d(\varphi_*\omega) = \int_{\mathbb{H}^m} d(\varphi_*\omega)
$$

$$
= \int_{\partial\mathbb{H}^m} i^*_{\partial\mathbb{H}^m} \varphi_*\omega = \int_{\partial\mathbb{H}^m} (\varphi_\partial)_* i^*\omega = \int_{\partial M} i^*\omega = \int_{\partial M} \omega \ .
$$

This proves the claim in this case.

(v) Finally, let $\{ (\varphi_j, U_j, \pi_j) \ ; \ j \in \mathbb{N} \}$ be a local system for M. Because $K := \operatorname{supp}(\omega)$ is compact, we can choose it so that there is a $k \in \mathbb{N}$ such that $\operatorname{supp}(\pi_j) \cap K = \emptyset$ for $j > k$. Letting $\omega_j := \pi_j\omega \,|\, U_j \in \Omega_c^{m-1}(U_j)$, it follows that $\omega = \sum_{j=0}^k \omega_j$. Hence from (iii) and (iv), we get

$$
\int_M d\omega = \sum_{j=0}^k \int_M d\omega_j = \sum_{j=0}^k \int_{U_j} d\omega_j = \sum_{j=0}^k \int_{\partial U_j} \omega_j = \sum_{j=0}^k \int_{\partial M} \omega_j = \int_{\partial M} \omega \ .
$$

Here we have used that $\{ (\varphi_{j,\partial}, \partial U_j, i^*\pi_j) \ ; \ j \in \mathbb{N} \}$ is a local system for ∂M with $\operatorname{supp}(i^*\pi_j) \cap K = \emptyset$ for $j > k$. ∎

Manifolds with singularities

For many applications, the assumption that M is a manifold is too restrictive. One would also like to apply Stokes's theorem to "piecewise smooth manifolds" such as cylinders or cones.

Were it not for "singularities", that is, edges, corners, pointy tips, etc., the sets above would be manifolds with boundary. In fact, these exceptions consist of sets that, relative to the boundary, are "thin". Thus it is to be expected that, as far as integration is concerned, the singularities make no difference.

We now introduce a class of "manifolds with thin singularities", which contains the examples above, and we show that Stokes's theorem also holds for these objects.

Suppose B is a closed subset of M with a nonempty interior. Then we denote by M_B the set of all $p \in B$ for which there is an open neighborhood V_p of p in M such that $B \cap V_p$ is an m-dimensional submanifold of V_p. Then M_B is an m-dimensional submanifold of M, the **support manifold** of B. The set $\mathsf{S}_B := B \backslash \mathsf{M}_B$ is called the **singular set** of B, and B is an m-dimensional **submanifold** of M

with singularities. Clearly S_B is closed in M.
By the **boundary** of B, we mean that of M_B;
that is, $\partial B := \partial M_B$. Do not confuse ∂B with
the topological boundary, $\mathrm{Rd}(B)$, of B in M.
Finally, we provide M_B with the orientation
induced canonically by M.

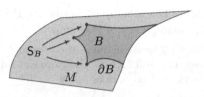

Let $\mathcal{H}^s$ be an s-dimensional Hausdorff measure on M (where M carries the
metric induced by $\mathbb{R}^{\overline{m}}$). Then we say that the singular set S_B of B is **thin** when
is an $\mathcal{H}^{m-1}$-null set. In this case, B is a **manifold with thin singular set**.

If we only require that the set M_B is a submanifold of class C^k for some
$k \in \mathbb{N}^\times$, then we naturally call B a C^k submanifold of M with singularities. Here
it suffices that M is a C^k manifold.

3.2 Examples   (a) Every m-dimensional submanifold of M that is topologically
closed in M — and thus in particular M itself — has a thin (indeed empty) singular
set.

(b) If the Hausdorff dimension has $\dim_H(S_B) < m - 1$, then S_B is thin.

Proof   This follows from the definition of $\dim_H$ in Exercise IX.3.5. ∎

(c) Suppose (J_k) is a sequence of intervals in $\mathbb{R}^{m-2}$ and $f_k \in C^{1\text{-}}(J_k, M)$ with
$\bigcup_{k=0}^\infty f_k(J_k) = S_B$. Then S_B is thin.

Proof   From Exercises IX.3.6(a) and (f), it follows that $\dim_H\big(f_k(J_k)\big) \le m - 2$. Therefore $f_k(J_k)$ is an $\mathcal{H}^{m-1}$-null set for every $k \in \mathbb{N}$, and the σ-subadditivity of the Hausdorff
measure finishes the proof. ∎

(d) (piecewise smooth domains)   Let Ω be a nonempty **domain in** M, that is,
a nonempty open and connected subset of M. Then $S_{\overline{\Omega}}$ and $M_{\overline{\Omega}}$, and therefore
also $\partial \overline{\Omega} = \partial M_{\overline{\Omega}}$, are defined. Let $\mathbb{B}_\infty^{m-1} = (-1,1)^{m-1}$ be the open unit ball in
$(\mathbb{R}^{m-1}, |\cdot|_\infty)$. We say Ω is a **piecewise smooth domain** in M there are finitely
many functions[1]

$$h_j \in C^1(\overline{\mathbb{B}}_\infty^{m-1}, M) \cap C^\infty(\mathbb{B}_\infty^{m-1}, M) \quad \text{for } 0 \le j \le n$$

such that

(i) $h_j \,|\, \mathbb{B}_\infty^{m-1}$ is a parametrization of a subset of $\partial \overline{\Omega}$ for $0 \le j \le n$;

(ii) $\partial \overline{\Omega} = \bigcup_{j=0}^n h_j(\mathbb{B}_\infty^{m-1})$;

(iii) $\mathrm{Rd}(\Omega) = \bigcup_{j=0}^n h_j(\overline{\mathbb{B}}_\infty^{m-1})$.

Then $\overline{\Omega}$ is an m-dimensional submanifold of M with a thin singular set, and we
set $\partial \Omega := \partial \overline{\Omega}$. More precisely, we have

$$S_{\overline{\Omega}} = \bigcup_{j=0}^n h_j\big(\mathrm{Rd}(\mathbb{B}_\infty^{m-1})\big) \quad \text{and} \quad M_{\overline{\Omega}} = \Omega \cup \bigcup_{j=0}^n h_j(\mathbb{B}_\infty^{m-1}),$$

---

[1]If $\overline{\Omega}$ is only a C^k submanifold of M with singularities, we call Ω a **piecewise C^k domain** in M.

and $\overline{\Omega}$ has a compact boundary $\mathrm{Rd}(\Omega)$. In addition, $\partial\Omega = \mathrm{Rd}(\Omega)$ if and only if the singular set of $\overline{\Omega}$ is empty.

Proof The mean value theorem and $M \hookrightarrow \mathbb{R}^{\overline{m}}$ imply that $h_j \in C^{1-}(\overline{\mathbb{B}}_\infty^{m-1}, M)$ (see Remark VII.3.11(b)). Therefore $h_j \mid \mathrm{Rd}(\mathbb{B}_\infty^{m-1})$ is also locally Lipschitz continuous, and the claim follows from (c). ∎

(e) Every open polyhedron in $\mathbb{R}^m$ with a nonempty interior is a piecewise smooth domain in $\mathbb{R}^m$. The boundary consists of "open" $(m-1)$-dimensional "faces". In other words, the singular set consists of all points that lie in "edges" of dimension $\leq m-2$. So, for example, the singular set of a cube in $\mathbb{R}^3$ consists the 12 edges and 8 corners.

Proof This follows from (d). ∎

(f) Let M and N be m- and n-dimensional, *topologically closed* submanifolds of $\mathbb{R}^m$ and $\mathbb{R}^n$, respectively. Then $B := M \times N$ is a submanifold of $\mathbb{R}^{m+n}$ with thin singular set. More precisely,

$$ \mathsf{S}_B = \partial M \times \partial N , \quad \partial B = (\mathring{M} \times \partial N) \cup (\partial M \times \mathring{N}) , \quad \mathsf{M}_B = (\mathring{M} \times \mathring{N}) \cup \partial B . $$

Proof It is easy to see that S_B, M_B, and ∂B are the given sets. From Exercise XI.1.8, we know that $\dim_H(\partial M) \leq m-1$ and $\dim_H(\partial N) \leq n-1$. Therefore Exercise IX.3.8 shows that $\dim_H(\partial M \times \partial N) \leq m+n-2$. Hence S_B is thin. ∎

(g) Let B be an m-dimensional submanifold of S^m with thin singular set and
$r > 0$. Then the cone $K(rB)$ of Example 2.7(d) is an $(m+1)$-dimensional submanifold of $\mathbb{R}^{m+1}$ with thin singular set.
To be more precise, (with $K(\emptyset, r) := \emptyset$)
we have

$$ \mathsf{S}_{K(rB)} = \{0\} \cup K(\mathsf{S}_{rB}) \cup \partial(rB) $$

and

$$ \partial K(rB) = \mathrm{int}(rB) \cup \{ t\,i(\sigma) \; ; \; 0 < t < 1, \; \sigma \in \partial(rB) \} $$

with $i : rB \hookrightarrow \mathbb{R}^{m+1}$.

Proof We leave it to you to justify the given representations of the singular set and the boundary of $K(rB)$. By $\dim_H(\partial B) \leq m-1$ and Exercise IX.1.4(b), $r\partial B = \partial(rB)$ is an $\mathcal{H}^m$-null set.

   As a closed subset of the compact set S^m, the set S_B is compact in $\mathbb{R}^{m+1}$. Let $\varepsilon \in (0,1]$ and $\rho > 0$. Because S_B is an $\mathcal{H}^{m-1}$-null set, there are open sets $O_0, \ldots, O_J$ in $\mathbb{R}^{m+1}$ such that $\mathrm{diam}\, O_j < \varepsilon$ and

$$ \sum_{j=0}^{J} [\mathrm{diam}\, O_j]^{m-1} < \rho . $$

Let $K \in \mathbb{N}^\times$ with $K\varepsilon < r \leq (K+1)\varepsilon$. Then the intervals

$$J_k := \big[k\varepsilon, (k+1)\varepsilon\big] \quad \text{for } 0 \leq k \leq K$$

cover the interval $[0, r]$ and satisfy $\operatorname{diam}(J_k) = \varepsilon$. Therefore

$$\{ Q_j \times J_k \; ; \; 0 \leq j \leq J, \; 0 \leq k \leq K \}$$

is a cover of $\mathsf{S}_B \times [0, r]$ in $\mathbb{R}^{m+1} \times \mathbb{R}$ with subsets of $\mathbb{R}^{m+1}$ for which

$$\operatorname{diam}(Q_j \times J_k) \leq \sqrt{2}\,\varepsilon \operatorname{diam}(Q_j) \leq \sqrt{2}\,\operatorname{diam}(Q_j) \leq \sqrt{2}\,\varepsilon \;.$$

With this we get

$$\sum_{j=0}^{J} \sum_{k=0}^{K} \big[\operatorname{diam}(Q_j \times J_k)\big]^m \leq (K+1) 2^{m/2} \sum_{j=0}^{J} [\operatorname{diam} Q_j]^m$$

$$\leq (r+1) 2^{m/2} \sum_{j=0}^{J} \varepsilon^{-1} [\operatorname{diam} Q_j]^m$$

$$\leq (r+1) 2^{m/2} \sum_{j=0}^{J} [\operatorname{diam} Q_j]^{m-1} < (r+1) 2^{m/2} \rho \;.$$

Because this holds for every $\varepsilon \in (0, 1]$ and every $\rho > 0$, we read off from Exercise IX.3.4(a) that $\mathsf{S}_B \times [0, r]$ is an $\mathcal{H}^m$-null set in $\mathbb{R}^{m+1} \times \mathbb{R}$. Clearly $K(\mathsf{S}_{rB})$ is the image of $\mathsf{S}_B \times [0, r]$ under the Lipschitz continuous map

$$\mathbb{R}^{m+1} \times \mathbb{R} \to \mathbb{R}^{m+1} , \quad (x, t) \mapsto tx \;.$$

Therefore Exercise IX.3.4(b) implies that $K(\mathsf{S}_{rB})$ is also an $\mathcal{H}^m$-null set. Now it follows that $\mathsf{S}_{K(rB)}$ is a thin singular set. ∎

(h) Let N be an m-dimensional manifold, and let $f \in \operatorname{Diff}(M, N)$. If B is a submanifold of M with thin singular set, then $f(B)$ is a submanifold of N with thin singular set.

Proof This follows easily from Remark XI.1.1(g) and Exercise IX.3.4(b). ∎

Stokes's theorem with singularities

We now generalize Theorem 3.1 to the case of C^2 manifolds with thin singular sets.[2] We first need a lemma. Here we denote by $\mathbb{B}^m(A, r)$ the open neighborhood of $A \subset \mathbb{R}^m$ with radius $r > 0$, that is,

$$\mathbb{B}^m(A, r) := \bigcup_{x \in A} \mathbb{B}^m(x, r) = \left\{ x \in \mathbb{R}^m \; ; \; \operatorname{dist}(x, A) < r \right\}$$

if $A \neq \emptyset$.

3.3 Lemma *Let K be a nonempty compact $\mathcal{H}^{m-1}$-null set in $\mathbb{R}^m$. Then for every pair $\varepsilon, r > 0$, there are open sets U and V and a $\chi \in \mathcal{E}(\mathbb{R}^m)$ such that*

$$K \subset\subset U \subset\subset V \subset\subset \mathbb{B}^m(K, r) \tag{3.6}$$

and

$$\chi \,|\, U = 0 \,, \quad \chi \,|\, V^c = 1 \,, \quad 0 \leq \chi \leq 1 \,, \quad \int_{\mathbb{R}^m} |\nabla \chi| \, dx \leq \varepsilon \,. \tag{3.7}$$

Proof We fix a $\psi \in \mathcal{E}(\mathbb{R})$ with $0 \leq \psi \leq 1$, let $\psi \,|\, [0, 1] = 0$ and $\psi \,|\, [2, \infty) = 1$, and set $\kappa := 2^m \operatorname{vol}(\mathbb{B}^m) \, \|\psi'\|_\infty$.

Because $\mathcal{H}^{m-1}(K) = 0$, we have $\mathcal{H}^{m-1}_\delta(K) = 0$ for every $\delta > 0$. Since K is compact, there are open set W_j for $0 \leq j \leq n$ with $K \subset \bigcup_{j=0}^n W_j$, $K \cap W_j \neq \emptyset$, and $\rho_j := \operatorname{diam}(W_j) < r/3$, and

$$\sum_{j=0}^n \rho_j^{m-1} \leq \varepsilon/\kappa \,. \tag{3.8}$$

We choose $x_j \in W_j \cap K$ and set $U_j := \mathbb{B}^m(x_j, \rho_j)$ and $V_j := \mathbb{B}^m(x_j, 2\rho_j)$ for $0 \leq j \leq n$. Then $U := \bigcup_{j=0}^n U_j$ and $V := \bigcup_{j=0}^n V_j$ are open and satisfy (3.6) because $U_j \supset W_j$ for $0 \leq j \leq n$.

Now we set

$$\chi(x) := \prod_{j=0}^n \psi\left(\frac{|x - x_j|}{\rho_j} \right) \quad \text{for } x \in \mathbb{R}^m \,.$$

Then χ belongs to $\mathcal{E}(\mathbb{R}^m)$ and satisfies $\chi \,|\, U = 0$ and $\chi \,|\, V^c = 1$. We also have

$$\nabla\chi(x) = \sum_{j=0}^n \psi'\left(\frac{|x - x_j|}{\rho_j} \right) \frac{x - x_j}{|x - x_j|} \frac{1}{\rho_j} \prod_{\substack{k=0 \\ k \neq j}}^n \psi\left(\frac{|x - x_k|}{\rho_k} \right)$$

---

[2] Our proof follows the ideas presented in [Lan95]. For another approach, see [HR72] and [AMR83].

for $x \in \mathbb{R}^m$, and therefore

$$|\nabla\chi| \le \|\psi'\|_\infty \sum_{j=0}^{n} \rho_j^{-1} \chi_{[\rho_j \le |x - x_j| \le 2\rho_j]} .$$

Then this and the translation invariance of the Lebesgue measure give

$$\int_{\mathbb{R}^m} |\nabla\chi| \, dx \le \|\psi'\|_\infty \sum_{j=0}^{n} \rho_j^{-1} \operatorname{vol}(2\rho_j \mathbb{B}^m)$$

$$= 2^m \operatorname{vol}(\mathbb{B}^m) \|\psi'\|_\infty \sum_{j=0}^{n} \rho_j^{m-1} = \kappa \sum_{j=0}^{n} \rho_j^{m-1} .$$

Because of (3.8), this implies the last statement of (3.7). ∎

We can now prove the advertised generalization of Stokes's theorem.

3.4 Theorem (Stokes's theorem with singularities)   *Let B be an m-dimensional submanifold of M with thin singular set, and let $\omega \in \Omega_c^{m-1}(M)$. If $\omega \,|\, \partial B$ is integrable, then $\int_B d\omega = \int_{\partial B} \omega$.*

Proof   (i) Suppose M is open in $\overline{\mathbb{H}}^m$ and $K := S_B \cap \operatorname{supp}(\omega)$. Because $\operatorname{supp}(\omega)$ is compact and S_B is closed in M, we know K is a compact $\mathcal{H}^{m-1}$-null set in $\mathbb{R}^m$. Therefore, it follows from Lemma 3.3 that there are a constant $\kappa > 0$ and open sets U_k and V_k, for every $\varepsilon > 0$ and $k \in \mathbb{N}^\times$, such that

$$K \subset\subset U_k \subset\subset V_k \subset\subset \mathbb{B}^m(K, 1/k) .$$

There is also a $\chi_k \in \mathcal{E}(\mathbb{R}^m)$ with

$$\chi_k \,|\, U_k = 0 , \quad \chi_k \,|\, V_k^c = 1 , \quad 0 \le \chi_k \le 1 , \quad \int_{\mathbb{R}^m} |\nabla\chi_k| \, dx \le \varepsilon .$$

In particular,

$$\bigcap_{k=1}^{\infty} (\overline{V}_k \cap \partial B) = \emptyset . \tag{3.9}$$

We set $\omega_k := \chi_k \omega$. Then ω_k belongs to $\Omega_c^{m-1}(M_B)$. Thus because $B \backslash M_B = S_B$ is a λ_m-null set, Theorem 3.1 gives

$$\int_B d\omega_k = \int_{M_B} d\omega_k = \int_{\partial M_B} \omega_k = \int_{\partial B} \omega_k \quad \text{for } k \in \mathbb{N}^\times . \tag{3.10}$$

With $\psi_k := 1 - \chi_k$, it follows that

$$\int_{\partial B} \omega - \int_{\partial B} \omega_k = \int_{\partial B} (1 - \chi_k)\omega = \int_{\partial B} \psi_k \omega .$$

Because $\omega \,|\, \partial B$ is integrable and $\mathrm{supp}(\psi_k) \subset \overline{V}_k$, (3.9) and Lebesgue's theorem (see Exercise 2.1) imply that $\left(\int_{\partial B} \psi_k \omega\right)$ is a null sequence. Therefore

$$\lim_{k \to \infty} \int_{\partial B} \omega_k = \int_{\partial B} \omega \,. \tag{3.11}$$

Further, Theorem XI.4.10(ii) gives the equality

$$\int_B d\omega_k = \int_B d\chi_k \wedge \omega + \int_B \chi_k \, d\omega \,. \tag{3.12}$$

Since $B \backslash \mathsf{M}_B$ is a λ_m-null set, we have

$$\int_B \chi_k \, d\omega = \int_{\mathsf{M}_B} \chi_k \, d\omega \quad \text{for } k \in \mathbb{N}^\times \,.$$

Because $d\omega \in \Omega^m(M)$ has compact support, $d\omega$ is integrable over M_B. Also $\chi_k(x) \to 1$ for $x \in \mathsf{M}_B$. Again using Lebesgue's theorem, we find

$$\lim_{k \to \infty} \int_B \chi_k \, d\omega = \int_B d\omega \,. \tag{3.13}$$

Now ω has the representation $\omega = \sum_{j=1}^m (-1)^{j-1} a_j \, dx^1 \wedge \cdots \wedge \widehat{dx^j} \wedge \cdots \wedge dx^m$ with $a_j \in \mathcal{D}(M)$. From this we read off that $d\chi_k \wedge \omega = b_k \omega_{\mathbb{R}^n}$, where

$$b_k := \sum_{j=1}^m a_j \partial_j \chi_k \quad \text{for } k \in \mathbb{N}^\times \,.$$

Therefore we get

$$\left| \int_B d\chi_k \wedge \omega \right| = \left| \int_M b_k \, dx \right| \le c \int_{\mathbb{R}^m} |\nabla \chi_k| \, dx \le c\varepsilon \quad \text{for } k \in \mathbb{N}^\times \,,$$

where c is a constant independent of k. As $k \to \infty$, results (3.10)–(3.13) now imply

$$\left| \int_B d\omega - \int_{\partial B} \omega \right| \le c\varepsilon \,.$$

Because this is true for every $\varepsilon > 0$, the claim is proved in this case.

(ii) Now suppose M is described by a single chart (φ, U). Then the claim follows from (i) by "moving down" to $\varphi(U) \subset \overline{\mathbb{H}}^m$.

(iii) Finally suppose M is arbitrary and $\{ (\varphi_j, U_j, \pi_j) \,;\, j \in \mathbb{N} \}$ local system for M. Then we see as in step (v) of the proof of Theorem 3.1 that it suffices to prove the claim for $\omega_j := \pi_j \omega$ and $B_j := B \cap U_j$ for $j \in \mathbb{N}$. Their validity in this case follows from (ii). We are done. ∎

3.5 Corollary

(i) *If ω is closed, then $\int_{\partial B} \omega = 0$.*

(ii) *$\int_M d\omega = 0$ if M is without boundary.*

We again point out that Theorem 3.4 contains the "regular case" of Theorem 3.1, namely, the case $B = M$.

3.6 Remarks	(a) (the one-dimensional case)	We consider a connected one-dimensional compact oriented manifold Γ. According to Theorem XI.1.18, Γ is either a 1-sphere embedded in $\mathbb{R}^m$ or diffeomorphic to $I := [0,1]$. Therefore Γ is an oriented smooth curve that is either closed or has an initial point A and an endpoint E. Because $\Omega^0(\Gamma) = \mathcal{E}(\Gamma)$ and $m := \dim(\Gamma) = 1$, we know $\omega \in \Omega^0(\Gamma)$ is a function on Γ. Then it follows from Remark 2.2(b) and Example VIII.4.2(b) (when one considers that the proof only uses the values of f and Γ) that

$$\int_{\Gamma} d\omega = \omega(E) - \omega(A) \, , \tag{3.14}$$

where we set $E = A$ in case Γ is an embedded 1-sphere. By stipulating that the volume measure of a 0-dimensional manifold is the 0-dimensional Hausdorff measure (the counting measure), and by providing the boundary $\partial\Gamma$, if nonempty, with the orientation given by $+1$ at E and -1 at A, we can write (3.14) in the form $\int_{\Gamma} d\omega = \int_{\partial\Gamma} \omega$. In the special case that Γ is the interval $[a,b]$, equation (3.14) is nothing other than the fundamental theorem of calculus. This shows that Stokes's theorem is a higher-dimensional generalization of — and is indeed based on — Theorem VI.4.13..

(b) Corollary 3.5(i) implies a higher-dimensional generalization of "half of" the fundamental theorem of line integrals, that is, the statement (i)$\Rightarrow$(ii) of Theorem VIII.4.4. The "second half" is likewise true in the general case (for example, Theorem XIII.1.1 in [Lan95]).

(c) (regularity)	An analysis of the proof shows that Stokes's theorem (with singularities) remains true if we only assume that ω belongs to $\Omega^{m-1}_{(1)}(M_B)$ and has compact support in M, and that $\omega \,|\, \partial B$ and $\omega \,|\, M_B$ are integrable. In addition, it suffices to assume B is a C^2 submanifold of M with thin singularities and that M is itself only a C^2 manifold. ∎

Planar domains

Suppose Ω is a piecewise smooth domain in $\mathbb{R}^2$. Then there are finitely many closed piecewise smooth curves $\Gamma_0, \Gamma_1, \ldots, \Gamma_n$ that are pairwise disjoint, free of self-intersections, and are such that $\mathrm{Rd}(\Omega) = \Gamma := \Gamma_0 + \cdots + \Gamma_n$. Here every

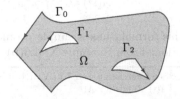

curve Γ_j is oriented by the outward normal of $\partial\overline{\Omega} \cap \Gamma_j$. This means that every Γ_j is oriented so that in traversing Γ_j, the part of Ω adjacent to Γ_j lies to the left. For short, we say Γ is the **oriented boundary curve** of Ω.

3.7 Proposition (Green–Riemann) *Let Ω be a bounded piecewise smooth domain in $\mathbb{R}^2$ with oriented boundary curve Γ. Also let X be an open neighborhood of $\overline{\Omega}$ in $\mathbb{R}^2$, and let $a, b \in C^1(X)$. Then[3]*

$$\int_\Gamma a\,dx + b\,dy = \int_\Omega \left(\frac{\partial b}{\partial x} - \frac{\partial a}{\partial y}\right) d(x,y) \ .$$

Proof Let $\overline{\Omega} \subset\subset U \subset\subset X$, and let χ be a cutoff function for U. Also define $\alpha := a\,dx + b\,dy$. Then α belongs to $\Omega^1_{(1)}(X)$, and $d\alpha = (\partial_1 b - \partial_2 a)\,dx \wedge dy$. Therefore $\omega := \chi\alpha$ lies in $\Omega^1_{(1)}(X)$, has compact support in X, and agrees on U with α. Because the line integral $\int_\Gamma \alpha$ exists, α is integrable over $\partial\Omega$, and

$$\int_\Gamma a\,dx + b\,dy = \int_{\partial\Omega} \alpha = \int_{\partial\Omega} \omega \ .$$

Because $\lambda_2\big(\mathrm{Rd}(\Omega)\big) = 0$, we have

$$\int_{\overline{\Omega}} d\omega = \int_{\overline{\Omega}} d\alpha = \int_\Omega d\alpha = \int_\Omega (\partial_2 b - \partial_1 a)\,d(x,y) \ .$$

Now the claim follows from Theorem 3.4 and Remark 3.6(c). ∎

3.8 Corollary *Under the assumptions of Proposition 3.7, we have the **Leibniz area formulas**[4]*

$$A(\Omega) := \lambda_2(\Omega) = \int_\Gamma x\,dy = -\int_\Gamma y\,dx = \frac{1}{2}\int_\Gamma x\,dy - y\,dx \ .$$

Proof As required, set $(a, b) := (0, X)$, $(a, b) := (-Y, 0)$, or $(a, b) := (-Y, X)$. ∎

3.9 Examples (a) Let Ω be a bounded piecewise smooth domain in $\mathbb{R}^2$ with oriented boundary curve Γ. In polar coordinates (r, φ), we have

$$A(\Omega) = \frac{1}{2}\int_\Gamma r^2\,d\varphi \ .$$

This formula has a simple geometric interpretation: From Example XI.4.6(b), we know that

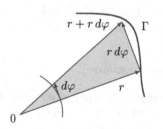

---

[3]See Exercises 6 and 7 in VIII.1 and Example VIII.4.2(a).
[4]$A(\cdot)$ stands for **Area**.

$d\varphi$ is the volume element of the unit circle S^1. Thus we can interpret $r\,d\varphi$ as the length of an[5] infinitesimal, positively oriented line segment which is tangent to rS^1. Then $r^2\,d\varphi/2$ can be interpreted as the area of the triangle whose vertices lie at the origin and the ends of the position vectors r and $r + r\,d\varphi$. The sum of these "infinitesimal" oriented areas, that is, the integral, then gives the total area.

Proof With the plane polar coordinate map f_2 (see Section X.8), we can verify that $f_2^*(x\,dy - y\,dx) = r^2\,d\varphi$. Therefore the claim follows from Corollary 3.8. ∎

(b) (Leibniz's sector formula) Let Ω be a piecewise smooth domain $\mathbb{R}^2$ whose boundary curve Γ is oriented and satisfies $\Gamma = \Sigma + \Gamma_0$, where Σ consists of line segments with endpoints at 0. Then

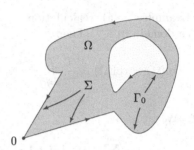

$$A(\Omega) = \frac{1}{2}\int_{\Gamma_0} x\,dy - y\,dx .$$

Proof It follows immediately from (a) that the integral over Σ vanishes. ∎

(c) (Cauchy's integral theorem) We identify $\mathbb{C}$ with $\mathbb{R}^2$ and consider a piecewise smooth domain Ω in $\mathbb{C}$ with oriented boundary curve Γ. Let X be an open neighborhood of $\overline{\Omega}$ in $\mathbb{C}$, and suppose $f : X \to \mathbb{C}$ is holomorphic. Then Cauchy's integral theorem holds,[6] that is, $\int_{\Gamma} f\,dz = 0$.

Proof We decompose $z = x + iy$ and $f = u + iv$ into their real and imaginary parts. This gives us $f\,dz = \alpha + i\beta$ with $\alpha := u\,dx - v\,dy$ and $\beta := u\,dy + v\,dx$. Then the Cauchy–Riemann equations imply

$$d\alpha = -(u_y + v_x)\,dx \wedge dy = 0 \quad \text{and} \quad d\beta = (u_x - v_y)\,dx \wedge dy = 0$$

(see Remark VIII.5.4(c)). Now the claim follows from Proposition 3.7 (or, by using a cutoff function, directly from Theorem 3.4). ∎

(d) Let[7] $B := \{ (x, y) \in \mathbb{R}^2 \; ; \; 0 \le x \le 1, \; 0 \le y \le f(x) \}$ with

$$f(x) := \begin{cases} 1 & \text{if } x = 0 , \\ 1 + x\sin(\pi/x^2) & \text{if } x \ne 0 . \end{cases}$$

Then $\mathsf{S}_B = \{(0,0), (0,1), (1,0), (1,1)\}$, and with $\omega = (x\,dy - y\,dx)/2$ we find $\lambda_2(B) = \int_B dx \wedge dy = \int_B d\omega < \infty$. Along the curve graph$(f)$, we have

$$2\omega = x\,dy - y\,dx = \big(xf'(x) - f(x)\big)\,dx = \left(-1 - \frac{2\pi}{x}\cos\frac{\pi}{x^2}\right) dx .$$

---

[5] See Remark VI.5.2(c).

[6] Compare Theorems VIII.5.5 and VIII.6.20, and note that Γ can now have multiple components.

[7] See Example II.B.10 in the fourth volume of [SW96].

Therefore $\omega \,|\, \partial B$ is not integrable, and the Leibniz area formula does not hold. This shows that Theorem 3.4 does require the assumption that $\omega \,|\, \partial B$ is integrable. ∎

Higher-dimensional problems

In the following examples, we consider generalizations of the results above to higher-dimensional cases.

3.10 Examples (a) (calculation of volumes) Let Ω be a bounded piecewise smooth domain in $\mathbb{R}^m$. Then

$$\operatorname{vol}(\overline{\Omega}) = \operatorname{vol}(\Omega) = \frac{1}{m} \int_{\partial\Omega} \sum_{j=1}^m (-1)^{j-1} x^j \, dx^1 \wedge \cdots \wedge \widehat{dx^j} \wedge \cdots \wedge dx^m \ .$$

Proof We set

$$\alpha := \sum_{j=1}^m (-1)^{j-1} x^j \, dx^1 \wedge \cdots \wedge \widehat{dx^j} \wedge \cdots \wedge dx^m \in \Omega^{m-1}(\mathbb{R}^m)$$

and $\omega := \varphi\alpha$, where φ is a smooth cutoff function for a compact neighborhood U of $\overline{\Omega}$. Then ω belongs to $\Omega_c^{m-1}(\mathbb{R}^m)$, agrees on U with α, and satisfies

$$d\omega \,|\, \overline{\Omega} = m\omega_{\mathbb{R}^m} \,|\, \overline{\Omega} \ ,$$

which follows from Example XI.3.7(b). Because $\mathcal{H}^{m-1}(\mathsf{S}_{\overline{\Omega}}) = 0$ and by Theorem 1.7(iii) and (iv), we see $\operatorname{Rd}(\overline{\Omega}) = \mathsf{S}_{\overline{\Omega}} \cup \partial\Omega$ is a λ_m-null set. Therefore we get

$$m \operatorname{vol}(\Omega) = m \operatorname{vol}(\overline{\Omega}) = \int_{\overline{\Omega}} dx = \int_{\overline{\Omega}} d\omega = \int_{\partial\Omega} \omega = \int_{\partial\Omega} \alpha \ ,$$

where the last equality follows from Stokes's theorem in the form of Theorem 3.4. ∎

(b) (spheres) For $\Omega = \mathbb{B}^m$, we again get the formula

$$\operatorname{vol}(S^{m-1}) = m \operatorname{vol}(\mathbb{B}^m) \ ,$$

which was already derived in Example 1.13(c).

Proof This follows with Example XI.5.3(c) from (a). ∎

(c) Let N be a nonempty compact hypersurface in $\mathbb{R}^m \setminus \{0\}$ such that every half ray going out from 0 intersects N at most once. Also let $K(N)$ be the cone with base N and tip 0. Then, in a (partial) generalization of the Leibniz sector formula, we have

$$\operatorname{vol}(K(N)) = \frac{1}{m} \int_N \sum_{j=1}^m (-1)^{j-1} x^j \, dx^1 \wedge \cdots \wedge \widehat{dx^j} \wedge \cdots \wedge dx^m \ .$$

Proof We check easily that $B := K(N)$ is an m-dimensional compact submanifold of $\mathbb{R}^m$ with thin singular set for which $\partial B = \mathrm{int}(N) \cup S$ and $S := \{ t\, i(p) \; ; \; p \in N,\ 0 < t < 1 \}$, where $i : N \hookrightarrow \mathbb{R}^m$. Example XI.4.13(b) gives

$$\alpha := \sum_{j=1}^{m} (-1)^{j-1} x^j \, dx^1 \wedge \cdots \wedge \widehat{dx^j} \wedge \cdots \wedge dx^m = v \lrcorner \, \omega_{\mathbb{R}^m}$$

with $v(x) := (x, x) \in T_x \mathbb{R}^m$. Then Remark 2.2(g) shows that $\alpha = (v \,|\, \nu) \omega_{\partial B}$. Because $v(p) \in T_p S$ for $p \in S$, we find $\alpha \,|\, S = 0$. Then because $d\alpha = m \omega_{\mathbb{R}^m}$, Theorem 3.4 gives

$$m \, \mathrm{vol}(B) = \int_B d\alpha = \int_{\partial B} \alpha = \int_{\mathrm{int}(N)} \alpha = \int_N \alpha$$

and therefore the claim. ∎

Homotopy invariance and applications

Let $I := [0, 1]$ and $r \in \mathbb{N}$, and suppose M is compact and without boundary. As in the considerations prior to Fubini's theorem for differential forms, we denote by $\widehat{\omega}(\,\cdot\,, p)$, where $\omega \in \Omega^{r+1}(I \times M)$ and $p \in M$, the 1-form-valued r-form on M induced by ω. It is defined by

$$\widehat{\omega}(\,\cdot\,, p)(v_1, \ldots, v_r) := \omega(\,\cdot\,, p)(\,\cdot\,, v_1, \ldots, v_r) \quad \text{for } v_1, \ldots, v_r \in T_p M .$$

Since the 1-form $\widehat{\omega}(\,\cdot\,, p)(v_1, \ldots, v_r)$ is continuous for every r-tuple $(v_1, \ldots, v_r) \in (T_p M)^r$ and is therefore integrable over I, it follows that

$$\int_I \omega := \left(p \mapsto \int_I \widehat{\omega}(\,\cdot\,, p) \right)$$

is a well-defined element of $\Omega^r(M)$ for every $\omega \in \Omega^{r+1}(I \times M)$. Therefore the linear map

$$K : \Omega^{r+1}(I \times M) \to \Omega^r(M) , \quad \omega \mapsto \int_I \omega \tag{3.15}$$

is defined. As in Section XI.3, we denote by i_ℓ the embedding

$$i_\ell : M \to I \times M , \quad p \mapsto (\ell, p)$$

for $\ell \in \{0, 1\} = \partial I$. Then Lemma XI.3.9 has a global generalization:

3.11 Lemma $K \circ d + d \circ K = i_1^* - i_0^*$.

Proof Because the statement is local with respect to M, it suffices to verify it in local coordinates. But this is exactly what Lemma XI.3.9 does. ∎

As an application of this lemma, we can now prove a higher-dimensional generalization of Proposition VIII.4.7, which was about the homotopy invariance of line integrals.

3.12 Theorem   *Let M and N be compact m-dimensional oriented manifolds without boundary. If $f_0, f_1 \in C^\infty(M, N)$ are homotopic, then*

$$\int_M f_0^* \omega = \int_M f_1^* \omega \quad \text{for } \omega \in \Omega^m(N) .$$

Proof   By assumption, there is an $h \in C^\infty(I \times M, N)$ such that $h(j, \cdot) = f_j$ for $j = 0, 1$. Now let $g := K \circ h^*$, where K is the map (3.15). Then, because d and h^* commute, we have

$$d \circ g + g \circ d = d \circ K \circ h^* + K \circ h^* \circ d = d \circ K \circ h^* + K \circ d \circ h^*$$
$$= (d \circ K + K \circ d) \circ h^* = i_1^* h^* - i_0^* h^* = f_1^* - f_0^* .$$

From $d\omega = 0$ for $\omega \in \Omega^m(N)$, it then follows that

$$f_1^* \omega - f_0^* \omega = (d \circ g)\omega - g \circ d\omega = d(Kh^*\omega) .$$

Therefore we get

$$\int_M f_1^* \omega - \int_M f_0^* \omega = \int_M d(Kh^*\omega) = 0$$

from Corollary 3.5(ii). ∎

In the following, we demonstrate several topological applications of Stokes's theorem.

3.13 Proposition   (hairy ball theorem)   *Every smooth vector field on an even-dimensional sphere has a zero.*

Proof   From Example VII.10.14(a), we know that $T_p S^m$ for $p \in S^m$ is the orthogonal complement of $\mathbb{R}p$ in $\mathbb{R}^{m+1}$. Therefore, we can regard $v \in \mathcal{V}(S^m)$ as a smooth map $v : S^m \to \mathbb{R}^{m+1}$ with $v(p) \perp p$ for $p \in S^m$. If v has a zero, then we can replace v by $p \mapsto v(p)/|v(p)|$. Therefore we can assume that $|v(p)| = 1$ for $p \in S^m$, that is, we can assume $v(S^m) \subset S^m$. From this, because

$$|\cos(\pi t)p + \sin(\pi t)v(p)|^2 = \cos^2(\pi t)|p|^2 + \sin^2(\pi t)|v(p)|^2 = 1 ,$$

we find that the map

$$h : I \times S^m \to S^m , \quad (t, p) \mapsto \cos(\pi t)p + \sin(\pi t)v(p)$$

is well defined. Remark XI.1.1(j) implies that h is smooth with $h(0, \cdot) = \mathrm{id}_{S^m}$ and $h(1, \cdot) = -\mathrm{id}_{S^m}$. Thus $f_0 := \mathrm{id}_{S^m}$ is homotopic to the **antipodal map** $f_1 := -\mathrm{id}_{S^m}$. Now Theorem 3.12 gives

$$\int_{S^m} \omega = \int_{S^m} f_1^* \omega \quad \text{for } \omega \in \Omega^m(S^m) . \tag{3.16}$$

Suppose m is even. Then $F := -\mathrm{id}_{\overline{\mathbb{B}}^{m+1}}$ is an orientation-reversing diffeomorphism of $\overline{\mathbb{B}}^{m+1}$ to itself. Now this and Remark XI.1.1(j) imply that $f_1 = F \,|\, S^m$ is also an orientation-reversing diffeomorphism of S^m to itself. Thus Remark 2.1(f) and Theorem 2.3 give the equation $\int_{S^m} f_1^* \omega = -\int_{S^m} \omega$, which, together with (3.16), shows that $\int_{S^m} \omega = 0$ for $\omega \in \Omega^m(S^m)$. However, this is in contradiction with $\int_{S^m} \omega_{S^m} = \mathrm{vol}(S^m) \neq 0$. ■

We can interpret a smooth vector field on S^2 as a (mathematically idealized) combing of a "hairy ball". Then Proposition 3.13 says that "a smoothly combed hairy ball has at least one bald spot".

We can also derive from Theorem 3.12 the fundamental Brouwer fixed point theorem.

3.14 Theorem (Brouwer fixed point theorem) *Every continuous map of $\overline{\mathbb{B}}^m$ to itself has at least one fixed point.*

Proof[8] (i) Let $f \in C(\overline{\mathbb{B}}^m, \overline{\mathbb{B}}^m)$, and suppose f does not have a fixed point. We consider the radial retraction

$$\rho: \mathbb{R}^m \to \overline{\mathbb{B}}^m , \quad x \mapsto \begin{cases} x & \text{if } x \in \overline{\mathbb{B}}^m , \\ x/|x| & \text{if } x \in (\overline{\mathbb{B}}^m)^c . \end{cases}$$

One verifies easily that ρ is uniformly continuous. Therefore the function $g := f \circ \rho : \mathbb{R}^m \to \overline{\mathbb{B}}^m$ is also uniformly continuous, and $g \,|\, \overline{\mathbb{B}}^m = f$. In particular, g does not have a fixed point. Because $g(\mathbb{R}^m) \subset \overline{\mathbb{B}}^m$, we have $|g(x) - x| \geq |x| - |g(x)| \geq 1$ for $|x| \geq 2$. Because $2\overline{\mathbb{B}}^m$ is compact, there is a $\delta \in (0, 1/2]$ such that $|g(x) - x| \geq 2\delta$ for $|x| \leq 2$. Therefore $|g(x) - x| \geq 2\delta > 0$ for all $x \in \mathbb{R}^m$.

Suppose $\{\,\varphi_\varepsilon \;;\; \varepsilon > 0\,\}$ is a smoothing kernel. Because g belongs to $BUC(\mathbb{R}^m)$, Theorem X.7.11 shows that there is an $\varepsilon_0 > 0$ such that $h := \varphi_{\varepsilon_0} * g$ satisfies the estimate

$$|h(x) - g(x)| \leq \|h - g\|_\infty < \delta \quad \text{for } x \in \mathbb{R}^m .$$

Therefore

$$|h(x) - x| \geq |g(x) - x| - |h(x) - g(x)| \geq \delta \quad \text{for } x \in \mathbb{R}^m .$$

Also, we find

$$|h(x)| = \left| \int_{\mathbb{R}^m} \varphi_{\varepsilon_0}(x - y) g(y) \, dy \right| \leq \|g\|_\infty \int_{\mathbb{R}^m} \varphi_{\varepsilon_0}(x - y) \, dy$$

$$= \|g\|_\infty \int_{\mathbb{R}^m} \varphi_1 \, dx \leq 1$$

for $x \in \mathbb{R}^m$. Finally, it follows from Theorem X.7.8(iv) that h is smooth. Therefore $h \,|\, \overline{\mathbb{B}}^m$ a smooth mapping of $\overline{\mathbb{B}}^m$ to itself that has no fixed point. In the next step, we show that this is not possible.

---

[8]For $m = 1$, the claim follows from the intermediate value theorem (see Exercise III.5.1).

(ii) Let $f \in C^\infty(\overline{\mathbb{B}}^m, \overline{\mathbb{B}}^m)$, and suppose f does not have a fixed point. Then for $x \in \overline{\mathbb{B}}^m$, we can define a half ray starting from $f(x)$ and passing through x. We denote by $g(x)$ the point where it intersects S^{m-1}. Then $g(x) = f(x) + t(x)(x - f(x))$ for $x \in \overline{\mathbb{B}}^m$, where $t(x)$ denotes the positive solution of the quadratic equation

$$|x - f(x)|^2 t^2 + 2t(x - f(x) \mid f(x)) + |f(x)|^2 = 1 .$$

From this, it follows that g belongs to $C^\infty(\overline{\mathbb{B}}^m, \overline{\mathbb{B}}^m)$ and satisfies $g \mid S^{m-1} = \mathrm{id}_{S^{m-1}}$.

We now consider the smooth map

$$h : I \times S^{m-1} \to S^{m-1} , \quad (t, x) \mapsto g(tx) .$$

Then $h_0 := h(0, \cdot) = g(0)$ and $h_1 := h(1, \cdot) = \mathrm{id}_{S^{m-1}}$. In other words, the identity on S^{m-1} is homotopic in S^{m-1} to the constant map h_0, that is, $\mathrm{id}_{S^{m-1}}$ is null homotopic in S^{m-1}. Because $h_0^* \omega = 0$ for $\omega \in \Omega^{m-1}(S^{m-1})$, we find from Theorem 3.12 the false statement that $\int_{S^{m-1}} \omega = \int_{S^{m-1}} h_0^* \omega = 0$ for $\omega \in \Omega^{m-1}(S^{m-1})$. This shows that every $f \in C^\infty(\overline{\mathbb{B}}^m, \overline{\mathbb{B}}^m)$ has at least one fixed point. ∎

Gauss's law

Unless we say otherwise, the sequel will use the following assumptions and conventions:

- $(\cdot \mid \cdot) := (\cdot \mid \cdot)_M$ is a Riemannian metric on M;
- B is an m-dimensional submanifold of M with thin singular set;
- $\nu := \nu_B$ is the **outward normal of** ∂B;
- $\mu := \lambda_M$ and $\sigma := \lambda_{\partial B}$.

Stokes's theorem (with singularities) implies immediately the divergence theorem. Because the divergence theorem is not usually formulated in terms of differential forms, it is perhaps the most well-known consequence of Stokes's theorem.

3.15 Theorem (Gauss's law, divergence theorem)   For $v \in \mathcal{V}_c(M)$ satisfying $(v \mid \nu) \in \mathcal{L}_1(\partial B, \sigma)$, we have

$$\int_B \mathrm{div}\, v \, d\mu = \int_{\partial B} (v \mid \nu) \, d\sigma .$$

Proof Remark 2.2(g) gives $v \lrcorner \omega_M = (v \,|\, \nu)\omega_{\partial B}$. Therefore we obtain from Theorem 3.4, because $d(v \lrcorner \omega_M) = \operatorname{div}(v)\omega_M$, that

$$
\int_B \operatorname{div} v \, d\mu = \int_B \operatorname{div}(v)\omega_M = \int_B d(v \lrcorner \omega_M)
$$

$$
= \int_{\partial B} v \lrcorner \omega_M = \int_{\partial B} (v \,|\, \nu)\omega_{\partial B} = \int_{\partial B} (v \,|\, \nu) \, d\sigma \ ,
$$

which finishes the proof. ∎

3.16 Remarks   (a) For $v \in \mathcal{V}_c(M)$, the assumption that $(v \,|\, \nu) \in \mathcal{L}_1(\partial B, \sigma)$ is automatically satisfied when either

(i) $B = M$

or

(ii) Ω is a piecewise smooth domain in M and $B = \overline{\Omega}$.

Proof   (i) is clear, and we leave (ii) to you. ∎

(b) (physical interpretation of the divergence)   Let $v \in \mathcal{V}(M)$ and $p \in M$. Then we have the relation

$$
\operatorname{div} v(p) = \lim_{\Omega \to p} \frac{\int_{\partial \Omega} (v \,|\, \nu) \, d\sigma}{\operatorname{vol}(\Omega)} \ . \tag{3.17}
$$

More precisely, this means that for every $\varepsilon > 0$ there is a neighborhood U of p in M such that for every relatively compact piecewise smooth domain Ω in M with $p \in \Omega \subset U$, we have

$$
\left| \operatorname{div} v(p) - \frac{\int_{\partial \Omega} (v \,|\, \nu) \, d\sigma}{\operatorname{vol}(\Omega)} \right| < \varepsilon \ . \tag{3.18}
$$

From Remark 2.2(h), we know that the quotient

$$
\frac{\int_{\partial \Omega} (v \,|\, \nu) \, d\sigma}{\operatorname{vol}(\Omega)} \tag{3.19}
$$

is the flux of the vector field v through $\partial \Omega$ per unit volume. In the special case $v := \rho w$, where ρ is the density and w is the velocity of a fluid in M, (3.19) represents the mass per unit time and volume flowing outward through $\partial \Omega$. Therefore (3.17) in this case measures how much mass is created or destroyed (depending on the sign of $\operatorname{div} v(p)$) per unit time at the point p. For this reason $\operatorname{div} v(p)$ is also called the **source** of the vector field v at point p. In particular, we say v is **source free** or **divergence free** if $\operatorname{div} v = 0$.

Proof By the continuity of $\operatorname{div} v$, there is to every $\varepsilon > 0$ a neighborhood U of p in M such that

$$
|\operatorname{div} v(q) - \operatorname{div} v(p)| \leq \varepsilon \quad \text{for } q \in U \ . \tag{3.20}
$$

Let Ω be a relatively compact piecewise smooth domain in M with $p \in \Omega \subset U$. Then we get with (3.20) the estimate

$$\left| \int_\Omega \operatorname{div} v \, d\mu - \operatorname{div} v(p) \operatorname{vol}(\Omega) \right| \leq \int_\Omega \left| \operatorname{div} v - \operatorname{div} v(p) \right| d\mu \leq \varepsilon \operatorname{vol}(\Omega) \ .$$

Now (3.18) follows from Gauss's law. $\blacksquare$

(c) (regularity) Gauss's law remains true when M is a C^2 manifold, B is a piecewise C^2 submanifold with thin singular set, and v is a C^2 vector field of M along B with $\operatorname{div} v \in \mathcal{L}_1(\mathsf{M}_B, \mu)$ and $(v \,|\, \nu) \in \mathcal{L}_1(\partial B, \sigma)$. $\blacksquare$

Green's formula

Let $f \in C^1(M)$. We denote by $\partial_\nu f$ the derivative of f in the direction of the outward normal of B, that is,

$$\partial_\nu f(p) := \big(\operatorname{grad} f(p) \,\big|\, \nu(p) \big) \quad \text{for } p \in \partial B \ .$$

We call $\partial_\nu f$ the **normal derivative** of f.

3.17 Theorem *Suppose Ω is a piecewise smooth domain in M with $\overline{\Omega} = B$ and $f, g \in \mathcal{E}(M)$. Defining the Laplace–Beltrami operator $\Delta := \Delta_M$ of M, we have*

(i) *(1. Green's formula)*

$$\int_\Omega f \Delta g \, d\mu + \int_\Omega (\operatorname{grad} f \,|\, \operatorname{grad} g) \, d\mu = \int_{\partial\Omega} f \partial_\nu g \, d\sigma$$

if f or g has compact support;

(ii) *(2. Green's formula)*

$$\int_\Omega (f \Delta g - g \Delta f) \, d\mu = \int_{\partial\Omega} (f \partial_\nu g - g \partial_\nu f) \, d\sigma$$

if f and g have compact support.

Proof (i) With $v := \operatorname{grad} g$ and because $\Delta = \operatorname{div} \operatorname{grad}$, the claim follows easily from Proposition XI.6.11(ii) and Theorem 3.15.

(ii) follows analogously from Proposition XI.6.11(iv). $\blacksquare$

3.18 Corollary
$$\int_\Omega \Delta u \, d\mu = \int_{\partial\Omega} \partial_\nu u \, d\sigma \quad \text{for } u \in \mathcal{E}_c(M) \ .$$

Proof Set $f := 1$ and $g := u$. $\blacksquare$

As an application, we derive a necessary condition for the solvability of Neumann boundary value problems.

Let Ω be a bounded **domain** in $\mathbb{R}^m$ with **smooth boundary**, that is, $\overline{\Omega}$ is a connected compact m-dimensional smooth submanifold of $\mathbb{R}^m$. Also let $f \in \mathcal{E}(\overline{\Omega})$ and $g \in \mathcal{E}(\Gamma)$ with $\Gamma := \partial\Omega$. By the **Neumann boundary value problem** for the Laplace operator in Ω, we mean the problem of finding the function $u \in \mathcal{E}(\overline{\Omega})$ satisfying the equations

$$-\Delta u = f \quad \text{in } \Omega , \qquad \partial_\nu u = g \quad \text{on } \Gamma . \tag{3.21}$$

Here $\Delta := \Delta_m$ is the m-dimensional Laplace operator, that is, we use the standard metric.

3.19 Proposition

(i) *In order to be solvable, the boundary value problem (3.21) requires that the* **compatibility condition**

$$\int_\Omega f \, dx + \int_\Gamma g \, d\sigma = 0$$

is satisfied.

(ii) *Two solutions $u, v \in \mathcal{E}(\overline{\Omega})$ of (3.21) differ by at most a constant.*

Proof   (i) is a consequence of Corollary 3.18.

(ii) It follows from the linearity of Δ and ∂_ν that $w := u - v$ satisfies the homogeneous equations

$$-\Delta w = 0 \quad \text{in } \Omega , \qquad \partial_\nu w = 0 \quad \text{on } \Gamma .$$

Thus it follows from the first Green's formula with $f := g := w$ that

$$\int_\Omega |\operatorname{grad} w|^2 \, dx = 0 .$$

From this we read that $\operatorname{grad} w = 0$, which implies $w = \operatorname{const}$, as we know from Remark VII.3.11(c). ∎

Obviously every constant function is a solution of the **homogeneous** Neumann problem

$$-\Delta u = 0 \quad \text{in } \Omega , \qquad \partial_\nu u = 0 \quad \text{on } \Gamma .$$

This implies that the Neumann boundary value problem (3.21) never has a unique solution. That is, if u is a solution of (3.21), then so is $u + c\mathbf{1}$ for every $c \in \mathbb{R}$.

The boundary conditions of the Neumann problem can be modified to give another important boundary value problem, the **Dirichlet problem**. This is the task of finding a $u \in \mathcal{E}(\overline{\Omega})$ such that

$$-\Delta u = f \quad \text{in } \Omega , \qquad u = g \quad \text{on } \Gamma .$$

In contrast to the Neumann problem, the Dirichlet problem has at most one solution, as the next result shows.

3.20 Proposition *The homogeneous Dirichlet problem*

$$-\Delta u = 0 \quad \text{in } \Omega , \qquad u = 0 \quad \text{on } \Gamma \tag{3.22}$$

has only the trivial solution $u = 0$.

Proof If $u \in \mathcal{E}(\overline{\Omega})$ solves (3.22), then it follows from the first Green's formula with $f := g := u$, and that u solves the Dirichlet problem, that $\int_\Omega |\operatorname{grad} u|^2 \, dx = 0$ and therefore $u = \text{const}$. Because $u \,|\, \Gamma = 0$, this means $u = 0$. ∎

The theory of partial differential equations proves that both the Dirichlet and Neumann problem are solvable, though the latter case requires the compatibility condition to be satisfied.

The classical Stokes's theorem

As usual, we assume $\mathbb{R}^3$ carries the standard metric. As a special case of the general Stokes's theorem, we now prove its classical version.

3.21 Theorem (Stokes) *Suppose X is open in $\mathbb{R}^3$ and M is an oriented surface in X. Also let $\mathfrak{t}$ be the positive unit tangent of ∂B, that is, of ∂M_B. Then for $v \in \mathcal{V}_c(X)$ with $(v \,|\, \mathfrak{t}) \in \mathcal{L}_1(\partial M, \lambda_{\partial M})$, we have*

$$\int_B \operatorname{curl} v \cdot d\vec{F} = \int_{\partial B} v \cdot \vec{ds}$$

and therefore

$$\int_B (\operatorname{curl} v \,|\, \nu) \, dF = \int_{\partial B} (v \,|\, \mathfrak{t}) \, ds .$$

Proof Example XI.4.13(b) implies

$$\operatorname{curl} v \lrcorner \omega_{\mathbb{R}^3} = (\operatorname{curl} v)^1 \, dy \wedge dz + (\operatorname{curl} v)^2 \, dz \wedge dx + (\operatorname{curl} v)^3 \, dx \wedge dy .$$

Thus Remark 2.2(e), definition (XI.6.24), and Theorem 3.4 give

$$\int_B \operatorname{curl} v \cdot d\vec{F} = \int_B \operatorname{curl} v \lrcorner \omega_{\mathbb{R}^3} = \int_B d(\Theta v) = \int_{\partial B} \Theta v = \int_{\partial B} v \cdot \vec{ds} ,$$

where the last equality follows from Remark 2.2(c). The second part of the claim is a consequence of Remarks 2.2(c) and (e). ∎

3.22 Remark (physical interpretation of
the curl)   The integral

$$\int_\Gamma (v \mid \mathsf{t})\, ds$$

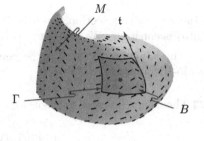

is called the **circulation** of the vector field v
along $\Gamma := \partial B$.   In the fluid model of
Remark 2.2(h), $\int_\Gamma (\rho v \mid \mathsf{t})\, ds$ is a measure
of the total mass transported along the
curve Γ per unit time.

For $p \in M$, we have

$$\lim_{B \to p} \frac{\int_{\partial\Omega}(v \mid \mathsf{t})\, ds}{\mathrm{vol}(B)} = (\mathrm{curl}\, v \mid \nu)(p)\,, \qquad (3.23)$$

where the limit is understood as follows: For every $\varepsilon > 0$, there is a neighborhood U
of p in M such that

$$\left| \frac{\int_{\partial\Omega}(v \mid \mathsf{t})\, ds}{\mathrm{vol}(B)} - (\mathrm{curl}\, v \mid \nu)(p) \right| < \varepsilon$$

for every piecewise smooth domain B in M with $p \in B \subset U$. The limit on the left
side of (3.23) is called the **circulation density** of the vector field v at the point p
with respect to the $\nu(p)$-axis (this is, the axis pointing in the $\nu(p)$-direction).

We now choose for M an oriented
plane through the point p such that $\nu(p)$
is the positive normal of M. Also let B_r
be a disc in M with center p and radius
$r > 0$, with Γ_r the oriented boundary
of B_r. Then

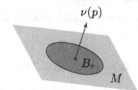

$$\lim_{r \to 0} \frac{1}{r^2 \pi} \int_{\Gamma_r} (v \mid \mathsf{t})\, ds = (\mathrm{curl}\, v \mid \nu)(p)\,. \qquad (3.24)$$

Because $\int_{\Gamma_r} (\rho v \mid \mathsf{t})\, ds$ is the amount of fluid transported along the oriented circle Γ_r
per unit time, (3.24) says that the component of $\mathrm{curl}\, v(p)$ along the unit vector $\nu(p)$
is equal to the circulation density with respect to the $\nu(p)$-axis. For $\mathrm{curl}\, v(p) \neq 0$,
it follows from the Cauchy–Schwarz inequality that

$$(\mathrm{curl}\, v \mid \nu)(p) \le |\mathrm{curl}\, v(p)| = \left(\mathrm{curl}\, v \,\Big|\, \frac{\mathrm{curl}\, v}{|\mathrm{curl}\, v|} \right)(p)\,.$$

Therefore the circulation density is largest at the point p when taken with respect
to the $(p + \mathrm{curl}\, v(p)\mathbb{R})$-axis.[9] On these grounds, $\mathrm{curl}\, v$ is also called the **vorticity
vector**. If $\mathrm{curl}\, v = 0$, then v is said to be **curl free**.

Proof   By Theorem 3.21, (3.23) follows in analogy to the proof of Remark 3.16(b). ∎

---

[9]See Remark XI.6.15(c).

The star operator and the coderivative

In the rest of this section, let g be a pseudo-Riemannian metric on M, so that $\text{sign}(g)$ is constant. We also set $(\,\cdot\,|\,\cdot\,)_M := g$.

The following theorem is the general form of the divergence theorem in the pseudo-Riemannian case.

3.23 Theorem (divergence theorem)   *For $v \in \mathcal{V}_c(M)$,*

$$\int_M \text{div}\, v\, d\lambda_M = \int_{\partial M} *\Theta v \ .$$

Proof   By definition (XI.6.13) (which defined the divergence) and Remark 2.2(i), this is an immediate consequence of Stokes's theorem (Theorem 3.1). ∎

Let $r \in \mathbb{N}$ with $r \leq m$. For $\alpha, \beta \in \Omega_c^r(M)$, we set

$$[\alpha \,|\, \beta]_M := \int_M \alpha \wedge *\beta \ .$$

3.24 Remarks $[\,\cdot\,|\,\cdot\,]_M$ is a nondegenerate symmetric bilinear form on $\Omega_c^r(M)$. If g is a Riemannian metric, then $[\,\cdot\,|\,\cdot\,]_M$ is a scalar product on M.

Proof   We know from Remark XI.5.11(a) that

$$\alpha \wedge *\beta = \beta \wedge *\alpha = \text{sign}(g)(\alpha \,|\, \beta)_{g,r}\omega_M \ .$$

The claim follows from this and the fact that $(\,\cdot\,|\,\cdot\,)_{g(p),r}$ is an inner product on $\bigwedge^r T_p^* M$. ∎

(b) Clearly $[\alpha \,|\, \beta]_M$ is defined when $\alpha \wedge *\beta$ is an integrable m-form on M. In particular, this is the case when α and β belong to $\Omega^r(M)$ and the intersection of their supports is compact. ∎

From Stokes's theorem, we easily get the *general Green's integral formula*:

3.25 Proposition   *Let $1 \leq r \leq m$, $\alpha \in \Omega_c^{r-1}(M)$, and $\beta \in \Omega_c^r(M)$. Then*

$$[d\alpha \,|\, \beta]_M + [\alpha \,|\, \delta\beta]_M = [\alpha \,|\, \beta]_{\partial M} \ .$$

Proof   From the product rule for the exterior derivative, it follows that

$$d(\alpha \wedge *\beta) = d\alpha \wedge *\beta + (-1)^{r-1}\alpha \wedge d*\beta \ .$$

Remark XI.5.9(d) shows $d*\beta = (-1)^{r+1}*\delta\beta$. Thus we get

$$d(\alpha \wedge *\beta) = d\alpha \wedge *\beta + \alpha \wedge *\delta\beta \ .$$

Now Theorem 3.1 (Stokes's) implies

$$\int_M d\alpha \wedge *\beta + \int_M \alpha \wedge *\delta\beta = \int_{\partial M} \alpha \wedge *\beta$$

and therefore the claim. ∎

3.26 Corollary *Suppose M compact and without boundary. Then we have the* **duality formula**

$$[d\alpha \,|\, \beta]_M = -[\alpha \,|\, \delta\beta]_M \quad \text{for } \alpha \in \Omega^{r-1}(M) \text{ and } \beta \in \Omega^r(M) \ .$$

In the case of a Riemannian manifold, the duality formula says that $-\delta$ is the operator formally adjoint to d with respect to the inner product $[\cdot \,|\, \cdot]_M$.[10]

The formulas above are the starting point for the topological exploration of manifolds; these we must relegate to further courses or other books in differential geometry and global analysis.

Exercises

1 Suppose M is compact and without boundary, and let $\omega \in \Omega^{m-1}(M)$. Show that $d\omega$ has a zero.

2 Let $\rho \colon \mathbb{R}^3 \setminus \{0\} \to S^2$ denote the radial retraction, and let $\sigma := \rho^* \omega_{S^2}$. Show σ is closed but not exact. (Hints: Example XI.4.13(c); consider $\int_{S^2} \sigma$.)

3 Let M be without boundary and Riemannian, and suppose $f \in \mathcal{E}_c(M)$ has $\Delta f \geq 0$. Show that f is constant. (Hints: Show first that $\Delta f = 0$, and then consider $\Delta(f^2)$; Green's formula.)

4 Suppose (e_1, e_2, e_3) is the canonical basis and (x, y, z) are the Euclidean coordinates of $\mathbb{R}^3$. Also let M be a compact three-dimensional submanifold of $\mathbb{R}^3$ with $\Gamma := \partial M$ and outward normal ν. Prove **Archimedes's theorem**, that is,

$$\int_\Gamma z\nu \, d\lambda_\Gamma = \text{vol}(M)e_3 \ .$$

Physical interpretation: We regard M as a body immersed in a fluid whose density is $\rho = 1$ and whose surface is the (x, y)-plane. Because $z < 0$ in the fluid, $\rho z \, d\vec{F}$ is the force (= pressure $\rho|z|$ in the direction of the inward normal, times the (infinitesimal) area element dF) that is exerted on the fluid at the point $p \in \Gamma$. Then because

$$\int_\Gamma z\nu \, d\lambda_\Gamma = \int_\Gamma z \, d\vec{F} \ ,$$

Archimedes's theorem says that the resulting force acts in the direction of the positive z-axis and is equal to the mass of the body: *The buoyant force is equal to the weight of*

---

[10]In the sign conventions that are typically used in geometry, the coderivative δ is formally adjoint to d.

the fluid displaced. Eureka.
(Hints: $\int_\Gamma z\nu^1\, d\lambda_\Gamma = \int_\Gamma z\, dy \wedge dz$ etc.; Stokes's theorem.)

5   Adopt the assumptions of Exercise XI.6.5. Using Gauss's law, find the integral form of Maxwell's equations. For example, show

$$\int_{\partial M} E \cdot d\vec{F} = 4\pi \int_M \rho\, dx$$

for every relatively compact, piecewise smooth domain M in Ω with outward normal ν, that is, *the flux of the electric field through a closed surface is proportional to the total charge it contains.*
Show that the differential and integral versions are equivalent.

6   Let Ω be a piecewise smooth bounded domain in $\mathbb{R}^3$, and let $p_1, \dots, p_k \in \Omega$. Calculate

$$\sum_{j=1}^k \int_{\partial\Omega} \partial_\nu \left(1/|x - p_j|^3\right) d\sigma(x) \,.$$

(Hint: Exercise X.3.6 and Corollary 3.18.)

7   Suppose M is a nonempty compact hypersurface in $\mathbb{R}^{m+1}\setminus\{0\}$ such that every half ray from 0 intersects M at most once. Also let

$$K_\infty(M) := \left\{\, t\,i(p) \;;\; t \in \mathbb{R}^+,\ p \in M \,\right\}$$

with $i: M \hookrightarrow \mathbb{R}^{m+1}$ be the (infinite) cone consisting of all line segments from the origin that intersect M. Finally let $\rho: \mathbb{R}^{m+1}\setminus\{0\} \to S^m$ be the radial retraction. Prove that

$$\int_M \rho^* \omega_{S^m} = \mathrm{vol}_{S^m}\left(K_\infty(M) \cap S^m\right) \,.$$

Remark $\mathrm{vol}_{S^m}\left(K_\infty(M) \cap S^m\right)$ is the **solid angle** of the cone $K_\infty(M)$.
(Hints: Examples XI.4.13(c) and 3.10(c); Stokes's theorem.)

8   Show that every closed differential form on S^2 is exact.
(Hint: Recall Lemma 3.11, and study the proof of Theorem XI.3.11.)

9   Let B be a compact m-dimensional submanifold of M with boundary, and let f be a smooth map from ∂B to a manifold N. Show these:

(a) If there is a smooth map $F: B \to N$ with $F\,|\,\partial B = f$, then $\int_{\partial B} f^*\omega = 0$ for every closed form $\omega \in \Omega^{n-1}(N)$.

(b) In the case $M = \mathbb{R}^m$ and $N := \partial B$, there is no smooth $F: B \to \partial B$ such that $F\,|\,B = \mathrm{id}_{\partial B}$, that is, there is no smooth retraction of B onto ∂B.

(Hints: (a) With $F = (F^1, \dots, F^n)$, consider the form $F^1\, dF^2 \wedge \cdots \wedge dF^n$; use Stokes's theorem.)

10   Prove that if M is without boundary, then the restriction of the Laplace–Beltrami operator Δ_M to $\mathcal{E}_c(M)$ is symmetric in $L_2(M, d\lambda_M)$, that is,

$$(\Delta f\,|\,g)_{L_2(M,d\lambda_M)} = (f\,|\,\Delta g)_{L_2(M,d\lambda_M)} \quad \text{for } f, g \in \mathcal{E}_c(M) \,.$$

11   Let $M := H^2$ be the hyperbolic plane and $v \in \mathcal{V}_c(M)$. Determine the explicit form of the divergence theorem (Theorem 3.15) in the following cases:

(a) the parametrization by polar coordinates (Example XI.5.5(k));

(b) the Poincaré model;

(c) the Lobachevski model;

(d) the Klein model.

12 Suppose M is without boundary and Riemannian, and let $\omega \in \Omega_c(M)$. Show the equivalence of

   (i) $\Delta\omega = 0$ and

   (ii) $d\omega = \delta\omega = 0$.

13 Let M be a Riemannian manifold. Prove that for $\alpha \in \Omega_c^{r-1}(M)$ and $\beta, \gamma \in \Omega_c^r(M)$ the following statements hold:

   (i) $\int_M [(d\alpha \mid \beta)_r + (\alpha \mid \delta\beta)_{r-1}] = \int_{\partial M} \alpha \wedge *\beta$;

   (ii) $\int_M [(d\beta \mid d\gamma)_{r+1} + (\beta \mid \delta d\gamma)_r] = \int_{\partial M} \beta \wedge *d\gamma$.

(Hint: Consider $d(\alpha \wedge *\beta)$.)

14 Suppose M is compact and without boundary. Show that the Hodge–Laplace operator with respect to the inner product $[\cdot \mid \cdot]_M$ is symmetric, that is,

$$[\Delta\omega_1 \mid \omega_2]_M = [\omega_1 \mid \Delta\omega_2]_M \quad \text{for } \omega_1, \omega_2 \in \Omega^r(M) .$$

References

[Ada75]   R.A. Adams. *Sobolev Spaces*. Academic Press, New York, 1975.

[Ama95]   H. Amann. *Gewöhnliche Differentialgleichungen*. W. de Gruyter, Berlin, 1983, 2. Aufl. 1995.

[AMR83]   R. Abraham, J.E. Marsden, T. Ratiu. *Manifolds, Tensor Analysis, and Applications*. Addison-Wesley, London, 1983.

[Arn84]   V.I. Arnold. *Catastrophe Theory*. Springer Verlag, Berlin, 1984.

[Art93]   M. Artin. *Algebra*. Birkhäuser, Basel, 1993.

[BG88]   M. Berger, B. Gostiaux. *Differential Geometry: Manifolds, Curves, and Surfaces*. Graduate Texts in Mathematics #115. Springer Verlag, New York, 1988.

[BS79]   J.M. Briggs, T. Schaffter. Measure and cardinality. *Amer. Math. Monthly*, **86** (1979), 852–855.

[Con93]   L. Conlon. *Differentiable Manifolds. A First Course*. Birkhäuser, Boston, 1993.

[Dar94]   R.W.R. Darling. *Differential Forms and Connections*. Cambridge Univ. Press, Cambridge, 1994.

[Dug66]   J. Dugundji. *Topology*. Allyn & Bacon, Boston, 1966.

[Els99]   J. Elstrodt. *Maß- und Integrationstheorie*. Springer Verlag, Berlin, 1999.

[Fal90]   K. Falconer. *Fractal Geometry. Mathematical Foundations and Applications*. Wiley, New York, 1990.

[Flo81]   K. Floret. *Maß- und Integrationstheorie*. Teubner, Stuttgart, 1981.

[Fol99]   G.B. Folland. *Real Analysis*. Wiley, New York, 1999.

[HR72]   H. Holmann, H. Rummler. *Alternierende Differentialformen*. Bibliographisches Institut, Mannheim, 1972.

[Koe83]   M. Koecher. *Lineare Algebra und analytische Geometrie*. Springer Verlag, Berlin, 1983.

[Lan95]   S. Lang. *Differential and Riemannian Manifolds*. Springer Verlag, New York, 1995.

[Mil65]   J.W. Milnor. *Topology from the Differentiable Viewpoint*. The Univ. Press of Virginia, Charlottesville, 1965.

[PS78]   T. Poston, I. Stewart. *Catastrophe Theory and its Applications*. Pitman, Boston, 1978.

[Rog70]   C.A. Rogers. *Hausdorff Measures.* Cambridge Univ. Press, Cambridge, 1970.

[RS72]   M. Reed, B. Simon. *Methods of Modern Mathematical Physics, I–IV.* Academic Press, New York, 1972–1979.

[Rud83]   W. Rudin. *Real and Complex Analysis.* Tata McGraw-Hill, New Delhi, 1983.

[Sch65]   L. Schwartz. *Méthodes Mathématiques pour les Sciences Physiques.* Hermann, Paris, 1965.

[Sch66]   L. Schwartz. *Théorie des Distributions.* Hermann, Paris, 1966.

[Sol70]   R.M. Solovay. A model of set theory in which every set of reals is Lebesgue measurable. *Ann. Math.*, **92** (1970), 1–56.

[SW96]   U. Storch, H. Wiebe. *Lehrbuch der Mathematik, 4 Bände.* Spektrum Akademischer Verlag, Heidelberg, 1996.

[Yos65]   K. Yosida. *Functional Analysis.* Springer Verlag, Berlin, 1965.

Index

Printed in the United States
By Bookmasters